SOIL and WATER CONSERVATION POLICIES and PROGRAMS

Successes and Failures

Edited by

Ted L. Napier, Ph.D.
Silvana M. Napier, Ph.D.
Jiri Tvrdon, Ph.D.

SOIL
AND WATER
CONSERVATION
SOCIETY

CRC Press
Boca Raton   London   New York   Washington, D.C.

Library of Congress Cataloging-in-Publication Data

Soil and water conservation policies and programs : successes and failures / edited by Ted L.
Napier, Silvana M. Napier, and Jiri Tvrdon.
    p. cm.
    Proceedings of an international conference convened at the Czech Agriculture
University in Prague, Czech Republic in September of 1996.
    Includes bibliographical references.
    ISBN 0-8493-0005-3 (alk. paper)
    1. Soil conservation—Government policy—Congresses. 2. Water
conservation—Government policy—Congresses. I. Napier, Ted L.
S622.2.S63 1999
333.76'—dc21
                                                                                      99-047523
                                                                                       CIP

No claim to original U.S. Government works
International Standard Book Number 0-8493-0005-3
Library of Congress Card Number 99-047523
Printed in the United States of America 1 2 3 4 5 6 7 8 9 0
Printed on acid-free paper

Preface

This book is the product of an international conference convened at the Czech Agricultural University in Prague, Czech Republic in September of 1996. The conference was organized to provide a forum for the presentation and discussion of soil and water conservation policies that have been implemented in North America, Europe, and Australia. Contributors and conference participants were challenged to debate the strengths and weaknesses of soil and water conservation programs and policies that have been implemented under a variety of socioeconomic, political, and cultural traditions. There were more than 30 countries represented at the meeting. Each conference participant contributed observations from his/her discipline and society. Hopefully, this text will contribute to resolving some of the conservation problems identified by program participants.

The organizers wish to thank several organizations for their contributions to the conference. Without the generous economic support of the Farm Foundation and the Natural Resources Conservation Service/United States Department of Agriculture this project would not have been possible. Special thanks is extended to the Czech Agricultural University administration and faculties for hosting the conference. Dr. Jon Hron, rector of the university, was a very strong supporter of the conference. We would also like to thank The Ohio State University for its contributions to the project. We wish to extend our thanks to many universities and government organizations that financially sponsored speakers and/or conference participants.

Ted L. Napier
Silvana M. Napier
Jiri Tvrdon

About the Editors

Dr. Ted L. Napier is a social scientist specializing in natural resources studies and technology transfer. He is a full professor at The Ohio State University Department of Human and Community Resource Development and the School of Natural Resources. Professor Napier's areas of academic specialities are the socioeconomics of natural resources development, statistics, research methodology, and technology transfer. Prior to joining Ohio State in 1970, he served as associate director of research for the AFL-CIO Appalachian Council. He has presented professional papers and has consulted with university staff and government officials in 20 different countries. Professor Napier has served in numerous leadership roles on regional and national research committees. He has successfully completed 35 externally funded projects during his tenure at The Ohio State University. Professor Napier has received several national awards for his research contributions to environmental studies. He was recently inducted as a fellow by the Soil and Water Conservation Society. He is president of the All Ohio Chapter of the Soil and Water Conservation Society and the features editor of the *Journal of Soil and Water Conservation*. He is also on the editorial board of the *Journal of Applied Research and Public Policy*.

Dr. Silvana Camboni Napier is an associate director in the Office of Grants and Contracts at The Ohio State University Research Foundation. She is an adjunct assistant professor in the Department of Human and Community Resource Development at The Ohio State University. She is a resources sociologist with extensive training in communication and journalism. Dr. Napier specializes in adoption and diffusion research with emphasis on technology transfer of agricultural technologies and techniques. She has published numerous articles focused on the adoption of soil and water conservation and has presented papers at many professional meetings throughout the United States and at several meetings abroad.

Dr. Jiri Tvrdon is a senior faculty member in the Faculty of Economics at the Czech National Agricultural University in Prague, the Czech Republic. His primary responsibilities are teaching and research. Dr. Tvrdon is well known in Europe for his empirical analyses of the Czech agricultural system. He was frequently consulted by officials within the Czech Ministry of Agriculture to recommend approaches for restructuring the Czech Republic following independence. Dr. Tvrdon was one of the first Czech scholars to recognize environmental problems associated with large-scale agriculture and has been active in promoting environmental studies within the Czech Republic.

Table of Contents

SOIL and WATER CONSERVATION POLICIES and PROGRAMS

Successes and Failures

1

Introduction

Ted L. Napier, Silvana M. Napier, and Jiri Tvrdon

Degradation of land and water resources via soil erosion is a universal problem in all geographic regions of this planet. While all land is subject to soil erosion due to forces of wind and water, the greatest proportion of environmental degradation due to displacement of soil is the result of human manipulation of land resources to produce food and fiber for human populations. Whenever human beings remove ground cover to produce agricultural products, soil loss is nearly always increased. This is especially true when highly erodible land is used for agricultural purposes.

Societies throughout the world have attempted to reduce degradation of water and land resources via use of conservation policies. Public policies to protect soil and water resources have employed a host of incentives and disincentives designed to motivate landowner/operators to adopt and use available conservation technologies and techniques. Some conservation policies have been successful, while others have failed to achieve objectives. Unfortunately, the reasons for successes and failures of conservation policies remain obscured, because factors affecting policy outcomes have seldom been examined.

In September 1996, an international symposium, which recognized the inadequacies of existing information concerning why some conservation policies succeed and others fail, was organized to examine soil and water conservation initiatives in different social, economic, and political environments. The conference, held at the Czech Agricultural University in Prague, was organized to provide a forum for the discussion of soil and water conservation policy initiatives that have been employed in Australia and in several other high-scale societies in North America, in Western Europe, in Eastern Europe, and in Central Europe. Chapter authors in this volume were commissioned to discuss specific conservation initiatives in their country of residence in the context of successes and/or failures of the policy approaches examined.

The chapters are grouped by geographic region and by country. The first two chapters outline the major problems associated with soil displacement on a global scale. The following section examines soil and water conservation policy initiatives in North America. The next group of chapters is from Western Europe, Eastern Europe, and Central Europe. The concluding section is devoted to soil and water conservation policies in Australia. The final chapter is devoted to a summary of major conclusions derived from extensive conference discussions and chapters published in this volume.

Soil erosion as a global problem

The chapter by Hurni outlines the major environmental problems associated with soil degradation on a global scale. He observes that approximately one-third of all agricultural land has been degraded by soil erosion and that about 84 percent of all degradation is the result of water and wind. Hurni notes that most technology transfer programs designed to address soil erosion problems are often not successful in lesser-scale societies, because poor people cannot purchase required inputs. He also suggests that many conservation technologies are not appropriate for lesser-scale societies, and that many conservation policies that have been successful in high-scale societies are not appropriate for people living in lesser-scale societies.

The chapter by Paoletti discusses some of the macrolevel causes of soil degradation. He suggests that reliance on a very limited number of animals and plants to provide food and fiber for the planet's populations has contributed to the development and use of agricultural production systems that degrade soil and water resources. He argues that use of a larger variety of plants and animals for food and fiber production would reduce the land required for production of feed grains and for grain stocks for human consumption. He also notes that deemphasis of large animal production would reduce pressure on land resources for grazing purposes.

Soil and water conservation policies in North America

The second section of this volume examines soil and water conservation policies in the U.S. The U.S. has a relatively long history of public policies designed to address environmental problems associated with soil erosion and subsequent water pollution. While there have been many successes, there have also been some failures.

Several authors note that national soil conservation policies began during the Dust Bowl Era when dust storms destroyed millions of acres of farmland in the Midwest. Since that time, a host of public policies have been developed and

4

implemented to reduce soil erosion and water pollution from agricultural sources to socially acceptable levels. While dust storms no longer darken the skies in the U.S. as they did in the 1930s, many environmental problems associated with soil erosion from cultivated farmland remain problematic.

The chapter by Johnson and the contribution by Weber and Margheim outline the role of the Natural Resources Conservation Service/U.S. Department of Agriculture (NRCS/USDA) in the development and implementation of soil and water conservation policies in the U.S. These authors correctly note that many of the early successes associated with motivating landowner/operators to adopt and use soil and water conservation production systems were the result of public conservation policies implemented by the NRCS/USDA (formerly known as the Soil Conservation Service/USDA) and other federal and state conservation agencies. These authors suggest that present social, political, and economic environments have changed the nature and complexity of soil and water conservation issues in the U.S. They recognize that new policy initiatives and programs will be required to adequately address soil and water conservation problems in the next millennium. Both chapters recognize that past initiatives will probably not be adequate to resolve future soil and water conservation problems in the U.S.

The chapter by Conrad offers the perspective of implementing soil and water conservation policy at the state level. He discusses a number of the problems associated with implementing Conservation Title programs at the local level. He notes that the Conservation Title mandated many programs but did not authorize sufficient human and economic resources to effectively implement them at the state level. Conrad observes that the quasi-regulatory role mandated by the Conservation Title was also difficult for many NRCS/USDA personnel to perform. He suggests that new policy initiatives will be required to adequately address conservation issues in the future. Conrad agrees with Weber and Margheim and with Johnson that future conservation policies will contain elements of coercion in the form of regulation.

Schnepf offers a number of observations about the role of private organizations in the development and implementation of soil and water conservation policies in the U.S. He observes that public conservation policies have been significantly affected by private conservation groups and by several professional associations. He asserts that future conservation policies will continue to be influenced by these groups. Schnepf argues that rapidly changing socioeconomic and political conditions within the U.S. require close cooperation among the various actors in the conservation policy arena.

The two chapters by Napier and Napier are less positive about the successes of contemporary soil and water conservation policies and programs in the U.S. While these contributors recognize the many successes achieved in the U.S. via

5

voluntary conservation efforts, they question the use of such approaches in the future. The Napiers argue that soil and water conservation policies in the U.S. have been inconsistent and in many instances ineffective in permanently addressing soil and water problems associated with production agriculture. They suggest that the Conservation Title of 1985 and subsequent conservation policies will not be able to reduce degradation of soil and water resources to socially acceptable levels without major redirection of conservation policies and programs. Napier and Napier suggest that regulatory approaches will have to be emphasized in the future or nonpoint pollution will not be controlled. They also assert that environmental policy will probably dictate agricultural policy in the future.

The Esseks et al. chapter partially supports the position advanced by the Napiers. These authors recognize many environmental benefits achieved via the quasi-regulatory approach used in the Conservation Title initiatives of 1985 and 1990. However, they note that considerable evidence suggests that many landowner/operators do not believe they will be forced to comply with Conservation Compliance provisions of the Conservation Title. The authors argue that NRCS/USDA needs to balance its desire to protect the rights of landowners to make land use decisions on their own land with its role as enforcer of national conservation policy.

The Hoag et al. chapter extends the criticisms offered by Esseks et al. when the authors observe that protection of soil and water resources cannot be entrusted to private landowners. Hoag et al. state that landowners are primarily concerned about private self interests and not pollution control. They assert that many landowner/operators do not adopt soil and water conservation production systems because such production systems are not profitable. Soil loss is perceived by many landowner/operators as posing no threat to land productivity during the lifespan of the owner. The authors argue that voluntary conservation initiatives have not been very successful in the U.S., because landowners have adopted production systems that maximize farm-level profits even when the production practices are erosive. Hoag et al. suggest that regulatory approaches will be required in the future.

The Hughes et al. chapter is critical of impacts of the Conservation Reserve Program on reducing wind erosion damage in New Mexico. The authors note that wind erosion is a serious environmental problem in the western portion of the U.S. and that the CRP was not very effective in addressing such environmental problems. A significant contribution of the Hughes et al. chapter is the observation that national conservation policy cannot adequately address local environmental situations. Empirical evidence from New Mexico strongly suggests that the Conservation Title was not cost effective in the context of wind erosion and that future conservation policy must consider location of land resources and land uses in addition to aggregate levels of soil loss.

The last chapter in the U.S. section raises the question of greater government control of pesticide use in the U.S. Manale reports that new standards for pesticides in foods are more restrictive than in the past and will probably require reassessment of weed management strategies for reduced till and no-till agricultural production. He argues that pesticides that have been widely used in the past will no longer be approved for use. Soil and water conservation production systems that rely heavily on pesticides to make them profitable will probably no longer be viable options for landowner/operators. Manale suggests that new pesticide standards will act as an incentive for agriculturists to place more emphasis on integrated pest management in the future.

Soil and water conservation policies in Canada

Soil and water conservation is an important issue in Canada. Provincial and national governments in Canada have developed and implemented soil and water conservation policies and programs. Some of the initiatives have been successful while others have not achieved desired objectives.

Cressman et al. examine trends in government soil and water conservation policies in Canada and discuss issues that they believe have not received adequate attention. The authors note that degradation of soil resources is a serious problem in Canada and that many approaches have been used to address conservation problems.

The wide diversity of policy approaches used in Canada is attributed to the fact that provinces have jurisdiction over natural resources within their boundaries. The federal government may intervene only in matters involving more than one province or when governments of other countries are involved. The authors suggest that many conservation policies are designed to achieve short-term objectives and cannot be employed to address long-term problems. They argue that future soil and water conservation policies should emphasize more involvement of local people, long-term programming, greater involvement of the private sector, and greater use of science in decision-making.

Stonehouse attributes much of the land and water degradation now occurring in Canada to the adoption and use of technology-intensive production systems by landowner/operators throughout the country. He suggests that agricultural policy has increased the pressure on farmers to employ degrading production practices. The author asserts that many conservation production practices are not profitable and that farmers use erosive production systems that will maximize farm profits. Stonehouse discusses the Land Stewardship Program and argues that the initiative has achieved some success; however, he asserts that the program has failed to achieve some objectives. The author suggests that off-site damages associated with agricultural pollution have not been adequately addressed in the Land Stewardship Program and that public funds have not been used in the most efficient manner.

The chapter by Vaisey et al. examines the Permanent Cover Program (PCP) that was implemented in western Canada during 1989. The authors observe that many of the benefits associated with the PCP are the same as the Conservation Reserve Program in the U.S. Farmers received about $50 per hectare to enroll land in permanent pasture. While many farmers enrolled land in the conservation program, the authors believe that the economic situation at the time the program was implemented was the reason many farmers participated in the program. Low grain prices combined with relatively high prices for cattle encouraged landowners to convert land to permanent pasture. The authors are uncertain that such a program would be implemented with the same level of success today. Given the high price of grain and the low price for cattle, the authors submit that higher government payments would be required to motivate farmers to enroll cropland in such a program today.

Soil and water conservation policies in Europe

The third section of the volume focuses on soil and water conservation policies in Europe. While soil erosion has been observed in Western and Northern Europe for many decades, concern for agricultural pollution was not perceived to be an important environmental issue until after World War II. Concern for soil loss and water pollution from agricultural sources increased substantially during the 1970s, because European agriculturists adopted technology intensive production systems that increased displacement of chemical-rich topsoil from cultivated cropland.

Fullen outlines the history of soil and water conservation in England and Western Europe. He suggests that the adoption of chemical fertilizers removed the necessity for leaving agricultural land fallow for extended periods of time, which made it possible for landowner/operators to cultivate cropland every year. Such farm production systems increased soil displacement. The author also notes that adoption of technology-intensive farm machinery combined with shifts in the types of crops produced further exacerbated soil loss on agricultural land. The author suggests that U.S. soil conservation policy significantly affected European soil conservation policies. However, he suggests that many conservation policies are not effective, because European countries do not have the institutional structures to implement conservation policies. Fullen strongly recommends that institutional structures such as the NRCS/USDA be formulated in European countries. He also argues that national soil surveys and targeting of environmentally sensitive areas need to be high priorities for environmental action.

The three chapters by Dubgaard, Schou, and Vedeld and Krogh focus attention on environmental problems in Denmark and Norway. The attempts to use policy instruments to address agricultural pollution in both countries have been quite similar.

Dubgaard and Schou discuss several environmental issues in Denmark. Both authors assert that the major agriculturally induced environmental problem is water pollution from agricultural fertilizers. Dubgaard suggests that the major reason for environmental degradation is the intensification and expansion of agriculture within Denmark. He notes that agricultural production increased 60 percent from 1960 to 1993. Fertilizer application rates per hectare more than tripled during the same time period primarily due to a shift to grain production. Dubgaard discusses several policy approaches to address agricultural pollution. He notes that recent initiatives include an eco-tax which signals a movement toward a national policy in which the polluter pays principle will be applied to agriculture. The author posits that much stronger economic incentives will have to be used to motivate farmers to adopt agricultural production systems that will be required to maintain rural landscapes in a manner defined as desirable by society.

Schou notes that about 95 percent of all drinking water in Denmark is from groundwater sources, and that contamination of groundwater by farm chemicals poses a potential threat to human health. About two-thirds of all nitrate pollution in Denmark is from agricultural sources, therefore, control of nonpoint pollution is essential to protect groundwater resources. Government policies have been formulated to address nonpoint pollution and the approach used in Denmark is primarily command and control. Many argue that government price support programs have encouraged farmers to maximize farm production, and that future policies should reduce farm subsidies. The author also suggests that cross-compliance policies should be implemented that require use of conservation production systems. Danish conservation policy designed to control nitrate pollution has emphasized taxes to increase the price of fertilizer, under the assumption that a higher cost of fertilizer will result in lower application rates.

Vedeld and Krogh examine soil and water conservation problems and policies in Norway. They argue that the primary cause of agriculturally induced pollution in Norway is the widespread adoption of technology-intensive production systems. Such production systems have contributed to the displacement of chemical-rich topsoil that has contaminated surface and groundwater resources. The authors observe that the most significant environmental problem in Norway is water pollution from agricultural sources. They discuss the agronomic approach for reducing pollution from agricultural chemicals. For example, agronomists argue that the solution for nutrient pollution is careful application of chemicals. If the plant is more efficient in taking up fertilizers, then fewer nutrients will be available to pollute waterways and groundwater. Economists disagree with this approach and argue that farmers will not reduce fertilizer application if reduced application rates pose a threat to farm-level profits. The authors assert that the major problem for policy formation is achieving a balance between loss of farm production and an improvement in environmental quality. Vedeld and Krogh submit that good science will be less significant in the development of conservation policy in Norway than political acceptability of the actions prescribed.

Land and water conservation problems and policies in Germany are examined by Weingarten and Frohberg, Frielinghaus and Bork, and Kindler. While all agree that environmental degradation is a significant issue in the country, the approaches to resolve environmental problems differ considerably.

Weingarten and Frohberg discuss environmental policies that have been implemented in Germany in recent years. They note that soil erosion is not a significant environmental problem in Germany due to flat topography and low rainfall; however, nutrient pollution from agricultural sources is a major issue. Pesticides are also an issue of concern, with atrazine and its metabolites constituting the most frequently observed pesticide in groundwater. The authors discuss several conservation initiatives developed by the European Union and by the German government to address agricultural pollution problems. They observe that practically all soil and water conservation policies employ command and control approaches, even though economic incentives have been used in a few instances. Conservation policies have been implemented to control surface application of sewage sludge, land use practices, fertilizer application rates, surface application of manure, and farm production practices designed to protect landscapes. The authors suggest that future conservation policy will demand clarification of what water resources will be protected. They submit that conservation policies will differ significantly if all groundwater resources are targeted for protection or if only water resources used for drinking purposes are targeted for protection.

Frielinghaus and Bork discuss environmental problems in the former East Germany. Unlike the situation in the U.S. and Canada and in other countries in Europe previously discussed, degradation of farmland and water resources by farm chemicals does not appear to be a serious issue. When erosion is problematic in the former East Germany, it is a function of intensive use of land resources by farmers. The major problem in the former East Germany, from the perspective of the authors, is soil compaction due to use of heavy farm equipment. Machine tracks become pathways for water movement and compaction prevents infiltration of water. The authors suggest that government policies have acted as barriers to adoption of conservation practices at the farm level. Frielinghaus and Bork posit that lack of government subsidies to adopt conservation production systems, combined with a lack of awareness among landowner/operators of the problems associated with soil erosion, have acted as important barriers to an adoption of conservation production systems.

Kindler addresses the problem of land conversion from extensive to intensive uses. The author argues that land conversion in Germany poses a serious threat to future landscapes and human life-styles. She argues that a tax should be imposed on land sales for nonagricultural purposes. Funds secured from the development tax would be paid by those who propose to change land uses from agricultural uses. Economic resources derived from the tax would be used to mitigate any environmental damages caused by land conversion.

Soil and water conservation problems and policies in Austria are discussed by Klik and Baumer. Contamination of water resources by farm chemicals is the major environmental issue in Austria. Groundwater contamination is a serious problem because 98 percent of all water for human consumption is secured from groundwater supplies. Contamination of groundwater supplies would pose a serious threat to public health. Local and federal conservation policies are very important factors for improving and protecting water quality. There is now a national program to monitor pollution of surface and groundwater resources. Water pollution standards have been established, and the government is responsible for ensuring that water quality is protected. The federal government also has responsibility for determining if land can be used for agricultural purposes. States have primary responsibility for protecting land resources, and considerable variance exists among Austrian states relative to public conservation policies. The authors observe that most conservation production practices and structures are too costly for farmers to adopt. Thus, many farmers have elected not to adopt conservation production systems.

Sapek and Sapek outline soil and water conservation problems and policies in Poland. The primary conservation issue in Poland is pollution of surface and groundwater by nutrients. The authors state that communistic regimes emphasized maximization of farm output with little or no regard for financial and/or environmental consequences. Nutrient content of animal manure was not considered in the decision-making process about use of chemical fertilizers at the farm level, and the price of chemical fertilizers was very low under communistic rule. These factors resulted in over application of nutrients. The environmental impact of excessive use of fertilizers was contamination of surface and groundwater resources. Policies and programs have been implemented in recent years to reduce the rates of fertilizer application at the farm level and to develop animal waste control structures. The Sapeks note that removal of animal waste dumps will not result in immediate improvement in water quality, because it takes long periods of time for nutrients to be leached from soil or to be consumed by plants. The authors recognize that the most important contributors to water pollution in Poland are municipal and industrial wastes. Point pollution is a very serious problem which must be addressed in Poland.

Soil and water conservation initiatives in Lithuania are examined by Foster and Budvytiene in the context of the Karst Zone in the northern portion of the country. The authors observe that a restructuring of the country and the fragmentation of large state farms have dramatically increased the number of farm production units and dispersed the production units over the landscape. The dispersion of point sources of pollution has reduced the threat of further pollution of water resources from agricultural sources. However, farmers must implement more conservation practices to achieve further reduction in pollution potential. The authors note that household and municipal wastes must be controlled in addition to animal wastes. Production systems being diffused in the Karst area are organic farming systems and sustainable farming systems. It is anticipated that future soil and water conservation policy initiatives will emphasize these production systems throughout Lithuania.

Kostadinov et al. discuss soil and water conservation policies and programs in Yugoslavia. They note that soil erosion is a very serious problem in the country and that conservation efforts have been in existence since the late nineteenth century. There were no laws governing soil erosion in Yugoslavia until 1930 when the Service for Erosion and Torrent (stream flow) Control was authorized. The first soil erosion control programs in Yugoslavia were focused on stream control and channeling to protect railroads from erosion damage. Between 1967 and 1977, national conservation efforts to protect soil resources were terminated with the repeal of the Service for Erosion and Torrent Control law. The authors note that emphasis in recent years has been on biological control due primarily to the lack of economic resources to invest in erosion control measures at the national level. Kostadinov et al. present statistics to show that investment in soil and water conservation efforts is correlated highly with national income. As the national income decreases, investments in soil and water conservation also decrease. The authors submit that conservation policies have been developed and implemented on a crisis basis to protect areas experiencing severe environmental degradation. The result has been inconsistent public policies. Soil and water conservation policies and programs will remain fragmented until economic resources become available to adequately fund conservation efforts.

Hron introduces the discussion of soil and water conservation problems in the Czech Republic by noting that many environmental problems within the country had their genesis during the socialist domination of the country. He observes that excessive use of agricultural chemicals, combined with overemphasis on agricultural production, resulted in degradation of soil and water resources throughout the former Czechoslovakia. Hron is optimistic that changes in the government system, which began in 1989, will result in laws that will protect soil and water resources from further degradation. He argues that public conservation policies should be targeted to specific land uses and that preferential tax policies should be used to protect land from degradation.

Dolezal et al. note that soil erosion in the Czech Republic is due to the movement of water. Wind erosion is of little consequence. They also note that all major waterways within the country originate in the Czech Republic, which means that environmental problems associated with soil erosion are generated within the country by Czechs. These authors confirm the observation made by Hron that fertilizer application rates during the socialist regimes were problematic, because there were no incentives for farmers to reduce fertilizer use. Dolezal et al. observe that recent soil conservation efforts in the Czech Republic have been focused on prevention of cropland conversion to nonagricultural uses. Land protection policies are based on the assumption that reduction in agricultural production will protect land resources from degradation. It is reasoned that if agricultural production can be significantly reduced, farmers will concentrate production on the fertile soils and retire marginal lands. Consideration is being given to compensating farmers who convert cropland to permanent pasture. Like a number of previous authors, Dolezal et al. observe that most conservation production systems are not profitable and are not adopted by land operators who are concerned about profitability of farm production systems. They believe that critical environmental areas should be protected by new environmental policies. A major issue that must be decided before effective environmental policies can be implemented is the question of land ownership adjacent to rivers. Stream bank erosion cannot be reduced until ownership is determined, because people will not invest in erosion protection programs unless they own the land.

Tvrdon suggests that protection of endangered species needs to be considered in new environmental policies in the Czech Republic. Wildlife is a very important natural resource within the Czech Republic and must be protected for future generations. Tvrdon argues that agriculture is the primary source of nonpoint pollution. However, an even more significant source of pollution is from point sources. Most sewerage within the country is not treated. Tvrdon observes that investment in the environment has increased tenfold since 1989. While the percentage of total gross domestic product is very small (about 2.7 percent), the increase is very dramatic in the context of the many pressing social problems within the country that compete for investment resources. The author argues that ignoring environmental problems will have severe socioeconomic consequences for the development of the country in the near future.

Lapka et al. examine conservation policy from the perspective of preserving rural lifestyles, in addition to protecting land from environmental degradation. They suggest that human values are seldom reflected in national conservation policy, even though conservation policy can affect the viability of rural communities. The authors argue that rural communities were basically destroyed in the Czech Republic by the emphasis on large collective farms under socialistic rule. Many small farmsteads were combined into large production units which resulted in changing the composition of most rural communities in the Czech Republic. Farmers were asked to comment about the adequacy of existing conservation policies in the Czech Republic, and they observed that contemporary policy and programs are totally inadequate to meet their needs. Farmers are concerned about maintaining the economic viability of the community rather than conservation. Economic survival is more important than protecting soil and water resources. Farmers also observed that conservation subsidies are not targeted on the appropriate agricultural products and adoption of conservation production systems is not profitable. Many farmers observed that transaction costs are very high relative to the economic benefits received. Finally, farmers argued that most subsidized programs are implemented without concern for the quality of the practices which will result in adoption of many conservation practices that provide very little improvement in the environment. Farmers wanted conservation programs that improved farm income.

Soil and water conservation policies in Australia and New Zealand

Soil erosion has been a serious environmental problem within Australia for decades. In recent years, a number of soil and water conservation policies have been implemented. While several of these efforts have achieved extensive attention in the world community, several of the contributors to this volume question the utility of existing conservation initiatives within the country.

Hannam emphatically argues that soil conservation has failed within Australia as evidenced by the increase in land degradation during the past 60 years. He states that soil conservation policy in Australia requires extensive change with greater emphasis on ecological sustainable land management strategies. Hannam notes that the increasing scale of agriculture, expanded acreage under cultivation, and an intensive use of land resources for agricultural production resulted in increased degradation of land resources via soil erosion. The author faults existing soil conservation policy as being too pro-agricultural production. Conservation practices are emphasized that maintain high levels of production rather than protection of environmental quality. Hannam implies that agricultural policy is the driving force behind soil conservation policy and that the two are not compatible. He asserts that government agencies are more concerned about economic development of the country than preservation of soil resources. Until this orientation is changed, effective soil conservation policies and programs will not be possible in Australia.

Bradsen discusses the Landcare program which has received considerable attention as an effective method of achieving soil and water conservation objectives. The author outlines the history of the social movement and how the approach has influenced conservation programs and policies in Australia. While emphasizing that Landcare raised the awareness of soil conservation issues in the country, Bradsen notes with concern how little impact the awareness approach has achieved in terms of reducing the incidence of land abuse. He observes that the "land ethic" approach constitutes the core of the Landcare program, however, he notes that profitability forms the basis of farm-level decision-making. Farmers will not adopt any conservation practice unless it is profitable for them to do so. Bradsen feels that the Landcare program falls short of what is needed to adequately address soil conservation problems in Australia.

Cary agrees with many of the positions advanced by Bradsen about the Landcare program. Cary notes that economic compensation is required to motivate farmers to adopt conservation production systems and that such approaches are not common in Australia. He observes that conservation programs in Australia place emphasis on the on-farm benefits of soil conservation, however, many soil conservation production systems are not profitable in the short term and often not in the long term. The result is that farmers do not adopt conservation production systems. The author discusses the gap between attitudes and behavior. He observes that people can have very positive attitudes toward something and not act on those attitudes even though they may wish to do so. He asserts that generating positive attitudes toward the environment will not be an effective means for motivating farmers to adopt conservation production systems in Australia. Cary posits that future conservation policy should include economic incentives to adopt conservation production systems. If economic incentives are not authorized in future soil conservation policies, it is highly likely that regulatory approaches will have to be employed to motivate landowner/operators to adopt and use conservation production systems in Australia.

The chapter by Ewing supports the positions advanced by Bradsen and by Cary concerning Landcare in Australia. Ewing argues that the Australian government is similar to the U.S. in that it is very unwilling to invoke legislation to control soil erosion. Preference is given to voluntary approaches that emphasize attitude change in the form of a land ethic. Ewing notes that success of the Landcare program has been assessed in terms of the number of community groups participating in the program; however, she argues that no causal linkage has been established between program participation and adoption of appropriate land management practices. Ewing discusses the findings from a case study of a community that became involved with Landcare. She observes that the Landcare program may have positively affected perceptions of community but probably did not achieve its environmental objectives.

Farrier examines water policy in the Murray-Darling River Basin in Australia in the context of protecting the resource for present and future generations. He notes that about 75 percent of the water used in Australia is used for irrigation and about half of the irrigation water is used for pasture. Irrigated pasture accounts for only about 20 percent of the production. Farrier posits that this is not a very efficient use of water. He also notes that environmental degradation is closely correlated with irrigation in the Murray-Darling Basin. Reduced stream flow, saltation of land resources, and many other problems result from overuse of the resource. The major problem associated with development and implementation of water policy in the river basin is that irrigators have used the resource for many years and to implement effective conservation policy will demand that rights of present irrigators be changed. Landowners resist attempts to reduce their access to water resources even though water has been overallocated. The author argues that regulation by the state is necessary to control illegal use of water resources and to legitimize transfer of water rights.

Bettjeman examines conservation policies in New Zealand. He notes that soil erosion is a serious problem in New Zealand due to the removal of forests and native bushes by native peoples and Europeans. Bettjeman observes that the government abolished financial assistance for the production of agricultural products in 1985, and a number of changes were made in the tax system that reduced subsidies to farmers. In 1987, the government also eliminated subsidies for soil conservation, flood control, and drainage. These actions represent the government's policy of polluter pays principle. The author notes a number of approaches that the government employs to force farmers to employ conservation practices. The most significant policy is the requirement that farmers obtain permits for a wide range of farming activities combined with monitoring to ensure compliance. All of the costs of implementing these conservation policies are borne by the landowner. New legislation is being developed to control the use of fertilizers and other farm chemicals. The author concludes that even though progress has been achieved in the advancement of national environmental goals, farmers still employ short-term profit-making criteria when making land management decisions.

Conclusions

Napier et al. summarize the major conclusions drawn from the chapters presented at the meeting. They also include several observations made during the discussions and informal group settings that occurred during the conference. The authors conclude that soil and water conservation policies are essential if land and water resources are to be protected for future generations. Napier et al. caution that experiences with conservation policies in other countries should act only as a guide for other societies engaged in the process of development and/or implementation of soil and water conservation policies. What has been shown to be successful in one society may not be successful in another. What has failed in one society may be successful in another. The role of policy-makers in every society is to assess soil and water conservation policy options in the context of the relevance of the policies to the social, economic, cultural, and political institutions of the country.

2

Soil Conservation Policies and Sustainable Land Management: A Global Overview

Hans Hurni

The dilemma of policy implementation in poorer countries

Observation of the damage caused by soil degradation processes reveals that soil degradation does not discriminate between countries or continents. The Global Assessment of human-induced Soil Degradation (GLASOD) showed that soil degradation damage in one form or another occurs in virtually all countries of the world (Oldeman et al. 1990). One-third of the world's agricultural soils, or roughly 2 billion hectares of land, are reported to be affected by soil degradation. Water and wind erosion account for 84 percent of this observed damage, while other forms like physical and chemical degradation are responsible for the rest. Some forms of soil degradation have been caused by industrialization and urbanization. Most damage, however, is the result of inappropriate land management in all farming systems, from subsistence to mechanized farming. This is an agricultural dilemma of global proportions. Figure 1 shows the global pattern of degradation, although types and severity have been neglected in this summary of the GLASOD map.

While soil degradation does not discriminate between different countries, distinctions must be made between nations when assessing their capabilities to cope with degradation. It appears that rich countries have much more potential to cope with their degradation problems, particularly at the policy level. Could they not transfer their knowledge and experience with soil conservation policies and strategies to less favorable countries? Unfortunately, the usual "transfer-of-technology" paradigm appears not to function here. The dilemma of policy implementation is thus particularly acute in poorer countries.

This chapter focuses on case studies in countries where more than 50 percent of the population is employed in agriculture. Fifty-seven percent of the world's population lives in such countries. A further 29 percent lives in countries where between 10 percent and 50 percent of the people are employed in

Figure 1. Global extent of human-induced soil degradation

Source: Generalized from Oldeman et al. 1990.

Figure 2. World distribution of per capita GDP and employment in the agricultural sector

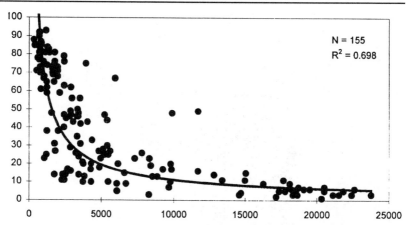

X = Gross domestic product in US$ per capita per year (1992)
Y = Percentage of population employed in the primary sector (agriculture, mining; 1992)

Statistical source: UNDP 1995. Source: Hurni et al., 1996.

agriculture. Globally, about 2.6 billion people, or 40 percent of the world's population, live and work in agriculture. Figure 2 presents an interesting correlation between people employed in the primary (basically agricultural) sector of a country, and the GDP[1] of that country. Low-GDP countries are defined as countries where the GDP per capita per year is less than U.S. $5,000. Out of 151 countries included in this overview, 93 can be classified in this low-income category, while 36 more have a medium-range GDP, between U.S. $5,000 and $15,000. Only 22 countries are in the high-GDP category, above U.S. $15,000.

Experience in soil and water conservation, as partly assessed in the global program WOCAT[2] under the auspices of the World Association of Soil and Water Conservation (WASWC), has shown that despite successes, most low-GDP countries have problems coping with their soil degradation problems.

Is there a possibility to learn from the small group of high-GDP countries—for example by transferring their policies in soil conservation? The author is optimistic about the fact that certain elements of these policies are worthy of consideration in local contexts, but doubts that a transfer of whole packages or systems will be feasible. Our basic hypotheses are that:

• there is no specific blueprint for a simple transfer of sustainable land management policies from one country to the other;
• national economic and political environments are the primary determinants

H. Hurni, October 1994.

Soil degradation in a barley cultivation area at 3300-3600 m asl in Northern Ethiopia.

in assessing opportunities and constraints related to sustainable land management; and

- the local context (household, community, and farming environment) is the ultimate determinant for adopting or rejecting land husbandry systems.

Examples of policy implementation problems in low-GDP countries

The first case study, from Ethiopia, demonstrates the impossibility of national implementation of an integrated soil conservation policy in a very low-GDP country, where threatening environmental degradation processes persist. This lack of policy implementation prevents soil conservation projects and programs from having a meaningful impact on the rural situation. In Ethiopia, long-term and widespread soil degradation resulted from traditional land use in rainfed agricultural and livestock mixed systems over many centuries. Favorable rainfall patterns in a highland situation are responsible for the fact that this country, which is situated in the Sahelian zone of Sub-Saharan Africa, has population densities more than ten times greater than other Sahelian countries, and a total population exceeding that of all other Sahelian countries together. The low productivity of subsistence farming systems, coupled with marginal mountain agriculture and a very low level of development in government and the private sector, results in high vulnerability to famine (see photo above).

22

H. Hurni, May 1991.

Swidden agriculture in secondary vegetation successions in Eastern Madagascar.

Consequently, foreign assistance in the form of food aid was used for more than two decades in "food-for-work programs," representing an investment of probably more than U.S. $500 million altogether. The impact of these programs, however, was ultimately much less than expected. It is estimated that less than 30 percent of all soil conservation measures were maintained by farmers over the years. Several factors account for this very modest success: an insufficient and often misused extension system, lack of a clear land use and ownership policy, population pressure on dwindling land resources, and the wrong approach to soil conservation. One of the main causes of these deficiencies is the fact that over 80 percent of the population is employed in subsistence agriculture. This leaves virtually no national economy with secondary and tertiary sectors which could provide meaningful services in all aspects of life. Land resource modeling and the application of scenarios have shown that increased investments in the four major sectors "environment," "population," "agriculture," and "livestock" are needed, apart from economic development, in order to adequately cope with degradation and land scarcity (Hurni 1993).

Another case study was done on land titling in Madagascar, where international interests have to be weighed against local traditions and practices (Brand 1996; Hurni et al. 1996). The eastern Malagasy escarpment is characterized by shifting cultivation along a dwindling strip with natural primary forest and much secondary fallow vegetation (see photo above). Land use is communal, and

23

clans decide on swidden areas in their territories on an annual basis.

A new national policy, which has been institutionalized, allows immigrants to buy individual land holdings which can be titled. The national environmental action program, which was launched in 1989 under the auspices of the World Bank to protect the last remaining primary forest zones of Madagascar, in line with the tropical forestry action plan and the biodiversity convention, pursues a strategy of formalizing landownership in the buffer zone around these forests with individual titles. Near the Andasibe Reserve, however, local communities successfully forced titling authorities not to register individual land, but to assign land to whole villages, introducing group property rights. Some villages with large numbers of immigrants preferred the individual titling approach. Early warning about the titling process led to accelerated forest clearing by the land users trying to increase their land holdings along the forest line.

In Kenya, private land titling is well established and an important commercial factor. Overuse of the agro-ecologically favorable moist-to-wet highlands, however, led to migration of small-scale farmers into the semiarid and arid parts of the country. In Laikipia District, large-scale, former colonial ranches were sold to small-scale agriculturalists who try to engage in rainfed farming, but with much less success than in their former high-potential areas. The result is an increasing conflict over water, which is exploited for drinking and irrigation purposes, and significant downstream effects for lowland users and wildlife (Wiesmann 1997). Land security as a single strategy to stimulate increased investment in soil conservation and better land husbandry was largely a failure in the semiarid areas because farming, as such, is at risk. However, private rates of discount for any agricultural investments are high, and households apply multiple strategies of on-farm and off-farm activities within their large family networks. The policy of "privatizing" large-scale farms and subdividing them into small holdings certainly did not increase sustainability in Laikipia District, while the overall productivity of the land was increased only at the expense of downstream inhabitants.

In Zimbabwe, about half of all agricultural land is still owned by about 4,000 formerly colonial farm families. The other half of the land is of generally lower quality and is divided among 8 million small-scale subsistence farmers. Sometimes, these two systems exist side-by-side (see photo, right), revealing vast differences in land use. While the former system is generally more environmentally sustainable, with better vegetation cover, greater biodiversity, and even rare wildlife species, the latter has a higher agricultural productivity and sustains high population densities.

Nevertheless, no government policy in recent years has changed this paradoxical system, for example, by subdividing large-scale farms as a matter of policy. The reason behind this inertia is rather obvious, as the state has a great interest in maintaining the large-scale farms because they are commercial, export-oriented, and produce for the growing markets within and outside the country. Small-scale farms, on the other hand, barely feed their growing populations at subsistence level. A system based on equity in access to land would thus undermine national security, while the present situation of land scarcity among smallholder functions as a resource-taxing system in favor of the state and the townspeople.

In Thailand's northern hills and mountains, two conflicting international interests have coexisted for the past 30 years. On the one hand, opium production was stimulated by attractive world market prices, while international pressure to ban production steadily mounted. This situation led to strict land use policies and a virtual ban on shifting cultivation practices in the mountain zones, because this was the common land use system used by highland ethnic groups to grow opium, as well as most agricultural subsistence crops. As a result, opium production was driven farther away, mostly into the hill areas in the countries around Thailand, and highland agriculture is now allowed only on permanent fields, mostly near roads for better control. This strict government regulatory policy could only be implemented and enforced due to the economic growth of the country and an enhanced institutional structure. On the environmental side, however, the ban on shifting cultivation has considerably accelerated soil degradation on the sloping rainfed agricultural land because fallow regen-

H. Hurni, February 1992

Abrupt land-use changes due to property in Zimbabwe. On the right side is a large-scale livestock farm; on the left side are small-scale settlers.

eration is no longer possible in this system (cf. Hurni 1982, or Hurni and Nuntapong 1983).

In Venezuela, the overall growth of the national economy resulted in agricultural improvements in many areas including sustainable land management. Examples are the refined irrigation systems with trickle irrigation for particular cash crops such as coffee on sloping land, or the irrigation systems on valley floors. Despite this modernization and sustainable intensification, small farmers were further forced to cultivate marginal lands where no appropriate technologies against erosion exist for the present land use systems. A change of land use, in conclusion, is not a regulatory measure guided primarily by land use policy; it must represent an economic opportunity at the farm level. This principle of subsidiarity applies to all low and medium GDP countries.

Finally, a positive example can be found in the policy introduced by a private company in 1992 in Sri Lanka, near the town of Kandy in the central highlands. Here, tobacco is grown by contract farmers on small plots, usually in very steep hilly areas, which are exposed to accelerated soil erosion due to a highly erodible cropping pattern with wide spacing at the early stages of plant growth. The company, alerted by the long-term consequences of this poor land use practice, introduced a simple hedgerow system for tobacco plantations with all its contract farmers applying the simple policy that only those farmers who have introduced the system could sell their product at harvest time. Although at the time of observation, in November 1993, the quality of the introduced technology was still shaky, it was nonetheless an astonishing feature and a real inspiration for other large-scale agricultural companies in low GDP countries to adopt similar policies.

Figure 3. Intervention levels and activities in sustainable land management

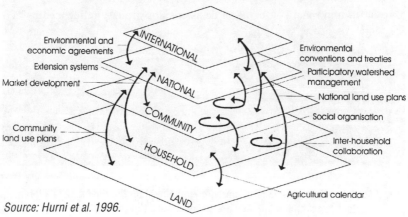

Source: Hurni et al. 1996.

26

A multilevel stakeholder approach to sustainable land management

In a multilevel stakeholder approach, principles relating to each level of intervention and the interactions between these levels must be compatible (see Figure 3). The basic level is the land, where good land husbandry should be promoted with measures which ensure vegetative cover, maintain favorable soil structure, allow for appropriate water flows, and optimize organic matter. Sustainable land use should be ensured by households and communities. They can do this only with an enabling institutional environment, at the community level and above, which promotes viable social, economic, and legal institutions, allows for integration at catchment levels, and employs participatory approaches in accordance with the principle of subsidiarity in decision-making. National and international levels, finally, are playing increasingly important roles through market development, land use plans, and conventions and agreements.

From the above approach and the examples described, the following can be concluded with regard to policy approaches in different countries:

- Policy development is a need at multiple levels of intervention, from international to national, and from institutional to private sector enterprises.
- A thorough assessment of sustainable land management options needs to include a multilevel stakeholder analysis and an analysis of economic viability, technical feasibility, ecological sustainability, and social acceptability of both technologies and approaches in sustainable land management.
- The transferability of policy frameworks between countries must be seen in light of the above assessment by using a conceptual framework which is interculturally sensitive, interdisciplinary in nature, and inter-generational in vision.
- Bear in mind that conservation decisions are strongly influenced by economic considerations, and that the different measures require thorough cost-benefit analyses (C.A.).
- Different rates of discount in C.A.s should be weighed against each other. Private rates of discount favoring individual profitability and farm prosperity are usually higher than social rates of discount which reflect national and intragenerational security, and these are again higher than sustainable rates of discount which reflect international and intergenerational security. The latter two rates, however, will require substantial commitment by nations and the international community, respectively.
- A particular challenge for the present generation in low-income countries will be to adopt an approach of waiving possible present-day benefits in favor of investments to protect the interests of future generations. It has been

shown that economic growth will only be stimulated if people are ready to make such a sacrifice (Grundlach 1995). Current population dynamics, however, tend to discourage such an approach because of the growing needs generated by population increase in low-income countries.

- In all decisions relating to sustainable land management, the principle of subsidiarity will have to be followed strictly. This means allowing decision-making at the lowest possible level of actor empowerment, with no decisions to be taken by higher institutions unless this is necessary to protect stakeholder interests.

Acknowledgments

I wish to thank all the hosts who facilitated research visits and studies in countries in which case studies reported on in this paper were conducted. Particular acknowledgment is made to the Swiss Agency for Development and Cooperation (SDC) which finances many partnership cooperation activities of the Center for Development and Environment (CDE) in numerous countries. Finally, I would like to thank Dr. Ted Wachs, DE editor, and Dr. Ted Napier for organizing the Prague workshop.

Abbreviations

C.A.	*Cost-Benefit Analysis*
CDE	*Center for Development and Environment, University of Berne, Hallerstrasse 12, 3012 Berne, Switzerland*
GDP	*Gross Domestic Product (usually given in U.S. $ per capita per year)*
GLASOD	*Global Assessment of human-induced Soil Degradation, UNEP and ISRIC*
ISRIC	*International Soil Reference and Information Center, Wageningen*
UNEP	*United Nations Environment Programme, Nairobi*
WASWC	*World Association of Soil and Water Conservation, Ankeny, Iowa, USA*
WOCAT	*World Overview of Conservation Approaches and Technologies. WASWC and CDE, Berne.*

Notes

1. GDP: Gross domestic product. It is usually given in U.S. $ per capita per year. GDP is an indicator derived from current economic factors but excluding many parameters such as external costs, impacts of economic activities on natural resources, human well-being, and other criteria which may be relevant in judging the sustainability of a national system.
2. WOCAT: The World Overview of Conservation Approaches and Technologies is a global framework for the evaluation of soil and water conservation initiated by WASWC in 1992. It is executed by a consortium of institutions and regional networks and coordinated by CDE. The current focus is on Africa, while other continents have joined the program more recently.

References

Brand, J. 1996. Unpublished data. Projet Terre-Tany, implemented by CDE in cooperation with FOFIFA, Antananarivo, Madagascar.

Grundlach, E. 1995. Humankapital als Motor der Entwicklung. Ein neuer Ansatz der neoklassischen Wachstumstheorie. Entwicklung und Zusammenarbeit, Vol. 36, No. 10, pp. 261-266.

Hurni, H. 1992: Soil erosion in Huai Thung Choa - Northern Thailand. Concerns and constraints. In: Mountain Research and Development, Vol. 2, No. 2, pp. 141-156.

Hurni, H. and S. Nuntapong. 1983. Agroforestry improvements for shifting cultivation systems: Soil conservation research in Northern Thailand. In: Mountain Research and Development, Vol. 3, No. 4, pp. 338-345.

Hurni, H. 1993. Land degradation, famine, and land resource scenarios in Ethiopia. In: World Soil Erosion and Conservation. Cambridge Studies in Applied Ecology and Resource Management, Cambridge University Press, pp. 27-61.

Hurni, H. with the assistance of an international group of contributors, 1996. Precious earth: From soil and water conservation to sustainable land management. International Soil Conservation Organisation (ISCO) and Center for Development and Environment (CDE), Berne, p. 89.

ISRIC: International Soil Reference and Information Center, Wageningen, the Netherlands.

Oldeman, L.R., V.W.P van Engelen, and J.H.M Pulles. 1990. The extent of human-induced soil degradation. Annex 5. Wageningen: ISRIC, and UNEP.

UNDP. 1995. Human development report 1995. Oxford and New York: Oxford University Press (based on 1992 statistics).

Wiesmann, U. 1997. Sustainable regional development in rural Africa: Conceptual framework and case studies from Kenya. CDE, Berne.

3

What Are the Alternatives to Improve Landscape Quality and Biodiversity?

Maurizio G. Paoletti

In Western countries we are facing the decreased importance of rural landscapes for production of commodities such as crops and livestock. Such commodities are increasingly surpluses. After having strongly supported increased productivity and having forgotten scarcity and famine, Western countries are looking at rural landscapes as environments that can be used to recreate millions of peoples that reside in large, crowded, sometimes nearly unlivable, towns. Most of those citydwellers have abandoned rural areas and rural lifestyle over the last two or three generations. We are also focusing our concerns on the misuse and side effects of technologies that have permitted this increase in productivity—which is dependent on fossil energy and contributes to water pollution, soil erosion, food contamination, decreased water availability, loss of biodiversity, and an increase of simplified rural landscapes. To face these concerns, strategies are under development to assess environmental sustainability of rural activities and to promote landscape transformation and remediation (Jordan 1993; Van Straalen and Krivolutsky 1996; Pimentel 1997).

In larger undeveloped tropical areas, however, the situation is just the opposite. Forested areas are increasingly transformed into rural areas from a poor unfamiliar peasantry that sometimes collapses soon after having reared their lot of forest. Other adopted conventional forms of agriculture based on alien crops in a monocultural design soon sell their "reclaimed" forest land to cattle growers who further modify the original forested landscape into a poorer savanna landscape managed by fire, as is the case especially in Brazil and in some parts of Indonesia and Africa (Jordan 1987; De Miranda and Mattos 1992; Paoletti and Pimentel 1992; Heywood and Watson 1995).

Most of the domestication that today permits high productivity and these current trends was first developed in one restricted area of the Middle East, the Fertile Crescent. Some of the plants and animals developed in this region are as follows: wheat, barley, oat, rye, beef, pork, sheep, and goat. This area is not forested yet and is now semidesert. This area has developed, as well, three important religions worldwide (Christian, Hebrew, Muslim). Cows, sheep, goats, and pork have migrated worldwide with wheat, barley, and oats. Few other key crops come from Asia (rice) or from South America (corn, potato, tomato). Small seeds, perennial crops such as perennial grains, small animals, and minilivestock have all been less developed since the beginning; they are in a certain way synonyms of underdevelopment. In general, they are not listed among the top priorities in the agenda of International Agencies and Research Centers.

We will discuss landscape structure and problems linked to the development of larger domesticated animals and of annual grain crops. We will discuss this in relationship with biodiversity maintenance and promotion, especially in the tropics. The following will also discuss alternatives that are feasible especially in locations in which forest and rich biodiversity are still a dominant situation.

The focus on our resources: Biodiversity

There are at least two points which amaze the researcher—how many beetle and insect species are living on the planet and how few plant and animal species we currently consider as our possible food. In Western countries, for instance, insects as well as most small invertebrates are still considered inedible in spite of the evidence that shows insects to be the large majority of living organisms (Defoliart 1997; Paoletti and Bukkens 1997). Approximately 90 percent of world food for people comes from just 15 plant and 8 animal species (Wilson 1988); however, the use of biodiversity is incredibly different among different human groups. In Java, small farmers cultivate 607 crop species in their gardens which have an overall species diversity comparable to deciduous subtropical forests (Dover and Talbot 1987; Michon 1983). Natural forests in Sumatra are similar to the cultivated ones in number of species and sometimes diversity (Foresta and Michon 1993). In Swaziland, Africa, 220 wild plant species are commonly consumed (Ogle and Grivetti 1985). Andean farmers cultivate many clones of potatoes, more than 1,000 of which have names (Clawson 1985). It has been estimated that about 10,000 plants have been used worldwide for human consumption (Esquinas-Alcazar 1993)—far more than plants currently found in crop indexes or in Western countries. For instance, Rehm (1994) says in Gottingen they have listed about 5,000 species of agronomic plants, but in his book he lists only 2,454 plant species. Most of these biodiversity resources are known to "primitive" local peoples living in a marginal situation, and sometimes these local resources are semidomesticated; such is also the case with small rodents in South America, Asia, and Africa (Hardouin Stievenart

1993; Feer 1993; Paoletti and Bukkens 1997). Some plants are also semidomesticated in the West; for example, the pistic in Western Friuli (Italy) which is a dish prepared with up to 52 wild plants that are still collected in spring in a few mountain villages (Paoletti et al. 1995).

Most of the few cultivated plants producing grains (wheat, corn, rice) are annual, and need tillage, fertilization, and weeding, which sometimes deeply affects the sustainability of the agroecosystem. The long history of breeding has selected larger grains over small ones (Zohary and Hopp 1993) in most countries such as North and South America (NRC 1989; Smith 1995) and Africa (BSTID 1996) (Figure 1). Most areas designated for crop and livestock have lost their forests and have developed taxes, towns, and religions. Grains in most countries are the key base for taxation (Bray 1984). Selection (by domestication) of larger annual grains and large mammals is linked to loss of forests and possibly biodiversity. But can small animals (minilivestock) and polyannuals or small grains better build a more biodiverse environment in a mosaic landscape pattern? Are crop and livestock dimensions linked to landscape pattern and biodiversity structure as well?

Two cases of tropical areas that still have their forests

The Sago culture in New Guinea: The Asmat case

The Asmat are a Papua group of about 70,000 people. Their territory (ca. 30,000 km^2) is located in the southwestern part of Irian Jaya (Indonesian New Guinea). It extends from the coast of the Arafura Sea to the interior foothill region. The wide alluvial swamps are covered with tropical rain forest and crossed by a dense network of rivers. The Asmat population's diet is based 80 percent on sago, a food obtained from the starchy pith of a palm (*Metroxylon rumphii* and *M. sagu*), which grows spontaneously in the humid forest, or is sometimes cultivated, for instance, in Papua New Guinea by the natives (Ulijaszek and Poraituk 1993). Other sago species are used for the same purpose in Malaysia (Brosius 1993). To get the edible part of the sago, fell the palm, crush the stipe's pith with a kind of adze, and then wash it in a trough. The liquid is filtered, and the starch is then collected in a draining pan, made out of the plant's leaves. Following Peters (cit. in Townsend 1974) 100 g of raw sago contain 27 g water, 71 g carbohydrates, 0.2 g proteins, 0.3 g fiber, 30 mg calcium, 0.7 mg iron, traces of fat, carotene, thiamin, and ascorbic acid. There are also discreet quantities of phosphorus and potassium. Although sago is high in calories, its plastic value is limited, and the diet has to be supplemented with the products of fishing and hunting and insect collection. The Asmat collect different kinds of insect as well, but in particular a palmworm (Coleopptera,

33

Figure 1. Wild and domesticated seeds

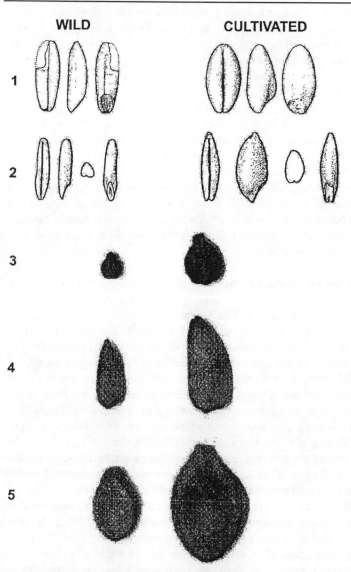

WILD   **CULTIVATED**

1. Wild and cultivated elnkorn wheat. 2. Wild and cultivated emmer wheat.
3. Marsh elder. 4. Sunflower. 5. Squash (from different sources).

Curculionidae, *Rhynchophorus ferrugineum*). The larvae develop quickly on sago palms. In preparation for the Pir-Jimi or "larvae feast," the larvae of palmworms are "grown" in the fallen trunks of sago palms and then gathered in large quantities up to several kilos a few weeks later and consumed in the course of the ritual (Sowada 1986).

In addition, many other insect species are collected on trees, bushes, and soil. Sometimes invertebrates support up to 30 percent of the meat diet requirements obtained by fishing and hunting. This is a great difference from our Western models, which comprise large ruminants with intensive food-production systems. Asmat and other New Guinea peoples, as well as many Amerindian tribes, still live in forested areas with local diversified resources. In Southern Vietnam, for instance, similar palmworm larvae (collected in bamboo shoots) are eaten by children and villagers, especially near the forested areas in which ethnic minorities live (personal observations, October 1996). It is important to focus on traditional uses of local biodiversity and minilivestock in order to maintain and promote the local knowledge systems that are suddenly lost when these groups come into contact with Western civilization.

The Incas agriculture case

In the Andean highlands, the Incas developed a rich and diversified agricultural system by horticultural adaption. They had available at least 50,000 species of plants, some of which have built the basis of current Western agriculture (corn, potato, tomato, pepper, bean, and tobacco). Most traditional agroecosystems were polycultures (a mixture of different, associated plants); however, they also respected the relationship and distribution of plants in the landscape pattern. The "lost Incas' crops" could become part of our sustainable future if more work were dedicated to refining these plants for use in modern agroecosystems (NRC 1989; Hernandez and Leon 1994). For instance, potatoes cultivated in Western regions belong only to one species, *Solanum tuberosum*, among the 235 described species of *Solanum* living in the Andean Region, of which at least 7 are cultivated species by Native Americans (Hawkes 1990). In fact, the rich diversity of potatoes has been used in breeding for insect resistance: the insect-repellent *Solanum berthaultii* with *S. tuberosum* (Tingey 1991) has been crossed, and promising results were obtained.

Amazonian forested areas

There are few references to local crops in the lowland Amazonian landscapes (Posey 1993) and their possible future. In these areas, little or no archeological evidence exists regarding domestication of plants like yucca (*Manihot esculenta*), tupiro *(Solanum sessiliflorum),* pijiguao (*Bactris gasipaes*), cucurito

35

(*Attapea maripa*), and others (see Hernandez and Leon 1994). Most crops of slash and burn agriculture are polyannuals, and sometimes the crops are highly poisonous, containing compounds such as cyanogenic glycosides in yucca, or alkaloids in tupiro leaves. Locally evolved technologies have been developed to detoxify the starchy roots of yucca after harvest; even if varieties with low glycoside content are known, their use is not considerably developed against any reasonable evidence of an easier way of production and consumption. Defense from parasites, strategies against neighbor enemies (who cannot use it as food because of poisoning if eaten raw), different taste, and higher productivity are among the possible, but not clear-cut, reasons for this permanence of poisonous yucca (McKey and Beckerman 1993; Dufour 1993). We must also stress that in horticultural communities, polyannual plants have been dominant compared to the abundant annuals, trees, and bushes; also, the semidomesticated status of many plants is latent in the Amazon (Posey 1993).

Evolution of our crops

Most of our crops and animals have been domesticated in eight areas of the world, for which sufficient documentation exists (Figure 2) (Smith 1995). In particular, one relatively small region in the Middle East (the Fertile Crescent), over a period ranging from 8,000 to 12,000 BC, produced most of the crops and animals we are accustomed to (Smith 1995, 1996; Zohary and Hopf 1993) (Figure 3). Today, we face rural landscapes in most countries that have been based on plants such as wheat, barley, rye, oats, and especially animals such as cows, sheep, pigs, and goats. To express need of food sufficiency the FAO has produced the logo *"fiat panis,"* which shows a head of wheat. Food is bread and bread is wheat—not only in figurative terms. Most of our science and our landscape is based on such annual cereals.

There is evidence that the wild antecedents for both animals and grain seeds were smaller; therefore, domestication has increased the size of most animals and plants and their grains (Figure 1).

Larger animals, especially ruminants, need larger pastures—and in general large animals and forests are incompatible. In fact, the landscapes in most Mediterranean areas in which sheep and goats have been developed have turned into degraded pasture or into poor "savanna" managed by fire. Some have suggested that the Mediterranean peoples invaded the New World because of the loss of sustainability due to overgrazing and degradation of the natural vegetation. It has also been suggested that most civilizations have collapsed by having generated improper use of the land, in general based on few crops, few animals, and poor soil erosion control (Carter and Dale 1974).

36

Figure 2. Centers of origin of domesticated plants

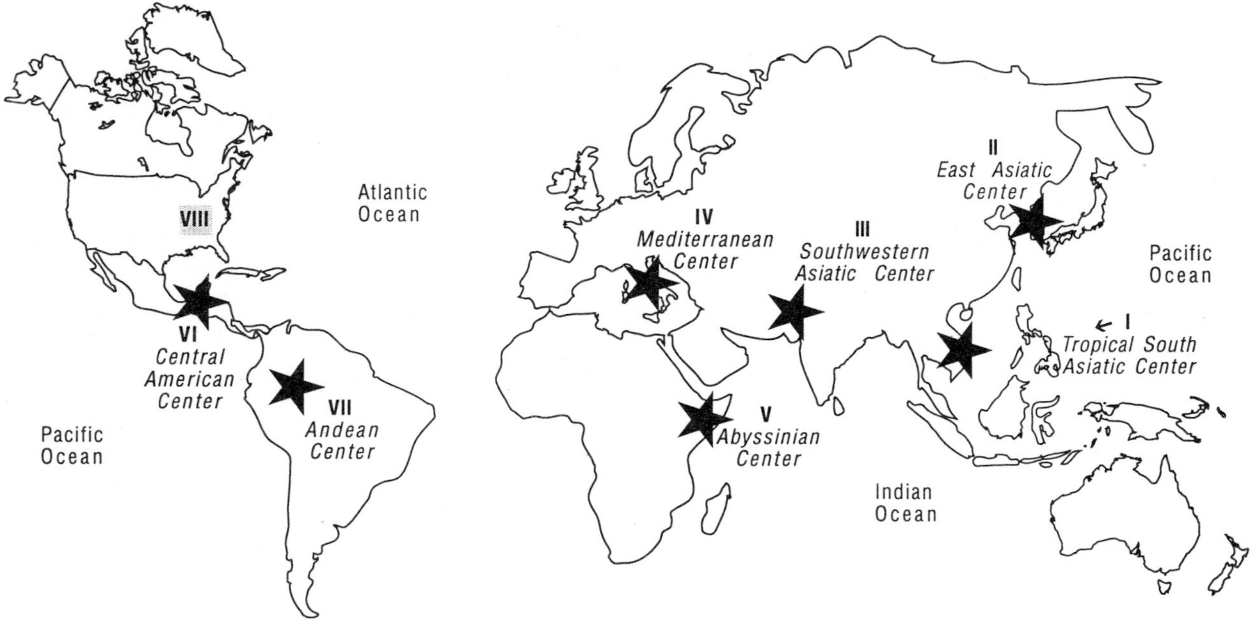

Vavilo's final map, published in 1940, showed seven centers of origin of domesticated plants.

Figure 3. Approximate time periods when plants and animals were domesticated in the seven primary centers of agricultural development

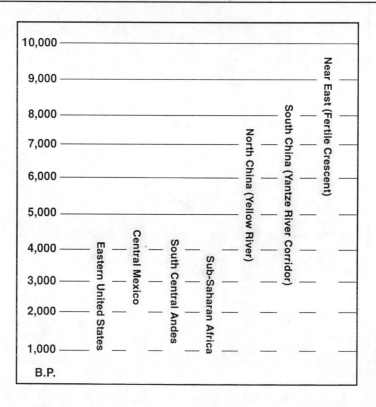

Small animals and polyannuals for a more sustainable landscape?

Then the question is: Is it possible to develop small grains, small animals, and polyannuals? And if so, which kind of sustainability could we perform? On which kind of landscape? And would this be a more suitable landscape? What links are interwoven between small animals and plurality of plants in our diet?

It has been reported that most of the lost crops of Africa have smaller grains than domesticated ones (BSTID 1996). The same has been observed for the current key grains and their wild relatives (Smith 1995). The same trend has been documented for the key cereals and their wild relatives in the Fertile Crescent (Zohary and Hopf 1993). Most wild relatives of crops have developed metabolites and semiochemicals for their defense against predators and pathogens, but most of these natural defenses have been deleted and reduced in the domestication process. However, in tropical forests some cultivated plants still have consistent toxicity that is eliminated only after harvest as is the case of casaba (*Manihot esculenta*) or quinoa (*Chenopodium quinoa*) in South America. Both are detoxified after harvest. Cows, sheep, and goats have also been enlarged in the domestication process (Smith 1995). Small animals, including insects, have been eliminated from the diet of "civilized" countries (Paoletti and Bukkens 1997). When Guajibo Indians, in Amazonas, were asked about palmworms (*Rhynchophorus palmarum*), they indicated that they considered the worms a gourmet snack, but the insect larvae is not an item to propose to Western tourists! The collapse of small vertebrates (insects) versus the large vertebrates (cows) is clear, especially in systems in which forests are disappearing from the landscape. In this situation, it appears clear that the alternative is technology use and forest retention. It has been observed that in the densely populated Yangtze river basin in subtropical China, only a few buffaloes are needed in the rural landscape. The buffaloes are not used for meat production but only for tillage and seeding operations. In India, taboo guards the cattle against meat consumption as well.

Discussion

Landscape structure and problems, including water depletion and pollution, soil erosion, biodiversity crisis, and pesticide residues, are only some of the issues we have resolved from the past. Specialized scientific branches have also allowed this trend of fewer plants, fewer animals, and less diversity in agroecosystems. The approach of science and technology to problem solving seems contaminated by doubts as well, and a cooperative holistic system approach seems better fitted to deal with complex problems related to the environment (Funtowicz and Ravetz 1993). Our reality can be measured by continuous comparison and discussion, coeducation or reeducation, and reallocation of trends and ideas as suggested by simulated strategies (G. Galoppin, personal communication, May 1996). Sustainability of further options to be implemented in our landscapes is not yet fully understood. Questions remain such as "What happens if we exchange our current annual grains for perennial grains?" (Jackson 1991; Beker et al. 1992); "What happens if we substitute our few crops with a more diversified set of plants and small animals, the minilivestock?" (Giampietro et al. 1997; Paoletti and Bukkens 1997).

Bioindication techniques can assess better alternative options to be played in different practical scenarios, such as the decreased use of pesticides in rural areas forecast by some countries such as the Netherlands, Sweden, and Canada (Pimentel 1997). We know, however, that the main trend is to reduce the hours of labor per capita allocated in agriculture in our modern societies and to reduce rural populations. Currently, genetically engineered crops try to fit these trends without any strong environmental and ecological reflection. For instance, the new engineered crops appear as results of lab work and marketing conveniences far more than as a trend to change highly subsidized farming systems. Crops genetically engineered to be resistant to target herbicides, such as glyphosate, seem better suited to some farmer's needs such as security against weeds, but pose serious environmental and possibly economic concern (Giampietro 1994; Paoletti and Pimentel 1996). It is time for sophisticated computerized tools to try to support options and opportunities that promote the highest possible level of biodiversity. If biotic diversity is considered the key to evaluate sustainability, then implementation of perennial crops, multiple crops, and small livestock as alternatives to current trends have to be considered priorities. For instance, if much is known about cows, pigs, sheep, and goats, then nothing (or only anecdotal knowledge) is available for minilivestock, including insects (Feer 1993).

Why not choose from the large number of small and diverse wild organisms to revise our landscape and our production system? What could be the relationship among small organisms as potential livestock and our potential landscape? This relation has to be modeled and analyzed carefully, but these alternative options have to be developed soon. The evidence from the forested areas is that small minilivestock do not compromise the forest itself as the larger livestock do.

Local populations that have these plants and minilivestock in their cultural knowledge and practical use, must be considered as target of this co-participative process of landscape resource use and sustainable promotion. Possibly small scale, local projects have to be linked with basic research (sometimes strongly lacking) and adequate policies. Evaluation and promotion of local sufficiency must be linked with promotion of marketing in order to establish new perspectives.

References

Becker, R., D. Meyer., P. Wagoner, and R.M. Sunders. 1993. Alternative crops for sustainable agricultural systems. Agriculture Ecosystems and Environments, 40(1-4): 265-274.

Bray, F. 1984. Biology and biological technology: Agriculture. In J. Needham (ed.), Science and Civilization in China. Cambridge Univ. Press, New York, 6(2), p.724.

Brosius, P.J. 1993. Contrasting subsistence ecologies among Penan foragers, Sarawak (East Malaysia). In C.M. Hladik et al., Tropical Forests, People and Food. The Parthenon Publ. Group, New York, pp. 515-522.

BSTID (Board on Science and Technology for International Development). 1996. Lost crops of Africa. National Academic Press, Washington D.C., 1:1-383.

Carter, V.G., and T. Dale. 1974. Topsoil and civilization. University of Oklahoma Press, Norman, 292 pp.

Clawson, D.L. 1985. Harvest security and intraspecific diversity in traditional tropical agriculture. Economic Botany, 39:56:67.

Dover, N., and L. Talbot. 1987. To feed the earth: Agroecology for sustainable development. World Resources Institute, Washington, D.C.

Dufour, D.L. 1993. The bitter is sweet: A case study of bitter cassava (*Manihot esculenta*) use in Amazonia. In C.M. Hladik et al., Tropical Forests, People and Food. The Parthenon Publ. Group, New York, pp. 575-588.

Esquinas-Alçazar, J.T. 1993. La diversidad genetica como material basico para el desarollo agricola. In J. I. Cuberto and M. T. Moreno, La Agricultura del Siglo XXI, Ediciones Mundi Prensa, Madrid, pp. 79-102.

Feer, F. 1993. The potential for sustainable hunting and rearing of game in tropical forests. In C.M. Hladik et al., Tropical Forests, People and Food. The Parthenon Publ. Group, New York, pp. 691-708.

Foresta, H., and G. Michon. 1993. Creation and management of rural agroforests in Indonesia: Potential applications in Africa. In C.M. Hladik et al., Tropical Forests, People and Food. The Parthenon Publ. Group, New York, pp. 709-724.

Funtowicz, S.O., and J.R. Ravetz. 1993. Science for the post-normal age. Futures, 25 (7): 739-754.

Giampietro, M. 1995. Sustainability and governance: Checking the mutual compatibility of socio-economic and ecological dimensions of human development. Proposal for the European Program on Environment and Climate. National Nutrition Institute, Rome.

Giampietro, M., M.G. Paoletti, S. Bukkens, and H.A.N. Chunta. 1997. Biodiversity in Agriculture for sustainable future. Agricul. Eco. and Environment. 62(2-3): 77-262.

Hardouin, J., and C. Stievenart. 1993. Invertebrates (minilivestock) farming. CTA (Center of Technical Cooperation), Wageningen, p. 223.

Hernandez, B., and J. Leon. 1994. Neglected crops. 1942 from a different perspective. FAO, Rome, 341 pp.

Heywood, V.H., and R.T. Watson. 1995. Global biodiversity assessment. Cambridge Univ. Press, New York, 1140 pp.

Jackson, W. 1991. Development of perennial grains. The 18th international conference on the unity of sciences. Seoul, Korea, August 23-26, 1991.

Jordan, C.F. 1987. Amazonian rain forests: Disturbance and recovery. Springer-Verlag, New York.

Jordan, V.W.L. 1993. Agriculture scientific basis for codes of good agricultural practice. Commission of European Communities, Brussels, p. 165.

McKey, D., and S. Beckerman. 1993. Chemical ecology, plant evolution, and traditional Manioc cultivation systems. In C.M. Hladik et al., Tropical Forests, People and Food. The Parthenon Publ. Group, New York, pp. 83-112.

NRC (National Research Council). 1989. Lost crops of the Incas: Little-known plants of the Andes with promise for worldwide cultivation. National Academy Press, Washington, D.C., p. 415.

Paoletti, M.G. 1997. IPM practices for reducing fungicide use in fruit crops. In D. Pimentel ed. Techniques for reducing pesticide use: Economic and environmental benefits. J. Wiley & Sons Ltd., New York, pp. 343-378.

Paoletti, M.G., and S. Bukkens. 1997. Minilivestock: Sustainable use of biodiversity for human food. Ecology of Food and Nutrition (s.i.), 36 (2-4): 91-346.

Paoletti, M.G., A.L. Dreon, and G.G. Lorenzoni. 1995. Edible weeds "pistic" found in W. Friuli (N.E. Italy). J. Econ. Bot., 49(1): 26-30.

Paoletti, M.G., and D. Pimentel. 1992. Biotic diversity in agroecosystems. Elsevier, Amsterdam, 356 pp.

Paoletti, M.G., and D. Pimentel. 1996. Genetic engineering in agriculture and the environment, BioScience, 46 (9): 665-673.

Pimentel, D. (ed.) 1997. Techniques for reducing pesticide use: Economic and environmental benefits. J. Wiley and Sons Ltd., New York.

Posey, D.A. The importance of semi-domesticated species in post-contact Amazonia: Effects of the Kayapo Indians on the dispersal of flora and fauna. In C.M. Hladik et al., Tropical Forests, People and Food. The Parthenon Publ. Group, New York, pp. 63-71.

Rehm, S. 1994. Multilingual dictionary of agronomic plants. Kluwer Academic Publishers, Dordrecht, 286 pp.

SITCPGR. 1996. The state of the world's plant genetic resources for food and agriculture. FAO, Rome, p.336.

Smith, B.D. 1995. The emergence of agriculture. Scientific American Library, Washington, D.C.

Smith, B.D. 1996. The origins of agriculture in the Americas. Evolutionary Anthropology, pp. 174-184.

Ulijaszek, S.J., and S.P. Poraituk. 1993. Making sago in Papua New Guinea: Is it worth the effort? In C.M. Hladik et al., Tropical Forests, People and Food. The Parthenon Publ. Group, New York, pp. 271-280.

Van Straalen, N.M., and D. Krivolutsky. 1996. Bioindicator systems for soil pollution. Nato ASI Series, Kluwer Academic Publishers, Dordrecht, p. 261.

Zohary, D., and M. Hopf. 1993. Domestication of plants in the old world. Clarendon Press, Oxford, p. 278.

4

The Role of the Natural Resources Conservation Service in the Development and Implementation of Soil and Water Conservation Policies in the United States

Paul W. Johnson

It is important to put the work of the Natural Resources Conservation Service (NRCS) into perspective. The agency was formed in the 1930s as the Soil Conservation Service (SCS). We recognized then that we could not achieve real conservation by putting all of nature into national parks, wildlife refuges, and wilderness.

Three eras

There have been three eras in the U.S. conservation movement. First, there was the protection of special places in national and state parks, forests, refuges, and wilderness, which began over a century and a half ago. Then, with the publication of Rachel Carson's *Silent Spring* in the 1960s, the age of environmental protection began. The "common" became global, and we began to recognize the interconnectedness of natural resources in ecosystems—from the smallest spade of earth to the earth as a whole. The third era arose three decades earlier with the Dust Bowl in the 1930s when we came to understand that private landowners held the world in their hands.

These three conservation eras—protecting special places, regulating things that foul the common, and being good stewards on private land—may have begun as individual elements of history, but they are as intertwined today as the ecosystems they aim to preserve. It is not possible, in my view, to achieve our obviously strong national conservation goals by focusing only on one or two of the three elements.

From the beginning, SCS provided technical assistance to individual farmers and ranchers and other landowners. Hugh Hammond Bennett, the first chief, put together a workforce of technicians—biologists, engineers, chemists, soil scientists, botanists, hydrologists, social scientists, and the like—and got them functioning as interdisciplinary teams. Team members were to help land users do well by their land after the ravages of the Dust Bowl. Bennett's message to farmers was this: "We will work side-by-side with you and not tell you what to do. The health of your land is your responsibility." The programs of the 1930s, particularly the demonstration program in the Driftless Area of Iowa, Minnesota, and Wisconsin, have allowed me to keep my farm in northeastern Iowa both productive and environmentally healthy.

We learn as we go. In the beginning, issues were simple. We knew, or thought we knew, that if we merely saved soil we could sustain farm productivity. Now things have changed. Farms are much more than food and fiber factories. My farm produces water, and it can do it well or poorly. All farms are like mine. The produce is not simply what is measured in the nation's commodity markets. Clean water, clean air, healthy soil, beautiful land, and habitat for the wildlife we share the earth with are commodities, too.

Today, 12,000 NRCS employees will get up and go to the field to work with farmers and ranchers. This is the most underestimated work in the world. These people, trained as natural resource experts just as in Bennett's day, work one-on-one with millions of landowners and land managers across the country. They help these landowners and land managers achieve their goals in an environmentally and economically sustainable fashion.

Regulation versus a voluntary approach

Certainly, there is a role for regulatory authority. The 1985 Farm Bill for the first time demanded that agricultural producers conserve soil and protect wetlands in return for commodity price supports and other federal farm program benefits, and that linkage has made a great deal of difference in the health and appearance of the American landscape over the past decade. But that situation has now changed. As we reduce farm program benefits, like the 1996 Farm Bill did, there is a question about whether and how much regulatory authority will ultimately remain. If it does, it probably will come from issues that go beyond individual land parcels, to such recognized public needs as clean water, endangered species, and biological diversity. There will be regulatory approaches—which I believe are and will be necessary—because ecosystems do not restrict themselves to individual farms and ranchers or to the political niceties of town, country, or state boundaries. Natural resource issues simply are not confined by the limits of human-drawn boundaries.

But regulation, as I've said, cannot do it all. Private land—nearly 70 percent of the U.S., exclusive of Alaska—is an important source of environmental protection—for salmon, migratory songbirds, and waterfowl; for clean air and water; for the biodiversity in a much broader sense. "Government," Aldo Leopold wrote, "cannot own and operate small parcels of land, and it cannot own and operate good land at all . . . It is the individual farmer who must weave the greater part of the rug on which America stands."

The voluntary approach is viable. We in NRCS have not been proactive enough in its use. Traditionally, we have provided service only to those who walked through our doors, but the NRCS knows who is not doing well on the land. We must approach those individuals and encourage them to do better. Regulation will be needed for setting standards that apply to and beyond the farm or ranch to ensure a healthy environment for us all. Wetland regulations are an example. We simply must keep wetlands on the landscape, and without policies, many farmers would drain them. But we must also get farmers to protect wetlands because they want to. That is where voluntary and regulatory programs come together.

Voluntary, locally driven programs—with standards set by the Environmental Protection Agency (EPA), the states, and local government and with conservation districts providing leadership and landowners providing initiative—will be successful.

Four steps

Four steps are needed to make conservation work on private land. The first step is a comprehensive inventory of the land. What natural resources comprise the landscape?

The second step is an assessment of the land's health, its condition and trends. Is the health of land getting better or worse? These farm appraisals must go beyond soil erosion and, as Leopold put it, "read the land." The assessments must look at the whole farm or ranch. However, Leopold's suggestion for reading the land accurately is no simple task. Land—soil, water, air, plants, and animals—is a marvelously complex and dynamic system that often changes in ways too subtle to perceive. Trying to understand that—the static and the changing—is what resource assessment is all about.

The third step is planning. Once a landowner knows his or her natural resource base and has set goals, both personal and business, it is possible to plan the best use and management of that natural resource base—in its entirety—to achieve the goals. As Leopold put it, "A farmer who conserves his soil but drains his marsh, grazes his woodlot, and extinguishes the native fauna and flora is not practicing conservation in an ecological sense. He is merely conserving one component of land at the expense of another." NRCS provides planning assistance to landowners, but it does not do planning. A landowner does the planning.

The fourth step is technical—and sometimes financial—assistance for conservation structures and management practices, including vegetative buffer strips, wildlife habitat, and conservation tillage, among others. Farmers can also obtain assistance for long-term protection of ecologically sensitive or highly erodible land or for protection of farmland or wildlife habitat.

If these four steps are taken, I am convinced that a voluntary approach can work in conserving our natural resources on private land. To those who disagree, I would say only this: Landowners simply will not bow to an exclusively regulatory approach, nor can we as a nation buy all of the land or development rights on the land we might want to protect.

As a farmer, I want to do what's right for my land. And I will. I do not, however, want to be fenced in, ordered about, told what to do and how to do it. Most farmers and ranchers feel this way. Voluntary, incentive-driven approaches must do the lion's share of private land conservation; private landowners, with help from the rest of us, must make the difference. People—farmers and ranchers—do acts of kindness for neighbors every day that we do not get paid for. The same must be true for conservation. Conservation, then, is not a limit on freedoms, but a protector of those freedoms.

We must promote the conservation ethic. We need a stewardship commitment from landowners. From the rest of us we need a commitment to help landowners maintain the nation's environmental health, with financial support when needed and appropriate. After all, as Leopold wrote, "Some components are of economic importance to the community, but of dubious profit to the individual owner." This includes most marshes, most cover on streambanks and steep slopes, and most windbreaks.

The research imperative

There is one other need I must mention. That need is research. We have not challenged the research sector adequately on conservation issues. The focus has been on and remains on food and fiber production. It is time for the research community to turn its attention to conservation. We must have a workable soil quality index. We need information on how to bring about and protect organic matter buildup. We have fairly good information on filter strips and other buffer strips and where and how they work, for example, but not much about why. We have scarcely begun to understand the ways in which the land can be productive on its terms and nature's terms. We have not done much with new conservation technologies either. Across the board, good solid research and the transfer of its results are absolute necessities for conservation.

We also need to do a better job of sharing history, natural history and human history alike. We in NRCS and our partners should be telling people what they have on their land, its natural resource base and its production potential. We should also be talking about opportunities to use the land to perform multiple functions. Natural history exists. The question is how to put it into laptop computers and share it with farmers.

To make the voluntary approach work, all of us must help. As the one agency in our national government with primary responsibility for conservation on private land, NRCS must speak for the land and its health, and I believe our employees are committed to doing that. But others—landowners, users, and city dwellers—must likewise take some responsibility to see that the land stewardship task is accomplished.

5

Conservation Policy in the United States:
Is There a Better Way?

Thomas A. Weber and Gary A. Margheim

Natural resource conservation policy in the United States has many dimensions and obviously does not emerge in a vacuum. It evolves within a complex set of forces that influences choices made by resource users. Its boundaries extend beyond the nation's borders—since U.S. policies are clearly influenced by what happens elsewhere.

Humans, in their quest for economic development and improvement of conditions affecting life, must come to terms with the realities of the limitation of natural resources and the carrying capacity of ecosystems, and must also take account of modern conservation. As such, conservation is basic to human welfare and indeed to human survival. There are many well documented cases, over the history of humankind, of the destruction of the ecosystem by humans resulting in the ruin of civilizations. Regardless of background, nationality, type of government, or political concern, and even regardless of socioeconomic status, humans are directly affected by natural resource conservation—whether or not they recognize it at this time.

Additionally, a discussion of public policy dealing with conservation on private agricultural lands—the focus of this chapter—must be tempered with an understanding of political systems, the concept of private property rights, and social and cultural values of the society.

This paper will address the development and maturity of conservation policy in the U.S. and will provide a synopsis of lessons learned from this policy over time. Such lessons should be useful to others as they craft conservation policy.

Defining conservation

In this chapter, conservation refers to the direction of human activities involving the use and management of natural resources in order that the greatest sustainable benefit to present human generations will be achieved, while main-

51

taining the potential to meet the needs and aspirations of future generations. A central part of conservation, then, is managing natural resources in
such a way that the options for use of the resources are maintained for
future generations.

Conservation policy in retrospect

Historically, agricultural policies have focused on production and economic
issues which evolved independently from resource conservation and environmental policies. The institutions that developed and implemented agricultural
policies frequently differed from those responsible for natural resource and conservation policies affecting agriculture. This is not to suggest that environmental quality and natural resource conservation are new issues in American farm
policy. On the contrary, the need to conserve soil to maintain land productivity
was recognized in colonial times and was advocated by such early leaders as
Thomas Jefferson. But the focus and public influence on natural resource concerns has changed over time relative to those of agricultural interests.

Conservation policies affecting agriculture in the U.S. may be broadly characterized as evolving through five stages: the age of apparent natural resource
abundance, before 1930; a period of resource conservation for agricultural interests, 1930 to 1960; an era of diverging agricultural and environmental policy,
1960 to 1985; the era of improved consistency between policies for agricultural
production and conservation, 1985 to 1995; and the stewardship transition era,
the present into the future.

Natural resource abundance—Prior to the 1930s, land and water resources,
the vast share which had never been tapped for agricultural use, were considered abundant. This view was reflected in major public policies aimed at development of the natural resource base to accelerate economic development and,
not necessarily, to sustain or protect it.

The Homestead Act of 1862 heavily subsidized private ownership of federal
land to encourage population dispersal and create an economic development
base in the western U.S. Likewise, the Reclamation Act of 1902 authorized
federal subsidization of water resource development in which the government
constructed numerous irrigation projects and provided water for irrigated farming at prices below their real value.

Conservation policy for agriculture—As America entered the Great Depression of the 1930s, the economic vitality of American agriculture was devastated
and it became apparent that agricultural resources were limited. Public awareness of soil erosion and water availability was heightened by the severe drought
of that period. Many production and conservation policies were established to

keep agriculture viable. Conservation policies were voluntary in nature and relied on the use of incentives (cost-share, technical assistance, etc.) to accomplish goals, but they were not linked to production incentives. In fact, conservation policy was seen and used as a vehicle to deliver economic and production assistance to producers. In its origins conservation policy was seen to primarily benefit the producer. The real change in modern time is that conservation is increasingly seen as benefiting the public.

It was during this era (1935) that the Soil Conservation Service (SCS), now the Natural Resources Conservation Service (NRCS), was created. SCS's popularity was its voluntary approach to conservation, its ability to provide technical assistance for conservation measures, and the fact that most assistance enhanced the productive capacity of the farm. A landowner simply called the local office, explained his or her problem with soil erosion or drainage, and an SCS employee would visit the farm and provide technical advice. In addition, a payment from the U.S. Department of Agriculture (USDA) was often sent to the farmer for the public share of financial support in solving the resource problem. During this period federal agricultural policies were developed that ensured a stable and adequate supply of food and fiber at reasonable prices while generally maintaining farmers' incomes and well-being.

Diverging policy interests—The environmental awareness growth and movement of the 1960s, and thereafter, provided the growing "public will" to reform attitudes and actions of people towards the natural resource base. It spawned society's first broad policy contained in the National Environmental Policy Act to reorient the priorities and responsibilities of government, and to establish a mechanism to fully consider the adverse impacts of federal actions on the resource base. It stimulated public pressure for conservation and environmental policies aimed at enhancing general welfare rather than strictly meeting the needs of agricultural interests.

However, during this same period the boom and bust grain cycle of the 1970s occurred. When the Soviet Union and China began importing grain in large amounts, the federal government encouraged planting "fence row to fence row." Producers were provided irrigation water at subsidized rates. A significant drawdown of groundwater supplies resulted from bringing new land under irrigation for crop production. Drainage of wetlands was common to allow greater production. This occurred at the same time federal payments were being made to landowners to prevent soil erosion and water depletion, and improve wildlife habitat. It is no wonder that questions about consistency between conservation and farm program policy arose.

Converging policy interests—The boom collapsed and agriculture entered into the farm financial crisis of the early 1980s. Public concern for the plight of the small farmer was particularly high but at the same time was often at odds with its equally vocal demands for a clean, environmentally sensitive agriculture. A force that initiated the process for rectifying these inconsistencies was the first National Conservation Program (NCP) released by the USDA in 1982. The NCP did several things. First, it established soil erosion reduction as USDA's highest natural resource conservation priority and provided recognition of the increasing concern about the magnitude of the off-site costs of soil erosion. Second, it called for more consistency and linkage between the USDA programs in meeting conservation and production objectives. Lastly, it called for increased involvement of people in natural resource management decisions.

The Food Security Act of 1985 (FSA) implemented many of the recommendations contained in the NCP with a series of new conservation programs and conservation policy. Two FSA provisions in particular—the highly erodible land conservation and the wetland conservation—made the availability of selected USDA farm program benefits contingent upon compliance with specified conservation provisions. Under the highly erodible land provision a producer was required to have a conservation plan fully implemented on highly erodible cropland by January 1, 1995, or lose eligibility for USDA program benefits. The wetland conservation provision extended this loss of benefits to producers converting wetlands, after December 23, 1985, for agricultural production. In addition, FSA provided authority for a Conservation Reserve Program (CRP) under which USDA was authorized to remove from crop production up to 45 million acres of environmentally sensitive cropland and place it in permanent cover through the use of multi-year rental contracts in lieu of commodity payments. The Food, Agriculture, Conservation, and Trade Act of 1990 (FACTA) continued this policy direction of consistency and added a Wetland Reserve Program (WRP) under which USDA was authorized to restore 1 million acres of wetlands through the use of wetland easements.

Stewardship transition—The Federal Agriculture Improvement and Reform Act of 1996 (FAIRA) ushered in a major transition period not only for agricultural production policies but for conservation policy as well. It represents the start of a movement from conservation programs being driven by commodity programs to one in which the concern for the sustainability of the natural resource base will be the major factor in determining policy. Under FAIRA, deficiency payments for wheat, feed grains, cotton, and rice are replaced with seven year "market transition" contracts covering the 1996 through 2002 crop year. Payments are fixed in advance and gradually decline over the seven years. Farmers can plant any mix of crops, except for fruits and vegetables, and receive annual payments. They must comply with the highly erodible land and wetland conservation provisions of FSA to be eligible for such payments.

At the same time, FAIRA sends a powerful message and demonstrates a renewed commitment by the agricultural community and the public to environmental protection. Provisions to extend the Conservation Reserve Program and the Wetland Reserve Program represent the cornerstone of this commitment and ensure the continued viability of these two important conservation programs. The wetland and highly erodible land conservation provisions of FSA were also maintained, but were amended to provide increased flexibility without sacrificing the significant progress that the American farmers had made in protecting wetlands and other sensitive lands.

The Act also authorizes a number of new initiatives that will enhance and improve natural resources on America's private lands. It provides up to $2.2 billion in new funds for a variety of conservation programs that will allow for more flexibility in the delivery of USDA financial and technical assistance that farmers and ranchers need in managing their operations in a conservation oriented manner. This increased commitment of federal dollars for conservation, while most budget items are decreasing, is recognition that agriculture produces more than commodities—it produces environmental benefits from which all of society benefits.

Past accomplishments in conservation

Agriculture has made major conservation progress over the last 60 years with tremendous improvements since 1985. This progress is attributed in large part to the unique conservation delivery system in the U.S. The USDA, more than any other major federal agency, is linked to the local and state level through its field offices and relevant conservation districts and associations of conservation districts. In particular, locally sponsored or elected conservation districts have made it possible for the USDA to operate programs that are responsive to the unique and specific needs of people and communities at the local level. These efforts have been guided by national priorities that serve the nation.

The mix of voluntary programs and policies, for incentives as well as disincentives established by FSA and FACTA, have allowed farmers and ranchers to achieve enormous progress in improving the nation's natural resources on private lands. At the same time, many of the adverse off-farm environmental effects of agriculture have been reduced. Farmers have adopted practices that have significantly changed agriculture and the world we live in for the better. For example, data from the USDA and other sources indicate that:
- Since 1985 more than 1.7 million conservation plans that were developed under the highly erodible land conservation provisions of FSA, covering more than 140 million acres, have been implemented. Coupled with the CRP,

soil erosion on cropland has declined by a third. In fact, we are rebuilding the soils on many acres of the nation's most important farmland.
- Wetland losses due to agriculture are down by more than 80 percent from the 1970s to an estimated 80,000 acres per year. This is less than the acreage that is proposed for wetland restoration under the WRP each year. With WRP, not only are we achieving a "no net loss" of wetlands due to agriculture activities, but we are near realizing a net increase.
- Wildlife habitat improvement has increased the population of pheasants and other upland birds and mammals, waterfowl, songbirds, and other wildlife species nationwide.
- Other significant improvements in the landscape from the adoption of conservation practices on non-highly erodible land have occurred driven in part by conservation compliance.

These achievements are important for the environmental gains they represent. However, additional benefits have resulted that:
- confirmed that public policy incorporating incentives and disincentives can have a positive impact on the resource base without substantial negative impacts to the agricultural economy,
- proved that the existing conservation delivery system can be utilized to effectively deliver a host of conservation programs,
- effectively changed landowners' operations to benefit the natural resource base,
- expanded the realm of public involvement in, and concern for, conservation policy and its relationship to agricultural productivity.

Agriculture's changing role—conservation policy implications

The investment in, and the policies for, natural resource conservation within the U.S. in the next decade and beyond will be affected by a series of factors. These factors, as discussed below, will identify the major features of the conservation policy landscape and the opportunities that landscape brings for refining such policy.

Rural environment—Agriculture's place in the rural setting and the total food system has changed significantly over recent time. Farm commodity prices and income support programs in the U.S. began at a time when the farm population made up 25 percent of the total population. The nation was mostly rural and based on agriculture. Today, the farm population is less than 3 percent of the total population and nearly 25 million fewer people live on farms today than in the 1930s. Farming's importance in the rural economic base has also

56

changed. Less than 20 percent of U.S. counties now depend on farming for at least one-fifth of their economic base. An important implication of these changes is that the case for policies based on the unique characteristics of agriculture and rural communities has diminished significantly over the past 60 years and must now reflect urban and other interests.

Today there are 2.4 million farms that are highly diverse in their characteristics and very difficult to characterize broadly. One of the most striking features is the concentration of production among them. Only 28 percent of the farms—those selling more than $40,000 in agricultural products annually—account for more than 86 percent of all agricultural output. Consequently, no single policy is likely to be equally applicable to this diversity.

Dependence on foreign trade—A significant change for agriculture, and one with important policy implications, has been the extent to which the U.S. has come to depend on trade. In 1970, agricultural exports were $7.3 billion and imports were $5.8 billion, resulting in a trade surplus of $1.5 billion. Exports were only 14 percent of farm cash receipts, but export growth has been phenomenal. In fiscal year 1997 we expect exports to be $60 billion. Long-term projections for exports are $66 billion the first year of the twenty-first century. The farm sector is clearly no longer isolated from economic trends at home or abroad. It has a large stake in the pace at which developing countries grow, how the Third World manages its debt, the size of the U.S. budget deficit, the value of the dollar, and achieving sustainability in the natural resource base.

Expanding productive capacity—By conventional measures, productivity and output of U.S. agriculture have increased dramatically since World War II. Based on recent research, productivity grew at an average annual rate of 2.2 percent in the 15 years immediately after World War II, 1.0 percent in the 1960s, 1.7 percent in the 1970s, and 1.2 percent in the 1980s. There are no inherent technical reasons why these rates could not be maintained well into the twenty-first century. Even though commodity prices have been at historic highs, the growth in productive capacity, in the U.S. and in other countries in the next decade or two, may well perpetuate the long-term decline in real prices of agricultural commodities. If this happens, the need to adjust an expanding productive capacity consistent with demand for farm products will continue to be central to policy issues.

Conservation ethic—As the human population expands in numbers and activities, concerns have increasingly magnified over the quality of our land and water resources, and the sustained yield of citizen benefits from this resource base. A continuing stream of news articles on excessive soil erosion, degrading surface and ground water, loss of wetlands, conversion of rangeland to cropland, loss of prime agricultural land to nonagricultural uses, and degradation of wildlife habitat

57

emphasizes how natural resources are viewed, used, and managed. The result has been a recognition that as a society there are moral and ethical standards by which policies should be formulated in addition to economic criteria. The clear duty of government, which is the trustee for unborn generations as well as for its present citizens, is to watch over and, if need be, use legislation to protect the exhaustible natural resources of the country. The pragmatic issue is what types of policies do we want for the future: ones which preserve the benefits of the current conservation delivery system; are sustainable for the natural resources in the long run; and are consistent with equity goals?

However, caution must be exercised on building conservation policy entirely on ethical and moral arguments. The era of the ubiquitous stewardship sermon is drawing to a close. Basic and applied research, coupled with a political mandate for policy effectiveness, have produced significant advances in specifying the causes, processes, and consequences of natural resources degradation. The conservation arsenal of policy tools must evolve beyond the ethical argument to one predicated on solid scientific fact.

Local government—The final trend affecting conservation policy in the U.S. is the shift of conservation-related responsibilities from the federal government to state and local government. Local control of conservation programs is consistent with prevailing political philosophies, and conservation districts are a logical choice to assist in framing and implementing local decisions. A debatable issue is the capability, and in some instances the desire, of local government to assume these responsibilities. Capacity building of these local institutions must occur in the future if they are to assume an increased role in addressing natural resources issues. The trend toward local control also implies new responsibilities for local officials including the politically unpopular tasks of monitoring and enforcement. State and local governments must be prepared to exercise leadership regardless of the popularity of these activities.

Lessons learned

Through the evolution of conservation policy in the U.S., several lessons have been learned which, when coupled with the changing role of agriculture, provide insight into conservation policy direction for the future. These lessons include the following:

- delegate as much discretion as possible for policy implementation to the local level,
- maintain some reasonable consistency along geographic/political boundaries in the application of policy, yet allow for flexibility to address state and regional differences,

- encourage voluntary, decentralized, and incentive-driven programs with the flexibility to permit adoption of new science and to tailor programs for site-specific conditions and needs are strongly supported by the American public,
- ensure that policy decisions must be supported by credible facts, data, and scientific research,
- recognize that conservation policies of the past may have actually penalized the "good" conservationist while rewarding those not using their natural resources within their capability,
- advance the position that the conservation delivery system of the USDA, through local conservation districts, provides an effective delivery system for conservation programs,
- acknowledge that awareness of the agricultural sector to conservation has been greatly heightened and has been effective in changing the behavior of a substantial number of farmers in managing their operations,
- involve a broad diversity of interests, not just the agricultural community, to ensure a major stake in conservation policy,
- facilitate historic relationships between the NRCS and private landowners so the relationships are not adversely affected due to a perceived regulatory role of the agency,
- recognize that a key role for the federal government in the conservation of natural resources on private land is to provide for consistent and scientific-based technical standards and guidance in conjunction with oversight for quality control.

Conservation policy for a new century

Population growth, economic progress, and technological progress will continue to fundamentally reshape our world. The picture that emerges is that of a diverse, science-driven, highly productive agricultural sector marked by increasing economic gains, ever-closer links to other sectors of the domestic economy, and growing dependency on export markets to absorb its products. These changes will place staggering pressure on the nation's natural resource base accompanied by unprecedented public demand for responsible stewardship of these resources.

Within the context of future conservation policy, there will be those who feel the natural resources are being damaged, but have little direct control over the source of the problem. They may seek to require that landowners sustain the natural resource base. This is the straightforward approach of regulation—"thou shall not permit destruction of our natural resources." The individual must comply regardless of cost, or face the legal consequences. However, national regulation of agricultural private lands is unlikely. Enforcement would be almost impossible, and national standards difficult to define. In addition, more and more states will enact their own regulations, mostly to reduce the off-farm damages of soil erosion, animal wastes and chemical use. Regulatory activities at the state level are well documented and will play an increasingly important role in the context of future conservation policy.

A softer approach to mandatory conservation policy includes a wide array of measures ranging from pure voluntary through incentives (financial, technical assistance, etc.) and disincentives (USDA eligibility linked to conservation requirements). This policy defines a greater level of responsibility for the landowner to provide adequate protection of the natural resource base without the iron fist of regulation. It is built on a premise that production agriculture and increases in farm income are compatible with the concerns we all share for a clean and healthy environment. Achievements of the past, particularly over the past decade, support this premise.

Integrated management of natural resources designed to achieve uses of the resource base on a sustainable foundation, rather than a degrading one, is being advanced through realignment of a number of management policies. These initiatives are being supported broadly by the public, as witnessed in a variety of polls, and through efforts of coalitions encompassing a broad spectrum of organizations. For example, in 1995 and 1996 the NRCS conducted a series of public listening forums to obtain public input on future conservation policy. Several crosscutting issues raised at all the forums provide insight into future conservation policy. These were:

- emphasizing that natural resource issues are complex and no one institution is likely to have the resources to address them effectively—thus the idea of partnerships within and across federal, state and local levels of governments; between the public and private sector; and among private-sector interests,
- increasing program flexibility built upon the contention that "one shoe" does not fit all; that problems differ from locale to locale, in subtle if not obvious ways; and that landowners need flexibility to respond to changing circumstances,
- utilizing a locally driven approach to conservation program administration—not a top-down approach,
- recognizing that natural resource issues transcend field, farm, and other human-drawn boundaries. Consequently, solutions to natural resource problems should be addressed through a locally led large area or watershed approach to conservation planning,
- basing conservation policy and program administration on fact derived from scientific research,
- providing recognition for the prevention of natural resource problems rather than policies based solely on solving existing problems. This concept, known

60

as "environmental credits," would recognize and reward the actions of producers who voluntarily apply conservation systems on their land. Landowners would earn environmental credits which could be applied toward offsetting taxes, could be redeemed for cash, or even traded on the open-market.

Policy implications of this integrated approach will place more emphasis on a locally led conservation effort. Local people will determine the "state of land" within their individual community and build on this knowledge to establish conservation goals and performance measures. These measures will be the report card that provides the accountability to the American public for any federal funds invested in solving community-defined natural resource problems. Inherent in this approach is the ability of landowners to look at the land differently, capitalize on its many functions, and to determine how they can best contribute to natural resource sustainability.

Conclusion

The context of future conservation policy in the U.S., as it does today, will span the policy spectrum from voluntary, to incentives, to disincentives, to regulation. The mix between the delivery strategies (i.e., voluntary, incentives, disincentives, or regulation) may change to accommodate what the public and political system envision as the "best" way to achieve sustainability of the natural resource base. However, the foundation of any conservation policy in the U.S. will continue to be voluntary conservation programs delivered through local conservation districts. These programs will increasingly be built upon expanded conservation partnerships and increased emphasis on a locally led approach to solving problems.

Flexibility and cooperation as well as interdisciplinary and ecological approaches will be tomorrow's watchwords when it comes to twenty-first century policy. In the U.S., it is important to remember that it is the people on the land who must manage the natural resource base. Aldo Leopold said it well, "All the acts of government, in short, are of slight importance to conservation except as they affect the acts and thoughts of citizens." The challenge of conservation policy is to affect those acts and thoughts.

6

Implementation of Conservation Title Provisions at the State Level

Daniel Conrad

This chapter summarizes experiences with implementation of the Conservation Title at the state level. The Natural Resources Conservation Service (NRCS), formerly the Soil Conservation Service, U.S. Department of Agriculture, was charged with much of the responsibility of implementing the Conservation Title. The Conservation Title in this chapter involves Title 12 of the 1985 Food Security Act (FSA) and further amended by the 1990 Food, Agriculture, and Conservation Trade Act (FACTA).

Barriers to the implementation of the conservation provisions

There were a number of problems encountered at the state level in the implementation of these provisions from 1986 through 1995. The problems included lack of adequate technical assistance, clear definitions, coordination among agencies, communication, resistance on the part of land users to accept quasi-regulatory roles, and many other problems. Some of the most significant problems were as follows:

- The USDA alienated some farmers in the wetland and HEL determination process. This alienation is harmful in that alienated farmers will not actively seek NRCS technical assistance on "other lands/farms" for fear that they will be found not to be in compliance. The "other lands" on their farms also need conservation treatment such as erosion control, pasture and hayland treatment, water quality improvement, animal waste treatment and distribution, etc. Some farmers feel that the wetland determination process is a "taking issue," or private property issue—the taking of land without adequate compensation (i.e., when a wetland is determined, the use of the land for crop production is reduced and/or eliminated).

- The implementation of wetland and HEL provisions appears to have moved the NRCS to a more regulatory mission. Officially, the Conservation Provisions are voluntary—voluntary in that the farmer has an option to participate in the 20-plus USDA financial assistance programs. Once a farmer participates, then the Wetland and HEL Provisions are applicable as to whether or not he/she receives USDA financial assistance benefits. NRCS field staffs were and are sometimes viewed as "regulators." It is difficult for field staffs to be viewed in this manner because traditionally NRCS agents "wore white hats" in the community. Some soil and water conservation districts "distance" themselves from the NRCS activities to avoid that same image. Some districts were threatened with reduced "appropriations" from local units of government because of the "regulatory perception." This was especially true when a local farmer was found "not using an approved conservation system."
- The quality of HEL conservation planning suffered during the implementation of these provisions. The outright requirement that conservation plans be developed on participating farms with HEL and with limited NRCS resources made it necessary to develop conservation plans without "going out on the farm." Many conservation plans were developed in the office or in group settings—thus reducing the quality of planning. Many younger NRCS employees have not experienced "total resource identification or planning experience." We were encouraged by law to provide planning assistance only on the cropland resource base and to consider only soil erosion—not woodland, pastureland/hayland, water quality, animal waste, or wildlife resources. In order to comply with the "substantial erosion reduction" requirement, the level of soil erosion protection was reduced in order to prevent farming enterprises from going out of business. The quality of conservation planning suffered because of a lack of resources (people and time to plan out on the land with farmers).
- Our traditional water resource programs, conservation operations program, water quality programs suffered because all available resources, people, and equipment had to be redirected to the implementation of the Conservation Provisions. This was necessary in order to assist farmers with conservation planning and application in order that they meet the Conservation Provision requirements. Some of our employees are not as technically competent in other natural resource areas as they previously were because of the focus on cropland erosion issues only.
- Another barrier to the successful implementation of the provisions was the fact that "NRCS tried too much in too little of time space." Our resources suffered both in well-rounded technical training and in attention to other

resource problems. Our attention to water quality, flood control, water resource planning, wildlife management, and erosion on other lands suffered. We simply did not have adequate resources (i.e., money, people, and equipment). A much better job of land treatment planning and application could have been achieved had adequate resources been made available.

- In the race to implement the Conservation Provisions, NRCS alienated certain soil and water conservation districts and land users. NRCS did not have time to carry out soil and water conservation district priorities such as conservation education, treatment of pastureland and hayland, erosion on woodlands, water quality initiatives, and other items. Districts traditionally set conservation priorities within their district boundaries. We had to ignore many of those priorities because of the need to assist farmers with keeping them eligible for the USDA financial assistance programs. We also ignored other farmers that had limited or no HEL or wetland needs. In Ohio, the largest soil erosion problem from a gross erosion standpoint is on Soil Capability System IIe soils. These erosion problems simply were left unattended. NRCS also alienated some districts because of the quasi-regulatory approach. Districts want to maintain voluntary. The Conservation Provisions were viewed by many as "regulatory."

- The huge workload associated with the implementation of the Conservation Provisions simply "stressed out" some of our employees. A few actually resigned from the day-to-day stress of dealing with huge numbers of farmers. The "enforcement actions" also brought on additional stress. Although agency support systems were put into place, stressful conditions did exist. The stress was brought on from the expectation that the job needed to be done without adequate resources (employees, equipment, and money).

- During the implementation of the Conservation Provisions, the NRCS "lost some other conservation opportunities." For instance, a Stewardship Incentive Program (SIP) and Flooding Mitigation Program were developed without NRCS traditional involvement. NRCS was simply too busy implementing the Conservation Provisions to notice. Some of our traditional programs could have been enhanced, but are now being carried out by other agencies.

- The documentation of HEL, CRP, and wetland adverse decisions slowed the agency's actions. Large numbers of farmer appeals were filed with NRCS. Large numbers of agency resources were directed to reviewing and deciding on these appeals. The documentation files increased tenfold. It was estimated that one appeal decision through all levels of appeal could command about $5,000 of resources. While we were documenting, other conservation work went unattended. It also became necessary to document compliance with the Conservation Provisions. Status reviews were implemented for checking compliance with the HEL, CRP, WRP, and wetland provisions. Many additional resources were expended carrying out these duties. Compliance documentation is necessary to report to Congress, other agencies, and private individuals whether or not NRCS is complying with the Conservation Provisions.

- NRCS changed some conservation standards. The pressure to meet the "substantial erosion reduction" effort led to NRCS reducing our erosion standards for the HEL provisions. Although this author now believes it was necessary to successfully implement these provisions, at the time it was perceived that "we were caving into public pressures."

- Some of the public perceived that NRCS actions were not "independent enough," especially in the appeal process. The agency reviewed the adverse decisions itself without an "outside independent organization" reviewing the same decisions. The National Appeals Division (NAD) was created and the USDA Reorganization Act of 1994 in a response to this national sentiment. Now appeals are reviewed at higher levels than the field office by other organizations.

- The implementation of the Conservation Provisions brought in the administration of some "oddities." For instance, small tobacco fields are covered under the HEL provisions. Tobacco in the U.S. is a commodity crop where USDA financial assistance is extended. The average tobacco field in Ohio is 1.5 acres in size with a total of 7,000 units. Very little erosion is lost from these small fields, but nonetheless, a conservation plan is required. There is no minimum size to the protection of a wetland in the Wetland Provisions. Therefore, NRCS finds itself sometimes using tremendous resources in upholding wetland determinations on wetlands the size of 0.1 or 0.2 acres.

- The implementation of these Conservation Provisions also impacted some private industry. It is well documented that some small seed, fertilizer, and chemical agricultural dealers in rural areas were impacted by the CRP. Farmers who place their cropland into the CRP for 10 years receive an annual rental payment. They do not have to plant agricultural commodity crops, thus no fertilizer, seed, pesticides, etc., are needed annually. Therefore, some small businesses suffered when large amounts of CRP lands were placed into the reserve. The USDA has a policy that no more than 25 percent of the county's cropland can be put into the CRP.

- The Conservation Provisions caused much confusion among farmer clients brought on largely by "lack of sufficient communication." Farm magazines would highlight stories that often would be unique to certain farmers and not relevant to the majority. In other instances, the media would release stories about the "details" of the Conservation Provisions before the implementing agencies could get the correct information to field staffs. Communication problems also existed with local people.

- There was also a lack of knowledge on the farmers' part in implementing the Conservation Provisions. Farmers didn't know in many cases which areas the USDA would determine to be a wetland. Farmers thought wetlands were equivalent to "water inundated on the soil surface year around." In fact, wetlands were and are determined on farms with much less hydrology being present on site. Another instance to highlight the "lack of knowledge" item was that many farmers didn't understand how much crop residue after planting was necessary in their planned conservation system. Many farmers believed that if they used "conservation tillage equipment" they were meeting the conservation tillage standard. The number of "field cultivations," "speed of the tillage operation," and the type of "tillage shank" all play a part in successfully applied conservation tillage. Lack of knowledge among many farmers was a large barrier in "getting them up to speed" in order to meet Conservation Provision standards.

- Implementing Swampbuster was very difficult, especially in the Midwest U.S. where drainage of excess surface and subsurface water was an ingrained culture of the people. Not improving agricultural drainage goes against the "grain" of many farmers since they were raised to improve drainage which will increase crop yields. The Swampbuster stopped some "drainage maintenance" or drainage improvement on identified wetland areas in order for them to be eligible for USDA financial benefits.

- The lack of skills among NRCS field staffs was also problematic. NRCS had to improve skills associated with wetland determinations, grass seedings, native grass seedings, management of grasses for erosion control over a long period of time, weed control in CRP, wetland restorations, mitigating wetland values, etc. The initial lack of skills at all field locations was problematic which required very intensive training programs.

- The problem of dealing with a huge structural conservation practice workload was a barrier. In Ohio, during the final implementation years of 1993 and 1994, the rate of installing grassed waterways was about 100 to 150 times as great compared to the pre-Conservation Provision period. We had to automate survey and design equipment, encourage farmers to schedule their practices earlier in the implementation period, and train conservation contractors to plan and lay out more of the practices. This workload caused a lot of stress among our field personnel as well. NRCS did not have enough field staff technical assistance to deal with this huge workload.

- The largest barrier to the implementation of these provisions was that the "ongoing" soil and water conservation program suffered. The provisions directed NRCS with scarce resources to work only on HEL. Much of the other cropland, not determined as HEL, still had an erosion problem. For instance, in Ohio, the majority of the gross erosion problem exists on USDA Classification System IIe soils. Almost none of these soils were identified as HEL cropland. Therefore, the conservation treatment on these soils went "untouched." Other soil and water conservation priorities of local importance also were scaled back because all of the available resources were directed to HEL cropland. Pastureland and rangeland conservation treatment was also neglected because of the direction toward treating HEL cropland.

- NRCS did not involve a sufficient number of "advisory groups" representing the environmental and agricultural communities in implementing these provisions. There were efforts to meet with certain agricultural constituency groups to determine acceptable "evaluation criteria" or standards to be imposed.

It was difficult to implement the Conservation Provisions at the state level.

Benefits in implementation of these provisions at the state level

The USDA/NRCS has experienced many benefits from the implementation of the Conservation Title at the state level. The benefits outdistance the barriers with the implementation of the Conservation Provisions at the state level. Now let us list the many strengths from the implementation of the Conservation Provisions at the state level:

- NRCS serviced many new clients. The Conservation Title encouraged many farmer clients who participated in other USDA financial assistance programs to "practice soil and water conservation." It is estimated that nearly 20 percent of the farmer clients had never voluntarily worked with the NRCS through soil and water conservation districts. Since HEL determinations and conservation plans and/or wetland determinations were required on applicable lands, these farmers were encouraged to seek such assistance. Some farmer-clients became acutely aware for the first time that conservation technology exists to significantly reduce soil erosion (i.e., managing crop residues and/or using cover crops to increase erosion control benefits). Additionally, many of these farmers/farms had the worst soil erosion problems in the area. These lands were basically "untreated" for sound technical land treatment to curb soil erosion. On these farms most of the structural works of improvement, conservation tillage, or cover crops were first planned and applied. Extensive soil erosion was reduced in the U.S. from these farming enterprises. It must be remembered that never before in the U.S. was conservation coupled to receiving other USDA financial assistance program benefits.

- NRCS assisted farmer-clients to significantly reduce soil erosion on large acreages of HEL cropland. In certain states, it is estimated that up to 65 percent of existing cropland was covered under the HEL provisions.
- The volume of cropland treated to "obtain a substantial reduction" in soil erosion contributed to millions of tons of soil erosion reduction. For instance, in Ohio nearly 20 percent of all cropland was treated, or about 2.1 million acres, with a "soil savings" of about 10 million tons. That amount of soil savings is equivalent to about 500,000 "dump trucks" of soil savings—soil being kept on the field. In other states with larger HEL workloads, the acres of cropland and resultant soil reduction was significantly larger. This massive cropland treatment effort was the largest concentrated effort in a short time period in the history of U.S. agriculture. In Ohio, the average "soil savings" was about five tons/acre with the before treatment at 9.3 tons/acre/year and the after at about 4.7 tons/acre/year. The tolerable soil loss level was at an average of 4.2 tons/acre/year.
- NRCS gained support from many nonfarm groups. Implementation of the Conservation Title brought support from many local, state, and national environmental groups and organizations. With the treatment of HEL and specifically through the CRP and WRP, millions of acres of land were improved for wildlife habitat, erosion reduction, and water quality benefits. Pheasants Forever, Ducks Unlimited, and other private organizations praised the effort to "set aside" land for improvement of wildlife habitat for upland game and songbirds. In Ohio, the pheasant populations and waterfowl populations have increased significantly. Hunting groups have enjoyed these new levels of wildlife populations. Public and private lake associations have praised the CRP and WRP where targeted efforts to apply grasses, trees, wetland restorations, and riparian establishments in upstream watersheds have reduced pesticide, nutrient, and sediment loadings in their waterbodies. Private land conservancy groups have utilized these programs to target streams for wildlife, native plant preservation, and recreational opportunities along certain stream corridors. Much support has come from nonfarm groups from the Conservation Title.
- The NRCS, Farm Service Agency (FSA), Rural Development (RD), U.S. Fish and Wildlife Service (USFWS), U.S. Army Corps of Engineers (COE), and the Extension Service (ES) learned to be more "team oriented" because of the Conservation Title. This occurred because of necessity, the rules, and regulations which established federal agency roles and responsibilities for implementation. All USDA agencies had to coordinate and share information much more readily than in past history. It was necessary for the FSA to share compliance and producer data and vice versa with NRCS in order to

carry out the provisions. Cropping history data and status of self-certification relative to the HEL and Wetland provisions was paramount for NRCS in carrying out our status review responsibilities. NRCS had to coordinate with the RD agency (formerly the Farmers Home Administration) in servicing their farm loan clients with respect to these provisions. The Ohio State University ES cooperated closely in informational and educational activities with the NRCS. NRCS "teamed up" with the COE, USFWS, and the Environmental Protection Agency in devising the first ever wetland mapping conventions (off-site mapping procedures utilizing photography) in training all field agency personnel in wetland determinations/criteria per the 1987 Wetland Jurisdictional Manual. This effort was a big chore, but it has strengthened NRCS in the wetland determination business on U.S. farmland. Not only were federal agencies critical for implementing the provisions, but NRCS found new friends in the private sector to assist with implementation. NRCS teamed with the farm equipment and chemical companies to encourage conservation tillage—a predominant practice to help farmers meet the HEL provisions. Private wildlife organizations, such as Pheasants Forever, Ducks Unlimited, and private hunting organizations assisted in the implementation of the Wetland Reserve Program and wetland provisions. Private contractors were trained to do simple conservation practice survey, design, and layout. In all, NRCS and USDA cooperating with local, state, and private organizations became much more team oriented. This effort assisted the USDA to get more conservation on the land than ever before in its history.

- Another benefit gained from the implementation of the Conservation Provisions was the ability of NRCS to implement "innovative methods" which helped to achieve conservation planning and application. NRCS contracted out to soil and water conservation districts (they hired retired NRCS and/or soil and water conservation district employees) the job of completing status reviews, the compliance checking component of the Conservation Provisions. We also contracted out the job of assisting with the CRP planning and application to soil and water conservation districts. Group meetings were held in about every field office to "group plan" individual HEL conservation plans with farmers. Group planning meetings were necessary to achieve the completion of nearly 44,000 HEL conservation plans in the time period mandated under the Federal Conservation Provisions. NRCS also trained conservation land improvement contractors and a few individual land users to survey, design, layout, and checkout simple conservation practices. This became a necessity when a record number of grassed waterways were planned in HEL conservation plans. Nearly four times the normal rate of grassed waterways were installed in 1993-1994 per year compared to the period before the Conservation Provisions in Ohio (i.e., about 2,000 acres of grassed waterways were installed in 1994 compared to an average level of 400 acres/year in the years 1980-1985). Private industry "teamed up" with NRCS and the Ohio State University Extension Service to develop better conservation equipment and weed control chemicals/applications to increase the use of no till from 20 percent application on cropland in before 1985 years to nearly 50 percent/year today.

- The USDA and NRCS became proficient in the planning and establishment of grasses, especially for long-term management (i.e., 10 years). Prior to the CRP, NRCS had limited technical expertise in pasture and hayland planting and management. Implementation of the CRP forced the agency to develop technical expertise in all locations to assist farmers with grass establishment. New technical knowledge and experience were gained in the Midwest for the planting and management of native grasses. Nearly 350,000 acres of CRP lands were established to grasses from 1985 to the present.

- An added benefit from the implementation of the Conservation Provisions was the favorable media coverage relative to soil and water conservation. The public and farmers became keenly aware of conservation measures. Conservation tillage, cover crops, cropping sequence, and wetland restoration and management became "familiar household terms" to millions of Americans. Even private industry highlighted conservation measures to market their products (i.e., equipment industry highlighted their no-till planters and other reduced conservation tillage equipment).

- The Wetland Provisions or USDA Swampbuster "slowed" the decline in wetland losses in the U.S. Prior to the Swampbuster, the decline in wetland losses was great. NRCS National Resources Inventory (NRI) has shown that the decline in wetland losses was almost zero, and in some cases, wetland restorations with the CRP and the WRP were actually increasing wetland numbers. In Ohio, prior to the Wetland Provisions, Ohio had about 850,000 acres of wetlands. The loss during these provisions was only about 10,000 acres, and much of those losses occurred in the early 1980s prior to full implementation. Nearly 3,500 acres of wetlands will be restored to date with the WRP—that is new wetland areas on agricultural areas. The USDA Swampbuster also "assisted the Section 404 Wetland Protection Program" in making nonfarmer/nonagricultural lands more sensitive to wetland protection. These provisions greatly increased the public's perception that wetlands serve a vital link in the ecosystem.

- NRCS became proficient in wetland restoration and mitigation. Utilizing the agency's expertise in agricultural land drainage, the thrust to wetland restoration became a simple task. All that was needed was to reverse the land drainage process on hydric soils—and a wetland is restored! Shallow dikes, blockage of subsurface drains, and drainage ditches are the basic practices needed for restoring America's wetlands on hydric soils. In Ohio to date nearly 700 acres of wetlands have been restored on CRP and 3,700 acres on WRP lands with the implementation of these Conservation Provisions. NRCS also helped other agencies such as the U.S. Fish and Wildlife Service and the Corps of Engineers to gain expertise in soil hydrology and hydric soils, an important concept in wetland restoration. NRCS has even introduced wetlands as a vehicle in treating liquid wastes for farm animal waste systems and treating acid water runoff from abandoned mineland areas.

- Another benefit of the implementation of the Conservation Provisions was the ability to provide technical assistance to farmers in the areas of crop residue management and cover crops. Crop residue management is a key to reducing soil erosion on cropland. Farmers began measuring crop residues after planting, thus taking measures to "keep residues on the surface after planting." They reduced the number of tillage trips across fields, better distributed residues from harvesting operations, and purchased/adapted tillage equipment which left more residues on the surface. Cover crops became familiar with farmers as a means of applying cover crops after harvest on lands with low crop residue amounts. The application of cover crops became normal with the introduction of "fly-on techniques." These methods also increase crop yields since more soil moisture was preserved for the commodity crops. Ohio's conservation tillage with more than 30 percent crop residue on the surface after planting increased from 20 percent in 1985 to 50 percent in 1995.

- The private land improvement contractors benefited from the large number of conservation practices constructed during the implementation of the Conservation Provisions. In Ohio, it is estimated that approximately 3,400 more acres of grassed waterways were installed as a result of the provisions. Grassed waterways and critical area seedings were necessary to control gully erosion on HEL fields. In addition, nearly 4,000 acres of wetlands were restored or mitigated requiring the construction of shallow water dikes, structures on blocking subsurface drains, and/or open ditches. Land drainage improvement increased on prior converted cropland areas in order to prevent "reversion of areas to farmed wetlands." The Conservation Provisions provided a "boost" in construction activities to the land improvement contractors; however, NRCS has difficulty dealing with this large structural workload.

- One of the largest strengths to the implementation of the Conservation Provisions was the public recognition to the standards and specifications contained in the Field Office Technical Guide (FOTG). The FOTG located in all NRCS field offices contains the time proven standards and specifications for planning and application of conservation practices. No other document in the U.S. or world provides the "technical details" on how to plan and apply conservation practices. They have also been refined for over 60 years based on proven experience and application. The public was referred to the use and application of the FOTG. A majority of farmers found the standards and specifications contained in the FOTG to be very practical and technically sound to apply.

- NRCS became more efficient and effective in documenting conservation facts—a helpful method for "upholding appeals." Farmers are allowed to appeal "adverse decisions" made by NRCS in the HEL, CRP, WRP, and Wetland Provisions. Therefore, it became necessary for the NRCS to become proficient to document "why decisions were made."
- NRCS also adopted and refined "new technologies" in the administration of the Conservation Provisions. The wetland off-site mapping conventions, a tool to make wetland determinations more precise and efficient for millions of acres of pastureland/hayland and cropland. NRCS used satellite imagery, aerial photography, and remote sensing equipment to complete wetland determinations. These tools were refined which helped NRCS to complete the nearly 35,000 wetland determinations in Ohio with client confidence.
- Finally, another strength gained from the implementation of the Conservation Provisions was the establishment of the state technical committee process. This process requires that federal, state, and local agencies and private environmental/farm business organizations participate in the review and development of technical inputs into these Conservation Provisions. Therefore, these groups and agencies have an opportunity to improve technical aspects of these programs. Improved technical applications and communications among agencies and constituent organizations are a product of this committee process. An improved "buy in" is achieved from the process from governmental agencies and agricultural and environmental constituent organizations.

We cannot leave this subject of implementation of the Conservation Provisions without noting the "changes or adoptions" of new policies made by the NRCS policy-makers that have further improved the implementation at the state level. You could also say these were the lessons learned from the earlier days of implementation at the state level.

- NRCS has established a stronger policy of providing technical assistance for farmer-clients to help them correct deficiencies before they are determined "not using an approved conservation system." We now have the opportunity to observe deficiencies in the HEL systems and to provide them assistance such as revision of the conservation plan, or providing increased time to correct observed deficiencies. This effort overcomes the issue of otherwise alienating the client from seeking our assistance for fear of being "determined not using an approved conservation system" and ultimately losing USDA benefits. It is a policy of trying to be more farmer friendly. This approach is also much more acceptable by local soil and water conservation districts. This concept also attempts to nullify the perception that the NRCS is quasi-regulatory. This approach also encourages a voluntary approach to conservation application and reduces agency employee stress by fewer client appeals.

- The NRCS has adopted a "holistic conservation planning approach" where the attempt is to focus on all of the resource needs on the farm in an environmentally safe approach. This concept takes us away from the narrow approach of focusing only on cropland erosion. The concept also encourages the broad use of technical skills of our soil conservationists and partners to recommend multiple alternatives to treating all resource concerns on the farm such as nutrient, manure, pesticide, soil erosion, grazing issues, wildlife concerns—all resource concerns. This concept strengthens the skills and knowledge of the conservation planner and makes for wise land use and conservation.
- The USDA Reorganization Act of 1994 shifted the appeals process to an independent agency thus making for a more "impartial review of technical appeals." Some may argue that the system has made things more difficult, but it has provided a more "impartial" and independent body to hear client concerns on Conservation Provision appeals.
- Farmer-clients now have a much clearer understanding about the minimum requirements of the Conservation Provisions. A focused informational and educational program was implemented throughout USDA which has greatly overcome the "lack of knowledge or miscommunication syndrome" with our clients.
- In regard to the Swampbuster Provisions, Congress has assisted in making the many wetland provisions more flexible for farmers. Wetlands are still wetlands, but the provisions now allow greater flexibility in mitigation of wetland values for farmers. This policy initiative reduces the farmer-client stress in regard to wetland conservation rules.
- The USDA Wetlands Reserve Program was also enacted in the 1990 Farm Bill which provided a tremendous opportunity for farmers to be compensated for some of the previously determined cropped wetlands (when included in adjacent conservation easement areas). The policy initiative also

encourages the restoration of additional wetlands with voluntary compensation—a very popular program initiative from farmers and the environmental community. It's a win-win situation.

- Today, the farmers' knowledge has been greatly improved over previous years in the application of approved conservation systems. The private sector has played a huge part in this effort with informational and educational opportunities. Private machinery companies have greatly improved planting and land preparation tools that directly support successful application of HEL conservation systems. Farmers have been better educated in the advantages of maintaining and restoring wetlands.

- The technical skills of NRCS and partner agency employees have been greatly improved in regard to HEL and wetland conservation systems. Additional training programs were implemented in the use of the RUSLE, wetland determination techniques, off-site wetland mapping procedures, measurement of crop residues, improved computer hardware and software applications, automated engineering design and drawing software systems, and the planning and management of grasses, especially native grasses, etc. These training programs have greatly improved agency personnel skills. Increased training and the addition of new and improved tools have increased output per employee as well.

- Finally, we have improved the quality of policy-making and procedural development with better acceptance through the policy adoption and use of state technical review committees. A broad spectrum of agricultural and environmental interests are brought together in this approach to review and make recommendations to USDA agencies for improvement in technical applications in these Conservation Provisions.

7

The Role of Private and Professional Organizations in the Development of Soil and Water Conservation Policy

Max Schnepf

This is a conference about soil and water conservation policy successes and failures. Over the past 15 years, the work of private, nonprofit organizations in U.S. policy-making must be judged a success by any measure. During this period, these groups collectively have become America's soil and water conservation policy think tank.

Many of the innovative conservation policies adopted by Congress in the last three farm bills, for example, had their genesis in the private sector. The list includes the Conservation Reserve Program and the sodbuster, conservation compliance, and swampbuster policies in the Food Security Act of 1985; the Wetlands Reserve Program, Water Quality Incentives Program, and important amendments to the CRP and compliance policies in the Food, Agriculture, Conservation and Trade Act of 1990; and the Farmland Protection Program and Flood-Risk Reduction Program in the Federal Agriculture Improvement and Reform Act of 1996.

Important policy ideas from the private sector appear in other legislative proposals as well, including the Clean Water Act amendments and pesticide law reform. But the farm bill has been the principal focus of the sector's efforts so far as soil and water conservation policy is concerned.

A recent phenomenon

What most people forget is that the involvement and success of private, non-profit conservation groups in soil and water conservation policy matters is a relatively recent phenomenon. Except for two organizations founded in the 1940s specifically to address soil and water conservation policy issues—the Soil and Water Conservation Society and the National Association of Conservation Districts—such issues drew only fleeting attention from most private-sector groups until the 1970s. About that time, the consequences of a series of events converged, resulting in a synergism within the sector that has produced the rather impressive legislative record of the past decade and a half. Those events, among others, included the following:

- Publication of Rachel Carson's *Silent Spring*.
- The bitter controversy over the environmental impacts of the Small Watershed Protection and Flood Prevention Program.
- Earth Day and the spate of environmental legislation that followed, including the Federal Water Pollution Control Act with its 208 areawide nonpoint pollution control planning process.
- The sudden increase in world demand for U.S. agricultural commodities in the early 1970s and the plea by then-Secretary of Agriculture Earl Butz for farmers to plant fence row to fence row.
- The legitimization of soil and water conservation as a national policy issue by virtue of passage of the Soil and Water Resources Conservation Act of 1977, which called for the periodic assessment of soil and water resources nationally and subsequent development of a national conservation plan.
- A study by the General Accounting Office showing how ineffectively public cost-sharing funds were being used for soil and water conservation purposes.

Three categories of organizations

The private, nonprofit sector, of course, is anything but a homogenous group of member and other organizations with common interests. In reality, the sector consists of three general categories of organizations, each of which has its particular agenda.

First are those professional societies that have a direct stake in soil and water conservation matters. Included are such discipline-oriented groups as the Soil Science Society of America and the American Water Resources Association, along with the multidisciplinary-oriented Soil and Water Conservation Society.

Second is the proliferating assortment of advocacy organizations from which has come much of the really innovative policy thinking.

A third category of private, nonprofit organizations that I will not deal with in any detail here, but which deserves mention, is the substantial group of philanthropic foundations that fund a significant portion of the policy work carried on by the advocacy groups in particular, as well as some professional societies. I mention this category of organizations only to call attention to its evolving role in soil and water conservation policy activities. Until recently, the foundation community generally was content to support those advocates in the private, nonprofit sector with innovative policy proposals. In recent years, however, foundations have become more activist in nature, actually soliciting interest in the public and private sectors alike from groups willing to carry out projects the foundations themselves want done. This may have something to do with what I will call the "thin green line." More on this later.

All private, nonprofit groups, whether professional societies or advocacy organizations, exist for two fundamental purposes. First, they attempt to do things for members collectively that the members cannot do for themselves individually. Most of these are educational in nature. Examples include the publication of a journal or magazine, sponsorship of an annual meeting, or support of a certification or other professional credentialling program. Second, all private, nonprofit groups are expected to represent their members' interests where it counts: in policy-making circles, for example.

A new activism

From a policy-making point of view, all professional societies and advocacy groups essentially conduct three types of activities. These are (1) educational; (2) advocacy; and (3) evaluation-type activities. Not all organizations undertake activities in each of these three categories. Neither do they always pursue the same types of activities within a particular category. In fact, professional societies and advocacy groups tend to inject themselves into the policy-making process at different points. Professional societies, for example, have been generally cautious about becoming directly involved in policy making, choosing instead to play it safe and make the one important contribution they can make to the process, that of ensuring the integrity of the science that undergirds all sound policy-making. This cautiousness is slowly evaporating, however, and more and more, one finds professional groups willing, at the very least, to more forcefully represent their members' interests in the policy-making arena by articulating legislative ideas based on the natural resource principles they adhere to.

A good example of this new activism among professional societies was the comprehensive policy statement prepared prior to the 1996 Farm Bill debate by the American Fisheries Society. This policy statement described agriculture's impacts on aquatic ecosystems and what policy initiatives could be pursued that would minimize or eliminate those impacts. A second example is the American Society of Agronomy's participation in a congressional fellow program.

Professional societies, it seems to me, have far more to offer the policy process than what they have contributed thus far. The written policy statement and congressional fellow program are but two examples. Professional organizations could do much more to foster policy-relevant research and to educate policy-makers, environmental advocates, the media, and the public on the state of the science surrounding important soil and water conservation issues. They could also more vociferously advocate the role of science and their respective professions in soil and water conservation policy matters. Finally, they have something to offer with respect to the evaluation of conservation policy initiatives, not only in terms of how program evaluations are conducted, but also in terms of the credibility that is ultimately attached to the outcomes of such evaluations.

Work by the Soil and Water Conservation Society in the past decade and a half exemplifies what is possible. Between 1979 and 1995, the Society sponsored four national conferences on soil and water conservation policy, making a concerted attempt in the process to bring a spectrum of interests into the discussions and ensuring that new policy ideas were identified and debated. These discussions and debates were continued in the pages of the Society's *Journal of Soil and Water Conservation*. The Society also conducted two national surveys of farmers regarding the Conservation Reserve Program, the results of which have been used extensively for both program administration and policy-making purposes. In cooperation with the U.S. Department of Agriculture and other interests, the Society also conducted the first national evaluation of experience under the Conservation Title of the Food Security Act of 1985. Findings from this 3-year project were the basis for important amendments incorporated into the 1990 Farm Bill.

In contrast to the cautious approach to policy making exhibited by professional societies, advocacy groups are quick to enter the policy-making fray, engaging, without hesitation, in conceptualizing policy, writing and debating legislative proposals, supporting appropriations, engaging in rule-making, and critiquing program implementation and evaluation efforts. It is in this sector where advocacy groups have worked over the past 15 years, essentially unimpeded, to be innovative, to serve as a catalyst for change, pulling the science, the public, and policy-makers along.

In addition to policy-related activity, advocacy groups are particularly good at networking. Several have developed nationwide webs of activists who have proved extremely influential in the policy-making process. One association of sustainable agriculture advocates, for example, claims at least one or more members in every congressional district in the U.S.

Collaboration begets synergism

Real synergism occurs, of course, when policy-making becomes a truly collaborative process, when professional and advocacy groups individually and jointly serve as conveners, bringing multiple interests together to discuss issues, to translate on-the-ground experience into sound public policy proposals, to correct policy when the policy-makers or program administrators don't get it right the first time through.

It is precisely a synergism of this sort that has existed through much of the 1980s and into the 1990s, accounting for the substantial innovation in soil and water conservation policy making that has characterized the past three U.S. farm bills. It all began with a clear articulation of soil and water conservation issues by the professional and advocacy communities conjunctively; with a willingness to air new and controversial policy ideas at their meetings and in their literature, both books and magazines; with a commitment by some of the major advocacy organizations to allocate, for the first time, staff and other necessary resources to influence soil and water conservation policy-making; with a willingness on the part of many of the important professional and advocacy organizations to work in 1985 toward a common soil and water conservation policy agenda; with a willingness to enter into private sector-led evaluations of federal program performance; and with an inclination to work to correct program shortcomings and to supplement workable programs with additional program innovations in subsequent farm bills.

The upshot of this largely privately led collaborative effort is a series of conservation policies that exact some measure of stewardship from those agricultural producers who enjoy the benefits of federal farm programs and that provide a means of moving voluntary conservation efforts along in significant fashion with such new conservation tools as the Environmental Quality Incentives Program; a broader, much more conservation-oriented CRP; more flexible conservation compliance and swampbuster policies; and perhaps most importantly, the funding of conservation through the Commodity Credit Corporation, which previously funded only price support and related programs.

A thin green line

Leadership in soil and water conservation policy is making ebbs and flows, of course. The past 15 years represents a high point for private sector interests in the U.S. Whether this sector will continue to exert the level of influence it is capable of is a matter of conjecture at the moment, which brings me back to my earlier reference to the "thin green line."

Despite the substantial number of professional societies and advocacy groups in the U.S., much of the thinking and leadership in the private sector on soil and water conservation policy issues in recent years has come from relatively few groups and individuals. Some of those individuals and organizations have diverted their attention to other issues of late, making what in reality has been a thin green line even thinner. This may well signal a shift in soil and water conservation policy making back to the public sector, where it all began 60 years ago with the innovative thought and action by Hugh Hammond Bennett and colleagues in the U.S. Department of Agriculture. There clearly was evidence of this in the 1996 Farm Bill debate, which produced a much more proactive role throughout for administration and congressional interests.

On the other hand, with world grain stocks at near-record low levels and continuing pressures on limited land resources in the U.S. and elsewhere, the course of events may prompt private and professional groups to extend well into the future their collective track record of success in soil and water conservation policy development since the early 1980s.

8

Soil and Water Conservation Policy Within the United States

Ted L. Napier and Silvana M. Napier

Salary and research support that made this chapter possible were provided by the Ohio Agricultural Research and Development Center of the Ohio State University, and the Management Systems Evaluation project funded by the U.S. Department of Agriculture. The positions advanced in this chapter are solely those of the authors and do not represent official positions of the funding agencies.

The evolution of soil and water conservation policy in the United States

Soil erosion has been perceived to be an environmental problem within the United States for more than 200 years. During the Colonial Period, enlightened conservationists advanced the position that soil resources should be protected and recommended farming practices that would reduce soil loss on cultivated cropland (Napier 1994a; Rasmussen 1982). Unfortunately, most land operators during that time period ignored concerns expressed by conservationists, because most farmers did not perceive soil erosion to be a threat to their long-term economic well-being.

One of the primary reasons that landowner/operators in the 1600s, 1700s, and 1800s did not view soil erosion as being a significant environmental issue was the availability of free land on the frontier. Large tracts of land were available at little or no economic cost to anyone willing to clear the land and willing to live on the farmstead for a few years. Virgin forests and grasslands could be transformed into productive farms by anyone willing to expend physical labor required to prepare the land for cultivation.

Land operators during this period were not concerned about using erosive farm production systems, because degraded land could be abandoned and new fields could be carved from free land on the frontier. Since there were no incentives to conserve soil, landowner/operators often exploited agricultural land without regard for future productivity of land resources and/or rights of future

83

generations to have access to fertile land and clean water.

Situational conditions that influenced how agricultural land was perceived and used in the U.S. changed drastically during the early 1900s. The close of the western frontier eliminated access to free land that could be transformed into productive farms. Land operators could no longer deplete land and move to the frontier to begin anew. Another important factor was the Great Depression which rendered many landowner/operators poverty-stricken. The situation was further compounded by an extended drought in the late 1920s and early 1930s which contributed to the development of dust storms that threatened the future productive capacity of U.S. agriculture (Rasmussen 1982).

The destruction of future agricultural productivity of several million acres of agricultural cropland during the Dust Bowl years, combined with the economic deprivation of many landowner/operators, created a situation which political leaders could not ignore. Congresspersons responded to the crisis by formulating soil conservation policies designed to prevent further destruction of agricultural lands. Authorization of public funds to implement soil conservation programs was justified on the basis of protection of future agricultural productivity that was being threatened on millions of acres of farmland in the Midwest.

The first national soil conservation initiative established to address soil erosion was the creation of the Soil Erosion Service (SES) within the U.S. Department of the Interior in 1933. For the first time in U.S. history, soil erosion was nationally defined as a threat to the future well-being of the society, and government action was taken to ensure that research was conducted on the causes and consequences of soil displacement. The SES was also commissioned to provide technical assistance to agriculturalists, regardless of need and ability to pay.

The SES was short-lived, because national concern for soil erosion resulted in the formation of the Soil Conservation Service (SCS) in 1935 via authorization of the Soil Conservation Act (Rasmussen 1982; Rasmussen and Baker 1972). The SCS replaced the SES as the primary public agency commissioned to address soil conservation issues in the U.S., and the new agency was placed within the U.S. Department of Agriculture. The emphasis on technical assistance and research was continued by the SCS.

Shortly after the authorization of and implementation of the SCS, Congress authorized the Soil Conservation and Domestic Allotment Act of 1936. This legislation legitimized government payments to farmers who were willing to adopt less erosive agricultural production systems. Farmers were provided economic assistance to shift production away from commodities that contributed to degradation of soil and water resources. This subsidy program was justified on the basis that public funds were being used to reduce soil loss. It was argued that public funds were being wisely invested, because the future productivity of land resources was being protected. Critics of this policy, however, suggested that the program was nothing more than an ill-disguised government subsidy program designed to stabilize agricultural commodity prices and to protect farm income. Such approaches were perceived by many to be counter to the "free market" system.

The Agricultural Conservation Program (ACP), which was the name often used for set-aside programs shortly before and immediately following World War II, was used to divert cropland from production without consideration of environmental impacts. All farmland was defined as being eligible for enrollment. This policy made it possible for land that was not highly erodible to be retired from production. Critics of the ACP argued that soil and water protection was secondary to maintenance of commodity prices and farm income. The legislation was modified in 1977 to ensure that land enrolled in the ACP program was being threatened in some manner by soil erosion (Rasmussen and Baker 1979). This change in public policy was the beginning of targeting limited soil and water conservation resources to farmland contributing disproportionately to environmental degradation.

The next major policy initiative formulated to address soil erosion and water pollution problems in the U.S. was the Soil Bank program authorized in 1956. The Soil Bank was a federal program designed to control the supply of agricultural commodities, even though it was often justified to taxpayers on the basis of conservation benefits (Harmon 1987). Farmland could be diverted from agricultural production for 10 years, and farmers were paid not to produce specific commodities. No requirement was made to ensure that land bid into the Soil Bank program was highly erodible or in need of erosion control. Landowners could enroll all or portions of their farmland regardless of erosion potential. Farmers often enrolled the least productive land in the Soil Bank and intensified agricultural activities on cropland that remained in production. As a result of this practice, food and fiber output on many farms was not substantially reduced by the set-aside program (Napier 1990a; Rasmussen 1982).

While soil loss was reduced from land enrolled in the Soil Bank, the program was subject to intensive criticisms from many segments of society. Many landowner/operators enrolled entire farms in the Soil Bank program and basically retired from farming for the duration of their contracts. Farmers who enrolled all of their land in the Soil Bank recognized that they could make more money from leasing land to the government than they could from the production of agricultural products. Critics of the Soil Bank questioned the legitimacy of government policy that paid landowners not to work. Opponents of the program also noted that rural communities often suffered considerable economic deprivation when a significant portion of local land operators enrolled large tracts of farmland in the Soil Bank. Long-term retirement of cropland adversely affected local communities because purchase of agricultural inputs was substantially reduced when farmland was retired on a long-term basis. The effect of declining local purchases of farm inputs was a decline in the economic viability of local economies.

Soil Bank policies were subject to even more severe criticism after the program was terminated, because many of environmental benefits achieved by public investment in conservation initiatives were negated when enrolled land was returned to production. Much of the Soil Bank enrolled land was returned to production using the same farming systems that had been used prior to the implementation of the program. No provision had been made to ensure that landowner/operators would maintain conservation structures and/or conservation production systems on set-aside land. In a very short time period, land formerly enrolled in the Soil Bank was being degraded by soil erosion.

While a number of federal and state conservation initiatives were introduced between the late 1960s and the early 1980s, practically all policies authorized to control soil erosion were governed by the same philosophical underpinnings and assumptions used in previous soil conservation policies and programs. Practically all soil conservation policies and programs implemented during this time period emphasized voluntary participation in conservation initiatives, provision of information and education to potential clients, and some form of government cost-sharing (Harmon 1987; Helms 1990; Napier 1990a; Napier 1990b; Rasmussen 1982).

The most significant characteristic of public soil and water conservation policies and programs conceived and implemented from 1933 to 1985 was the emphasis placed on voluntary participation by landowner/operators. Public conservation policies for the first 50 years avoided use of regulatory approaches. Public conservation policies were formulated on the basis that landowner/operators possessed almost absolute rights to land resources. It was assumed that property owners had the right to use land resources in any manner deemed appropriate to secure economic returns to investment (Napier 1990a; Napier 1994a). It was assumed that society did not have the right to affect on-farm production decisions without "just compensation" to the landowner. Public conservation policies emphasized incentives to motivate landowner/operators to adopt soil and water protection systems rather than use of regulatory approaches that employed some form of coercion.

Soil conservation programs developed and implemented prior to 1985 placed primary emphasis on the provision of information and education. It was reasoned that land operators were ignorant of the adverse impacts of highly erosive production systems on the environment. It was assumed that provision of information would make farmers aware of the damage being done to the environment as a result of the use of inappropriate farming systems. Proponents of the information-education approach suggested that, once farmers were made aware of the environmental costs of using erosive farming systems, they would modify production practices to reduce soil loss. Opponents of this intervention strategy, however, suggested that lack of adoption of conservation practices at the farm level was not a function of landowner/operators being unaware of adverse environmental consequences associated with use of erosive production systems. Critics of the information-education model argued that farmers were often well informed of erosion problems on their land and that farmers did not adopt soil erosion control systems because they were aware that highly erosive production systems were often the most profitable even in the long term. Critics of the information-education approach argued that farmers could not be expected to adopt farm production systems that would not enhance farm-level income. Critics posited that until economic barriers to adoption of conservation production systems were removed, landowner/operators would never voluntarily adopt conservation protection systems at the farm level.

Research focused on returns to investment in soil conservation at the farm level during the late 1970s and early 1980s revealed that arguments advanced by critics of the voluntary approach had validity (Halcrow et al. 1982; Lovejoy and Napier 1986; Mueller et al. 1985; Napier 1990a; Napier 1990b; Napier et al. 1994). Investments in information and education did not appear to be justified using adoption of soil conservation as the criterion for making assessments. Access to information and education were shown to be very poor predictors of adoption behaviors at the farm level (Camboni and Napier 1993; Lovejoy and Napier 1986; Napier et al. 1994; Napier and Johnson 1996; Swanson and Clearfield 1994).

A second major component of most soil conservation programs implemented within the U.S. has been the use of economic incentives to motivate land operators to adopt soil and water conservation production systems. Monetary incentives have been shown to encourage landowners to participate in soil conservation programs (Napier and Forster 1982) because such subsidies reduce risks associated with adopting conservation production systems. Proponents of this policy approach have argued that landowners should not be expected to internalize all of the costs of protecting the environment. Proponents have observed that land operators are much more willing to implement conservation programs if risks associated with adoption are reduced by public subsidies. Critics of these types of policy approaches argue that land operators should not be permitted to export a portion of the production costs via pollution of waterways to nonfarm populations. Opponents of public policies that legitimize economic subsidies suggest that it is not appropriate for society to bear production costs so that individual land operators can reap larger profits.

While the voluntary-education-subsidy approach can be severely criticized from many perspectives, it must be recognized that substantial reduction of soil loss from cropland occurred during the 1930s using this approach (Napier 1990a; Napier 1990b; Rasmussen 1982; Rasmussen and Baker 1972). However, it must also be recognized that these types of policies have become less effective over time. Proponents of the education-information-subsidy policy approach have failed to recognize that land operators in the U.S. have become well informed of the causes of soil loss and are aware of the many methods of resolving soil erosion problems. Lack of awareness of environmental problems and of potential technical solutions has not been a significant issue for several decades. Most modern agriculturalists are very well educated and informed. Ignorant farmers do not stay in business very long in the U.S. Unfortunately, conservation policies continue to be developed and implemented using information-education-subsidy strategies long after such approaches have been shown to be basically ineffective (Napier 1990a).

After 50 years of public policies and billions of dollars of investment in conservation programs that emphasized the information-education-subsidy approach, environmental degradation remained a significant environmental issue in the U.S. in the early 1980s. Conservationists, concerned citizens, and political leaders began to question the return to investment in soil conservation policies and programs in the 1980s (Napier 1990a; Napier 1990b; Napier 1994a; Napier 1994b). Concerned citizens and organized environmental groups began to reexamine existing soil conservation policies and intervention strategies. These actions provided the political support base for the development and authorization of the Conservation Title of the Food Security Act of 1985 (Napier 1990b). This legislation was destined to substantially change soil and water conservation policies and programs in the U.S.

New policy approaches for soil and water conservation within the United States

Conservation Title of the 1985 and 1990 Farm Bills

The Conservation Title authorized by the Food Security Act of 1985 ushered in a new era of soil conservation policies and programs in the U.S. For the first time in U.S. history, the right of nonfarm populations to influence farm-level production decisions without use of incentives was legitimized. The Conservation Title introduced elements of cross-compliance that required landowners who participated in conservation programs authorized by the legislation to comply with contractual agreements, or suffer loss of access to federal farm programs and other penalties. Furthermore, the legislation significantly constrained the rights of landowners to use certain types of farm production systems on highly erodible cropland. Landowner/operators who continued to operate highly erodible cropland without implementation of approved conservation farm plans were subject to penalty. Also, highly erodible cropland that had not been cultivated within specified time periods prior to the authorization of the Conservation Title of 1985 could not be put into production without implementation of an approved farm plan (Napier 1990b). These elements of the Conservation Title were revolutionary, because property rights were changed by legislative action.

Even though there were elements of coercion contained within the Conservation Title, the legislation included a number of components that were consistent with previous conservation policies. One of the most highly acclaimed conservation initiatives within the Conservation Title was the Conservation Reserve Program (CRP). The CRP was designed to retire highly erodible cropland from production for a 10-year period. The mechanism used by the CRP to retire highly erodible cropland from production was government purchase of cropping rights for the duration of lease agreements. In many respects, the CRP was a modified version of the Soil Bank program of the 1950s. The only major difference between the Soil Bank and the CRP was the stipulation that land retired by CRP must comply with other components of the legislation when brought back into production when contracts expire. This action was taken to ensure that conservation farming systems would be employed on set-aside land when the land was brought back into agricultural production.

Landowners were permitted to enroll highly erodible cropland in the CRP using a bid system at or below government established maximum rental values. Production of agricultural products on CRP-enrolled land was forbidden during the 10 years the government leased the land. These implementation policies were designed to minimize economic costs and to maximize soil savings. Use of other implementation policies would have significantly changed the outcomes of the program (Napier 1990b; Reichelderfer and Boggess 1988). If the CRP had placed primary emphasis on protecting water quality, then land that would have been eligible for enrollment would have been significantly different, and the land retired by the program would have been distributed differently. Also, if implementation policies had permitted production of nonerosive crops and/or animal production, the economic costs of the program would have been reduced, and the impact on local community infrastructures would have been substantially different (Napier 1990b).

Another significant component of the Conservation Title is Conservation Compliance. Conservation Compliance mandates that owners of highly erodible cropland must have developed and fully implemented an approved conservation farm plan by January 1995. If owners of highly erodible cropland elect not to comply with the legislation, they were subject to penalties.

The Conservation Title was initially conceived and implemented using "T" as the goal to be achieved. The expression *T by 2000* was the theme used to characterize the legislation. This policy expectation was changed, because it was argued by many agricultural interests that such a stringent goal was unrealistic. Opponents of the use of T as the goal to be achieved argued that use of such a low level of soil loss would force many landowner/operators of highly erodible land out of production agriculture. Critics of T argued that it would be too costly to implement farm production systems that would meet such an expectation.

While it is probably true that use of T would have permanently removed some highly erodible land from agricultural production, such a policy would have resolved the erosion problem. It is highly likely that the economic impacts of the removal of a small portion of highly erodible land permanently would have been negligible in the long term.

Other elements of the Conservation Title that have important policy implications for future soil conservation initiatives in the U.S. are Swamp Buster and Sod Buster. Even though these elements of the Conservation Title are only indirectly focused on soil conservation, they have established policy precedents that will have long-term effects on conservation initiatives in the U.S. Both of these components of the Conservation Title impose significant constraints on the rights of landowners relative to use of land resources. These components of the Conservation Title have successfully challenged absolute rights of landowners to use land resources in any manner the property owner deems appropriate. Landowners can no longer drain wetlands or bring land into production without consideration of the environmental impacts. In essence, the Conservation Title legitimized the right of nonowners to influence farm-level decision-making about farm production systems that can be legitimately used.

1996 Farm Bill

The conservation provisions of the Federal Agriculture Improvement and Reform Act of 1996 (FAIRA) have introduced several interesting policy initiatives that will significantly affect future soil and water conservation policies in the U.S. Several agricultural provisions of the 1996 Farm Bill will undoubtedly affect soil and water conservation initiatives within the U.S., particularly those focused on commodity programs.

Several conservation programs authorized by the two previous Farm Bills have been reauthorized in modified form by the 1996 FAIRA. The CRP and the Wetland Reserve Program (WRP) from previous farm bills have been combined with the Environmental Quality Incentives Program (EQIP) to form the Environmental Conservation Acreage Reserve Program (ECARP). Activities enacted under ECARP will be focused on watersheds, multistate regions, or regions of special environmental sensitivity. The goal of ECARP is to maximize environmental benefits and to facilitate adoption of relevant conservation practices by producers. Soil and water conservation resources will be targeted to areas that are defined as having priority for assistance. Advisory committees will be formulated to aid in the selection of regions that will receive conservation resources.

ECARP has considerable potential to be successful. However, there are pitfalls that may substantially reduce environmental benefits. Targeting of conservation resources should result in better utilization of limited human and economic resources. However, the criteria used to select "sensitive regions" and to assess impacts will be critical in the determination of success or failure of conservation efforts associated with ECARP. It will be very difficult for persons involved in the selection process not to be affected by "vested interests." Committee persons commissioned to select areas to receive conservation resources will find it difficult not to lobby for designation of places close to their community or for watersheds populated by constituents. It will also be difficult for committee persons to select resources that are particularly deserving of protection. It will be difficult to select protection of soil resources over wetlands or wildlife habitat. It will be difficult for committees to choose target areas that will maximize improvement in water quality when investment in other areas will maximize soil savings. Selection committees will hold extensive power relative to determination of where public resources will be expended, and it will be very difficult to formulate committees that can effectively represent all public interests.

Conservation provisions of FAIRA contain elements of cross-compliance that will penalize landowners for noncompliance with provisions of conservation contracts. Unfortunately, these penalties may be of very little consequence, because commodity programs are being phased out and provisions have been made within FAIRA for early withdrawal from conservation contracts. Most producers are not concerned about farm income in the near term, because commodity prices are so high. It is highly likely that farmers will have at least 3 to 5 years of high commodity prices and that they will not be concerned about minor penalties associated with violation of conservation contracts. It is also highly likely that market pressures to maximize production will motivate land operators to produce at the highest level possible and to be much less concerned about conservation of soil and water resources.

The conservation provisions of the 1996 FAIRA suggest that considerable interest remains in protecting soil and water resources. However, the means of achieving this policy end may not be contained within the legislation. Unless implementation of the policy objectives of FAIRA is completed in a very proficient and equitable manner, it is highly likely that market forces will overwhelm the nearly powerless conservation programs contained within the legislation. There do not appear to be major economic resources authorized by FAIRA to adequately subsidize landowners, and there are practically no disincentives to coerce landowners to practice good conservation of soil and water resources.

Conclusions

Contemporary soil conservation policies have evolved from totally voluntary approaches conceived and implemented in the 1930s, to a mixture of voluntary and regulatory approaches employed in the late 1980s and 1990s. Rights of resource owners have been constrained, and rights of society have been expanded. While use of economic and other forms of government subsidies remain effective mechanisms for implementing soil conservation policies, the use of coercion and regulation have emerged as useful policy instruments to motivate land operators to implement soil and water protection systems at the farm level.

It is highly likely that future soil conservation policies will continue to employ both voluntary and regulatory approaches. It is highly probable that regulatory approaches will be used more extensively in the future, because voluntary approaches have been shown to be relatively ineffective in the recent past. With the exception of lease programs associated with the CRP, voluntary conservation programs have not been very successful. The use of regulation will probably be accelerated by society's inability to exert pressure on landowners to comply with conservation expectations by denying access to farm programs that are being phased out of existence. New policy approaches must be developed in the near-term or many of the advances made by investment of billions of dollars of public funds in Conservation Title programs in the last decade will be lost by pressures to maximize farm production.

References

Camboni, S.M. and T.L. Napier. 1993. Factors Affecting Use of Conservation Farming Practices in East Central Ohio. Agriculture, Ecosystems and Environment 45: 79-94.

Halcrow, H.G., E.O. Heady, and M.L. Cotner. 1982. Soil Conservation Policies, Institutions, and Incentives. Ankeny, Iowa: Soil and Water Conservation Society Press.

Harmon, K.W. 1987. History and Economics of Farm Bill Legislation and Its Impacts on Wildlife Management and Policies. In Impacts of the Conservation Reserve Program in the Great Plains. J.E. Mitchell (ed.). Denver, Colorado: USDA Forest Service. pp. 105-108.

Helms, D. 1990. New Authorities and New Roles: SCS and the 1985 Farm Bill. In Implementing the Conservation Title of the Food Security Act of 1985. Ted L. Napier (ed.). Ankeny, Iowa: Soil and Water Conservation Society Press. pp. 11-25.

Lovejoy, S.B. and T.L. Napier. 1986. Conserving Soil: Insights from Socioeconomic Research. Ankeny, Iowa: Soil and Water Conservation Society.

Mueller, D.H., R.M. Klemme, and T.C. Daniel. 1985. Short- and Long-Term Cost Comparisons of Conventional and Conservation Tillage Systems in Corn Production. Journal of Soil and Water Conservation 40: 466-470.

Napier, T.L. 1990a. The Evolution of U.S. Soil Conservation Policy: From Voluntary Adoption to Coercion. In Soil Erosion on Agricultural Land. J. Boardman, I.D.L. Foster, and A. Dearing (eds.). London: John Wiley & Sons, Ltd. pp. 627-644.

Napier, T.L. 1990b. Implementing the Conservation Title of the Food Security Act of 1985. Ankeny, Iowa: Soil and Water Conservation Society Press.

Napier, T.L. 1994a. The Potential for Public-Private Partnerships in Ecosystems Management. In Ecosystems Management: Status and Potential. Washington, D.C.: U.S. Government Printing Office. pp. 243-249.

Napier, T.L. 1994b. Regulatory Approaches for Soil and Water Conservation. In Agricultural Policy and the Environment. Louis E. Swanson and Frank Clearfield (eds.). Ankeny, Iowa: Soil and Water Conservation Society Press. pp. 189-202.

Napier, T.L. and E.J. Johnson. 1996. Conservation Initiatives Within the Darby Creek Hydrologic Unit of Ohio: An Assessment of Impacts. Paper presented at the 1996 annual Soil and Water Conservation Society meeting. Keystone, Colorado.

Napier, T.L. and D.L. Foster. 1982. Farmer Attitudes and Behaviors Associated With Soil Erosion Control. In Soil Conservation Policies, Institutions, and

Incentives. H.G. Halcrow, E.O. Heady, and M.L. Cotner (eds.). Ankeny, Iowa: Soil and Water Conservation Society Press. pp. 137-150.

Napier, T.L., Silvana M. Camboni, and Samir A. El-Swaify. 1994. Adopting Conservation on the Farm: An International Perspective on the Socioeconomics of Soil and Water Conservation. Ankeny, Iowa: Soil and Water Conservation Society Press.

Rasmussen, W.D. 1982. History of Soil Conservation, Institutions, and Incentives. In Soil Conservation Policies, Institutions, and Incentives. H.G. Halcrow, E.O. Heady, and M.L. Cotner (eds.). Ankeny, Iowa: Soil and Water Conservation Society Press. pp. 3-18.

Rasmussen, W.D. and G.L. Baker. 1972. The Department of Agriculture. New York: Praeger Press.

Rasmussen, W.D. and G.L. Baker. 1979. Price Support and Adjustment Programs from 1933 Through 1978: A Short History. Washington, D.C.: U.S. Government Printing Office.

Reichelderfer, K. and W.G. Boggess. 1988. Government Decision Making and Program Performance: The Case of the Conservation Reserve Program. American Journal of Agricultural Economics 70 (1): 1-11.

Swanson, L.E. and F.B. Clearfield. 1994. Agricultural Policy and the Environment: Iron Fist or Open Hand. Ankeny, Iowa: Soil and Water Conservation Society.

Swanson, L.E., S.M. Camboni, and T.L. Napier. 1986. Barriers to Adoption of Soil Conservation Practices on Farms. In Conserving Soil: Insights from Socioeconomic Research. Stephen B. Lovejoy and Ted L. Napier (eds.). Ankeny, Iowa: Soil and Water Conservation Society Press. pp. 108-120.

9

Future Soil and Water Conservation Policies and Programs Within the United States

Ted L. Napier and Silvana M. Napier

Salary and research support that made this chapter possible were provided by the Ohio Agricultural Research and Development Center of the Ohio State University and the Management Systems Evaluation Area project funded by the U.S. Department of Agriculture. The positions advanced in this chapter are solely those of the authors and do not represent official positions of the funding organizations.

As noted in the previous chapter (Napier and Napier 2000), recent modifications in the economics of the agricultural industry and agricultural policies within the United States will significantly change the trajectory of future soil and water conservation policies and programs. While the specific directions of conservation policies and programs are uncertain at this time, it is highly likely that regulatory approaches will become much more significant for motivating land operators to adopt and to continue use of conservation production systems (Napier 1994a). It is also highly likely that state and federal governments will play a smaller role in the implementation of soil and water conservation policies than in the past, and new actors will assume greater responsibility for implementing soil and water conservation policies.

Success in predicting the nature of future soil and water conservation policies and programs is predicated on the validity of assumptions made about the social, economic, and political environments that will be in existence at the time future soil and water conservation policies are formulated and implemented. Also, the validity of future predictions is usually much higher in the near term than in the long term, because knowledge of the factors affecting conservation policies and programs is much greater in the near term. Therefore, predictions made about future directions of soil conservation policy in this chapter are differentiated in terms of near-term predictions and long-term predictions. The assumptions used to make predictions about future soil and water conservation programs and public policies in this chapter are as follows: 1) society will continue to value soil and water conservation and will politically support policies designed to protect soil and water resources; 2) the economy will remain viable and will continue to expand at the same rate as present; 3) public economic and human resources allocated to support soil and water conservation programs will continue to be stable; 4) environmentalists will maintain sufficient political influence to affect public conservation policy; 5) federal commodity programs will be phased out of existence and will not be replaced by some other price support system; 6) grain prices will remain relatively high for at least 3 to 5 years; 7) the goals to be achieved by conservation policies and programs will shift periodically as priorities of environmental problems change; and 8) food and fiber scarcity within the U.S. will not legitimize abolition of environmental constraints on agricultural production.

Near-term soil and water conservation policies and programs

In the near term, it is highly likely that soil and water conservation policies and programs will remain similar to those authorized by the 1985, 1990, and 1996 Farm Bills.

Contained within the Federal Agriculture Improvement and Reform Act (FAIRA) of 1996 are several soil and water conservation initiatives that are similar to those created in 1985 and reauthorized in 1990. A modified version of the original Conservation Reserve Program (CRP) was reauthorized by the 1996 FAIRA. The continuance of a modified CRP was supported by many segments of the agricultural community, because it authorizes government lease of cropping rights at local rent values. Public conservation policies that generate income are highly valued by U.S. landowner/operators, especially those that pay landowner/operators not to farm.

The CRP has been perceived by many publics to have generated win-win outcomes for society and for landowner/operators. Landowner/operators benefit from government rent payments, and society benefits by improvement in environmental quality. The major negative aspects of the CRP program are the high economic costs of leasing highly erodible cropland and the adverse economic impacts on local rural economies associated with long-term retirement of large tracts of agricultural land (Napier 1990a).

It is highly likely that major emphasis of the 1996 CRP will be shifted from reduction of soil loss to improvement in water quality. Such an emphasis shift should increase the cost of enrolling highly erodible cropland. Retirement of highly erodible cropland via a CRP designed to improve water quality should increase the per acre cost over previous CRPs because the qualitative aspects of land required to achieve improvement in water quality are much different from those of land that must be retired to reduce soil loss. Achievement of improved water quality can best be accomplished by retirement of highly erodible cropland in more humid areas of the Midwest, East, and Northwest rather than in more arid regions of the Midwest and the West. Cropland in the more humid regions of the U.S. is much more productive than in the arid West and High Plains where a major portion of the land in the original CRP was enrolled (Napier 1990a). As the productivity of cropland increases, the bid price required to motivate landowners to enroll eligible land in set-aside programs concomitantly increases. If commodity prices remain high for an extended period of time, the bid price necessary for landowner/operators to enroll highly erodible cropland in the CRP may hasten the demise of these types of conservation policies and programs because the economic costs may become prohibitive.

It is highly likely that politically popular programs such as the CRP will be continued into the future as long as economic resources are available from public sources for allocation to set-aside programs. While the environmental goal to be achieved by set-aside policies and programs will probably change over time (for example, emphasis on improvement of water quality rather than soil savings), leasing of production rights is a very popular and effective approach for addressing soil and water conservation problems at the farm level. If economic resources from public sources become scarce and/or competition for access to these programs becomes intense, it is highly probable that use of lease agreements will be discontinued because the political cost of continuing such programs will be very high.

Participation in soil and water conservation programs is significantly affected by the policy objectives to be achieved. Assuming that improved water quality will be the primary goal to be achieved by the 1996 CRP, distribution of land enrolled in the new CRP will be considerably different from previous CRP initiatives. A major portion of the land bid into the original CRP came from the arid West and portions of the High Plains (Napier 1990a), because the policies employed to implement the original CRP were designed to "minimize cost" and to maximize "soil savings" (Reichelderfer and Boggess 1988). The outcome of the "least cost-soil savings" implementation policy was the enrollment of large tracts of relatively inexpensive cropland in the West and High Plains. Much of the highly erodible cropland in the humid Midwest and East was not enrolled in the program, because the bid price established by the government was below local rent values for highly productive land. Much of the highly erodible land in humid areas of the U.S was not enrolled in the previous CRPs. Highly erodible cropland in humid regions of the U.S. were not attracted into the program until the last rounds of bidding when bid prices reached parity with local rent values. If improvement of water quality is emphasized by conservation policies contained within the 1996 Farm Bill and much of the cropland enrolled in the revised CRP is located in the humid East and Northwest, then distributional impacts will be experienced in all geographic regions of the country. The political consequences of these changes will almost certainly be reflected in future conservation policy initiatives.

The social, political, and economic implications of a major shift of CRP enrolled land to humid regions of the Mississippi River system, the East, and the Northwest will be significant for all regions of the country. As land enrolled in the original CRP in the arid West comes back into agricultural production due to termination of contracts, local economies will reflect the increase in purchases of agricultural inputs. It is also highly likely that much of the land enrolled in the original CRP will be returned to agricultural production using similar production systems employed in previous decades and environmental degradation may be similar to before CRP rates even though the land will be operated under the constraints imposed by Conservation Compliance. Conversely, local economies in more humid areas where land will be enrolled in the revised CRP will be significantly affected by a decline in purchases of agricultural inputs. Many of the economic consequences identified in farming communities when the Conservation Title of 1985 was implemented (Napier 1990a) will be observed in geographic regions where large tracts of cropland will be enrolled in the 1996 CRP. Environmental benefits in the form of reduced soil loss and improved water quality will be realized in the regions where highly erodible cropland will be enrolled.

The political consequences associated with exclusion of landowner/operators in the High Plains and arid West who participated in the 1985 CRP from the 1996 CRP is yet to be determined. However, it is highly likely that considerable political pressure will be applied from vested interests in those regions to be compensated in some form for loss of federal payments. The transition of government funds from one geographic region of the U.S. to another will not occur without considerable conflict. Resolution of interregional conflict will probably affect future soil and water conservation policies and programs when political leaders attempt to satisfy constituents in arid regions of the country.

Evidence to date strongly suggests that a number of the regulatory mechanisms incorporated in the 1985 and 1990 conservation titles will be continued. Landowner/operators in the U.S. have accepted the fact that government will be involved in farm-level decision-making relative to the types of production systems employed. Use of penalties such as loss of access to federal farm programs for noncompliance with Conservation Title initiatives will be continued. Other coercive instruments designed to motivate landowners to adopt and to continue use of soil conservation production systems may be forthcoming. One reason that new disincentives may have to be created is the elimination of commodity programs. As commodity programs are phased out of existence, society will lose much of the leverage it possesses to encourage landowner/operators to conform to contractual conservation agreements.

The loss of commodity support programs will have little impact on soil and water conservation efforts in the U.S. if other incentives are created to protect soil and water resources from environmental degradation. Elected representatives of society possess the right to authorize the use of a host of incentives and/ or disincentives to motivate land operators to adopt and to use soil and water conservation production systems. However, it is highly probable that few of the more coercive methods will be politically popular with landowners who are subject to penalties associated with such approaches. Coercive methods are also not popular with elected officials who must assume the responsibility for authorizing use of these approaches. For these reasons, it is highly likely that future conservation policy initiatives, at least in the near term, will favor noncoercive methodologies to achieve conservation goals. One approach that is almost certain to be explored is the use of the private sector to assume greater responsibility for achieving soil and water conservation objectives (Lovejoy 1994; Napier 1994b; Napier et al. 1995). Once it is realized that environmental goals cannot be achieved with noncoercive mechanisms, it is highly likely that more coercive approaches will be incorporated in future conservation policies.

The targeting of limited conservation resources to highly sensitive lands by the 1996 Environmental Conservation Acreage Reserve Program (see discussion of ECARP in the previous chapter) is highly desirable from the perspective of securing the highest environmental return per public dollar allocated for environmental protection. Assuming that implementation procedures can be developed that are equitable and effective in selecting the most deserving environmental problems and project sites, society should benefit from better utilization of limited economic and human resources available for use in protecting soil and water resources. While there are many political and social forces that may prevent limited conservation resources from being allocated to the most critical environmental problems (Napier and Napier 2000), the use of targeting appears to be justified.

While elements of regulation exist within the 1996 Farm Bill, the conservation policies contained within the legislation are clearly oriented toward the use of voluntary participation of landowner/operators of highly erodible and environmentally sensitive land. There is very little emphasis placed on the use of regulation to control landowner behaviors. Supporters of the 1996 conservation policy initiatives appear to be convinced that education-information-partial subsidies will result in increased use of soil and water conservation production systems at the farm level. Unfortunately, research to date brings such an assumption into serious question (Lovejoy and Napier 1986; Napier et al. 1994; Napier and Johnson 1996). While landowner/operators will undoubtedly participate in subsidy programs and enroll land in long-term lease arrangements when it is profitable to do so, environmental benefits associated with such approaches are often short-lived. When it becomes more profitable to employ production systems that threaten soil and water resources within designated "environmentally sensitive" areas, landowner/operators will do so. Society will be basically powerless to prevent degradation of land and water resources. This scenario is almost certain to be realized if grain prices remain as high as they are in the summer of 1996, if commodity programs are eliminated, and if bid prices for set-aside land do not increase adequately to make it profitable for landowner/operators to retire land from production.

Long-term soil and water conservation policies and programs

It is highly likely that greater emphasis will be placed on regulatory approaches to motivate landowners to adopt soil and water conservation production systems in the long-term. If substantial improvements are made in remote sensing technologies that make it possible to monitor individual fields at very reasonable cost, it is almost certain that regulatory approaches will become the primary mechanism used to control soil erosion at the farm level (Napier 1994b; Napier 1990b).

The primary reason that command and control approaches have not been used extensively in the past to reduce soil erosion on highly erodible cropland in the U.S. is that it is expensive to monitor nonpoint pollution (Napier 1990b; Napier 1994a). When monitoring of individual farmsteads for soil erosion becomes economically feasible, society will undoubtedly use regulatory approaches to stop agricultural pollution. The cost of economic subsidies and many environmental educational programs to motivate landowner/operators to adopt and to continue use of conservation production systems will be eliminated. Cost savings from elimination of economic subsidies and education programs will be employed to justify use of such policy approaches.

It is becoming extremely difficult to argue from an ethical perspective that agriculture should be exempt from environmental norms of society. Proponents of regulatory approaches to achieve conservation goals argue that agricultural pollution should not be treated differently from any other source of pollution, such as urban sewerage treatment systems and industrial plants. When the majority of U.S. society embraces this orientation, some form of regulation will almost certainly be applied to agricultural land, because regulations were required to control point pollution. City governments and owners of industry did not voluntarily adopt pollution abatement technologies and practices. They were forced to internalize the costs of implementing pollution control systems. Control of point pollution required regulation combined with severe penalties, and it is highly likely that similar approaches will be required to adequately address nonpoint pollution from agricultural sources.

If effective monitoring of agricultural pollution can be achieved at reasonable cost, a host of mechanisms can be used to encourage landowner/operators to adopt soil and water conservation production systems. Some of the mechanisms that could be used to control degradation of soil and water resources are as follows: fines for violation of environmental standards; restricted access to agricultural chemicals; restrictions placed on timing of agricultural chemical applications; restrictions placed on the types of agricultural products that may be produced on specific soil classifications; purchase of cropping rights to highly erodible cropland; permanent retirement of highly erodible cropland; and many others. It is highly likely that many segments of society will support some or all of these approaches in the future, because such methods would be much less expensive to implement than information-education-subsidy approaches presently used (Napier 1994a).

Without effective and inexpensive means of monitoring pollution from individual farmsteads, it is highly probable that future soil conservation policies and programs will contain elements of government incentives combined with various types of regulations. However, it is doubtful that society will continue to invest in information-education approaches, because such mechanisms have been shown to be relatively ineffective for motivating landowner/operators to adopt soil and water protection practices at the farm level. Most U.S. farmers are well aware of soil erosion on their farms, they are knowledgeable of the causes and consequences of soil displacement, and they are knowledgeable of the many technologies and techniques that can be used to control erosion on highly erodible cropland. Provision of additional information to make farmers more sensitive to environmental problems has very marginal impact on conservation behaviors (Lovejoy and Napier 1986; Napier et al. 1994; Napier and Johnson 1998). Economic savings derived from reducing or eliminating information-education components of conservation programs could be used to finance more effective approaches on highly erodible cropland, such as the provision of economic resources to cost-share purchase of conservation technologies or construction of needed conservation structures; purchase of cropping rights on highly erodible cropland; payment for on-site monitoring and/or remote sensing to identify polluters; and the provision of technical assistance to make it possible for landowner/operators to adopt complex conservation production systems.

Another policy option that society may elect to implement in the future is the establishment of environmental standards without specifying how landowner/operators should meet environmental expectations. Such an approach would permit landowner/operators to satisfy environmental goals in any manner they wish to employ. Such an approach may have merit, since farmers have a long history of discovering ways to maximize farm-level profits under many constraints. To be effective, this type of approach will require extensive utilization of technical assistance because many conservation options will require knowledge not possessed by most farmers.

Unlike contemporary conservation policies that require the state to prove that specific farmers are not in compliance with environmental standards, the proposed option could place the burden of proof on individual farmers. Landowner/operators could be periodically required to demonstrate that production systems in use satisfy environmental standards. The cost of implementing such a program for the state would be significantly less because monitoring by the state would not be required. Inclusion of severe penalties within soil and water conservation policies, which authorize such an approach, would act as strong motivators for landowner/operators to comply with established environmental standards.

While it is not possible to specify the specific combination of incentives and disincentives that will be employed in future soil and water conservation policies and programs in the U.S., it is certain that soil conservation initiatives that emphasize voluntary participation, use of costly government subsidies, and information-education campaigns to make landowner/operators aware of environmental issues will be subject to much closer scrutiny in the future (Napier and Johnson 1998). It is highly likely that society will rely more heavily on command and control mechanisms where it is feasible to do so because such approaches tend to be less expensive for the taxpayer.

New actors in the development and implementation of soil and water conservation policy

While soil conservation programs have been perceived to be the primary responsibility of state and federal governments for more than sixty years in the U.S., serious consideration will probably be given in future soil and water conservation policies to modifying this orientation. An option that has received attention in recent years is use of the private sector to address soil erosion problems (Lovejoy and Napier 1986; Napier 1994b; Napier et al. 1995). It is quite likely that future public policies will be formulated to strongly encourage private sector participation in soil and water conservation initiatives and to de-emphasize government involvement in the implementation of conservation policies. It is also highly likely that future conservation policies will place a large proportion of the responsibility for achieving environmental quality goals on landowner/operators.

Present soil and water conservation policies and programs rely heavily on environmental ethics, environmental values, and on-site benefits to be derived from investments made in soil and water conservation to motivate land operators to adopt conservation production systems. This approach basically argues that when landowner/operators develop a strong environmental ethic and perceive that economic benefits will be forthcoming from adoption of conservation production systems, they will adopt recommended conservation production systems. This approach does not appear to be very useful when assessed in the context of changes in actual conservation behaviors at the farm level. The primary reason this approach fails is that most landowner/operators already value soil and water resources; they perceive themselves to be stewards of the land which they control, and they recognize that few on-site benefits can be achieved by additional investments in soil and water conservation at the farm level. Use of on-site economic returns to investments as a motivator of adoption of soil and water conservation production systems is particularly flawed, because it has been demonstrated repeatedly that adoption of most soil and water conservation production systems is usually not profitable to land operators in the short-term and often not in the long-term (Batte and Bacon 1995; Mueller et al. 1985; Putman and Alt 1987). It is very difficult to motivate land operators to adopt new production systems when the return to investment in the new systems is very low or nonexistent.

Many public conservation policies and programs fail to recognize that farmers are business persons and are primarily motivated by profits. Conservation policies and programs often fail because the importance of profit to farmers is underestimated and the return to investment in conservation at the farm level is overestimated. The introduction of new practices and/or technologies into existing production systems will always introduce higher levels of uncertainty and risk into farm business operations. If farmers do not expect substantially higher net returns as a result of changing production systems, they will not voluntarily adopt. Economic inducements are always required to make adoption of conservation production systems profitable. When the subsidies are withdrawn, farmers will return to the practices that produced profits even if the production systems contribute to environmental degradation. This implies that subsidies, once initiated, will always be required to motivate landowner/operators to employ conservation production systems. Unfortunately, recent evidence suggests that society will probably not continue to allocate adequate economic resources to keep conservation practices on land that must be operated using conservation production systems.

If future soil conservation policies and programs have any hope of permanently resolving soil and water resources problems within the U.S., it must be recognized that land operators adopt alternative production systems because they are more profitable than those presently being used. Until intervention strategies recognize that production agriculture is a business and that farm-level production decisions are made in the context of profitability, it is highly probable that future programs will remain ineffective.

The private sector is probably the best agent to implement a profitable approach to soil conservation. Research and development to produce technologies and techniques that are highly productive and environmentally friendly appear to be consistent with the goals of private business. There is and will remain a very strong demand for effective conservation technologies and techniques that are environmentally benign. Establishment of environmental standards by the government would encourage land operators to adopt relevant technologies and techniques developed and marketed by the private sector. Public policies that encourage government funding of private sector research and development of conservation farm technologies could facilitate this process. The emergence of private businesses to provide technical expertise to landowner/operators concerning agricultural and environmental issues could replace existing extension functions and would release massive amounts of economic and human resources for other uses.

Establishment of public policies to control the application of agricultural inputs could generate considerable demand for "contract farming." Authorization of conservation policies designed to reduce chemical inputs and to improve the timing of application could encourage landowner/operators to employ licensed commercial applicators to apply chemicals in environmentally approved ways. Input costs savings would partially pay for the service provided by commercial applicators.

Public policies could be formulated that establish liability for use of farm production systems that threaten the health and well-being of human and non-human populations. Such policies would further contribute to the involvement of contract applicators and contract agriculturalists. Insurance costs to protect individual farmers from lawsuits would probably more than compensate for use of contract farmers who would assume liability for any damages. If liability were extended to include agricultural input from manufacturers and dealers, they would become significant actors in on-farm decision making to ensure that they would not be subject to lawsuit.

It must be recognized that the establishment of soil and water conservation policies has many social, political, and economic consequences. Establishment of stringent environmental standards could practically eliminate small-scale agriculture in the U.S. Liability and penalties for violation of environmental norms could result in almost all farming being done by contract agriculturalists using "precision technologies." Farm technologies presently exist that make it possible to apply chemicals differentially within fields governed by the physical structure and variability of the land resource. Use of precision farming to meet environmental standards would not be feasible for most farmers, at least in the near-term, because such technologies are very expensive, and an effective use of such technologies necessitates human skills that are often beyond the level of most agriculturalists in the U.S. Therefore, specialized contract farmers could justify purchase of such equipment and make investment in the development of human skills needed to effectively use such technologies, if the demand for precision farming was sufficiently high to establish profitable businesses. Over time, however, it is highly likely that costs of technologies would decline and the human skills would be enhanced adequately for landowner/ operators to do their own precision applications.

If several of these policy options are implemented in the U.S., the nature of the agricultural industry will change and the institutions serving the industry will be drastically modified. The greatest benefactors will be the private sector and the public agencies that will assume modified roles. The principal losers will be traditional agricultural research and development actors, such as land-grant universities, that continue to produce technologies and techniques that often are not relevant to the needs of potential adopters. Government agencies specializing in the provision of education-information focused on soil and water conservation to agricultural populations will be phased out of existence or will be required to shift their efforts to other issues. For example, it is highly likely that the Extension Service will have a much less useful role to play in future soil and water conservation initiatives because farmers do not require "awareness information." Few extension agents possess the level of technical knowledge that will be required to satisfy the information needs of landowner/ operators operating businesses in the new conservation policy arena. Land operators now require very technical data for decision making, not general "sensitivity" information. In the future, it is highly likely that any education-information function required by land operators to make them aware of innovations in soil and water conservation will be much better provided by the Natural Resources Conservation Service/U.S. Department of Agriculture (formerly called SCS/USDA) than any other agency. NRCS/USDA field staff are much more competent to provide technical information and to act as technical consultants than professional communicators employed in the Extension Service. If NRCS/USDA elects to assume the new information role, it is highly probable that the agency will require additional training in communication skills so that technical information can be communicated in an understandable manner to potential adopters.

Conclusions

Soil and water conservation policies and programs have the potential to revolutionize the agricultural industry and related service agencies in the U.S. Society will soon determine the direction that future soil and water conservation policy initiatives will take, and it is highly likely the direction chosen will create significant concern among contemporary landowner/operators, faculties and administrators of land-grant institutions, agricultural service industries, and government agencies commissioned to provide technical services to landowner/operators. It is not unreasonable to predict that environmental policy may soon dictate agricultural policy and programs in the U.S. Unfortunately, most academicians, conservation agency personnel and administrators, politicians, and rank and file citizens with vested interests in the agricultural industry are not prepared to address the complexities associated with the changes that will be forced upon us in the near future.

References

Batte, M.T. and K.J. Bacon. 1995. Economic evaluation of three production regimes at the Ohio MSEA Project. In Clean Water-Clean Environment-21st Century Working Group on Water Quality. St. Joseph, Michigan: American Society of Agricultural Engineers, pp. 17-20.

Lovejoy, S.B. 1994. Meeting America's soil and water goals without regulation. In Agricultural Policy and the Environment: Iron Fist or Open Hand. Louis E. Swanson and Frank B. Clearfield (eds.). Ankeny, Iowa: Soil and Water Conservation Society Press, pp. 183-188.

Lovejoy, S.B. and T.L. Napier. 1986. Conserving soil: Insights from socioeconomic research. Ankeny, Iowa: Soil and Water Conservation Society Press.

Mueller, D.H., R.M. Klemme, and T.C. Daniel. 1985. Short- and long-term cost comparisons of conventional and conservation tillage systems in corn production. Journal of Soil and Water Conservation 40 (5): 466-470.

Napier, T.L. 1990a. Implementing the Conservation Title of the Food Security Act of 1985. Ankeny, Iowa: Soil and Water Conservation Society Press.

Napier, T.L. 1990b. The evolution of U.S. soil-conservation policy: From voluntary adoption to coercion. In Soil Erosion on Agricultural Land. J. Boardman, I.D.L. Foster, and J.A. Dearing (eds.). New York: John Wiley and Sons.

Napier, T.L. 1994a. Regulatory approaches for soil and water conservation. In Agricultural Policy and the Environment: Iron Fist or Open Hand. L.E. Swanson and F.B. Clearfield (eds.). Ankeny, Iowa: Soil and Water Conservation Society Press, pp. 189-202.

Napier, T.L. 1994b. The potential for public-private partnerships in ecosystems management. In ecosystems management: Status and potential. Washington D.C.: Congressional Research Service and the U.S. Government Printing Office, pp. 243-249.

Napier, T.L. and Eric J. Johnson. 1998. Impacts of voluntary conservation initiatives in the Darby Creek watershed of Ohio. Journal of Soil and Water Conservation 53(1) : 78-84.

Napier, T.L., Sam E. McCarter, and Julia R. McCarter. 1995. Willingness of Ohio landowner/operators to participate in a wetlands trading system. Journal of Soil and Water Conservation 50 (6): 648-656.

Napier T.L. and S.M. Napier. 2000. Soil and water conservation policy within the U.S. In Soil and Water Conservation Policies: Successes and Failures. Ted L. Napier, Silvana M. Camboni, and Jiri Tvrdon (eds.). Ankeny, Iowa: Soil and Water Conservation Society Press, pp. 89-100.

Napier, T.L., S.M. Camboni, and S.A. El-Swaify. 1994. Adopting soil conservation on the farm: An international perspective on the socioeconomics of soil and water conservation. Ankeny, Iowa: Soil and Water Conservation Society Press.

Putman, J. and K. Alt. 1987. Erosion control: How does it change farm income? Journal of Soil and Water Conservation 42 (4):265-267.

Reichelderfer, K. and W.G. Boggess. 1988. Government decision making and program performance: The case of the Conservation Reserve Program. American Journal of Agricultural Economics 70 (1): 1-11.

10

Policy Lessons From a Quasi-Regulatory Conservation Program

J. Dixon Esseks, Steven E. Kraft, and Douglas M. Ihrke

Acknowledgment: Funding for this research came from the Joyce Foundation, the Illinois Groundwater Consortium, The American Farmland Trust, and the Natural Resource Conservation Service. The chapter reflects the opinions of the authors and not the funding agencies. We would like to thank all the farmers, District Conservationists, and NRCS staff people who have made these studies possible.

Conservation Compliance is a quasi-regulatory program that was enacted as part of the Conservation Title (Title XII) of the 1985 Food Security Act (PL 99-198). Through conservation compliance, farmers operating highly erodible land (HEL) who want to remain eligible for U.S. Department of Agriculture (USDA) program benefits (e.g., commodity price supports, commodity loans, crop insurance, disaster relief, Conservation Reserve Program, and Farmers Home loans) are required to acquire and implement a Conservation Compliance plan that should reduce the level of soil loss on the HEL to an acceptable level. Through the use of survey interviews of farm operators with HEL and of district conservationists of the Natural Resources Conservation Service (NRCS), we have been able to study the implementation of conservation compliance over time. These studies have allowed us to identify a number of important policy lessons for the development and implementation of soil and water conservation policy.

As a policy, Conservation Compliance developed in response to a number of concerns coming together in the early 1980s: the use of the National Resources Inventory (NRI) identified geographical areas accounting for the largest proportion of soil erosion, skepticism regarding the effectiveness of voluntary measures to control soil erosion, the identification by Edwin Clark and others of the dollar value of off-site damage caused by soil erosion, and the existence of relatively large commodity support payments to farmers. Analysis of the

109

1977 NRI data indicated that for sheet and rill erosion, 25 million acres of cropland accounted for 43 percent (828 million tons per year) of the total tonnage of cropland sheet and rill erosion (American Farmland Trust 1984). The analysis of data from the 1982 NRI revealed that approximately 118 million acres of the 421 million acres of the country's cropland base were responsible for 70 percent of the sheet and rill erosion in the country (USDA 1989). The skepticism of a reliance on volunteerism for controlling soil erosion was captured in 1982 by Wittwer: "After 46 years and a $16 billion investment, no more than 25 to 35 percent of the nation's farmland is under approved conservation practices. Meanwhile, enormous losses of topsoil continue (16.5 tons per hectare per year for a total of 2.8 billion tons)." (Wittwer 1982, p. 209). Work by Clark and others for the Conservation Foundation indicated that the off-site damage caused by soil erosion had a monetary value ranging from $3.2 billion to $13 billion per year (Clark et al. 1985). Similarly, during the federal FY80 through FY85, USDA payments of program benefits to farmers increased from a low of $1.2 billion to a high of $9.3 billion in FY83. Because of (a) concern about the efficacy of voluntary programs to control soil loss; (b) the large payments of USDA benefits to farmers with HEL; (c) data indicating the magnitude of off-site damage from erosion; and (d) evidence that a relatively small proportion of the cropland base was responsible for the majority of the soil loss, Congress was ready to consider major reform of conservation policy during its deliberations for the 1985 Farm Bill. More specifically, they sought ways to significantly reduce the level of soil erosion on HEL.

As the debate over the 1985 Farm Bill proceeded, a consensus gradually emerged that farmers receiving USDA program benefits should have to meet an eligibility requirement involving the use of HEL in ways that would significantly reduce the level of soil erosion on this land. The result was Conservation Compliance, through which the NRCS was directed to work with farmers in the development and implementation of conservation plans for their HEL. Under the 1985 Food Security Act (the 1985 farm bill), farmers with HEL desiring to remain eligible for USDA program benefits were required to have a plan by January 1, 1990, and to have the plan fully implemented by the first of January 1995. NRCS (then called the Soil Conservation Service) was put in charge of working with farmers subjected to Conservation Compliance and of monitoring and enforcing the extent to which farmers were implementing their plans. Farmers that NRCS employees found out of compliance were reported to local offices of the Farm Services Agency (FSA) (formerly called the Agricultural Stabilization and Conservation Service or ASCS). Initial and amended rules under which Conservation Compliance was implemented were published in the *Federal Register* from 1986 to 1988. A controversial provision of these rules was the acceptance of Alternative Conservation Systems (ACS). Through ACS, the NRCS could develop Conservation Compliance plans for farmers that were not as demanding as plans designed with basic conservation systems to reduce soil loss to a tolerance level called "T." The ACS was designed so that farmers could still significantly reduce the rate of soil loss on their land without incurring large economic costs.

Before the passage of the 1985 Food Security Act, NRCS had worked only with those farmers voluntarily requesting help from the agency in dealing with their soil and water conservation problems. Essentially, the 1985 legislation transformed NRCS from an agency providing technical assistance to willing cooperators to an enforcement agency directed to provide technical assistance and enforcement for a quasi-regulatory program.

As part of Conservation Compliance, NRCS along with the FSA had to identify parcels of HEL to notify the owners and operators of the identified parcels. Once the parcels were identified and the owners and operators notified, the local employees of NRCS worked with the farmers in developing conservation plans.

In all, NRCS identified 149 million acres of HEL subject to Conservation Compliance. This program has been successful in reducing the overall level of soil loss in the U.S. With the application of the conservation plans, the erosion rates on the HEL dropped from an average of 17.4 tons per acre per year to 5.8 tons (Economic Research Service 1994). While presenting testimony in 1995 to the U.S. House of Representatives Appropriations Committee, Paul Johnson, Chief of NRCS, stated that NRCS had worked in the development of 1.7 million conservation plans covering 142 million acres resulting in a reduction of soil loss of more than 1 billion tons per year (U.S. House of Representatives 1995). In short, Conservation Compliance is having a major impact on reducing the level of soil loss. As a consequence of Conservation Compliance, NRCS employees have provided technical assistance to more farmers than they would have through the traditional voluntary approach to conservation planning.

Longitudinal studies of Conservation Compliance

With the passage of the 1985 Food Security Act, we started on a series of surveys on farmers with HEL. The first series of the telephone surveys, 1987-92, consisted of four that focused on selected watersheds in the Midwest dominated by HEL. Included were questions about both Conservation Compliance and the Conservation Reserve Program (CRP), another conservation provision of the 1985 Food Security Act (see Esseks and Kraft 1988, 1989, 1991, 1993).

The response rates were 80 percent or better. The second series consisted of four, 1992-1996, which, rather than focusing on a census of farmers in specific watersheds with HEL, involved sampling HEL tracts in the five Corn Belt states of Iowa, Illinois, Missouri, Indiana, and Ohio and then interviewing the operators or managers of the randomly selected tracts. For each study, the HEL tracts were chosen in two stages: first, 100 counties were randomly picked from across the Corn Belt so that each state's allotment from the 100 counties was proportional to the state's share of HEL tracts. From each of the selected counties, 15 tracts were selected at random and their operators surveyed. Phone interviews were conducted in 1992, 1993, 1995, and 1996 with response rates varying from 75 to 80 percent (Esseks, et al. 1993, 1994, 1995a and Esseks et al. 1996).

Finally in June 1994, a survey was conducted of a sample of NRCS District Conservationists (Esseks et al. 1995b). A mail survey instrument was used with a response rate of 91.8 percent. The design of the questionnaire reflected the results of four focus groups comprised of district conservationists. The purpose of the survey was to ascertain the district conservationists' perceptions of their work load associated with their expanded responsibilities under the 1985 Food Security Act.

Survey results and lessons for implementation

Our earlier studies conducted on a watershed basis yielded a number of results that lend themselves as lessons for implementation of soil and water conservation policy. First, the 1985 Food Security Act was signed into law on the December 23, 1985. The implementing agencies were instructed to have the first sign-ups for the CRP in March 1986 with subsequent sign-up periods in May and August. Surveys conducted in February and March 1987 of farmers with HEL, who would have been eligible for the CRP, indicated that a year after the program was initiated large proportions of the potential clientele were uninformed or misinformed about the program. According to our studies, this situation persisted for a number of years after the CRP was fully operational.

Data collected during surveys in the late 1980s as farmers approached the January 1, 1990 deadline for acquiring a Conservation Compliance plan indicated that farmers were similarly uninformed or misinformed about the program and the implications it would have for their farm operations. For example, 26 to 31 percent of the surveyed producers with HEL in a Wisconsin watershed were unaware of rather common mistakes that under Conservation Compliance would cause them to lose USDA benefits (Esseks and Kraft 1991). That is, farmers frequently lacked information so they were unable to make informed decisions about either participating in programs or informed assessments about

112

the implications of new conservation policies for their farm operations. These results suggested that the agencies implementing soil and water conservation policies and programs must take a vigorous, proactive approach to getting timely and accurate information about these programs into the hands of farmers who will have to make decisions vis-à-vis their participation in the programs. We have argued elsewhere that a useful model for implementing agencies to follow is that for industrial marketing (see Kraft et al. 1989 and Esseks and Kraft 1989). The important lesson is that the personnel of implementing agencies cannot assume that there will be a timely and adequate flow of information to the population of farmers potentially affected by the policies.

Similarly, the early studies conducted on the watershed basis indicated that farmers falling under Conservation Compliance had expectations regarding enforcement that were closely tied to the signals sent by NRCS through its implementation activities. That is, as farmers perceived the agency was less emphatic about implementing the program, their expectations about the extent to which monitoring and enforcement would take place fell. Similarly, as the leadership of the agency changed and a new message of fulfilling the requirements of Conservation Compliance was sent out, farmers' expectations regarding monitoring and enforcement changed reflecting an enhanced expectation of enforcement. The lesson is that as agencies implement conservation policies with more or less vigor, the expectations of the targeted farmers also change. The result can have a direct impact on the amount of conservation that is actually achieved, the extent to which the conservation practices will be maintained in the future, and the perceived legitimacy of the implementing agency.

Although not reflected directly in the data collected in our surveys, there is a third important lesson from this early period of Conservation Compliance implementation. While many environmental and conservation groups were critical of the development of the alternative conservation systems mentioned above, they probably had a positive impact on the acceptance of Conservation Compliance planning by the agricultural community. Given the NRCS experience with the development of conservation plans and the potential impact that the plans might have on farm profitability, the "agency decided that conservation plans could be based on systems that are economically and technically feasible for the farmers' local area" (Economic Research Service 1994, p. 184). This frequently meant that the conservation plans were less stringent regarding the level of soil loss reduction that the plan was designed to achieve. However, the apparent willingness of the agency to adopt an attitude of flexibility in implementing Conservation Compliance plans probably had an impact on farmers' favorable evaluations of the program as they approached the development of the 1995-1996 Farm Bill (Esseks et al. 1996). The important lesson for conservation policy is to determine the extent to which the rules under which a policy is implemented allow for flexibility in the design and application of conservation practices. The implementing agency must use this flexibility judiciously so that the greatest amount of conservation is achieved without undermining the effectiveness and legitimacy of the agency or the policy.

Survey data collected in 1992, 1993, 1995, and 1996 give us an opportunity to track farmers' perceptions to various aspects of Conservation Compliance across time. These perceptions provide the basis for deriving additional lessons. Often the best indicator of whether the targets of regulations will comply, and probably the most important policy issue affecting agency-clientele relations, is whether the clients believe that they are losing money because of the regulations. In the case of Conservation Compliance, for which the NRCS has expended vast amounts of field staff hours in developing Conservation Compliance plans and helping producers to apply them, a finding of many or most producers believing the plans were money-losers would be very disappointing for the agency.

Therefore, all four surveys asked the producers to assess whether their land's earnings after production costs decreased, increased, or did not change given the application of their Conservation Compliance plan. Table 1 shows that consistently about a quarter of our samples reported that their compliance plans were money-losers. Most of the remaining respondents reported no change in their earnings, while about a fifth to a quarter of the whole sample believed that they made money.

Though a minority, this 24-27 percent may have been numerous enough or otherwise sufficiently influential to persuade policy-makers to modify Conservation Compliance. For example, prior to passage of the 1996 Farm Bill, Kansas farmers complained directly to Representative Roberts, then Chair of the House Agriculture Committee, that Conservation Compliance needed to be made more farmer-friendly. Provisions of the enacted bill may reduce the frequency or impact of complaints about financial losses. One provision permits producers who are dissatisfied with practices in their Conservation Compliance plans to seek relief from the county FSA committees. Such relief depends on the committee's finding that applying the practices "would impose an undue hardship" (Section 315 of the Conference Report; see U.S. House of Representatives, 1996). Another provision permits producers who self-certify their compliance to change their conservation practices without need for NRCS approval as long as "the same level of conservation treatment" is achieved (Section 315).

Regression analysis indicates that among other factors, farmers who believed that NRCS would be fair in administering Conservation Compliance were less likely to feel that they would be losing money as a result of applying their

Conservation Compliance plans to their HEL. However, farmers who are participants in commodity programs have a higher likelihood of believing that Conservation Compliance will result in a loss of earnings. These farmers appear to be "reluctant conservationists" (e.g., farmers with Conservation Compliance plans solely to remain eligible for USDA program benefits). The 1996 Farm Bill created a new USDA program benefit for farmers, Market Transition Payments. Farmers with HEL who wanted to be eligible for these payments had to have a Conservation Compliance plan and implement it. Many farmers with qualifying base acres who had not recently participated in farm programs have

Table 1. Farmers' opinions regarding the effect of compliance plans on earnings

Applying CC practices will/ would:	Fall '92*	Fall '93*	Winter '95**	Winter '96***
			—%—	
Decrease earnings	27.3	25.5	25.3	24.5
Not change earnings	43.0	42.3	54.1	52.2
Increase earnings	29.3	28.5	19.1	21.3
Don't know/ won't answer	0.4	3.7	1.6	2.0
Total percent	100.0	100.0	100.0	100.0
Number of respondents	256	918	839	836

* Text of question in 1992 and 1993 surveys: "Here is a question about the financial effects of applying conservation practices listed in your compliance plan. Let's say you have been applying those practices a few years and are experienced in using them. After you gain or have that experience, will applying the practices have any effect on the land's earnings after production costs? Will applying the practices decrease earnings after production costs, not really change earnings, or will it increase earnings?"

** Text of question in 1995 survey: "Here is a question about the financial effects of applying the conservation practices listed in your compliance plan. Some farmers, for good reasons, have not applied the practices listed in their plans. That is their business. We are interested in your estimates of the financial effects if the practices listed in your plan were applied to your farmland. Does or would applying them decrease the land's earnings after production costs, not really change earnings, or increase earnings?"

***Text of question in 1996 survey: "We are interested in your estimates of the financial effects of applying the conservation practices in your compliance plan. In a year of normal weather and normal prices (such as corn at $2.30 and soybeans at $5.80), would applying those practices to your farmland decrease the land's earnings after production costs, not really change earnings, or increase earnings?"

signed up for Market Transition Payments. Many of these farmers have HEL and had not previously installed adequate conservation systems. They had a great potential to expand this pool of "reluctant conservationists." They presented NRCS with a challenge to apply the lessons the agency had learned about implementing Conservation Compliance over the past 10 years. The agency will have to determine how they will provide technical assistance to the farmers while monitoring how effectively the farmers are applying their conservation practices.

In addition to the questions about the perceived financial effects of compliance plans, we asked in all four surveys how the producers perceived NRCS' fairness in administering Conservation Compliance. The kind of fairness perception tapped in all surveys is reported in Table 2. The respondents were asked if they believed that NRCS would be fair to producers who were unable to apply Conservation Compliance practices due to circumstances beyond their control, such as flooding or drought. The percentage of interviews shows that those who selected the "very fair" option rose from 28.9 percent in 1992 to 43 percent in the fall of 1993, after the terrible flooding of that summer in the western Corn Belt. In response to the flooding, NRCS granted variances to farmers in the flood-affected counties. Subsequently the "very fair" percentage

Table 2. The expectations of respondents with compliance plans regarding the fairness of USDA's enforcement of conservation compliance*

Fairness of enforcement	Fall '92	Fall '93	Winter '95	Winter '96
			%—	
Not at all fair	3.5	4.0	5.7	6.2
Somewhat fair	22.3	15.5	18.8	18.2
Moderately fair	41.0	30.6	31.2	34.8
Very fair	28.9	43.0	40.3	37.9
Don't know/ won't answer	4.3	6.8	3.9	2.9
Total percent	100.0	100.0	100.0	100.0
Number of respondents	256	918	839	836

* Text of question: "Some farmers may be unable to apply compliance practices successfully due to circumstances beyond their control. It will not be intentional. For example, the crop residue levels may not be reached because of flooding or drought. Will the Natural Resource Conservation Service office serving your county be fair to this kind of producer who faces circumstances beyond his control? Will that office be not at all fair, somewhat fair, moderately fair, or very fair?"

declined to 40.3 percent in the 1995 study and then to 37.9 percent in the winter 1996. The 5.1 percentage point decrease between the 1993 and 1996 studies is statistically significant. Corn Belt producers' perceptions of NRCS fairness may improve as a result of another 1996 Farm Bill amendment, section 314, calling for "expedited procedures for the consideration and granting of a temporary variance" that "address[es] weather, pest, or disease problems" (U.S. House of Representatives, 1996, Section 314). As discussed above with the issue of flexibility, the implementing agency is placed in the position of "walking a tightrope" between being flexible in dealing with farmers coping with the vagaries of weather and pests and administering in the program so that the agency is not perceived as abdicating its monitoring and enforcement responsibilities.

Research on various regulatory programs, including a study we did on Conservation Compliance (Davis 1988; Donovan 1989; Erickson, Gibbs, and Jensen 1977; Esseks, Kraft, and Furlong 1997), indicates that program clients are more likely to comply if there is a high probability of being caught out of compliance. In other words, while many clients may comply for other reasons (such as because conservation practices in their plans improve their earnings and/or they believe in the goal of conserving soil), other clients need the negative incentive of probable detection and penalty if they fail to comply.

All four surveys asked the interviewed producers to estimate the likelihood of noncompliers being discovered. In both the 1992 and 1993 studies, about 31 percent of the sample chose the response option, "high likelihood," which was defined in the interview as something greater than a 50–50 chance (Table 3). In the winter 1995 survey, the percent choosing high likelihood dropped slightly to 26 percent; and a year later it decreased more significantly to 16.3 percent. Some of that drop may be attributable to a slight change in the question's wording (see Table 3).

We are concerned that if producers with these perceptions become more numerous, some of the impressive conservation gains attributable to Conservation Compliance may be in jeopardy. Not all operators of HEL are motivated by strong stewardship values. And, as our Corn Belt surveys consistently indicate, only about a fifth of the clientele believes that their earnings increase because of Conservation Compliance. Some part of the remaining 80 percent most probably need a credible threat of detection to deter them from neglecting their land's conservation needs.

To prevent producers from believing that noncompliance will escape detection, NRCS could increase its monitoring function or at least target its limited monitoring resources so as to deter abuse of the environmentally most important tracts of HEL. The risks of monitors being seen as the farmers' "enemy" should be reduced by provisions of the 1996 Farm Bill that allow a year's grace period before penalties are imposed. This amendment should enable the NRCS field staff persons to be encouragers of good conservation practices for that year's time. They need to be "policeman" only after that generously long period and only after the violators have been clearly warned.

In the Corn Belt between 1993 and 1995 one important kind of NRCS monitoring became less frequent rather than more common. Each year since 1991, NRCS selected random samples of HEL tracts with Conservation Compliance plans to be subject to status reviews. For each selected tract there was a determination by a district conservationist of whether the producer is "actively applying the approved conservation plan, and/or using an approved conservation system" (USDA 1992). In 1991 and 1992, all field offices were instructed to select random samples of 5 percent of their total HEL tracts for status reviews. Then beginning in 1993, the tracts were chosen at National Headquarters of NRCS in order to achieve estimates by state and to make it more likely that current participants in USDA programs were selected for review rather than producers who had dropped out of the programs. The result was status reviews were conducted on a 2 percent sample of all HEL tracts, and the number of

Table 3. The expectations of respondents with compliance plans regarding the likelihood of noncompliance being discovered by USDA

Likelihood of discovery	Fall '92*	Fall '93*	Winter '95*	Winter '96**
			%—	
Zero likelihood	0.4	0.7	0.7	1.7
Low likelihood	14.1	11.9	14.2	21.7
Moderate likelihood				
(50–50 chance)	52.7	52.0	57.4	58.6
High likelihood	30.9	31.9	26.0	16.3
Don't know/				
won't answer	2.0	3.6	1.7	1.8
Total percent	100.0	100.0	100.0	100.0
Number of respondents	256	918	839	836

* Text of question in 1992–95 surveys: "Before violations of conservation compliance can be penalized, they have to be discovered. In your county, how likely is it that USDA will discover that a producer has failed to apply a practice required by his compliance plan? In your opinion, is there a zero likelihood of being discovered, a low likelihood, a moderate likelihood (such as a 50–50 chance), or a high likelihood of violations being discovered?"
* Text of question in 1996 survey: "Before violations of conservation compliance can be penalized they have to be discovered. If a producer in your county failed to apply a practice required by his compliance plan, how likely is it that the USDA would discover that failure? In your opinion, is there a ...?"

status reviews per county fell. As noted, between our data collected in 1993 and 1995, there is a reduction in the proportion of farmers who have high expectation that noncompliance will be discovered. If, indeed, farmers with HEL are sensitive to the level of enforcement activities carried out by agency personnel, then the agency needs to maintain a credible level of enforcement if the gains from reducing erosion through Conservation Compliance are going to be maintained.

In times of tight budgets, USDA executives may look for program activities from which resources can be transferred to sustain technical assistance programs that are popular both with clients and field staff. Enforcement activities may be attractive targets for budgetary raids. However, for political as well as conservation reasons, the agency probably needs to maintain a credible level of enforcement effort.

For noncompliance to be deterred, potential violators may need to see a reasonably high likelihood of meaningful penalties being imposed, as well as a high enough chance of the violations being detected. All four of our Corn Belt surveys contained essentially the same question about the perceived likelihood of the respondent's local FSA committee voting to deny USDA benefits to a producer who intentionally did not "apply a practice required by his compliance plan." As Table 4 indicates, the percentage of respondents choosing the "high likelihood" option (i.e., greater than a 50-50 chance) increased from 44.1 percent in the 1992 study to 55.8 percent in 1995. Then there was a statistically significant 8.6 percentage-point decrease between the 1995 and 1996 surveys to 47.2 percent.

In the winter 1996 survey, 27.9 percent of the sample knew of someone in their counties having received a graduated penalty; 22.8 percent knew of someone having lost complete eligibility; and 36.7 percent were aware of at least one out of these two kinds of penalties being imposed in their counties. The remaining reported being unaware of penalties of either type, or they were "unsure." There may not be a credible deterrent effect unless at least someone locally is known to have been penalized.

Although the current national political climate does not favor regulatory approaches to solving environmental problems, Conservation Compliance may be treated as an exception. The 1996 Farm Bill's requirement about participating producers fulfilling their Conservation Compliance and swampbuster obligations is one of the justifications for guaranteed government payments over seven years under the Market Transition Payments. That justification is likely to be politically conspicuous if many producers receive such payments on top of relatively high commodity prices. Although the current congressional leadership is hostile to regulation, the masters of authorizing legislation and appropriations in a Congress in the future may have different values; or at the least they may expect USDA to have held the producer-recipients of this public largesse accountable for some protection of HEL and wetlands. Therefore, we would feel more optimistic about the long-term future of NRCS if the agency maintained a credible monitoring and enforcement effort for Conservation Compliance. Perhaps, given budgetary constraints and the current political opposition to regulation, the agency should engage in selective enforcement, for example, monitoring the tracts with high potential to degrade water quality when not adequately protected against soil erosion or tracts operated by individuals receiving large program benefits.

As discussed elsewhere in this volume, the enactment of Conservation Compliance in conjunction with the other conservation provisions of the 1985 Food Security Act challenged the agency culture of NRCS. Personnel of the agency

Table 4. The expectations of respondents with compliance plans regarding the likelihood of an intentional violator of conservation compliance being denied eligibility for USDA program benefits by the local FSA committee

Likelihood of lost eligibility	Fall '92*	Fall '93*	Winter '95*	Winter '96**
			%	
Zero likelihood	1.2	2.2	2.0	2.5
Low likelihood	7.4	8.6	7.2	9.8
Moderate likelihood (50–50 chance)	43.8	34.9	31.5	37.0
High likelihood	44.1	47.1	55.8	47.2
Don't know/ won't answer	3.5	7.3	3.6	3.1
Total percent	100.0	100.0	100.0	100.0
Number of respondents	256	918	839	836

* Text of question in 1992–95 surveys: "Let's say that the local Soil Conservation Service office found that a producer in your county failed to apply a practice required by his compliance plan. The office discovered this failure through visiting the farm or through inspecting aerial photographs. Let's also say that this failure was considered intentional. In your county, how likely is it that the county ASCS committee will deny such a producer USDA benefits like disaster payments and deficiency payments? In your opinion, is there a zero likelihood of benefits being denied, a low likelihood, a moderate likelihood (such as a 50–50 chance), or a high likelihood of benefits being denied?"

* Text of question in 1996 survey: "Let's say the local office of the NRCS found that a producer in your county failed to apply a practice required by his compliance plan. The office discovered this failure through visiting the farm. Let's also say...?"

120

who traditionally provided technical assistance to voluntary cooperators were placed in the position of assisting farmers in developing Conservation Compliance plans and then monitoring the extent to which the farmers implemented the plans. Farmers who did not implement were to report to FSA so that the benefits could be reduced or eliminated. As students of the conservation programs and of NRCS as an implementing agency, we were concerned about the impact of this agency change on personnel who had direct contact with farmers (i.e., District Conservationists). Consequently, we undertook a study of District Conservationists' perceptions of their work.

One area that the study concentrated on was the extent to which district conservationists were suffering from burnout (see Gabis and Ihrke 1995). Burnout has been defined as a syndrome of emotional exhaustion, depersonalization, and reduced accomplishments that occur among persons doing "people work" of some kind (Maslach and Jackson 1981). While district conservationists work with natural resources, a large portion of their work is interactive and people oriented. Increasingly, they have to deal with farmers experiencing problems with federal regulations. This puts strain on district conservationists, which increases the likelihood they will experience burnout. Individuals experiencing burnout tend to be less effective as employees.

Maslach has developed an index to measure the level of burnout a person is experiencing (Maslach and Jackson 1982). Through a series of questions, this Maslach Burnout Inventory (MBI) was applied to District Conservationists in the study. Table 5 presents the results. As the MBI increases, an individual is experiencing higher and higher levels of burnout and the associated impacts on job performance. Data from the Midwest in the table indicate that 61 percent of the District Conservationists were experiencing high levels of burnout (e.g., phases 6 through 8). In the Midwest, 31.8 percent of the responding District Conservationists were at phase 8 of the MBI. Regression analysis indicated that the MBIs were related to the District Conservationists' new conservation activities.

Subsequent analysis of the MBI data in conjunction with insights derived from focus groups of District Conservationists point to a number of lessons for the agency. The NRCS needs to support its field staff who are carrying out the farm-level planning, providing technical assistance to farmers, and enforcing conservation compliance. The agency needs to deal aggressively with the problems of high burnout and low morale if they are going to maintain their committed field staff. Examples of what the agency can do include providing counseling on stress relief, training in mediation and team building, and supporting District Conservationists when their enforcement decisions are appealed.

During the survey of the District Conservationists, they were asked their assessment of the impact of the new conservation programs on achieving long-term protection of resources. Table 6 presents the results. In the Midwest, 63 percent of the District Conservationists believed that more resource protection would be achieved through the new programs in the Food Security Act than through the traditional voluntary approach. Similarly, 88.7 percent of the Midwestern District Conservationists thought that the Food Security Act conservation programs were having a positive impact on the soil and water resources of the country. Consequently, while the agency was undergoing a profound change in its culture and operations, the employees directly involved with the farming community and with land conservation judged the new programs to be effective.

Summary and conclusions

Based on our longitudinal studies of farmers with HEL and a survey of District Conservationists who have been actively engaged in implementing Conservation Compliance and other conservation programs, we have been able to identify a number of important lessons for the implementation of conservation policy in the U.S. These lessons are discussed above. However, in summary there are a number of general lessons that reside "between the lines" in the preceding material and a few that bear being repeated. First, it is important that a conservation agency such as NRCS maintain its capacity to deal with farmers seeking technical assistance. This is especially true for those farmers seeking guidance on how to alter their conservation compliance plans so that they can fit with the changing business. Farm management is a dynamic process. Farmers respond to changes in commodity and input prices, weather, pest populations, technology, and new governmental policies. Consequently, the conservation compliance plans have to be flexible enough to permit the farmer to adapt

Table 5. Maslach Burnout Inventory phase categories: Distribution of district conservationists by phase

Phases of burnout	West	Midwest	South	Northeast	All regions	National norms
1–3	26.9	26.9	28.8	34.2	28.7	42.0
4–5	22.4	12.2	21.4	15.8	17.6	16.0
6–8	50.7	61.0	49.8	50.0	53.7	42.0
Total %	100.0	100.0	100.0	100.0	100.0	—
Total respondents	219	354	313	178	1064	—

them. And, farmers need to feel secure that there is a source of technical assistance they can go to for help. As a corollary, NRCS needs to maintain its capacity to educate farmers and input suppliers about the relative conservation effectiveness of different and new structural and nonstructural conservation practices.

A second lesson is that the NRCS must learn how to balance its concern about being perceived as flexible by the farmers with HEL while maintaining its credibility as an enforcer of conservation policy. If it diminishes its regulatory role, it risks undermining all of the gains in soil conservation Compliance has achieved while potentially being viewed as another client-captive agency of the USDA.

Third, NRCS as an agency needs to provide its field staff with the resources and training necessary to successfully implement the given policies. Increasingly, conservation programming and policies involve technically trained personnel carrying out interpersonal activities with farmers for which they have not been adequately trained. That is, conservation is a people-orientated activity with landscape consequences. District Conservationists need to be trained with this perspective in mind.

Through Conservation Compliance, much conservation has been achieved. As we enter the era of the 1996 Farm Bill with Market Transition Payments and a general population still concerned about the environment, NRCS will have to draw on its recent, past experience for developing strategies for the future. The above lessons are offered to aid in this process.

Table 6. District conservationists' opinions of the statement: "The Food Security Act will do more to achieve long-term resource protection than the voluntary program would have achieved."

Response options	West	Midwest	South	Northeast	All regions
			—%—		
Strongly agree	3.7	15.1	10.7	9.7	10.6
Agree	35.3	47.9	49.7	46.8	45.7
Disagree	29.4	19.9	19.8	19.4	21.7
Strongly disagree	18.8	6.4	10.4	13.4	11.3
Don't know	11.0	10.1	8.5	10.2	9.8
Didn't answer	1.8	0.6	0.9	0.5	0.9
Total %	100.0	100.0	100.0	100.0	100.0
Number of respondents	218	357	318	186	1079

References

American Farmland Trust. 1984. Soil conservation in America: What do we have to lose? Washington, D.C.: American Farmland Trust.

Clark, E.H., J.A. Haverkamp, and W. Chapman. 1985. Eroding soils: The off-farm impact. Washington, D.C.: The Conservation Foundation.

Davis, M.L. 1988. Time and punishment: An intertemporal model of crime. Journal of Political Economy. 96 (2): 383-390.

Donovan, D.M. 1989. Driving while intoxicated: Different roads to and from the problem. Criminal Justice and Behavior. 16 (3): 270-298.

Economic Research Service. 1994. Agricultural resources and environmental indicators. Agricultural Handbook no. 705. Washington, D.C.: U.S. Dept. of Agriculture.

Erickson, M.L., J.P. Gibbs, and G.F. Jensen. 1977. The deterrence doctrine and the perceived certainty of legal punishments. American Sociological Review. 42 (April): 305-317.

Esseks, J.D. and S.E. Kraft. 1988. Why eligible landowners did not participate in the first four sign-ups of the Conservation Reserve Program. Journal of Soil and Water Conservation 43: 251-256.

Esseks, J.D. and S.E. Kraft. 1989. Marketing the Conservation Reserve Program. Journal of Soil and Water Conservation 44: 425-430.

Esseks, J.D. and S.E. Kraft. 1991. Land user attitudes toward implementation of conservation compliance farm plans. Journal of Soil and Water Conservation 46: 365-370.

Esseks, J.D. and S.E. Kraft. 1993. Midwestern farmers' perceptions of monitoring for conservation compliance. Journal of Soil and Water Conservation 48: 458-465.

Esseks, J.D., S.E. Kraft, K. Sullivan, and M. Dellinger. 1994. Conservation compliance and producers in the Corn Belt. DeKalb, IL: Center for Agriculture in the Environment, American Farmland Trust.

Esseks, J.D., S.E. Kraft, E.J. Furlong, V.A. Krause, and B.L. Myers. 1995a. Conservation compliance and producers in the Corn Belt: Comparisons across 1992, 1993, and 1995 surveys, DeKalb, IL: Center for Governmental Studies, Northern Illinois University.

Esseks, J.D., S.E. Kraft, G.T. Gabris, E.J. Furlong, D.M. Ihrke, V.A. Krause, and B.L. Myers. 1995b. Implementing federal soil conservation and wetlands protection policies at the field level: A survey of a national sample of district conservationists of the USDA's Natural Resource Conservation Service. DeKalb, IL: Center for Governmental Studies, Northern Illinois University.

Esseks, J.D., S.E. Kraft, V.C. Clarke, and B.L. Myers. 1996. Corn Belt farmers' perceptions of the implementation of conservation compliance and the Conservation Reserve Program: Comparisons across 1992, 1993, 1995, and 1996 surveys—draft report. DeKalb, IL: Center for Governmental Studies, Northern Illinois University.

Esseks, J.D., S.E. Kraft, and E.J. Furlong. 1997. Why targets of regulation do not comply: The case of Conservation Compliance in the Corn Belt. Journal of Soil and Water Conservation 52(4):259-264.

Gabris, G.T. and D.M. Ihrke. 1995. Personnel management issues: district conservationists' levels of burnout, district conservationists' motivation, trust of supervisors, and their intentions about resigning, retiring or transferring. In Implementing Federal Soil Conservation and Wetlands Protection Policies at the Field Level: A Survey of a National Sample of District Conservationists of the USDA's Natural Resource Conservation Service. DeKalb, IL: Center for Governmental Studies, Northern Illinois University.

Kraft, S.E., P.L. Roth, and A.C. Thielen. 1989. Soil conservation as a goal among farmers: Results of a survey and cluster analysis. Journal of Soil and Water Conservation 44:487-490.

Maslach C. and S.E. Jackson. 1981. The measurement of experienced burnout. Journal of Occupational Behavior 2:99-113.

Maslach C. and S.E. Jackson. 1982. Maslach burnout inventory. Palo Alto, CA: Consulting Psychologists Press.

USDA. 1989. The second RCA appraisal: Soil, water, and related resources on nonfederal land in the U.S.—Analysis of conditions and trends. Washington, D.C.: U.S. Dept. of Agriculture.

USDA/Soil Conservation Service. 1992. National Food Security Act Manual, 2nd ed., 180-V-NFSAM, Amendment 7.

U.S. House of Representatives 1995. Agriculture, rural development, Food and Drug Administration, and related agencies—Appropriations for 1996: Hearing, 104th Congress, 1st Session, Washington, D.C.: U.S. Government Printing Office.

U.S. House of Representatives, 1996. Federal Agriculture Improvement and Reform Act of 1996—Conference Report, 104th Congress, 2nd Session, Rept. 104-494. Washington, D.C.: U.S. Government Printing Office.

Wittwer, S. 1982. New technology, productivity, and conservation, in soil conservation: Policies, institutions, and incentives. H.B. Helcrow, E.O. Heady, and M.L. Cotner (eds.). Ankeny, IA: Soil Conservation Society of America, pp. 201-215.

11

Is U.S. Soil Conservation Policy a Sustainable Development?

Dana L. Hoag, Jennie S. Hughes-Popp, and Paul C. Huszar

The definitions of sustainability can vary significantly (Gold 1994). At one end of the spectrum are the "sustain-me" definitions that cast sustainability narrowly on a particular concern such as environmental conservation (U.S. Department of Agriculture 1980), use of regenerative inputs (Rodale 1991), rural economic health, family farms, or economic health and the ability to feed the world (DowElanco 1994). At the other end of the spectrum is the "all-inclusive" approach, which incorporates everything, as the following definition from the Food, Agriculture, Conservation and Trade Act of 1990 (Section 1603, Title XVI) demonstrates:

> Sustainable agriculture is an integrated system of plant and animal production practices having a site-specific application that will, over the long-term: (1) satisfy human food and fiber needs; (2) enhance environmental quality and the natural resource base upon which the agricultural economy depends; (3) make the most efficient use of nonrenewable resources and on-farm resources and integrate, where appropriate, natural biological cycles and controls; (4) sustain the economic viability of farm operations; and (5) enhance the quality of life for farmers and society as a whole.

The third approach is intergenerational. A widely quoted definition given in 1987 by the World Commission on Environment and Development is that the needs of the present are met without compromising the ability of future generations to meet their own needs. Very similar definitions have been expressed by Dicks (1991) and Victor (1991), and adopted at the 1992 United Nations Conference on the Environment and Development in Rio de Janeiro.

Implications for resource management

We adopt the intergenerational view that the main principle of sustainability is to provide future generations with an equal or better quality of life than that of the present generation. This definition captures the essence of sustainability but it offers few guidelines about how resources should be managed to assure sustainability. How, for instance, does society even know that agriculture is not already sustainable without definitive guidelines or measurement endpoints? One indication is public concern about negative impacts of agriculture, such as pollution, resource degradation, and reduced quality of life in the surrounding communities (Grossi 1993; Rodale 1991; NRC 1993).

In effect, it is much easier to agree to be sustainable than it is to define or achieve it (Helmers and Hoag 1993; Schuh and Archibald 1993). Even the most ardent of supporters, such as the leadership at the Henry A. Wallace Institute for Alternative Agriculture, have become disillusioned about the prospect of "specifying the empirical content and characteristics of a sustainable agriculture" (Heller and Youngberg 1994, 2). But does this lack of detail mean that the concept of sustainability has no practical utility in guiding resource management as some have argued (Beckerman 1994), or does it help society generate goals and concepts that are helpful for resource management (Daly 1995)?

Hyde (1991) and Beckerman (1994) suggest that it is welfare that must be sustained, not resources. There are two opposing viewpoints about what sustainability requires. Weak sustainability requires only that this generation leaves the next generation with at least as much wealth as it has. The "current generation does not especially owe to its successors a share of this or that particular resource. If it owes anything, it owes ... access to a certain standard of living" (Solow 1986, 142). On the other hand, strong sustainability requires special attention for natural resources (or natural capital). Pearce and Atkinson (1993) argue that natural resources deserve more attention than other kinds of capital because there is uncertainty about how human activities impact the environment, because natural capital is hard or impossible to replace once it is lost, and because the environment might provide yet undiscovered services and products. These viewpoints are the compass for soil conservation policy; they dictate whether society's goal is resource conservation or welfare conservation.

Does sustainability require conservation?

Soil conservation or preservation is not required for weak sustainability, but it would certainly be part of resource management. It is clear that U.S. policy has given special attention to how conventional practices affect natural resources and the environment (NRC 1993; Grossi 1993; Riechelderfer 1991). This makes it easy to see why many assume conservation is sustainability. Nevertheless, conservation does not have a one-to-one relationship with sustainability objectives (Hoag and Skold 1996). Conservation has both beneficial and adverse impacts on other sustainability issues such as family farms, rural economies, or the environment. The Conservation Reserve Program (CRP) for example did conserve soil, but it impaired sectors of some rural economies and was a boon for others (Hughes, Hoag and Nipp 1995). In addition, sustainability programs must be economically feasible while conservation programs have no such limitation. Public cost-sharing has been justified for many conservation practices because they are not profitable for the landowner but provide public benefits (Riechelderfer 1991).

According to Hoag and Skold (1996), conservation and sustainable agriculture programs do share some common characteristics. Both rely on voluntary cooperation from producers for their implementation. They have similar and limited objectives, but may use very different approaches to achieve them. Both may have beneficiaries other than those who implement the programs. And, finally, both have suffered from severe budget limitations.

Soil conservation policy has not necessarily supported the strong sustainability position that no soil loss should occur.[1] Government programs have supported conservation, but these programs are not designed to eliminate soil losses nationwide. Soil conservation policies have tended to follow the pattern of current crises. Major policies were implemented after the devastation of the Dust Bowl in the 1930s. Since then, booms in crop prices have pushed producers to exploit soil resources to their limit, as occurred in the early 1970s, and in other periods crop surpluses produced a sense that resources were not scarce (Libby 1993). The primary issue driving much of the policy today may be government budgets, or lack of budgets.

Zinn (1993) predicted that the country was moving away from concern about erosion since no crises were eminent in the recent past or foreseen for the near future. In that same year, Libby (1993, 290) said "there is virtually no reliable evidence today of any 'natural resource limits of U.S. agriculture.' In fact, evidence is quite the contrary." There is no question that the government is pursuing weak sustainability where soil is concerned. It is allowing erosion to occur, implicitly because the gains from erosion exceed the losses (weak sustainability). The question remaining is how much loss is optimal? This will depend on the costs of erosion compared to the benefits.

Costs and benefits of conservation

Sustainability has been described in terms of three intertwined yet often competing objectives, *economic, environmental,* and *social welfare* (Cernea 1993; Munasinghe 1993; Rees 1993). Agricultural producers are primarily economic agents where the market dictates the rules by which they operate. However, market forces may not always lead to optimal resource allocation when there are externalities or public goods. The existence of market failures (environmental and social externalities) and public goods provide reason for considering government intervention for soil management. A major consideration for sustainability policy is therefore the extent of nonmarket impacts.

Economic

The economic dimension to sustainability is usually based on the profitability of a firm. Private benefits of erosion control include reductions in the damages of erosion on the farm (Davis and Condra 1989). Besides reducing the inherent productivity of the soil, nutrients may be lost when topsoil is eroded which requires higher input costs and increases the threat to the environment. The costs of erosion control are born in structures, such as conservation terraces, or in conservation practices, such as no-tillage or contour plowing. Farmers around the country have adopted conservation measures at varied rates, indicating that the net benefits are highly site specific.

In general, farmers have little incentive to control for erosion for three reasons. First, the average level of erosion across the United States is only 6.4 tons per acre per year (Pierce and Nowak 1995). At this rate it takes more than 25 years to remove an inch of topsoil (Crosson and Pierce 1984) and therefore the losses to productivity over a farmer's lifetime may be negligible. Without some sort of public intervention (incentives or disincentives), farmers have little or no reason to protect the soil endowment for future generations (Grossi 1993). Second, even when productivity losses due to erosion are evident, other inputs may be substituted into the production process to maintain economic viability at a cost lower than that of a conservation practice. Third, because society rarely holds the farmer responsible (via taxes or regulations) for erosion impacts offsite, he has no economic incentive to control them (Riechelderfer 1991).

Environmental

The economic component of erosion control focuses on production viability. Viability is determined by comparing the costs of conservation practices to the private (on-site) benefits of erosion control. These costs and benefits are captured in market prices. Environmental and social considerations often escape the viability question because they are felt off the farm (off-site), have negligible or no direct impact on production, and are not captured by market prices.

These off-site environmental benefits of erosion control have been enumerated in the literature. Environmental benefits include improved water quality, air quality, and habitat for fish and wildlife, as well as reduced sedimentation in lakes and rivers (Clark, Havercamp, and Chapman 1985; Ribaudo et al. 1994; Young and Osborn 1990). Related benefits include reducing roadway blockages, landscape damage, dust-related illnesses, interior and exterior building damages, as well as improved land stabilization and increased recreational opportunities (Piper 1985). The value of public erosion control benefits greatly outweighs the value of the on-site private benefits (Davis and Condra 1989; Huszar and Piper 1986).

Social welfare

Environmental impacts are not the only externality associated with farming. There are many nonenvironmental social concerns that are impacted by the way producers farm. For example, healthy family farms, some would assert, are better for a community than corporate farms. Conservation policies that promote bigger farms may seem more sustainable to consumers and less so to rural residents worried about declining populations in their communities. Other examples include food safety, animal welfare, and impacts on other economic sectors (such as equipment manufacturers).

The public trust

When conservation decisions are left to the market, farmers and ranchers are entrusted with managing U.S. soil resources. As explained in the previous section, market failures may lead to the misallocation of these resources and consequently to inter- and intratemporal inequities between sustainability objectives. We have identified three major objectives, all of which are not accounted for by the market. This alone may be sufficient reason for more public input about the way producers manage soil, but the sustainability literature contains three additional reasons why the public should have input about the management of nonrenewable resources: substitutability, uncertainty, and reversibility. A producer's endowment of soil represents only one of multiple inputs exploited in the production of agricultural commodities. There may be fundamental differences in the way the market and society value substitutability, reversibility, and uncertainty, and therefore differences in the viewpoint of which tradeoffs in soil for other inputs are appropriate.

Substitutability refers to the ease of substitution among inputs and illustrates how well one input may substitute for another when prices change or when one or more inputs become constrained. The ease of substitution is extremely relevant when one or more inputs to the production process are (or are becoming) scarce, since sustainability will hinge on how easily and effectively other resources can substitute for the scarce input. When substitutes are clean (i.e., pose no environmental threat) and readily available, there may be little cause for concern. However, conservation may be needed if input contribution is unique and not easily substitutable.

Reversibility and uncertainty are best explained when approached together. Reversibility pertains to the ability to revert back to a former input mix once others have been chosen. Uncertainty refers to any unforeseen circumstances (both positive and negative) that may either follow as a consequence of, or an impact on, production. Uncertainty arises with respect to all prices, input supply, output supply, profits, and environmental impacts. When a producer depletes the natural soil base and becomes dependent on one or more inputs that can compensate for the soil loss, the producer has limited the set of possible input combinations for the production process. If use of these substitute inputs later leads to unforeseen consequences, the producer may not be able to readjust the mix because the producer may have depleted the natural soil base. This becomes extremely important, especially when circumstances resulting from uncertainty are negative, for two reasons. First, in many cases these impacts do not limit themselves to the particular production process but may affect other sectors and the economy or the environment. Second, these impacts may be irreversible.

When input mixes dependent on substitutes for a natural resource lead to unforeseen negative consequences, it is likely that the magnitude of the negative impact will be much larger for society than it will be for the individual producer. Therefore, it is also likely that the risk that society attaches to depleting a natural resource will be higher than that for an individual. Individuals will be more likely to deplete a natural resource than society; therefore, if society places a high value on conserving soil then the government must take an active role in soil management.

Soil conservation policy in the United States

Soil conservation programs have been around since the Dust Bowl days of the 1930s. Technical assistance and cost-sharing incentives were established then to provide farmers with both the knowledge and a percentage of the implementation cost of a conservation practice. These incentives continue today through the U.S. Department of Agriculture. However, they have evolved to include both on- and off-site goals and objectives. In 1985, the major farm legislation in the U.S., popularly called the farm bill, established firm linkages between conservation and farm price and income supports. The environmental objectives survived the 1990 Farm Bill and while support seems to have wavered some, the 1996 Farm Bill also continued the tradition.

While there are many types of conservation policies, there are really only two basic conservation approaches. One is to completely set aside or retire land from production. In these cases, the land is usually put into a conservation cover that is suitable for wildlife. The other approach is to reduce erosion levels stemming from erosion. This may require special equipment, specific cropping patterns, intensive management or a combination of all three. Either of these methods reduce erosion, but one continues to allow production. In addition, it is often the poorest soils that are set aside in temporary land "retirement" programs.

Since the 1930s when national conservation programs were first initiated policy makers have chosen to apply a variety of policies, but all have relied on one or more of four tools, technical assistance, conservation cost share, buy outs, and set asides, as the means to attain their goals (Jeffords 1982; Laycock 1991; NRC 1993; Strohbehn 1986). These four tools amount to financial incentives or education and technical assistance. Policy makers can use disincentives, but have chosen to rely mostly on incentives for the adoption of soil conservation (Riechelderfer 1991). Regulations are rarely applied, and usually only at the state level (Riechelderfer 1991). Financial disincentives, such as erosion taxes, are virtually unused.

Studies have found that these policies alone may not even meet the goals of conservation, let alone sustainability. One reason is that incentives provided in other farm programs may negate the efforts of soil conservation programs. For example, many commodity programs reward farmers for high yields which are often achieved at the expense of soil loss. Nevertheless, it is more likely that sustainability is simply too amorphous, to define, and to accomplish. As noted by the National Research Council (1993):

> ... (soil) problems would be easier to solve if the most promising opportunities for improving farming systems were more clearly defined. Policies could then be developed to take advantage of those opportunities.

Sustainability, conservation, and policy

The discussion above emphasizes how difficult it is to define and measure sustainability. No model or entity will ever be able to be all inclusive, nor will there ever be complete consensus about what is and what is not sustainable. However, better attention can be paid to be more inclusive of the objectives and issues identified here and by others in the decision-making process. We have identified three objectives where consensus is building: economic, environmental, and social welfare. In addition, we identified three key issues where the market may differ from society: sustainability, reversibility, and uncertainty. Combined, these areas of consideration imply the following questions and resulting guidelines for soil management:

Should society pursue weak or strong sustainability? Strong sustainability implies a goal of no soil loss; weak sustainability implies a goal of maintained productive capacity.

If we pursue weak sustainability, how much soil erosion is optimal? When and where should soil be conserved? Important considerations include:

- Consider economic impacts, environmental impacts, and social impacts (inter- and intratemporal).
- Consider what is substituted for soil, the uncertainty of the substitution relationship and the availability of the substituted resources, and the reversibility of erosion damages.
- Consider investment in man-made capital, the extent of externalities (social and environmental), and the complementarity (substitutability) of inputs.

A sustainability model

Let us start with the case where the objective of sustainability is to maintain constant per capita consumption as suggested by Solow (1974a, 1974b) and Hartwick (1977). Assuming a single sector economy with one output (which can be consumed or invested), a constant returns to scale production environment, and homogenous natural and human inputs, Dixit, Hammond, and Hoel (1980), Hartwick (1977, 1978), Page (1977), and Solow (1974a, 1991) have shown that a society can sustain a constant stream of consumption when:

- the shadow price of the unextracted natural resource increases at a rate equal to the marginal product of reproducible capital (known as Hotelling intertemporal efficiency);
- the competitive rents from extracting natural capital are invested in physical capital (known as the savings-investment rule);
- physical capital is more productive than natural capital. That is, the production elasticity of physical capital exceeds that of natural capital.

Hartwick (1977, 1978) has illustrated mathematically via a growth model framework how per capita consumption may remain constant even as the natural resource base is depleted.[2]

Net output is a function of natural (nonrenewable) capital, man-made (in this example, physical) capital and labor.[3]

$$Q_t = F[K_t, Y_t, L_t] \tag{1}$$

where

Q_t = net output per capita (net of costs)

Y_t = natural capital, yielding a flow of services per unit of time in constant proportion to its size

K_t = man-made capital stock, yielding a flow of services per unit of time in constant proportion to its size

L_t = flow of labor

Using this well-known example, it can be shown that output can be maintained over time even if one of the inputs is degraded. Solow has stated in lectures and in this mathematical example that natural and man-made resources are fungible in a certain sense; therefore, natural capital and man-made capital may substitute for each other in the production process. Under this definition, sustainability does not require any particular endowment of capital or final goods be passed on to future generations. Instead it requires only that a general capacity to reproduce or to achieve a desired level of social welfare be conveyed to future generations.

An alternative viewpoint is that sustainability means that nonrenewable resources should be maintained over time. Pearce and Atkinson (1993, 1995) among others (Boulding 1973; Daly 1995; Jansson et al. 1994) contend that natural and man-made capital complement each other in the production process. In this complementary relationship, natural capital, which is not reproducible, is the limiting factor of production and therefore must be preserved over time in order for the production process to be sustainable.

Supporters of this second definition defend their position by invoking three arguments. First, uncertainty associated with the lack of full knowledge of the consequences of natural resource depletion should lead decision-makers to adopt a conservative position with regard to their use. As Pearce and Warford (1996) note, this is comparable to the notion of safe minimum standards advocated by Bishop (1978) and discussed by Lesser and Zerbe (1993). Second, natural resource depletion is permanent and any permanent change should be approached very slowly and carefully. Third, not only do natural resources provide inputs for a production process but they perform multiple functions in the environment. Resources should be preserved to ensure these other functions are fulfilled.

The arguments for weak and strong sustainability were addressed by Miller (1996) who showed that the two definitions are the same if there is no investment. That is, strong sustainability is required to maintain constant consumption. However, the corollary was that maintaining a constant natural resource base is not required when there is investment, depending also on many other conditions such as well-defined property rights. This can also be shown for the case where non-renewable resources are substitutes for other inputs rather than the more restrictive case of complementarity between inputs. Therefore, *a zero erosion rate is not required* if there are close substitutes for soil and if the capital purchased with the revenues from agriculture provides more income growth than the loss in soil stock.

A soil sustainability model

Soil can be easily cast into the Solow framework. Consider the following function. Where Solow expressed output as a function of natural capital, man-made capital, and labor, we express the equivalent version as crop output (yield) in any year t as a function of natural capital (soil quality, sq), other capital and variable inputs (X_1 and X_2).

$$Y_t = f(x_{1t}, x_{2t}, sq_t)g(t) \tag{2}$$

Crop yields, or our potential to maintain the ability to produce food, are dependent on the natural endowment of soil, but as soil quality is degraded by erosion, other capital or inputs can be substituted for it. Another expression, $g(t)$, is included at the end of the equation to indicate that the relationship of inputs to outputs is not constant over time. Erosion suppresses output while new technologies increase it. Hoag (1997) showed via a dynamic model how the impact of these two offsetting influences vary according to soil type and the substitutability of inputs. In addition, he showed that erosion changes the relationship between the substitution of soil for other inputs over time. Therefore, economic sustainability would be achieved by a profit-maximizing producer because he or she would allocate soil across time appropriately by responding to market forces, given his or her soil type and the price and availability of other inputs. A producer would also factor in risk associated with uncertainty and inability to reverse previous erosion. However, he or she may not value these risks the same as society if land prices fail to fully incorporate these fears.

Expression (2) also failed to account for externalities. Suppose that the use of an input, such as X_1, has an external impact (E) on the environment or society, or that soil quality (sq) had external effects as shown in expression (3):

$$E_t = h(x_{1t}, sq_t) \tag{3}$$

Sustainability would require that external costs and benefits are maximized simultaneously with economic sustainability.

Several observations can be made from this simple soil model. First, soil can be eroded, but its substitutability with other inputs will be highly site specific and will change over time. There will always exist a degree of uncertainty about the production process, especially about the availability of the substituted inputs and the reversibility of eroded soils. In addition, there will be external impacts such as eroded soil entering streams, and these impacts will change over space and time. A producer with a given soil may change the ability of the soil to buffer environmental impacts as erosion is allowed to occur.

Second, the substitutability of other forms of capital must be carefully monitored. Investments in man-made technologies have decreased the real cost of food from 25 to under 15 percent of per capita income in the past 25 years (Libby 1993). This indicates that investments have more than offset the depreciation of soil capital over the same period. Nevertheless, there is uncertainty about how much and for how long the movement of capital out of soil and into new technology can expand output. If consumption of soil (erosion) occurs too long before a net loss is realized, lost productivity may not be recaptured.

Third, technology can offset soil depreciation. However, technology can also increase the costs of erosion since it sometimes widens the gap between potential yield if erosion had not occurred and post-erosion yields. Finally, producers and society have different goals, and so too might different individuals in society. Therefore, putting sustainability into practice for conservation could prove extremely difficult.

Policy implications

In our introduction we asked the question whether conservation programs are implemented improperly or whether conservation is something different from sustainability? How can society design soil conservation programs that are sustainable without first addressing what is and what is not sustainable? Hoag and Skold (1996) suggest that conservation is not the same as sustainability. Society, they assert, will conserve soil because the cost seems worth the benefits, even though the problem is defined more narrowly than the holistic versions of sustainability. Therefore, one viewpoint about whether U.S. conservation programs are promoting sustainability is that sustainability is not the goal of conservation programming. Conservation programming is pursued to provide certain benefits, such as cleaner air and water, or more wildlife. Since conservation activities are often associated with other goals for sustainability, conservation policies will often have positive or negative impacts in other areas. However, these are not the intended targets.

An alternative view is that conservation programs have been improperly implemented. Efforts should be made to coordinate the complex web of sustainability objectives as decisions are made about conservation programs. This seems like a reasonable viewpoint providing expectations about the level of effort and results. It will be impossible to totally coordinate all activities in a way that makes everyone happy. Tradeoffs between goals will occur.

As policy-makers seek to be more inclusive, they must monitor externalities and work to find solutions that are not worse than the original problems. Do current programs address sustainability? What do the two major approaches to conservation policy accomplish? Hughes, Hoag, and Nipp (1995) showed that the Conservation Reserve Program had many varied impacts, resulting in winners and losers. Most analyses indicate that the program resulted in a net cost to society; nevertheless, given certain assumptions, one could show the program was a net social benefit. Setting aside land from production therefore provides mixed results, but may reduce sustainability by more than it increases it.

What about reducing the erosive impacts of production? Here again, it is difficult to show whether sustainability has or has not been improved. However, programs could be evaluated on a case-by-case basis by the criteria discussed in this paper. Many programs, for example, reduce environmental impacts and preserve productivity but do nothing for other social interests such as labor use. In addition, there appears to be no plan at all for defining an appropriate strategy for influencing producers to manage risk and reversibility in a manner consistent with social interests.

Conclusion

Sustainability is a difficult concept to define. However, it appears that the best choices for soil conservation itself are not that easy to determine either. Since conservation is a relatively simple concept, it will be infinitely easier to borrow concepts from the sustainability literature and use them to improve conservation programs than it would be to attempt to achieve sustainability. It would be difficult to argue that current programs in the U.S. already are sustainable as defined by the criteria used here. However, to the extent the market already accounts for externalities, weak sustainability may be achieved as policy-makers utilize the limited tools they have to make corrections where market failures exist. Libby (1993) found that contrary to repeated warnings about the sustainability of agriculture by many, there are no signs that output potential is threatened.

Notes

1. Soil loss is defined as no net loss, after considering any reformation.
2. Details of the mathematical proof via the growth model can be found in Miller-Sanabria (1996).
3. This example examines the substitutability of physical capital only natural nonrenewable capital. Solow shows how technology may also substitute for nonrenewable capital.

References

Beckerman, W. 1994. Sustainable development. Is it a useful concept? Environmental Values. 3(3) pp. 190-209.

Bishop, R. 1978. Endangered species and uncertainty: The economics of a safe minimum standard. American Journal of Agricultural Economics. 60(1) pp. 10-13.

Boulding, K. 1973. The economics of the coming of the spaceship Earth. In: Herman Daly (ed.) Toward a Steady State Economy. W.H. Freeman Press. San Francisco, CA.

Cernea, M. 1993. The sociologist's approach to sustainable development. Finance and Development. 30(1) pp. 11-13.

Clark, E., J.A. Havercamp, and W. Chapman. 1985. Eroding soils: The off-site impacts. The Conservation Foundation. Washington, D.C.

Crosson, P. and K. Pierce. 1984. Soil erosion and policy issues: Agriculture and the environment. Resources for the Future. Washington, D.C.

Daly, H. 1995. On Wilfred Beckerman's critique of sustainable development: Four parting suggestions for the World Bank. Ecological Economics 10(3) pp. 183-187.

Davis, B. and G. Condra. 1989. The on-site costs of wind erosion on farms in New Mexico. Journal of Soil and Water Conservation. 44 (4) pp. 339-343.

Dicks, M. 1991. What will be required to guarantee the sustainability of U.S. agriculture in the 21st century? Journal of Alternative Agriculture. 6. pp. 191-195.

Dixit, A., P. Hammond, and M. Hoel. 1980. On Hartwick's rule for regular maximum paths of capital accumulation and resource depletion. Review of Economic Studies. XLVII-3 (148) pp. 551-556.

Dow Elanco. 1994. What makes agriculture sustainable? In The Bottom Line on Agri-chemical Issues. Sustainable Agriculture Issue, Indianapolis, IN.

Food, Agriculture, Conservation and Trade Act of 1990, Subtitle B of Title XVI, Chapter 3.

Gold, M. 1994. Sustainable agriculture: Definitions and terms. USDA. National Agricultural Library, Special Reference Brief No. SRB 94-05, Beltsville, MD.

Grossi, R.E. 1993. A green revolution: Retooling agricultural policy for greater sustainability. Journal of Soil and Water Conservation. 48(4) pp. 285-288.

Hartwick, J.M. 1978. Substitution among exhaustible resources and intergenerational equity. The Review of Economic Studies. XLV-2 (140) pp. 347-354.

Hartwick, J.M. 1977. Intergenerational equity and investing of rents from exhaustible resources. American Economics Review. 67(5) pp. 972-974.

Heller, M. and G. Youngberg. 1994. Letter from the President and Executive Director. In the 1994 Annual Report, Henry A. Wallace Institute for Alternative Agriculture, Greenbelt, MD.

Helmers, G. and D. Hoag. 1993. Sustainable agriculture. In M.C. Hallberg, R. F. Spitze, and D.E. Ray, eds. Food, Agriculture, and Rural Policy into the twenty-first century. Westview Press, Boulder, CO.

Hoag, D. 1997. The intertemporal impact of soil erosion on non-uniform soil profiles: A new direction in analyzing erosion impacts. Agricultural Systems, 56(4) pp 415-429.

Hoag, D. and M. Skold. 1996. The relationship between conservation and sustainability. Journal of Soil and Water Conservation, 51(4) pp. 292-295

Hughes, J., D. Hoag, and T. Nipp. 1995. The Conservation Reserve: A survey of research and interest groups. Special Publication, Number 19, Council for Agricultural Science and Technology, Ames, IA.

Huszar, P. and S. Piper. 1986. Estimating the off-site costs of wind erosion in New Mexico. Journal of Soil and Water Conservation. 41(5) pp. 414-416.

Hyde, W. 1991. Social forestry: A working definition and 21 testable hypotheses. Journal of Business Administration. 20(1&2): Chapter 18.

Jansson, A., M. Hammer, C. Folke, and R. Costanza (eds.). 1994. Investing in natural capital: The ecological economics approach to sustainability. International Society for Ecological Economics. Island Press, Washington, D.C.

Jeffords, J.M. 1982. Soil conservation policy for the future. Journal of Soil and Water Conservation. 37(1) pp. 10-13.

Laycock, W. 1991. The Conservation Reserve Program: How did we get where we are and where do we go from here? In L. Joyce, J. Mitchell, and M. Skold (eds.) The Conservation Reserve—Yesterday, Today and Tomorrow. General Technical Report RM-203. USDA, Forest Service, Rocky Mountain

Forest and Range Experiment Station, Fort Collins, CO.

Lesser, J. and R.O. Zerbe Jr. 1993. What can economic analysis contribute to the sustainability debate? Contemporary Economic Policy. 13(3) pp. 88-100.

Libby, L.W. 1993. The natural resource limits of agriculture. Journal of Soil and Water Conservation. 48(4). pp. 289-294.

Miller, T. 1996. The economics of sustainable development in the Gran Chaco region of South America. Ph.D. Dissertation. Department of Agricultural and Resource Economics. Colorado State University. Fort Collins, CO.

Munasinghe, M. 1993. The economist's approach to sustainable development. Finance and Development. 30(1) pp. 16-19.

National Research Council (NRC). 1993. Soil and water quality: An agenda for agriculture. National Academy Press. Washington, D.C.

Page, T. 1977. Conservation and economic efficiency. John Hopkins University Press. Baltimore, MD.

Pearce, D. and G. Atkinson. 1995. Measuring sustainable development. In D. Bromley, ed. The Handbook of Environmental Economics. Blackwell Publishers, Cambridge, MA. pp. 166-181.

Pearce, D. and G. Atkinson. 1993. Measuring suitable development. Ecodecision, No. 9 p. 64.

Pearce, D. and J.J. Warford. 1996. World without end economics. Environmental and Sustainable Development.

Pierce, F.J. and P. Nowak. 1995. Soil erosion and soil quality: Status and trends. In: Proceedings of the 1992 National Resources Inventory Environmental and Resource Assessment Symposium. Washington, D.C., July 19-20.

Piper, S. 1985. The off-site economic costs of wind erosion in New Mexico. M. S. Thesis. Department of Agricultural and Resource Economics. Colorado State University. Fort Collins, CO.

Rees, C. 1993. An ecologist's approach to sustainable development. Finance and Development. 30(1). pp. 14-15.

Ribaudo, M., D. Colacicco, L. Langner, S. Piper, and G.D. Schaible. 1994 natural resources and users benefits from the Conservation Reserve Program. USDA. Economic Research Service. Agricultural Economic Report Number 627. Washington, D.C.

Riechelderfer, K. 1991. The expanding role of environmental interest in agricultural policy. World Agriculture. 40(1) pp. 18-21.

Rodale, R. 1991. Agricultural systems: The importance of sustainability. In J. D. Hanson, M.J. Shaffer, D.A. Bell, and C.V. Cole (eds.) Sustainable Agriculture for the Great Plains, Symposium Proceedings, USDA, Agricultural Research Service ARS-69. pp. 9-16.

Schuh, G.E. and S. Archibald. 1993. A methodological framework for the integration of environmental and sustainable development issues into agricultural planning and policy analysis in developing countries. Paper presented at a seminar sponsored by the Food and Agricultural Organization of the United Nations. Rome, Italy.

Solow, R. 1993. Sustainability: An economist's perspective. In Robert Dorfman and Nancy Dorfman, eds. Economics of the Environment, Selected Readings, Third Edition. W. W. Norton & Company, NY.

Solow, R. 1991. Sustainability : An economist's perspective. Paper presented at the Eighteenth J. Seward Johnson Lecture to the Marine Policy Center, Woods Hole Oceanographic Institution. Wood Hole, MA. June 14.

Solow, R. 1986. On the intergenerational allocation of natural resources. Scandinavian Journal of Economics. Vol. 88: p. 142.

Solow, R. 1974. The economics of resources or the resources of economics. American Economics Review. 64(2) pp. 1-13.

Strohbehn, R. (ed.). 1986. An economic analysis of erosion control programs: A new perspective. USDA. Economics Research Service. Agricultural Economics Report Number 560. Washington, D.C.

U.S. Department of Agriculture. 1980. Report and recommendations on organic farming. U.S. Government Printing Office, Washington, D.C.

Victor, P. 1991. Indicators of sustainable development: Some lessons from capital theory. Ecological Economics 4(3) pp. 191-213.

World Commission on Environment and Development. 1987. Our common future. Oxford University Press, Oxford 1987.

Young, C.E. and C.T. Osborn. 1990. The Conservation Reserve Program: An economic assessment. United States Department of Agriculture. Economic Research Service. Agricultural Economic Report No. 626. Washington, D.C.

Zinn, J. 1993. How are soil erosion control programs working? Journal of Soil and Water Conservation. 48(4) pp. 254-259.

12

Reducing Wind Erosion Damages and the Conservation Reserve Program

Jennie S. Hughes-Popp, Paul C. Huszar, and Dana L. Hoag

The Conservation Reserve Program (CRP) continues an approach to soil conservation in the United States which began after the Great Dust Bowl of the 1930s. This approach seeks to reduce soil erosion by either purchasing or renting highly erodible land in order to retire it from production. The Bankhead-Jones Act of 1935 authorized Land Utilization Projects to purchase and seed to grass submarginal land. By 1951, these projects had seeded approximately 1 million of the 6 million acres purchased for conservation purposes. Most of this land is still managed by the USDA Forest Service. The Soil Bank Act of 1956 provided payments to landowners to retire cropland for 3 to 10 years and to establish permanent cover on the land. The program peaked in the 1960–1961 period when about 28.7 million acres were under contract. Most Soil Bank land was plowed again during the 1970s in response to high wheat prices. This purchase/rent approach to soil conservation has been very expensive and of questionable effectiveness (Laycock 1991). The purpose of this chapter is to address the cost-effectiveness of the CRP, another rental program, in the particular case of wind erosion.

Wind erosion in the U.S.

Wind erosion is a significant problem in the western U.S. Total soil erosion on nonfederal cropland in the U.S. equals more than 2 billion tons per year, with approximately 950 million tons (44 percent) due to wind erosion and 1.2 billion tons (56 percent) due to water erosion. More than 770 million tons per year or 80 percent of all erosion in the western U.S. is due to wind erosion, which represents 36 percent of the total erosion from all sources in the U.S. (NRCS 1995).

Referring to Figure 1, the western U.S. is generally defined as consisting of the Pacific, Mountain, Northern Plains and Southern Plains production regions. As Table 1 indicates, within the western U.S., wind erosion is greatest in the

Figure 1. Map of U.S. farm production regions

Table 1. Wind and water erosion in the United States, 1992

Region	———Wind erosion———		———Water erosion———	
	(tons/yr)	(% of total erosion)	(tons/yr)	(% of total erosion)
USA	953,129,810	45.6%	1,191,680,330	54.4%
Appalachian	290,110	0.3%	108,047,040	99.7%
Corn Belt	39,082,010	9.0%	393,544,180	91.0%
Delta	0	0.0%	78,675,260	100.0%
Lake States	142,758,500	59.1%	98,656,760	40.9%
Mountain	283,347,500	81.3%	65,162,160	18.7%
Northeast	831,200	1.6%	52,243,210	98.4%
Northern Plains	171,384,070	47.5%	189,445,630	52.5%
Pacific	42,937,740	47.2%	47,993,970	52.8%
Southeast	0	0.0%	59,032,940	100.0%
Southern Plains	272,497,680	73.4%	98,879,180	26.6%

Source: Natural Resource Conservation Service, 1995

Mountain region accounting for more than 280 million tons per year or nearly 30 percent of all wind erosion, followed by the Southern Plains region with wind erosion of more than 270 million tons per year or nearly 29 percent of all wind erosion, and the Northern Plains region with wind erosion of more than 170 million tons per year or 18 percent of all wind erosion (NRCS 1995).

Conservation Reserve Program

The Conservation Reserve Program was established under Title XII of the Food Security Act of 1985 to reduce cropland soil erosion. Under the CRP, the U.S. government has entered into ten-year contracts to pay participants to convert highly erodible land to permanent cover.

Participants receive annual rental payments and half the cost of establishing permanent land cover in exchange for retiring highly erodible cropland for 10 years. More than 36 million acres were enrolled during the 1986-1992 period with annual rental payments averaging $50 per acre and soil erosion reductions averaging 19 tons per acre.

Table 2 summarizes cropland, CRP acreage, and rental costs by region. As can be seen, CRP acreage represents 10 percent of U.S. cropland, with the largest proportion of regional cropland enrolled in CRP located in the Mountain (18 percent), Northern Plains (11 percent), Southeast (12 percent) and Southern Plains (14 percent). The western U.S. accounts for more than 23 million acres of CRP land, representing 64 percent of total CRP land.

The annual rental costs of CRP contracts average nearly $50 per acre for a total cost of more than $1.8 billion. While average CRP rental costs in the west-

Table 2. Cropland, CRP acreage, and CRP rental costs, 1992

Region	Cropland (ac)	——CRP acres——		————Rental cost————	
USA	381,675,800	36,396,843	10%	$49.67	$1,808,069,203
Appalachian	19,724,000	1,158,124	6%	$53.97	$62,503,952
Corn Belt	87,876,200	5,603,333	6%	$74.26	$416,103,509
Delta	19,427,500	1,248,403	6%	$44.28	$55,279,285
Lake States	41,154,100	3,008,337	7%	$58.66	$176,469,048
Mountain	37,513,100	6,687,262	18%	$39.67	$13,419,380
Northeast	15,783,100	226,411	1%	$59.27	$265,283,684
N. Plains	86,983,800	9,664,110	11%	$46.00	$444,549,060
Pacific	20,572,400	1,765,294	9%	$49.55	$87,470,318
Southeast	14,299,600	1,692,580	12%	$42.69	$72,256,240
S. Plains	38,342,000	5,342,989	14%	$40.19	$214,734,728

Source: Osborn, Llacuna, and Linsenbigler; and Natural Resource Conservation Service

ern U.S. of $43.85 per acre are less than the $66.63 per acre in the eastern U.S., total CRP costs in the western U.S. of more than $1 billion per year exceed total CRP costs in the eastern U.S. of $796 million, due to the greater acreage enrolled. Within the western U.S., the rental cost of CRP is nearly $445 million in the Northern Plains, more than $265 million in the Mountain region, and nearly $215 million in the Southern Plains.

Conceptual framework

While the primary objective of the CRP when it was first established was to reduce soil erosion, its secondary objectives were to curb surplus commodity production, improve water quality, and enhance wildlife habitat. Interestingly, the benefits from the secondary objectives have received most of the attention (Barbarika and Langley 1992; Johnson et al. 1994; Ribaudo et al. 1990; Young and Osborn 1990). Also, most studies have focused on water erosion to the virtual exclusion of wind erosion (Lant 1991; Ribaudo 1989). The major contribution of this study is to include the benefits of reduced erosion for the case of wind erosion.

In order to conduct a benefit-cost analysis, the benefits and costs need to be identified and measured. We have measured costs as the sum of federal CRP expenditures, which include rental payments, conservation-practice cost sharing, and administrative expenses. The benefits of the CRP are improved air and water quality, improved crop productivity, savings in commodity program payments and enhanced wildlife habitat. To date, the only empirical measures of the off-site benefits of wind erosion are for the state of New Mexico (Huszar 1989). For this reason, the current study uses New Mexico as a case study.

Using benefit calculations from previous studies, this study determines if the total measured benefits outweigh the costs of the CRP in New Mexico by (1) determining the costs of the CRP; (2) subtracting previously determined estimates of benefits of the CRP from these costs; (3) calculating the off-site wind erosion reduction benefits of the CRP; and (4) determining if the off-site benefits are sufficient to cover the costs remaining after all previously measured benefits are subtracted.

CRP statistics for New Mexico

Annual CRP rental costs and erosion rates were collected from the Agricultural Stabilization and Conservation Service (ASCS) county offices in New Mexico during the fall of 1993. The cost share expenditure information, which includes relevant technical and administrative expenses, was obtained from Osborn, Llacuna, and Linsenbigler (1993).

As of the fall 1993, there are approximately 48,000 acres in more than 1,500

CRP contracts in New Mexico. Contracts are held in 19 counties. Most of the land enrolled is concentrated along the eastern border of the state. While 6 counties exceeded the 25 percent enrollment limit, 11 counties enrolled only 10 percent or less of their land. On average, wind erosion on CRP lands is reduced by more than 90 percent, from 41 tons per acre per year (t/a/y) to 3.69 t/a/y.

The average annual rental cost is approximately $18.3 million or $38 per acre. The one-time cost share payment is estimated at $28 million or $58 per acre. Using a 4 percent discount rate[1] the value of the rental cost over the life of the CRP equals $148.55 million or $307.64 per acre. The value of cost share payments, including technical assistance and administration, equals $24.92 million or $51.60 per acre. That is, total CRP costs for New Mexico have a value of $173.47 million or $359.24 per acre.

Value of secondary benefits

Secondary benefits include soil productivity, surface water and groundwater quality, wildlife habitat, and commodity program outlay savings.[2] Data from the National Ecological Research center shows that nearly one half of every dollar spent on the CRP is offset with savings in commodity programs—in other words, the present value of savings to commodity programs in New Mexico equals approximately $150 per acre (Ekstrand 1994). The estimated present value of regional economic benefits from changes in soil productivity, water quality, wildlife habitat and groundwater quality for the SCS's Mountain Region falls between $40 and $109 per acre, with a best estimate of $74 per acre (Ribaudo et al. 1990).

Combining the benefits from these studies (Ekstrand 1994; Ribaudo et al. 1990) yields per acre secondary benefits in New Mexico of $190 to $259, with a best estimate of $224. Subtracting these secondary benefits from the total costs yields a net per acre cost of the CRP between $100.24 and $169.24, with a best estimate of $135.24 per acre. Using the best estimate, off-site benefits of reduced wind erosion must equal $135.24 per acre in order to just break even.

Off-site wind erosion reduction benefits

Total CRP benefits in New Mexico equal secondary benefits plus off-site wind erosion reduction (OSWER) benefits. OSWER benefits are equal to the difference between off-site damages due to wind erosion with and without the CRP. The off-site damage function for wind erosion in New Mexico estimated by Huszar (1989) is

$$lnTC = 2.98 - 0.14X^{-2} - 0.09Y^2 + 0.01Y^3 + 2.40Z - 0.02W \qquad (1)$$

where TC = household costs from blowing dust per year, X = wind erosion rate (t/a/y), Y = household income level per year, Z = dummy variable for owning or renting residence, and W = years at present residence.

Equation (1) predicts that off-site damages of wind erosion depend upon both the level of erosion and the amount of property at risk. Moreover, since interior and exterior cleaning are major components of the off-site damages, attitudes of individuals toward the cleaning of blown dust will affect the level of damages incurred. The income level is used as a proxy for the value of property at risk. Whether individuals own or rent their homes and the number of years they have lived at their present locations are used as proxy measures of attitudes toward cleaning blown dust. Home owners are expected to expend more time on cleaning than renters, due to pride of ownership and protection of their investment. On the other hand, the more years individuals have lived in an area affected by wind erosion, the more tolerant of blowing dust they are expected to be and, thus, the less time they are expected to spend cleaning.

Income, residence, and population data needed for computing the value of the damage function were gathered from 1990 census reports (USDC 1992). County erosion rates without the CRP are computed from National Resource Inventory (NRI) data (SCS 1990), while erosion rates with CRP are calculated as a weighted average of NRI data and erosion rates reported on CRP contract. The difference between total damages with and without CRP represents the OSWER benefits per household. These per household benefits, computed for both owners and renters, are multiplied by the county owner and renter population levels, respectively, and then summed to derive an estimate of total county off-site benefits. County benefits are then summed to yield total OSWER benefits for New Mexico.

OSWER benefits accrue from the time the conservation practice is in place until the land is converted back to cropland. Garrison, et al. (1993) have estimated that 49.2 percent of CRP land will remain in permanent cover after contracts expire.[3] For illustrative purposes, in addition to a 50 percent conversion rate, we also consider the possibility that all CRP land will be converted to cropland immediately after contracts expire and the possibility that none will be converted. That is, benefits are calculated for three assumptions: (1) 100 percent conversion of CRP land to cropland upon expiration, which represents the "low estimate"; (2) 0 percent conversion, the "high estimate"; and (3) 50 percent conversion, which is considered to be the "best estimate."

Benefit-cost calculations

The results of the benefit-cost analysis are presented in Table 3. As shown, the best estimate of OSWER benefits is $99,136. Subtracting total costs from total benefits yields a net loss to the state of $65.82 million for the best estimate. Even when the high estimate for total benefits is assumed, the program still yields a net loss of more than $42 million.

On a per acre basis, CRP costs for New Mexico averaged $359.24. OSWER benefits were only $0.21 per acre in the best estimate case. Even when these benefits were added to the secondary benefits, the net benefits of the CRP in New Mexico are a negative $136.31 per acre.

Even when counting the most prevalent source of erosion in the West, the CRP in New Mexico did not come close to producing benefits in excess of costs. Relative to the costs, the benefits are practically insignificant. Two questions immediately arise from this observation: (1) Do the results make sense? (2) And if so, what could have been done differently to increase the net benefits of the program?

Analysis of results

While the CRP reduced overall wind erosion statewide and dramatically reduced wind erosion on program land, it likely had a relatively small impact on the dollar value of erosion damages for two main reasons. First, the CRP land represents only 21 percent of the state's total cropland and less than 1 percent of

Table 3. Cost-benefit analysis results

	Total benefits & costs ($ millions)			Per acre benefits & costs ($)		
	High	Low	Best	High	Low	Best
Rental costs			148.55			307.64
Cost share, tech assist.& admin.			24.92			51.60
Secondary benefits	124.89	91.94	107.55	259.00	190.00	224.00
OSWER benefits	0.151	0.0467	0.099	0.31	0.10	0.21
Net benefits[a]	(48.42)[b]	(81.48)[c]	(65.82)[d]	(100.27)	(168.75)	(136.31)

[a] Discrepancies in sums are due to rounding
[b] Assumes that 100% of land remains in permanent cover and the "high" estimate of secondary benefits
[c] Assumes that 0% of land remains in permanent cover and the "low" estimate of secondary benefits
[d] Assumes that 50% of land remains in permanent cover and the "best" estimate of secondary benefits

all land in the state. As a consequence, large reductions of wind erosion on CRP land were overwhelmed by continuing erosion on non-CRP land.

Second, the nature of the damage function is such that relatively large reductions in the erosion rate have small impacts on off-site damages until nearly all of the erosion is eliminated. As can be seen in Figure 2, if erosion is initially at a level corresponding with point A on the damage function, then a relatively large reduction in the erosion rate to point B will yield a very small reduction in off-site damages. Only once the erosion rate is reduced to a level within the steeply sloped portion of the damage function will further reductions in the erosion rate significantly reduce off-site damages. For example, if the erosion rate corresponds with point C, then a relatively small reduction in the erosion rate to a point D will significantly reduce off-site damages.[4] By simply selecting land with the greatest erosion rates, CRP reduced erosion rates by large amounts without significantly affecting the off-site damages.

On the other hand, the damage function likely under-represents health and recreation damages, as well as completely omits business and government damages from blowing dust (Huszar 1989). Moreover, the calculations simply assume that reductions of off-site wind erosion damages are only for those counties in which the wind erosion is reduced. The significance of these likely under-estimations can be examined by a sensitivity analysis.

Figure 2. Total off-site damages

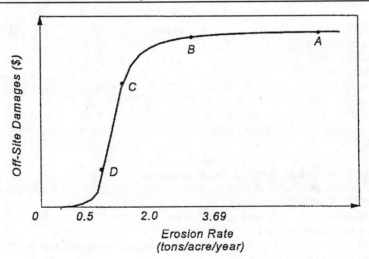

150

The high estimate shows OSWER benefits of $151,513. Corresponding estimates of costs net of secondary benefits are $48.42 million. In order for benefits to exceed costs of CRP, additional OSWER benefits would have to exist that are 320 times greater than those included in this study. Similarly, additional OSWER benefits would need to be 665 times greater for the best estimate and 1,744 times greater for the low estimate scenario. Additional benefits of these magnitudes seem highly unlikely, so that it must be concluded that the results are not very sensitive to possible underestimation of the off-site benefits of reduced wind erosion.

Alternative approaches

Greater benefits and, consequently, a positive net return from the CRP in New Mexico may have been achieved by selecting program land not on the basis of the greatest erosion rate, but based upon a combination of erosion rates at key levels of erosion, land location, and land uses.

Key levels of erosion—Analysis of the results of the damage function indicate that 95 percent of all off-site damages from wind result from erosion levels below 2 t/a/y. Beyond this rate, damages increase at a sharply decreasing rate and nearly level off beyond 3 t/a/y. That is, the steep portion of the damage curve shown in Figure 1 occurs between erosion rates of 0.5 t/a/y and 2 t/a/y. But the CRP in New Mexico has been targeted at highly erodible cropland which is located in areas with an average erosion rate of 15 t/a/y. Benefits can only be captured in these areas when two conditions are met.

First, erosion levels must be reduced below 2 t/a/y to capture a major portion of the benefits. Conservation practices, although effective, have on average been unable to bring erosion rates down to this level in New Mexico. Second, a large proportion of the land must be treated in order to effect the overall erosion rate. Many of the counties have enrollment levels below 10 percent of total cropland. Even large reductions of erosion rates on only 10 percent of highly erodible cropland will not significantly affect the overall county erosion rate.

Key locations—The main benefit of the CRP is the reduction of erosion costs, but for wind erosion a major component of these costs are off-site. In New Mexico, on-site damages amount to only about $10 million per year (Davis and Condra 1989), whereas off-site damages likely exceed $466 million (Huszar and Piper 1989). And a key factor in determining the magnitude of off-site damages is the proximity and density of population. In many counties, land enrolled is located in low population areas. This resulted in relatively low levels of OSWER benefits, so that the costs of reducing erosion on these lands greatly outweighed the benefits received by such small populations. Had implementors targeted areas with higher affected populations, total OSWER benefits would have been larger.

Key land uses—Huszar and Piper (1989) found that the largest source of off-site erosion damages in New Mexico is from rangeland erosion. Moreover, rangeland constitutes 52 percent of the area of the state. Since cropland constitutes only 3 percent of the land in New Mexico, reducing erosion on cropland alone has little potential for effecting the overall erosion rate. The effectiveness of the program could be increased by allowing the targeting of lands that have significant impacts on the overall erosion rate, regardless of the use.

Conclusions

The CRP is the first program since the Soil Bank to maintain environmentally sensitive land in vegetative cover for an extended period of time. Since its inception in 1985, the CRP has controlled a large proportion of federal funds allocated to conservation programs. The CRP has shown that it has the potential to produce benefits, but as this paper indicates for wind erosion, these benefits maybe at the expense of far greater costs.

Much of the discussion surrounding the renewal of CRP in the 1996 Farm Bill has been concerned with water erosion and water quality. Perhaps, as currently administered, the CRP is the appropriate mechanism for addressing these concerns. However, wind erosion in the western U.S. is also an important soil conservation issue and one which this study indicates is not effectively dealt with by CRP. Within the federal government's efforts to balance its budget and to address soil conservation problems, increasing the cost-effectiveness of the CRP must be addressed.

Notes

1. This is the long-term rate of return used by the ERS and it is used here in order to make our results comparable with previous studies (e.g., Ribaudo et al. 1990; Young and Osborn 1990).
2. There is debate as to whether including commodity program outlay savings overstates the benefits of the CRP. The general conclusions obtained from the results of this analysis, however, are not sensitive to inclusion or omission of these savings.
3. Since our study was completed, another survey of CRP contract holders indicated that 63 percent intend to return their land to production (Osborn, Schnepf, and Keim 1994). If correct, this would make our best estimate of benefits high, but the overall evaluation would not be changed.
4. An example of such a relationship might be house cleaning. A 50 percent

reduction of dust accumulations in the house from 10 centimeters (cm) to 5 cm per day would likely not greatly reduce the frequency or time costs of cleaning; the house would need to be cleaned daily. But a reduction from 1 cm to 1/2 cm per day may result in a more acceptable level of dust that only requires cleaning on a weekly basis.

References

Barbarika, A. and J. Langley. 1992. Budgetary and farm sector impacts of the 1985-1990 conservation reserve program. Journal of Soil and Water Conservation. 47(3):264-267.

Davis, B. and G. Condra. 1989. The on-site costs of wind erosion on farms in New Mexico. Journal of Soil and Water Conservation. 44(4): 339-343.

Ekstrand, E. 1994. Economist, National Ecology Research Center, U.S. National Biological Survey, Fort Collins, CO. Personal communication. January.

Garrison, C.O. et al. 1993. Recropping rates of conservation reserve program acreage. Presented at the Southern Agricultural Economics Association Annual Meeting, Tulsa, OK, February 3-7.

Huszar, P.C. 1989. Economics of reducing off-site costs of wind erosion. Land Economics 65(4): 333-339.

Huszar, P.C. and S.L. Piper. 1986. Estimating off-site costs of wind erosion in New Mexico. Journal of Soil and Water Conservation. 41(6): 414-416.

Johnson, R., E. Ekstrand, and K. Jon. 1994. The economics of wildlife and the CRP. Proceedings of Soil and Water Conservation Society Policy Options Conference: When Conservation Reserve Program Contracts Expire. Arlington, VA, February 10-11. pp. 45-51.

Lant, C.L. 1991. Potential of the conservation reserve program to control agricultural surface water pollution. Environmental Management. 15(4): 507-518.

Laycock, W. A. 1991. The conservation reserve program—how did we get where we are and where do we go from here? The Conservation Reserve—Yesterday, Today and Tomorrow. Edited by Linda A. Joyce, John E. Mitchell, and Melvin D. Skold. USDA Forest Service, Rocky Mountain Forest and Range Experiment Station, General Technical Report RM-203, Ft. Collins, CO.

Natural Resources Conservation Service (NRCS). 1995. Summary Report 1992 National Resources Inventory. USDA, Iowa State University, Statistical Laboratory.

Osborn, C.T., F. Llacuna, and M. Linsenbigler. 1993. The Conservation Reserve Program, Enrollment Statistics for Signup Periods 1-11 and Fiscal Years 1990-1992. USDA, ERS Statistical Bulletin Number 843, Washington, D.C.

Osborn, C.T., M. Schnepf, and R. Keim. 1994. How farmers plan to use the land after CRP contracts expire. Presented at the Soil and Water Conservation Society Annual Meeting, Norfolk, VA, August 7-10.

Ribaudo, M.O. 1989. Water quality benefits from the Conservation Reserve Program. USDA, ERS Agricultural Economics Report Number 606. Washington, D.C.

Ribaudo, M.O., D. Colacicco, L.L. Langner, S. Piper, and G.D. Schaible. 1990. Natural resources and users benefits from the Conservation Reserve Program. ERS Agricultural Economics Report No. 627. Washington, D.C.

Soil Conservation Service (SCS). 1990. Data for decisions: 1987 National Resource Inventory: New Mexico Data. U.S. Department of Agriculture, Washington, D.C.

U.S. Department of Commerce (USDC). 1992. 1990 census of population and housing: Summary population and housing characteristics New Mexico. Washington, D.C.

Young, C.E., and C.T. Osborn. 1990. The Conservation Reserve Program: An economic assessment. USDA, ERS Agricultural Economic Report No. 626. Washington, D.C.

13

United States' New Pesticide Law: Implications for Soil and Water Conservation

Andrew P. Manale

The Food Quality Protection Act of 1996 established new standards for pesticide residues in foods. Numerous pesticides registered before the stricter data requirements established in 1984 and now undergoing reregistration may not meet these stricter standards. Because some of these pesticides have been incorporated into soil conservation strategies as alternatives to soil tillage, the Act may lead to the need to reformulate weed management strategies in reduced or no-tillage systems. The stricter dietary standards may also result in the regional curtailment or national banning of the use of a subset of the pesticides that pose water quality problems in some parts of the country. Changes the Act makes in the registration process for pesticides important in Integrated Pest Management (IPM) and for minor uses may offset the loss of existing chemical pest management alternatives by encouraging and expediting the registration of alternative pesticides, particularly biological and biochemical pesticides.

Summary of the current and new laws

The Federal Insecticide, Fungicide and Rodenticide Act (7 U.S.C. 136 et seq.) (FIFRA) requires that any agent to be sold or distributed for the purpose of controlling pests must be registered by the United States Environmental Protection Agency (EPA). Should the pesticide be used on a food crop, the Federal Food, Drug, and Cosmetic Act (21 U.S.C. 301 et seq.) (FFDCA) mandates a food tolerance for a pesticide residue.[1] A tolerance is required only for the raw product if the chemical does not concentrate in processing. If it does concentrate, a separate tolerance is required for the processed food. However, if the chemical is a carcinogen, the Delaney Clause in FFDCA stated that no processed food tolerance was permitted, regardless of the actual risk or benefit of the pesticide. Fluidity among sectors of the fresh and processed food markets, which readily diverts foods intended for the fresh food market into processing,

155

makes enforcement of the separation of fresh and processed foods administratively impossible. EPA therefore adopted a policy of linkage: if it is used on a raw product that can be processed, it assumes that the raw product will be used in processing as well (USEPA 1996a).

Amendments to FIFRA enacted in 1988 required reregistration reviews for all pesticides and their associated food tolerances first registered before November 1984, the date on which the modern health standards for pesticides were implemented. Registrants submit data showing that the pesticides, when used in commonly recognized practices, do not pose an unreasonable risk to health or environmental safety. Pesticides that do pose some risk could nevertheless still be registered if there were counterbalancing health and welfare benefits associated with their use. In 1989, there were 600 active ingredients to be reregistered. There are currently 382 active ingredients undergoing reregistration in 20,000 products.

Pesticides grandfathered into the program, which represent the great majority of pesticide registrations, posed a dilemma for the Agency. Carcinogenic, generally water-soluble pesticides that posed only negligible risks could be still be registered for use on foods. On the other hand, pesticides that are generally fat-soluble and concentrate in processing was not allowed regardless of whether or not their carcinogenic dietary risk were negligible, and regardless of whether or not their benefits to society through avoidance of greater risks were deemed significant.

The new law removes the paradox by setting a uniform standard of "safe" for pesticide residues in both raw and processed foods. A pesticide residue is safe if the Administrator of EPA has determined that there is a reasonable certainty that no harm will result from aggregate exposure to the pesticide chemical residue, including all anticipated dietary exposures and other exposures for which there is reliable information. It also establishes a system of periodic review for all pesticide registrations on a 15-year cycle.

The "reasonable certainty of no harm" standard derives from the legislative history of the Food Additives Amendment of 1958 [P.L. 85-929], the amendment that created the section of pesticide law establishing the requirement for a food tolerance for processed foods [section 409 of FFDCA] [Vogt 1995]. Under this concept of safety, the administrator must take into account the probable consumption of the substance, the cumulative effect of the substance in the diet, any chemically related substances in the diet, and safety factors that are appropriate. For pesticides that pose risk of cancer and other adverse health effects at any level of exposure, "reasonable certainty of no harm" is generally understood to mean no more than one case of cancer occurring in a population of a million people (104th Congress 1996). Because of scientific uncertainties, a one-in-a-million risk is generally considered below the limit of the risks that scientists can reliably quantify (National Research Council 1987 and 1996). It can also mean zero risk. For chemical agents that cause adverse health effects only at levels exceeding a scientifically determinable concentration in foods, so-called "threshold toxicants," reasonable certainty of no harm means that exposure to the chemical residue through all anticipated routes of exposure will fall below the threshold with an ample margin of safety.

The law does provide for exceptions to the standard in cases where there are benefits that outweigh the risks. The exception, however, applies only to a non-threshold toxicant, for which the lifetime risk can be quantified. The use of the pesticide must protect consumers from adverse health effects greater than the dietary risk posed by the pesticide residue or where the pesticide is necessary to ensure domestic production of an adequate, wholesome, and economical food supply. This inclusion of a risk-benefit test makes the FFDCA language governing pesticide contaminants in foods consistent with the risk-benefit balancing of the FIFRA, which allows for registration of a pesticide if it can be used without "unreasonable risk to people or the environment, taking into account the economic, social and environmental costs and benefits of the use" [FIFRA section 2 (bb), i.e., its risks are outweighed by the societal benefits resulting from its use].

The Food Quality Protection Act, unlike FIFRA, caps the risks associated with the exception. The annual risk associated with a carcinogenic pesticide cannot exceed by a factor of ten the yearly risk of the acceptable risk standard, which, through regulation, is expected to be set at one additional case of cancer in a population of 1 million over a lifetime of exposure. Nor can the lifetime risk associated with exposure to the pesticide residue in food exceed by a factor of two the acceptable lifetime risk.[2] Finally, the tolerance issued under this exception must be reviewed after a period of 5 years to determine whether the conditions that permitted the exception still hold.

Perhaps more important for future registration decisions than the numerical risk caps is language in the bill that changes the procedures for assessing risk. The new law expands the kinds of data to be considered in assessing dietary risk to include cumulative and aggregate dietary and nondietary exposures to the pesticide and other substances with common mechanisms of toxicity and effects on the endocrine system. In addition, the law specifically calls for the administrator of EPA to determine that there is reasonable certainty that no harm will result to infants and children from aggregate exposure to the pesticide residue. If the data on exposure or toxicity to infants and children are not adequate, the law states that an additional tenfold margin of safety for threshold toxicants is to be applied to account for potential pre- and postnatal toxicity.

Pesticides that pose a significantly lower risk to human health from dietary exposure for the same or similar uses have priority in establishment or modification of tolerances. Furthermore, such lower risk pesticides may benefit from expedited tolerance reviews and registration procedures. Lower risk pesticides eligible for expedited registration include pesticides that reduce the risk of pesticides to public health, reduce the risks of pesticides to nontarget organisms, reduce the potential for contamination of ground water, surface water, or other environmental resources, and broaden the adoption of integrated pest management strategies or make such strategies more available or effective.

Loss of efficacious pesticides due to more stringent requirements for food tolerances may be offset by a new minor use program established by the new law which provides incentives for manufacturers to register pesticides for eligible uses. Pesticides eligible for the new program are those used on an animal, an agricultural crop, or for the protection of public health where the total acreage is less than 300,000 acres, or where there is insufficient economic incentive to support pesticide registrations. However, the following conditions apply: there must not be efficacious alternative pesticides registered for the use. If there are alternatives, they must pose greater risk to the environment or human health; the pesticide will play a significant role in managing pest resistance; or the pesticide will play a significant role in an IPM program. The program allows for extensions of the period of exclusive use of data that supports the registration of the pesticide for the minor use, extensions of time for the development of data for registration and reregistration of pesticides for a minor use, waiver of data requirements, and expedited registrations.

A new consumer right-to-know provision requires EPA to publish and distribute, to large retail grocers, information on pesticide residues for public display. The information discusses the risks and benefits of pesticide chemical residues in foods, lists actions that allow residues greater than negligible levels on certain types of foods, and contains recommendations on how consumers may reduce dietary exposures to pesticides. Economic analyses and consumer surveys suggest that this additional information will have only a modest impact on consumer food buying decisions that could affect agricultural production practices (Wohl 1994).

Finally, the law establishes a new screening program for estrogenic substances which must be implemented within three years. EPA must first develop the tests that it will require pesticide registrants to conduct to allow identification of estrogenic substances. The additional cost of testing could further dissuade registrants from maintaining existing chemical pesticide registrations or developing new chemical pesticides.

Pesticides at risk of cancellation under the new law

Pesticides that were at risk of being canceled or having restrictions put on their use because of Agency concern regarding carcinogenicity from dietary exposures, particularly where there are significant exposures that occur in more than one medium, are likely to remain at risk under the new standard. Some of these pesticides share common mechanisms of toxicity and are suspected of being reproductive toxicants or endocrine disrupters. The new focus on multiple exposures, risks to children and infants, and indication of possible endocrine disruption will likely replace Delaney as the driving force for cancellation of pesticides for dietary uses.[3]

Two classes of pesticides are particularly important to conservation—the triazine herbicides and the ethylene-bis-dithiocarbamate (EBDC) fungicides. As these chemicals undergo reregistration under the new law, they face the possibility that further restrictions will be imposed upon their use or that certain or all registrations and tolerances for food uses are canceled.

In November 1994, EPA announced a special review of the triazine herbicides atrazine, cyanazine, and simazine.[4] EPA conducts special reviews to determine whether or not new scientific evidence justifies a rulemaking to cancel, deny, or reclassify registration of a pesticide product that may cause unreasonable health or environmental effects. Special reviews tend to signal the agricultural industry that overall use of the pesticide should be reduced. It initiated the special review because of concern that they "may pose significant cancer risks to consumers who are exposed to residues in food and drinking water...EPA is also concerned about the risks to the environment from the large amount of triazines used (USEPA 1995a)." It estimates that the increased lifetime risk from use of the triazines on foods are one in 23,000 for atrazine, one in 34,000 for cyanazine, and one in 91,000 for simazine (USEPA 1995b).

The reasons stated for the Special Review represented a change in Agency policy. Until this announcement, EPA looked solely at the risks associated with the particular pesticide undergoing review. In response to the 1993 report by the National Research Council of the National Academy of Sciences, *Pesticides in the Diets of Infants and Children* (National Research Council 1993), the Agency adopted the recommendations of the Academy to estimate potential risk from exposure to several pesticides with common mechanisms of action and/or exposure via multiple routes. The Special Review for the triazine herbicides represented a test case for conducting a risk assessment in accordance with the best science, as determined by the Academy: "the triazines present an example where the Agency can address pesticides with commonality of mechanism as well as exposure through multiple pathways..."

Aggregate risk from exposure to all three triazines through multiple exposure pathways was given as the justification for its determination of unreasonable risk. The risk estimates for each triazine and for each exposure pathway are presented in Table 1.

In the announcement for the Special Review, the Agency expressed its concern over the potential ecological impacts of ground and surface water contamination resulting from the use of products containing the triazines. It decided not to include ecological effects as a trigger in the Special Review at the time but reserved the authority to do so in the future.

Between 0.1 and 3 percent of atrazine, for example, applied to fields is lost to the aquatic environment due to its mobility. At the lowest rates of loss (0.1 percent) and use (64 million pounds), 64,000 pounds of atrazine annually enter surface water supplies. At the higher rates of loss and use, some 2.4 million pounds annually contaminate water resources. Similar mobility of the three triazines suggest that the upper limits of triazine pollution of water resources approaches 3.3 million pounds annually.

Conducting a risk assessment according to the concept of aggregate exposures and risk is, however, not the same as making a regulatory decision that can withstand court challenges that it is inconsistent with the law. The new law incorporates this concept and provides the legal authority for the regulatory action to implement this risk policy. Thus, a regulatory decision based upon aggregate risk and multiple exposures is more likely to be enforceable under the new law than the old.

The Delaney Clause was not given as the reason for the triazine herbicides being put into Special Review. Hence, eliminating the Delaney Clause does not necessarily reduce the Agency's concern regarding these compounds. The current estimate of cancer risk for atrazine exceeds the one in a million lifetime risk threshold that is presumed to be set under the new "reasonable certainty of

Table 1. Triazines—Upper bound cancer risk estimates across several exposure pathways

Exposure pathway	Atrazine	Simazine	Cyanazine
Dietary	4.4×10^{-5}	1.1×10^{-5}	2.9×10^{-5}
Drinking water	4.2×10^{-6}	6.2×10^{-7}	9.7×10^{-6}
Occupational	7.7×10^{-3}	4.6×10^{-3}	6.6×10^{-3}
Residential	2.4×10^{-5}	N/A	N/A
Total:	$\mathbf{7.8 \times 10^{-3}}$	$\mathbf{4.6 \times 10^{-3}}$	$\mathbf{6.6 \times 10^{-3}}$

no harm" test (Table 1). And, even though the triazines are not used on foods that children consume to a greater proportion of their diets than adults (Ralson et al. 1995), they are present in environmental media to which children and infants may be exposed. By expressly calling for consideration of multiple exposures, common modes of toxicity, and endocrine disruption, the new standard raises the burden of proof that the triazines can be safely used.

On August 2, 1995, EPA announced that the manufacturer of cyanazine, Dupont Agricultural Products, had agreed to phase out the production of the chemical over 4 years (USEPA 1995b). The phaseout of cyanazine is likely to lead to greater dependence upon the other triazines, even though they are not direct substitutes in pest control strategies.

More recently, EPA published a Notice of Proposed Rulemaking on state management plans (USEPA 1996b) which, when implemented, requires states which wish to maintain the registration of the triazines and two other pesticides that have been found to contaminate ground water supplies to implement a plan for protecting its ground water resources from these pesticides. The authority for the rules derives from FIFRA section 3 (d)(1)(C) [7 U.S.C. 136a(d)(1)(C)] which is not affected by the amendments in the new law. In the plan, the state must detail how the state intends to carry out this commitment, using such measures as groundwater vulnerability assessments, groundwater monitoring, and direct management of pesticide use. The agency justified its action with the triazines on the basis of the frequent occurrence of these chemicals in groundwater supplies, the evidence of carcinogenicity, and the potential for transport to surface waters where they can have an adverse effect on aquatic ecosystems. Any success of these new rules in lowering overall exposures in water supplies should reduce the pressure to ban the triazines.

The EBDCs are the second major group of pesticides that face possible changes in the status of their registrations under the new law. The resolution of the Delaney Clause momentarily relieves the pressure to cancel their food tolerances. However, the new standard and the probable higher threshold for potential toxicants to children and infants may again put them at jeopardy of restrictions on their use.[5]

Impacts on soil and water conservation policy

Impacts on practices

Common agricultural practices on most farmland in the U.S. have had a major impact on soil and water quality (Office of Technology Assessment 1995). Agriculture ranks as the number one source of water quality impairment for rivers and lakes and number three for estuaries (USEPA 1995c). Agricultural pollutants include sediment, pesticides, nutrients, and pathogens. Pesticides have resulted in moderate impairment of 11 percent of the lakes surveyed, for example.

Current national agricultural policy on minimizing the adverse impacts of agricultural production is to provide a mix of carrots and sticks to encourage producers to adopt conservation practices, such as conservation tillage, that minimize the tilling of the soil to reduce soil erosion and to adopt IPM practices. Conservation tillage is defined as any tillage or planting system that leaves at least 30 percent of the planted soil surface covered by crop residue or leaves at least 1000 pounds of residue per acre during critical periods when wind can cause soil erosion (Office of Technology Assessment 1995). IPM uses scouting and other management information to target better a range of chemical and nonchemical treatments to control pests in an economically and ecologically sound manner. IPM can also lower pesticide applications and their associated costs. The major stick driving producers to use conservation practices is the Conservation Compliance provision of the Federal Agriculture Improvement and Reform (FAIR) Act of 1996. Producers of commodity crops grown on highly erodible land who do not adopt conservation practices face the possible loss of federal commodity program payments and other benefits. The carrots provided under Federal policy are technical assistance and educational programs and the incentive programs authorized under the research and education and conservation provisions of the FAIR Act.

The triazine herbicides play a major role in weed control for the major commodities under soil conservation practices. The triazine herbicides are used in reduced and no-till systems as an alternative to tilling the soil to control weed populations. In an average year, 60 million pounds are applied to roughly 65 percent of the corn acreage and about 3.3 million pounds to 60 percent of the sorghum acreage in the cornbelt states of the U.S.[6] Nationally, total atrazine use amounts to about 60 million pounds with 85 percent used on corn (USEPA 1994). Approximately 21 to 34 million pounds of cyanazine are used annually, of which 95 percent is used on corn. Approximately 20 percent of the total U.S. corn acreage, primarily in Iowa, are treated with cyanazine as either the sole active ingredient or in combination with other herbicides. About 1 to 2 million pounds of simazine are used on corn acreage annually.

In 1993, EPA conducted a study of the consequences for water quality of a ban on the triazine herbicides. It is illustrative of changes in agricultural practices that the loss of these preemergent herbicides could produce and the associated impacts on the environment, assuming that the rules for maintaining residue levels of the soil to prevent soil loss under the conservation compliance provisions of the 1996 Federal Agriculture Incentive Reform (FAIR) Act are not relaxed (USEPA 1993).

The study, which modeled the behavior of farmers regarding their choice of practices, found a significant shift to conventional postemergent herbicides and, in particular, a technological leap to the newly registered, low dosage sulfonylurea herbicides which are used at ounces per acre. Total herbicide use actually increases because the alternatives, with the exception of the sulfonylureas, employ higher application rates. Total loadings to surface waters, nevertheless, would not increase with the health risk from herbicide exposure actually decreasing, though there are exceptions for specific areas with certain soils. Significantly less pesticide would leach to ground water. However, the ecological impact is caveated because the long-term consequences for widespread introduction of the sulfonylurea herbicides into the environment are not well understood. The conclusion is that a ban on the triazine herbicides would result in only slight environmental and health benefit at considerable economic cost (because of the greater cost to producers of alternative weed management strategies) unless geographically targeted mitigation measures are employed.

A subsequent study conducted by EPA examined the economic and environmental consequences of targeted alternative conservation policies (Lakshminarayan 1996). The study used data from USDA's Management System Evaluation Area (MSEA) at Walnut Creek, Iowa, to model alternative policies to reduce not just soil erosion, but surface and ground water impacts as well. The results suggested that a ban on atrazine to address water quality concerns would lead to an increase in corn/soybean rotations and a decrease in land under conservation tillage with subsequent increase in sediment loadings to surface waters. The latter result illustrates the tradeoff between soil erosion and water quality. A win-win solution was found with a policy of "green payments" that compensated producers for a loss in income from a sustainable crop rotation.

A green payment program can be implemented under the conservation provisions of the 1996 FAIR Act and, in particular, the Environmental Quality Incentive Program. Under this program, producers can receive Federal subsidies for adopting practices that protect soil and water quality.

In wet weather, such as occurred in the Northern Great Plains for the past three years, crops like wheat and occasionally barley tend to be prone to fungal problems, such as scab, particularly when produced under the high residue conditions of conservation tillage (Salas 1991). In 1993, for example roughly 90 million bushels of wheat were lost to the epidemic. The *Fusarium* species that attacks small grains survives mainly in crop residue and not freely in the soil.

Effective control occurs through three means: (1) cultural control; (2) control by resistance; and (3) chemical control (Tekauz in *Proceedings*, 1994). Cultural control, where it involves tillage, may be inconsistent with residue man-

163

agement for conservation purposes. The chemical control method employs the field application of mancozeb, an EBDC fungicide. Roughly 1 million acres of wheat were treated with Mancozeb, for example, in 1993.

There are nonEBDC fungicides, which include Benlate, and Til, that can be used. But they would cost producers roughly $4 more an acre (Gianessi and Anderson 1995). Research on developing new varieties of wheat which are resistant to the disease (Frohberg 1994) may ultimately be the answer, but it will take time and money.

IPM policy

In September 1993, the Clinton Administration expressed the national goal of implementation of Integrated Pest Management (IPM) on 75 percent of crop acres by the year 2000. IPM is defined as a "sustainable approach to managing pests by combining biological, cultural, physical, and chemical tools in a way that minimizes economic, health, and environmental risks" (USDA 1993). IPM tends, over time, to reduce the overall use of pesticides and hence the levels of pesticides in the diet.

This measure puts greater pressure on pesticide registrants to reduce the overall level of chemical pesticide that contaminates the environment. Registrants can either help develop IPM for the uses of their chemical pesticides or else shift their investments in research and development to alternatives pesticides, such as biochemicals, and biologicals. The implications for water and soil quality are that pest control practices that are kinder to the soil will likely be used. Precision agriculture that employs the latest technology, including global positioning, to pinpoint crop needs and apply pesticides and nutrients only where needed can be expected to play an increasingly prominent complementary role to IPM in reducing overall input use (Office of Technology Assessment 1995).

Generally the most profitable uses of a pesticide are those associated with use on major field crops. Not surprisingly, over the course of EPA's reregistration program, the array of pesticides available for minor use crops has diminished. This loss of pesticides available for minor-use crops has inadvertently led to greater reliance and use of a small number of remaining pesticides registered for a particular use. This, in turn, has resulted in problems of pest resistance (Gianessi 1992).

The minor use provisions of the measure are designed to address this shortage of pest management alternatives by creating the incentives for manufacturers to make the necessary investment in data development to extend registrations now available for major commodity crops to minor crops. The availability of more pesticide alternatives assists the extension of IPM to more crops and adoption by more producers by reducing the overall reliance upon individual pesticide and the need to use more pesticide to overcome a pest infestation, thus reducing the likelihood that resistance develops.

Finally, another provision of the pesticide law supports IPM and overall efforts to introduce safer pesticides into agricultural use. This is the provision regarding fees to pay for the administrative costs of reregistering pesticides. Without sufficient funding for resources, the Agency cannot meet its commitments on reregistering pesticides and removing those pesticides that are unsafe. The availability of a pesticide for a crop production practice, even should it be risky, can dissuade manufacturers from committing the investment in research and development to introduce a safer alternative. The constraint on funds can also prevent the allocation of resources towards registering new pesticide alternatives that can be incorporated into IPM regimes.

Conclusion

Soil conservationists and the crop protection industry must not lull themselves into complacency in the belief that, with the passage of the new pesticide law eliminating the Delaney paradox, the pressure on pesticide registrations important for conservation tillage will subside. The industry must continue efforts to develop geographically targeted IPM strategies to reduce triazine use. In particular, there must continue to be resources devoted to the development of IPM weed control strategies in conservation systems that do not rely upon the triazines, that mitigate overall risk to health and the environment, and that support resistance management. Finally, in order to reduce disincentives for producers to adopt conservation practices in wheat production, agricultural research should make the development of scab-resistant strains a high priority.

Notes

1. A food tolerance is a numerical standard for the concentration of pesticide residues in a particular crop. For raw agricultural products, it is generally measured at the "farmgate," a point in time shortly after harvest.
2. A carcinogenic pesticide which exceeds, by a factor of ten, the presumed new risk standard of one additional case of cancer over a lifetime of exposure in a population of a million can remain in the marketplace under the caps for at least 14 years.
3. According to Dan Barolo, Director of the Office of Pesticide Programs, it will be at least 2 years before the agency will have new data requirements for endocrine disrupters. It therefore has no list of pesticides currently sus-

pected of mimicking hormones at the levels at which humans are exposed through their diet or environmental exposures. Nevertheless, there are chemical pesticides, because of extensive research on their impacts on animal systems, that are suspected of inducing hormonal changes in humans. According to Theo Colborn, Dianne Dumanoski, and John Peterson Myers, in *Our Stolen Future*, these pesticides include DDT and its degradation products, synthetic pyrethroids, triazine herbicides, ethylbisdithiocarbamate fungicides (EBDCs), and some dioxins. None of the chemicals identified in *Our Stolen Future* have been confirmed by EPA as posing a significant health threat to humans through endocrine disruption.

4. According to USEPA (58 FR 60412), "The Agency has also determined that a combined Special Review of atrazine, simazine and cyanazine is more appropriate than examining each individually. This determination is based on the following considerations: all three (1) are structurally related chemicals; (2) induce mammary tumors when fed to rats and are classified as Group C, possible human carcinogens; (3) degrade or metabolize to similar degradates/metabolites; (4) are generally similar in terms of environmental fate including relative persistence, leachability, run-off potential and possibly atmospheric transport; (5) are similar in toxicity to aquatic organisms and terrestrial plants; and (6) may serve as alternatives to each other for some situations."

5. USEPA is likely to examine with greater scrutiny any dietary risk resulting from mixtures of pesticides on foods because of their possible synergistic effects. See S. Stoney Simons Jr., "Environmental Estrogens: Can Two 'Alrights' Make a Wrong?" (Simons 1996), and Arnold et al., "Synergistic Activation of Estrogen Receptor with Combinations of Environmental Chemicals" (Arnold 1996).

6. Based upon data from Gianessi, Leonard P., and Cynthia Puffer, Herbicide Use in the United States, Resources for the Future, December 1990; and Gianessi, Leonard P., and Cynthia Puffer, Fungicide Use in U.S. Crop Production, Resources for the Future, April 1992, and subsequent communications with the authors.

References

Arnold, S., D.M. Klotz, B.M. Collins, P.M. Vonier, L.J. Guillette Jr., and J.A. McLachlan. 1996. Synergistic activation of estrogen receptor with combinations of environmental chemicals. In Science 272: 1489-1491.

Frohberg, R. 1994. Proceedings of the November, 1994, regional scab forum and research conference. Fargo, North Dakota.

Gianessi, L.P. and J.E. Anderson. 1995. National Center for Food and Agriculture Policy. Issue Brief. Resources for the Future.

Gianessi, L.P. and C.A. Patton. 1992. Reregistration of minor use pesticides: Some observations and implications. National Center for Food and Agriculture Policy. Resources for the Future.

Lakshminarayan, P.G., A. Bouzaher, and S.R. Johnson, Center for Agricultural and Rural Development, Iowa State University, and A. Manale, USEPA. 1996. Evaluation of soil and water quality policies: The case of the Iowa MESA site—Walnut Creek.

National Research Council. 1987. Regulating pesticides in foods. Washington, D.C.: National Academy Press.

National Research Council. 1993. Pesticides in the diets of infants and children. Washington, D.C.: National Academy Press.

National Research Council. 1996. Carcinogens and anticarcinogens in the human diet. Washington, D.C.: National Academy Press.

Office of Technology Assessment (U.S. Congress). 1995. Targeting environmental priorities in agriculture: Reforming program strategies. OTA-ENV-640. Washington, D.C.

Ralson, K., F. Kuchler, L. Unnevehr, and S. Crutchfield. 1995. Dietary intake of pesticide residues from fruits and vegetables: Lessons from the pesticide data program. Paper presented at the June 6-7, 1995 Economics of Reducing Health Risks from Food Conference. Washington, D.C.

Salas, B. 1991. Effect of tillage and cropping systems on root rot of wheat. NDSU Ph.D. Thesis. 230 pp. As reported by Robert Stack in Proceedings of the November, 1994. Regional Scab Forum and Research Conference. Fargo, North Dakota.

Simons, S.S., Jr. 1996. Environmental estrogens: Can two "alrights" make a wrong? Science, 272: 1488-1489.

Tekauz, A. 1994. Proceedings of the November, 1994, regional scab forum and research conference. Fargo, North Dakota.

USDA. 1993. USDA's pest management initiative. Publication of the Natural Resources Conservation Service. Washington, D.C.

USEPA. 1993. Office of Policy, Planning, and Evaluation. Agricultural atrazine use and water quality: A CEEPES analysis of policy options. EPA 230-R-23-008.

USEPA. 1994. Office of Pesticide Programs. Notice of special review of the triazine herbicides. 59 FR 60412 .

USEPA. 1995a. Office of Prevention, Pesticides, and Toxic Substances, Office of Pesticide Programs. Special review of triazine herbicides. April 25, 1995, packet distributed by the Office of Congressional Liaison.

USEPA. 1995b. Office of Communications, Education, and Public Affairs. August 2, 1995. Note to correspondents: Cyanazine pesticide voluntarily canceled and uses phased out.

USEPA. 1995c. Office of Water. National Water Quality Inventory: 1994 Report to Congress. Executive summary.

USEPA. 1996a. The Pesticide Coordination Policy: Response to petitions, [OPP-300409; FRL-4991-9].

USEPA. 1996b. Office of Pesticide Programs. Proposed rulemaking on state management plans. 61 FR 33260.

Vogt, D.U. 1995. Food additive regulations: A chronology. CRS Report for Congress.

Wohl, J.B. 1994. The effect of ambiguity on consumers' willingness to pay for pesticide-residue certification on applies. Doctoral dissertation for the Department of Agricultural Economics, Michigan State University.

104th Congress. 1996. House of Representatives. Food Quality Protection Act: Report to accompany H.R. 1627.

14

Soil Conservation Policy in Canada:
Adrift or in a State of Evolution?

David R. Cressman, Scott N. Duff, Paul H. Brubacher, and Jim Arnold

Canadian farmers contribute significantly to global food supplies. With only about 5 percent of Canada's total land mass having the capacity to support commercial agriculture (Dumanski et al. 1986), the robustness of this agricultural industry and the agroecosystem that supports it depend in large part on the health and productivity of fragile surface soil layers.

That soil degradation, including water and wind erosion, loss of organic matter, compaction, salinization, and acidification, are serious threats to long-term productive capacity and sustainability of the Canadian agrosystem is by now a well documented fact (Coote et al. 1981; Anderson and Knapik 1984; Rennie 1985). But the extent to which the processes of soil degradation have been arrested, and renovation of degraded soils accomplished, is ambiguous at best.

Throughout the late 1970s and the 1980s, the federal and provincial governments in Canada began to address soil degradation issues through public education, policy, and program instruments. Some authors have questioned the viability and effectiveness of policy directions in addressing soil degradation issues at the farm level (Duff et al. 1992; Cressman 1994). The early 1990s witnessed a basic shift in policy directions concerning soil degradation issues with the work of the Federal-Provincial Agriculture Committee on Environmental Sustainability (1990), and Canada's Green Plan (Government of Canada 1990).

This chapter summarizes and critiques some of these recent policy and program shifts and identifies current gaps in government responses to the soil degradation issue. Relevant policy and programs are reviewed relative to a brief list of suggested prerequisites for successfully addressing soil degradation and the sustainability of Canadian agricultural land and water resources. Map 1 provides the Canadian geographical context for the discussion that follows.

Map 1. Location map of Canada

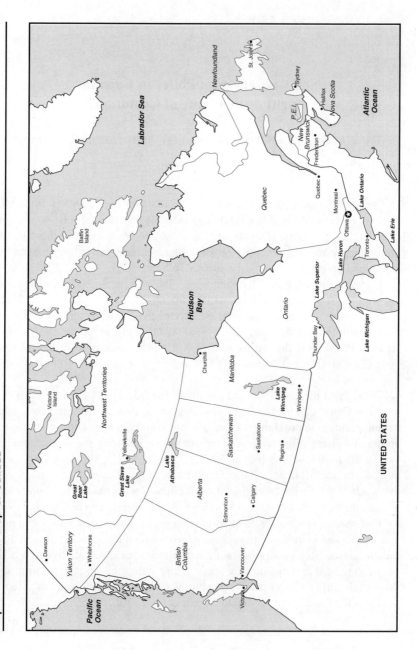

Figure 1. Government expenditures for agriculture conservation and development in Canada (1970–1995)*

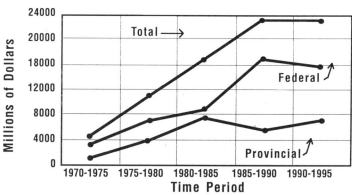

Time Period

*values are 5-year cumulative totals

Source: Environmental Perspectives—Studies and Statistics, Statistics Canada 1995

The jurisdictional context for remedial action programs

To appreciate the reasons for a wide diversity of approach in soil conservation policies and programs across the country, one needs to recognize the jurisdictional context within which resource management is carried out in Canada. Constitutionally, provincial governments have jurisdiction over natural resources within their boundaries. The federal government may only intervene in matters of a transboundary or binational nature, or through negotiated agreements with the provinces. Typically, most agreements with provinces involve the spending of federal tax dollars in the provinces. As natural resource programs are often equally funded by the provinces, negotiations about the nature of the programs and the methods of spending federal money are approached with vigor. Figure 1 illustrates the combined federal and provincial spending for agricultural conservation and development in Canada since 1970.

The federal government is also free to do research, and to collect and disseminate data within provinces. This authority is used to carry out some of the federal government's interests in conservation research and public education through direct federal programming. It is also the mandate under which most of the soil survey in Canada has been conducted and supports a national program to monitor soil quality (Acton et al. 1995).

The historical context for soil conservation programs in Canada

Nationally, soil and water conservation programs are very recent but they have historical roots which help explain their present character. A few of the more prominent programs that were initiated in the past are highlighted below.

The earliest institutional attempts at soil conservation are found in the Prairie Farm Rehabilitation Administration (PFRA) in the three prairie provinces (Manitoba, Saskatchewan, and Alberta). Established by the federal parliament in 1935 in response to the dust bowl of the 1930s, the PFRA was the first and only federal institution to deliver on-farm soil and water conservation services. The PFRA remains active, but its jurisdiction is restricted to prairie agriculture.

In response to severe flooding problems, and erosion of marginal lands that were unfit for continuous cultivation, the watershed-based Conservation Authorities were established under an act of the Ontario Legislature in 1942. Conservation Authorities were designed and implemented as municipal-provincial partnerships for dealing with a broad cross section of resource management problems. Though oriented more to water management issues, some conservation authorities have tried to address soil erosion and nonpoint source pollution problems in rural Ontario.

One example of soil conservation programming at the farm level is that of the Ontario Department of Agriculture and the Ontario Agriculture College (OAC). A farm planning service modeled after the work of the U.S. Soil Conservation Service was established in 1945. However, by the late 1950s, enthusiasm for the service waned. The ability of commercial fertilizers to substitute, cost effectively, for the plant nutrients lost because of soil erosion was heralded as the better way to farm.

Notwithstanding these and other smaller scale efforts across the country dealing with soil degradation problems, it took the environmental movement of the 1960s and 1970s to prod agriculture into taking seriously its deteriorating soil resource base. Specifically, it was concerned about the impact of cropland-derived sediment and agrichemicals in surface water systems that aroused public alarm in eastern Canada. The findings and the recommendations of the Pollution from Land Use Activities Reference Group about rural nonpoint sources of pollution in the Great Lakes (notwithstanding industrial point sources) forced agriculture to get on with remedial action programs (PLUARG 1978, International Joint Commission 1980).

Starting in 1983, the Province of Ontario developed subsidy programs to help farmers pay for measures to reduce farm-related water pollution. Eventually, staff was hired and trained to provide farmers with technical advice. Since then, several financial incentive programs have been carried out with enhanced networking among farmers, farm groups, government extensions, and agribusiness.

Following the conclusion of the PLUARG studies, the Canadian and U.S. governments agreed in 1986, under the terms of an annex to the 1975 Great Lakes Water Quality Agreement, to attack the problem of high phosphorus loads from rural nonpoint sources. In its response, Canada developed with Ontario a 5-year federal-provincial cost sharing program entitled "The Soil and Water Environmental Enhancement Program" (SWEEP). The objective of the SWEEP program was to achieve Canada's target reduction of 200 metric tonnes (220.4 tons) per year of phosphorus loading of Lake Erie by 1990 from nonpoint agricultural sources.

The biggest boost for soil conservation nationally came in 1984 when Herbert O. Sparrow, a Canadian senator and Saskatchewan farm operator, became so alarmed at the dust storms raging across the prairies that he roused his senate colleagues to hold a cross-country inquiry into the issue of soil degradation. Within 5 months, the Senate Standing Committee on Agriculture, Forestry and Fisheries completed its hearings and published its findings in "Soil at Risk: Canada's Eroding Future" (Senate Standing Committee on Agriculture, Fisheries and Forestry 1984).

Within 2 years of the release of "Soil at Risk," a conference, "In Search of Soil Conservation Strategies in Canada," was organized in April 1986 under the leadership of the Agricultural Institute of Canada. One tangible outcome of that conference was the call for an umbrella non-government organization, a National Soil and Water Conservation Council to act as a watchdog over governments and to be an advocate for soil conservation (Dumanski et al. 1986). In May 1987, Soil Conservation Canada (SCC), with Senator Sparrow as its first president, emerged as that umbrella organization.

Throughout this period, the PFRA continued to work within its mandate studying soil degradation issues and providing technical advice to farmers on soil conservation practices. Under several short-term economic regional development agreements (ERDAs) between the prairie provinces and the federal government, PFRA helped lay the groundwork for the national soil conservation efforts that were to follow.

In December 1987, the federal Minister of Agriculture announced Canada's first National Soil Conservation Program (NSCP). Designed as a 3-year, $150 million cost-share program with the provinces, the program resulted from a series of federal-provincial working group sessions. These were initiated as a result of agreement among the federal and provincial Ministers of Agriculture at their 1986 consultation that soil degradation was a significant issue and that it warranted further action.

For the program to come into effect, a series of lengthy negotiations were necessary. These focused first on generally worded accords, or statements of intent, and then on the specific agreements about the nature of the programs, the roles of the parties, and the administrative mechanisms for program execution. The federal government expected the provinces to follow its lead by injecting "new" money. However, several provinces resisted this expectation and insisted on including existing programs as part of their share of costs, which led to lengthy delays in reaching agreement with some of the provinces (Serecon and Ecologistics Limited 1993).

Characterizing the scope of the activities that were eventually agreed upon is complicated by the unique situations that had to be addressed in each province or region. However, the following six categories embrace all of the program activities. They are listed below, together with the relative proportion of the $150 million they represented:
- Technology development, 5.8 percent
- Education and awareness, 10.0 percent
- Conversion to alternate cover, 26.5 percent
- Research, monitoring, and survey, 10.7 percent
- On-farm soil conservation, 41.4 percent
- Other, 5.6 percent

The program objectives approved by the federal cabinet for the NSCP were "to promote the best use of the soil resource base, within practical economic limits and according to its capabilities, in order to sustain its productivity and thereby support the growth, stability, and competitiveness of the agri-food sec-

Table 1. Reduction in actual water erosion risk per hectare, 1981 to 1991

Province	Cultivated land in 1991 (million ha)	——Erosion reduction per hectare—— (percent)		
		Resulting from cropping practice	Resulting from tillage practice	Total
British Columbia	0.61	7	10	17
Alberta	11.06	5	8	13
Saskatchewan	19.17	5	3	8
Manitoba	5.06	6	9	15
Ontario	3.48	10	11	21
Quebec	1.65	3	3	6
New Brunswick	0.12	2	4	6
Prince Edward Island	0.16	-9	3	-6
Nova Scotia	0.11	-3	3	0
Canada	41.42	5	6	11

Source: Wall et al. 1995

tor, and to promote, where applicable, economic diversification in western Canada." Two aspects are noteworthy about the wording of these objectives. First, they suggest that soil conservation systems are constrained by on-farm economics. Lacking is an explicit rationale for public support of soil conservation efforts because of off-site environmental benefits to society. Second, the objectives imply that funding to farmers is an alternate mechanism for propping up farmers financially. As a consequence, universal financial support programs are likely to be favored over support to problems in target areas.

Current status of soil resources

Given the resources that have been devoted to soil and water conservation over the last 20 years, what do we really know about the health of Canadian soils? The answer is surprisingly little. In Canada, there has been no systematic national evaluation and inventory of the state of soil resources. As a result, there is no defined starting point from which to assess resource quality as there is with the National Resources Inventory in the U.S. (Kellogg et al. 1994).

One response to this situation has been the development and implementation of the Soil Quality Evaluation Program (SQEP) in Canada. Recommended in the 1986 National Agriculture Strategy, and implemented largely through Canada's Green Plan, the goal of this program is "to develop national capability to assess soil quality and associated environmental quality, as well as the effects of land use and management practices on these qualities" (Acton and Gregorich 1995, vol. IX).

A foundation of the SQEP has been the development of 23 benchmark sites across the country. Collection of baseline data from these sites was largely completed in 1995. It is expected that soil quality monitoring will continue for at least 10 years, and possibly longer until trends emerge. Assessment of soil health at regional and natural scales is based on a risk assessment approach, which relies on soil-landscape and census information to determine areas at risk from declining soil health.

In its first report, the SQEP provided an overview of trends in Canadian soil quality derived largely from existing data. While regional summaries are presented, the report finds that nationally, "soil health will continue to decline in areas of intensive cropping and marginal land where conservation farming methods are not used. Soil health is holding steady or improving in regions where conservation practices have been tailored to local problems of soil degradation" (Acton and Gregorich 1995, vol. III).

Table 1 illustrates estimated progress toward reduction of water erosion risk in Canada between 1981 and 1991.

Figure 2. Recent policy context for soil conservation in Canada

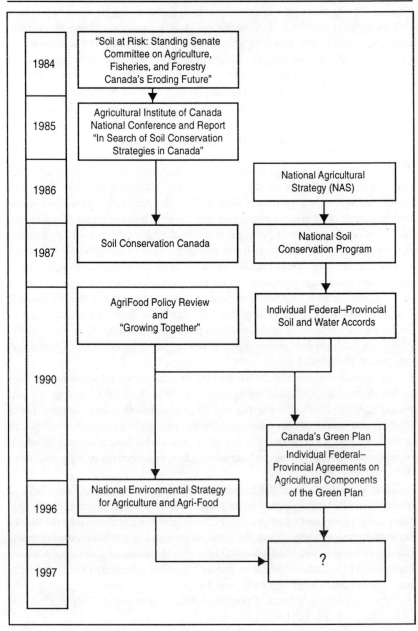

Recent policy

The last 6 years have seen shifts in the development of environmentally based agriculture policy and program planning in Canada. While agreements between the federal and provincial governments for cost-shared programming are still the order of the day, a key shift since the NSCP has been the encapsulating of agro-environmental initiatives within larger environmentally policy-based agendas. This has resulted in an apparent dilution of the agricultural agenda in the public eye. Figure 2 depicts the choronology and relationship of some of these major initiatives, which are discussed in the following section.

Growing together

In 1990, a Federal-Provincial Agriculture Committee on Environmental Sustainability was established under the Canada Agri-food Policy Review. The Committee developed a policy framework to the Federal and Provincial Ministers of Agriculture entitled "Growing Together" (Federal-Provincial Agriculture Committee on Environmental Sustainability 1990). Adopted by the ministers later that year, the framework became the Federal Government's first-ever policy on environmental sustainability and agriculture.

The overriding goal of the policy was to "maintain or enhance the natural resources that the agri-food sector uses or affects, while ensuring environmental economic and social integration" (Federal-Provincial Agriculture Committee on Environmental Sustainability 1990:1). The following objectives were adopted to further this goal:

- to conserve and enhance the natural resources that agriculture uses and shares (e.g., soils, wetlands, and wildlife);
- to be compatible with other environmental resources that are affected by agriculture (e.g., major lakes and rivers); and,
- to be proactive in protecting the agri-food sector from the environmental impacts caused by other sectors and factors external to agriculture.

Agricultural components of Canada's Green Plan

Canada's Green Plan—The Green Plan is the national response to the Brundtland Commission's challenge for global environmentally sustainable development plans and was developed through an extensive series of national consultations involving multiple stakeholders (Government of Canada 1990). Launched in December 1990, the Green Plan was a strategic policy initiative to guide the sustainable development of the Canadian economy. It also became the broad policy and programming envelope for Canada's next round of soil conservation and sustainable agriculture programming.

Building on the 1990 Report of the Federal-Provincial Agriculture Committee on Environmental Sustainability, the Green Plan became the implementing mechanism for this agricultural policy. Specific initiatives were identified under an agricultural component of the Green Plan including promoting soil conservation, promoting a clean water supply, integrating agriculture and wildlife, managing waste and pollution, and protecting genetic resources (e.g., native grasses). Soil and water conservation was now competing with other agro-environmental initiatives.

The Green Plan enabled the Minister of Agriculture, on behalf of the Government of Canada, to enter into cost-shared agreements with provincial governments. Such arrangements were negotiated for the agricultural regions of Canada. The specific issues to be addressed, and the relative importance given to each emerged from consultations with selected stakeholders in each province. The remainder of this chapter will focus largely on the agreements for Alberta, Saskatchewan, Manitoba, and Ontario.

Canada-Alberta Environmentally Sustainable Agriculture Agreement (CAESA)—The 5-year Canada-Alberta Environmentally Sustainable Agriculture (CAESA) Agreement was established in 1992. The original funding for the agreement was $44.12 million but was later scaled back to approximately $36 million. The CAESA program is being delivered through local and provincial producer organizations, agricultural service boards, Indian agricultural organizations, research institutes, and government agencies.

The following components are addressed under CAESA:
• Farm-based Program (65 percent of total budget) for adoption of sustainable agricultural practices, shelterbelts, and extension services.
• Research Program (13 percent of total budget) to provide technical support and applied research. (Federal researchers are the primary recipients of this program.)
• Resource Monitoring Program (12 percent of total budget) to monitor impact of agriculture on soil, surface water quality, and groundwater quality and supply.
• Processing-Based Program (8 percent of total budget) aimed at agribusiness.
• Public Awareness Program (3 percent of total budget)

Canada-Saskatchewan Agriculture Green Plan Agreement (CSAGPA)—The Canada-Saskatchewan Agriculture Green Plan Agreement (CSAGPA) was established in 1993. The original funding for the agreement was $46.6 million to be cost-shared equally between Saskatchewan and Canada. The value of the Agreement was subsequently reduced to approximately $41.7 million. The relative distribution of funding among the various programs under the Agreement was, however, maintained. The CSAGPA program is being delivered through local and provincial producer organizations, Agriculture Development and Diversification (ADD) Boards, research institutes, government agencies, and other agricultural and environmental groups.

The following program elements are being addressed under CSAGPA:

- Farm-based Program (45 percent of total budget) aimed to change the way agricultural-related resources are managed, and to promote the adoption of environmentally sustainable production practices.
- Research and Development Program (28 percent of total budget) supports practically oriented research projects in the areas of soils, crop pest management, water quality, and wildlife.
- Other Resource Management Programs (20 percent of total budget) include subprograms of Innovative Grazing and Pasture Technologies, Sustainable Water Management, Wildlife Enhancement and Shelterbelts.
- Innovative Partnerships Program (4 percent of total budget) provides flexibility to find innovative projects emerging priorities and technologies not addressed elsewhere in CSAGPA.
- Communications Program (2 percent of total budget).

Canada-Manitoba Agreement on Agricultural Sustainability (CMAAS)—The Canada-Manitoba Agreement on Agricultural Sustainability was established in 1993. The original funding for the Agreement, $20.8 million, was to be cost-shared equally between Manitoba and Canada. However, the value of the Agreement was adjusted to reflect Federal budget cuts. Total CMAAS funding now stands at $18.2 million (50 percent Federal, 50 percent Provincial contribution). The CMAAS program is being delivered through local and provincial producer organizations, government agencies, and the research community. Many of the producer organizations were created at local levels to help deliver previous soil conservation programs, and are thus providing some continuity in delivery.

The following seven program elements comprise CMAAS:

- Soil Resource Management (27 percent of total budget) includes subcomponents for Sustainable Cropping Systems, Permanent Practices, Equipment Support, and Soil Resource Monitoring.
- Water Resource Management (24 percent of total budget) includes subcomponents for Water Sustainability (efficiency), Water Quality Protection and Enhancement, and Special Projects.
- Integrated Resource Management (23 percent of total budget) includes conservation farm plans and other projects.
- Forage/Cover Crop Utilization and Livestock Management Program (10 percent of total budget).
- Integrated Pest Management (8 percent of total budget).

- Innovative Partnership Initiative (4 percent of total budget) provides flexibility to fund innovative projects on emerging priorities and technologies not addressed elsewhere in CMAAS.
- Conservation and Urban Awareness (2 percent of total budget).

Canada-Ontario Agreement on the Agricultural Component of The Green Plan (COAGP)—The Canada Ontario Agreement on the Agriculture Component of The Green Plan was established in September 1992. The Agreement was originally cofunded 50 percent by federal and 50 percent by provincial to approximately $64 million. Following budgetary adjustments, the total COAGP funding was reduced to approximately $49 million. Programs under the Ontario Agreement are being delivered largely through partnerships between federal and provincial agricultural agencies, producer groups and universities and associated institutions.

Programs implemented under COAGP include the following:

- Best Management Practices;
- Research;
- Technology Transfer;
- Rural Conservation Clubs;
- Stewardship Information Bureau;
- Environmental Farm Plans;
- Best Management Practices; and
- Wetlands/Woodlands/Wildlife.

National Environment Strategy for Agriculture and Agri-Food

Released in 1995, the National Strategy for Agriculture and Agri-Food resulted from another series of national consultations and workshops (Agriculture and Agri-Food Canada 1995). Stakeholders from approximately 100 groups representing government, producers, industry, academia, and special interests provided input.

The strategy recognized that, within individual provinces, there are soil and water conservation strategies and efforts being designed and implemented by governments, producer organizations, agribusinesses, and individual producers. The proliferation of these "independent" initiatives is addressed in the strategy. Building upon the past, and updating the 1990 strategy, the current strategy is "intended to frame these independent strategies within a national context and identify those challenges which cannot be dealt with effectively by a single province or industry group" (Agriculture and Agri-Food Canada 1995:1).

The strategy does not, nor was it intended to, address specific responses to agroenvironmental challenges. It does, however, call for the development of multilevel/jurisdictional action plans based on the following principles:

- economic and social viability;

- co-operation and partnerships;
- regional flexibility;
- local action;
- ecosystem-based approach;
- managed environments;
- stewardship responsibility;
- inclusive decision-making;
- proactive evaluation, measurement and reporting of performance; and
- long-term perspective for goals and objectives.

The Strategy was prepared for, and accepted by, the Federal and Provincial Ministers of Agriculture. It is widely circulated publicly and serves as a backdrop for deliberations on Green Plan programming (1997 and beyond).

Policy analysis

Almost 3 years ago the principal author of this chapter challenged Canadian agricultural policy-makers with provocative suggestions and actions felt to be necessary for significantly enhancing agricultural soil productivity and upgrading waterbodies impacted by agricultural nonpoint sources of pollution (Cressman 1994). These suggestions included

- committing to long-term federal and provincial programming arrangements, administered by a core of well-trained soil conservation professionals at the national level;
- targeting scarce financial resources to high priority problem areas for controlling off-site impacts. Specifically, policies and resultant programs should be applied on a watershed basis by working directly with those who farm significantly contributing areas;
- designing programs to be result-oriented and serve as the basis for regular program evaluations;
- ensuring that extension staff are teamed in a multidisciplinary atmosphere;
- explicitly seek and include the private sector (i.e., agribusiness) in policy formulation and implementing publicly funded research and development of sustainable farming practices and products; and
- embarking on a more open approach to sustainable agriculture policy and program formulation.

To this list, we would also add the need to design and undertake basic agricultural resource monitoring.

These points are equally valid today as prerequisites to effective soil and water conservation policy and programs. Consequently, they are used to assess Canada's policy and program initiatives as seen in the preceding section. Mid-term evaluations have been completed for most of the agricultural Green Plan Agreements

(Deloitte & Touche and Apogee Research 1994; The Advisory Group Inc. and Ecologistics Limited 1995, 1996; FT Ecologistics Limited and The Advisory Group Inc. 1996). While not complete, final assessments of the individual arrangements, these reviews provide a basis for assessing the effectiveness of general policy directions and implementation strategies. Insights gained from reviews of individual agreements mid-term, are next examined relative to the challenges above.

Long-term commitment

Money speaks louder than words. Funding for programs under the agricultural component of the Green Plan remains short term. While federal/provincial cooperative agreements were intended to run almost continuously over the last 15 to 25 years, the short-term and sometimes parochial nature of specific initiatives does little to foster identity, confidence, commitment, and vision within the agriculture and agribusiness community. The continuing presence and participation of PFRA in western Canada, has at least provided continuity of administrative structure, and has been instrumental in facilitating resource protection efforts. Notwithstanding some provincial initiatives, comparable presence of the federal government does not exist elsewhere in Canada.

Similarly, as experienced professionals accept early retirement or are shifted to nontraditional duties to deal with budget and deficit realities, the institutional knowledge base of the government is being weakened and depleted.

Federal-provincial agreements have, and are likely to remain, the significant conduit of program dollars to regional and local levels. However, in our political system long-term commitments are difficult to fulfill even when agreements are signed. Budgets remain vulnerable to cuts in the face of competing interests and priorities. Public opinion polls at the time of the Green Plan development, listed "environment" as the highest public concern. Environment is now surpassed by concern about "jobs." Consequently, we find ourselves sacrificing the well-being of our resource base for the sake of short-term political expediency. Having rolled policy about sustaining soil and water resources exclusively into a broad environmental envelope, we see diminished resources for this sector as relative support for the environment wavers; this despite a growing need to sustain our food production capacity.

Commitments to basic research that integrates and addresses concerns about the interrelationships between soil and water quality, agricultural productivity, economics, and the health of off-site aquatic and terrestrial species, need to be longer. By definition, research into finding solutions to soil and water degradation problems requires a longer time horizon than provided by most government funded efforts. In many cases, it can take up to a 2-year lead time to get a project up and running, especially where on-farm field work is required. Under the current funding mentality, it is difficult for researchers to plan for projects that can potentially yield accurate and meaningful results over the long term.

Targeting

Conservation policy researchers have, for decades, called for the targeting of policy initiatives for soil and water conservation in agriculture to those portions of the landscape that are most vulnerable to degradation (Culver and Seecharan 1986; Lovejoy et al. 1986; Duff et al. 1992; Cressman 1995; Stonehouse 1994).

There are very few examples of targeted approaches to soil and water conservation programming in Canada. One example is the Permanent Cover Program in the prairie provinces (Vaisey and Wettlaufer 1996). Canadian policy traditions are rooted in universality. This approach has relevance for social programs, but is not appropriate for soil and water resource-related agricultural programming.

The Green Plan did prioritize funding by resource issue, but no attempt was made to match program dollars to areas of the land base, and watersheds that are most degraded, fragile or susceptible. Policy-makers continue to miss the opportunity to maximize results by targeting dollars to geographic areas where problems are most acute.

Program evaluations

It is impossible to exploit and build on the successes of good programming and policy initiatives and avoid pitfalls without systematic and results-oriented evaluations. The federal government requires that program evaluations be completed for significant program areas. The agricultural components of the Green Plan are being evaluated under these provisions. However, to date, the evaluations have not been result-oriented in terms of actual impacts on the resource base. A major factor contributing to this deficiency is the fact that program goals and objectives are not generally stated in measurable terms, and, therefore, evaluation frameworks and questions tend to be vague and obscure.

There is a general lack of easily obtainable, reliable, socio-economic, and biophysical data at a scale to assess broad change. Therefore, evaluation results tend to focus more on process and administration than changes on the land and farmer behavior. As a result, conclusions from program evaluations are qualitative and opinion-based, and can only speculate on real program impact.

Multidisciplinary extension staff

A multidisciplinary approach is essential in addressing agricultural-related soil and water degradation issues and solutions. The building of strong teams of professionals is required within the farming community. The collection and sharing of experience and outlook in a multidisciplinary atmosphere, helps to ensure that all facets of problems are considered and addressed.

Multiple stakeholder consultations and the ensuing partnerships for Green Plan agreement implementation is providing a positive and multidisciplinary atmosphere. In western Canada, some provinces contract with trained and experienced extension staff to support activities. However, overall, there tends to be few established linkages to other areas, e.g., watersheds.

Private sector involvement

Agricultural environmental policy in Canada has tended to ignore the integration within the agriculture economic sector. Very little has been done to engage the agribusiness sector in exploring cooperative strategies for soil conservation products and systems on the farm. Policy-makers continue to miss a significant opportunity. For example, site-specific farming is driven by private sector interests with evidence of input cost savings and environmental benefits achieved under cash crop farming systems. Farmers have generally seen conservation till-plant systems as a soil enhancing moisture retention systems that sustain increased yields and increased net returns, mainly through reduced inputs. Other services traditionally government based are now undertaken by agribusiness which include crop scouting and soil testing.

There is a fundamental shift occurring in the way Canadian farmers are seeking, accessing, and utilizing technical information on their farms. Agribusiness tends to be the most important and frequently utilized source of the technical information for farmers across the country. This trend will only increase as government is forced to reduce programs and services.

The Green Plan Agreements have, for the first time, sought and received significant input from producers and producer organizations in both the design and implementation of programs. There is almost a unanimous agreement among administrators that this feature has contributed to high rates of participation for most on-farm programming components.

The Environmental Farm Plan (EFP) process in Ontario provides a proactive example where producers have been significantly involved in program delivery. Spearheaded by a coalition of farm organizations and funded under the Green Plan, the process promotes development of a farm-specific environmental "plan" following a comprehensive inventory of on-farm environmental risk. It is noteworthy that soil and water conservation/degradation issues are only one of a large set of environmental risks.

Between 9,000 and 10,000 farmers in Ontario completed plans by program termination (March 1997), which is a similar number of farmers who took advantage of a much richer provincial grants program (up to $30,000/producer) during the late 1980s/early 1990s (Rudy 1996). However, until systematic and detailed evaluations of these initiatives are undertaken, there is little evidence to support some of the claims made regarding farm uptake and the actual effectiveness of the program in reducing environmental impact from agricultural practices and operations.

Producer-driven associations with environmental protection agendas appear to be gaining momentum across Canada, even as overall government funding of environmentally sustainable agriculture programs may be on the decline. The groups are contributing often significantly and informally to soil and water conservation initiatives. Examples of these groups include Alberta Conservation Tillage Society (ACTS), Keystone Producers Association in Manitoba, Innovative Farms Association of Ontario (IFAO), and the Atlantic Farmers Council.

An open approach to policy formulation and implementation

Some progress has been made in many areas of the country with respect to achieving an open approach to policy formulation and implementation. The implementing agreements of agricultural components of the Green Plan were subject to extensive consultations involving multiple stakeholders. However, this still remains a selective process in that most policy consultation exercise are generally by invitation. In cases such as Ontario, stakeholder consultations excluded many producers and resource professionals who had given leadership to earlier soil and water conservation initiatives. This was not the case in western Canada.

The Green Plan agreements also appear to have built in a degree of flexibility in implementation. For example, the inclusion of "innovative partnerships" as program components in many of the agreements provide for non-traditional initiatives to be funded and emphasize the development of non-traditional partnerships.

There still remain, however, some restrictions in the way that government dollars are spent and allocated which may be limiting conservation efforts, especially where impacts cross provincial boundaries.

Monitoring

Systematic monitoring of the state of the agricultural resource base, including affected off-farm resources is essential to accurately gauge improving conditions related to the use of soil and water conservation practices.

As noted earlier, the Soil Quality Evaluation Program (SQEP) is the designated national initiative for monitoring and assessing soil health over time. The SQEP is founded on 23 benchmark sites designed to be representative of the main landforms in Canada's most important agricultural regions. While this effort is yielding, and should continue to yield, valuable scientific insights related to the impact of conservation practices on soil properties and general health, it will ultimately provide little direct utility to resource managers. Models like the Natural Resources Inventory (NRI) in the U.S. are an example of methods that provide resource managers accurate information at 5-year intervals.

National strategy and framework

The National Environment Strategy for Agriculture and Agri-Food is a significant initiative. Key approaches addressed in the strategy include:
- recognizing differences in scale, jurisdiction, and requirements of different approaches for each;
- understanding that local solutions are best achieved locally;
- guiding principle for specific action plans; and
- encouraging policy direction that builds upon the past.

Summary

The Canadian agriculture and agri-food sector has realized progress in its approach to environmental sustainability and stewardship since 1990 when the Federal-Provincial Agriculture Committee on Environmental Sustainability issued the "Growing Together" report. Most noteworthy are the following areas where progress is clearly evident:
- a more inclusive approach to policy and program formulation with participation from a broad cross-section of interests both within the agri-food sector and across sectors;
- increased awareness within the agri-food sector of issues relating to resource sustainability and of the relation between agricultural activities and the quality of the environment; and
- results from applied research that has helped adapt conservation systems for use under a variety of local conditions.

While policy directions appear to be farsighted in focus and provide continuity with previous documents, program design and implementation are still based on short-term (e.g., 4 to 5 years) commitments. Unless there is a strong political will to engage in real progress over the long term, progress in arresting and reversing agricultural-related soil and water degradation problems will be marginal.

Considerations for the future

By the end of 1997 Canada and its provinces spent in excess of $147 million on environmentally based research, demonstrations, and projects under the umbrella of the agriculture component of the Green Plan.

Discussions are currently under way concerning the shape of future programming in the area of agricultural resource sustainability. As is evident from the National Strategy for Agriculture and Agri-Food (1995),soil and water conservation is very much in competition with a host of other agroenvironmental issues for public attention and action. It appears the policy makers have yet to grasp the fundamental significance of soil and water resources in sustaining the very agroecosystems they wish to exploit in enhancing future outputs from the sector.

Nevertheless, the National Environment Strategy for Agriculture and Agri-Food represents a significant step forward in public policy. The framework and the principles for local action plans embody attributes that have been long suggested for making national and provincial policy more relevant to local conditions and needs. It also suggests that the role of governments will be changing, i.e., less direct government programming and more support of local initiatives.

What remains to be emphasized in the implementation of such policy and in the interests of cost-effective long-term protection and enhancement of agricultural soil and water resources is the following:

- more openness to partnerships with agribusinesses, recognizing the strong existing links between agribusiness, information, land use, and adoption by producers of particular farming practices;
- more attention to economic barriers (perceived and real) that slow the rate of adoption of soil conserving practices; more research on the economics of conservation farming systems;
- program goals and objectives that are stated in measurable terms and mechanisms for measuring, collecting, and analyzing the data that are built into program implementation plans;
- greater emphasis on the collection and maintenance of baseline data on soil conditions, land use, and other agricultural resources;
- development of a system for collecting data on a few indicators from many farms where the field information can be easily gathered by the farmer, the personal benefits to the farmer are self evident, and the farmer receives an annual status report while governments acquire a substantial data base on changing soil conditions in Canadian agriculture; and
- acceptance of the principle of local participation and decision-making by all government personnel whose principle mode of operation has been "control."

Ultimately the success of soil and water conservation efforts will be determined by the will of society to confront the issues and to "invest" in their resolution. The will and determination for governments to play their role most effectively will be governed by the degree of leadership exercised at the political and senior bureaucratic levels. But not since the rallying cries of "Senator Herb Sparrow" have Canadians experienced such leadership nationally. Some exceptions can be noted in several provinces, but without national leadership the future of our agricultural soil and water resource base will not be secure.

Participation in the global economy, and the securing of stable trade relationships within the international community currently creates employment in the Canadian agri-food sector, but we will clearly be "adrift" if our commitment to sustainable agriculture—research, practice, and technology development—wavers in the long-term. Food security and remaining "competitive" are fundamentally dependent on the health, resilience, and productivity of the land resource base. As Senator Sparrow has so effectively advocated in the past, we must all, producers and consumers alike, be active stakeholders in promoting sustainable resource use if we hope to realize a secure future in a quality environment.

References

Acton, D.F., and L J. Gregorich (eds.). 1995. The health of our soils: Toward sustainable agriculture in Canada. Agriculture and Agri-Food Canada. Publication 1906/E. Minister of Supply and Services Canada. Ottawa, Ontario.

Agriculture and Agri-Food Canada. 1995. National environment strategy for agriculture and Agri-Food. Prepared for the Federal and Provincial Ministers of Agriculture. August. Ottawa, Ontario.

Anderson, M., and L. Knapik. 1984. Agricultural land degradation in western Canada: A physical and economic overview. Regional Development Branch, Agriculture Canada, Ottawa, Ontario. p. 138.

Coote, D.R., J. Dumanski, and J.F. Ramsay. 1981. An assessment of the degradation of agricultural lands in Canada. Land Resource Inst. Contribution No. 118. Res. Branch, Agriculture Canada, Ottawa, Ontario. p. 86.

Cressman, D.R. 1994. Remedial action programs for soil and water degradation problems in Canada. In Ted L. Napier, Silvana M. Camboni, and Samir A. El-Swaify (eds.), Adopting Conservation on the Farm: An International Perspective on the Socio-economic of Soil and Water Conservation. Soil and Water Conservation Society, Ankeny, Iowa.

Culver, D., and R. Seecharan. 1986. Factors that influence the adoption of soil conservation strategies. Canadian Farm Economics 20(2):9-13.

Deliotte & Touche, and Apogee Research. 1994. Phase I evaluation of Canada-Ontario agreement on the agricultural component of the Green Plan. Report prepared for Agriculture and Agri-Food Canada. June.

Duff, Scott N., D. Peter Stonehouse, Don J. Blackburn, and Stewart G. Hilts. 1992. A framework for targeting soil conservation policy. Journal of Rural Studies 8:399-410.

Dumanski, J., L.J. Gregorich, V. Kirkwood, M.A. Cann, J.L.B. Culley, and D.R. Coote. 1994. The status of land management practices on agricultural land in Canada. Centre for Land and Biological Resources Research, Agriculture and Agri-Food Canada, Ministry of Supply and Services Canada, Ottawa, Ontario. Technical Bulletin 1994-3E.

Dumanski, J., D.R. Coote, G. Lucink, and C. Lok. 1986. Soil conservation in Canada. Journal of Soil and Water Conservation 41:204-210.

Federal-Provincial Agriculture Committee on Environmental Sustainability. 1990. Growing together: Report to ministers of agriculture. June. Ottawa, Ontario.

FT·Ecologistics Limited, and The Advisory Group Inc. 1996. Interim evaluation of the Canada-Manitoba agreement on agricultural sustainability (CMAAS). Report prepared for the Agreement Management Committee, CMAAS. May.

Government of Canada. 1990. Canada's Green Plan. Minister of Supply and Services Canada, Ottawa, Ontario.

International Joint Commission. 1980. Pollution in the Great Lakes basin from land use activities. IJC Report to the Governments of the United States and Canada. Windsor, Ontario.

Kellogg, Robert L., Gale W. TeSelle, and J. Jeffery Goebel. 1994. Highlights from the 1992 National Resources Inventory. Journal of Soil and Water Conservation. 49:521-527.

Lovejoy, S.B., J.G. Lee, and D.B. Beasley. 1986. Integration of social and physical analysis: The potential for micro-targeting. Conserving Soil: Insights from Socioeconomic Research. S.B. Lovejoy and T.L. Napier (eds.). Soil and Water Conservation Society, Ankeny, Iowa.

PLUARG. 1978. Environmental management strategy for the Great Lakes System. Final Report to the International Joint Commission, Windsor, Ontario.

Rennie, D.A. 1985. Soil and water issues and options in Canada. Paper presented to the Canadian Agricultural Outlook Conference, Ottawa, Ontario.

Rudy, Harold. 1996. Environmental farm plan coordinator, Ontario Soil and Crop Improvement Association, Guelph, Ontario. Personal communication September 5, 1996.

Senate Standing Committee on Agriculture, Fisheries, and Forestry. 1984. Soil at risk: Canada's eroding future. Ottawa, Ontario. p.120.

Serecon Management Consultants Inc., and Ecologistics Limited. 1993. Evaluation of the National Soil Conservation Program. Report prepared for Program Evaluation Division, Agriculture Canada. March.

Statistics Canada. 1995. Environmental perspectives 2: Studies and statistics. Catalogue 11-528E, No. 2 - Occasional. Minister of Industry, Science and Technology, Ottawa. June.

Stonehouse, D. Peter. 1994. Canadian experiences with the adoption and use of soil conservation practices. Adopting Conservation on the Farm: An International Perspective on the Socio-economic of Soil and Water Conservation, Ted L. Napier, Silvana M. Camboni, and Samir A. El-Swaify (eds.). Soil and Water Conservation Society, Ankeny, Iowa. pp. 369-395.

The Advisory Group Inc., and Ecologistics Limited. 1996. Interim evaluation of the Canada-Saskatchewan Agriculture Green Plan (CSAGPA) Agreement: Findings report. Report prepared for Agreement, Management, Implementation, and Evaluation Committees of CSAGPA. May.

The Advisory Group Inc., and Ecologistics Limited. 1995. Interim evaluation of the Canada-Alberta Environmentally Sustainability Agriculture (CAESA) Agreement: Final report. Report prepared for Evaluation Committee, CAESA. June.

Vaisey, J., and R.J. Wettlaufer. 1996. The permanent cover program: Is twice enough? Paper presented at conference on Soil and Water Conservation Policies: Success and Failures. Prague, Czech Republic. September 17-20, 1996.

Wall, G.J., E.A. Pringle, G.A. Padbury, H.W. Rees, J. Tajek, L.J.P. van Vliot, C.T. Stashnoff, R.G. Eilens, and J.M. Cossette. 1995. Erosion. In The health of our soils—Toward sustainable agriculture in Canada; D.F. Acton and L.J. Gregorich (eds.). Centre for Land and Biological Resources Research, Research Branch, Agriculture and Agri-Food Canada, Ottawa, Ontario. pp. 61-76.

15

A Critical Assessment of the Ontario Land Stewardship Program

D. Peter Stonehouse

Compared with many other countries in the Organization for Economic Co-operation and Development (OECD), and particularly with the United States, Canada has suffered serious agriculture-related resource degradation problems only in recent decades (Stonehouse 1994). This is true especially of eastern Canada, where agriculture was characterized by mixed livestock-cropping systems and long crop rotations with a high proportion of forage and cereal grain crops until the early 1960s. Since then, pervasive technological progress and intensification of production techniques have led to increasing specialization, separation of livestock from cropping enterprises, shorter crop rotations, and greater dependence upon off-farm inputs such as synthetic pesticides, fertilizers, feed additives, and animal hormones (Kay and Stonehouse; Miller et al.). In turn, these trends have resulted in severe natural resource degradation problems, epitomized by erosion and compaction of soils and pollution of waterbodies by sediment, plant nutrients, pesticide residues, and bacteria from livestock manure (Dumanski et al.; Miller).

Degradation problems in Ontario

Among the provinces of eastern Canada (those lying east of the prairies), Ontario figures most prominently in the agricultural industry. Accounting for some 65 percent of the total farm receipts in eastern Canada (nearly 24 percent of total Canadian farm receipts), Ontario is the eastern region's most important producer of beef cattle, poultry, and eggs. It also is the second most important province in the production of hogs and dairy products, after Quebec (Statistics Canada). Ontario also accounts for almost all of Canada's (as well as the eastern region's) output of soybeans, tobacco, and tree fruits other than apples, nearly two-thirds of its corn, and more than 40 percent of Canada's output of vegetables, apples, and floriculture and nursery industry products (Statistics

Canada). At the same time, Ontario produces only some 5 percent of Canada's hay and less than 5 percent of its cereal grains.

Ontario therefore features prominently in the production of intensive row crops, relative both to other provinces in Canada and to other crops such as hay and cereal grains. Such intensity of crop production has emerged only since the early 1960s, (Figure 1), concomitantly with the intensification trend throughout eastern Canada. Recent trends in Ontario indicate continued increases in land allocated to bean crops, mostly at the expense of cereal grain and hay crops, but also of corn, especially for fodder purposes (Figure 1). Acreages of corn appeared to have peaked in Ontario in the early 1980s. The overall trend in farmland use has nevertheless been unmistakable. Intensive row crops have been supplanting hay and cereal grain crops.

One positive consequence of this intensification process has been the increasing volume and value of agricultural output in Ontario (Statistics Canada). This has been achieved despite a steadily declining area of land in agriculture, from some 18.5 million acres in 1961 to about 13.5 million acres in 1991 (Statis-

Figure 1. Trends in Ontario cropland use, 1960–1994

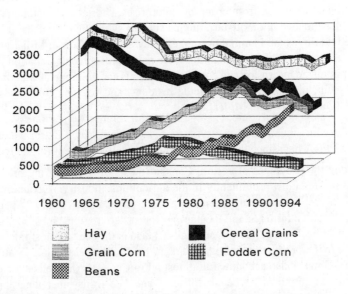

Cereal grains - includes winter wheat, oats, barley, mixed grains, spring wheat and rye
Beans - includes soybeans and dry white beans

tics Canada, Census). A negative consequence has been the degradation of the natural resource base and environment. Separation of livestock from crop enterprises has left high density livestock farms with a manure disposal and plant nutrient surfeit problem. In contrast, specialized cash-cropping farms have suffered from soil organic matter depletion, soil erosion and compaction, and increasing needs to import synthetic fertilizers and pesticides. Both types of specialized farms have contributed to on-farm and downstream watercourse pollution from sediment, manure, and/or fertilizer nutrients, bacteria and pesticide residues (Kay and Stonehouse; Miller et al.).

From the farmer's viewpoint, both the positive and negative consequences have been rational and justified. On-farm economics have very much favored specialized and intensive production systems over those that are more conservation-oriented (Zantinge et al.; Baffoe et al.; Stonehouse et al. 1987; Stonehouse et al. 1988). Very few conservation practices have been found to be profitable, in and of themselves (Turvey; Stonehouse 1995), and of those that have transpired to be profitable, they are not necessarily so under all circumstances (Stonehouse 1991). Nor do farmers need to be aware or to take account of the damages to the resource base and environment off the site (i.e., externalities) because of market failure or imperfections (Stonehouse and Bohl; Stonehouse 1996). At the same time, government farm policies throughout Canada have accentuated and encouraged higher production and productivity, often at the expense of resource and environmental conservation (Fox et al.; Stonehouse 1994).

Responses to degradation problems

The Canadian prairie region has experienced public intervention through the Prairie Farm Rehabilitation Act (PFRA) since the 1930s (Anderson and Knapik). Designed as a combined regulatory and voluntary compliance vehicle for conserving and stabilizing prairie farmlands eroded by wind or water action, the PFRA has, since the 1980s, become the delivery agent for the federal government's part of joint federal-provincial agreements on resource conservation in the three prairie provinces (Stonehouse and Bohl).

In Canada, resource conservation and environmental protection are provincial jurisdictions. However, the federal government has often taken the lead in encouraging greater conservation efforts. The government has done this through offers of financial assistance to participating provinces as the focal point of federal-provincial agreements.

Due to the rising incidence of and concerns about resource degradation in eastern Canada, and in British Columbia, public intervention became more prevalent in these parts of the country in the mid-1980s (Stonehouse and Bohl). While most public programs have comprised federal-provincial agreements, typically using a voluntary compliance approach, some have been independent provincial initiatives. One such independent thrust was initiated by the province of Ontario in 1987. Known as the Land Stewardship Program (LSP), this program is the focus of the present study.

Ontario's Land Stewardship Program (LSP)

The LSP was selected for scrutiny in this study because of the importance of Ontario's agriculture to the national agri-food industry, because more than half the class one farmland in Canada lies in southern Ontario's agricultural belt, because resource degradation problems had been particularly marked in Ontario by the 1980s, and because the LSP represented the most conspicuous, comprehensive, and concerted effort to date to use public intervention to address degradation problems in Ontario.

History of LSP

Farmer awareness and concerns about degradation during the 1970s and early 1980s engendered experimentation with various conservation practices by the most innovative, but this revealed how unprofitable most practices were to most farm operators. Subsequent research has confirmed that most conservation technologies are not profitable in Canada (Stonehouse 1995), and that selected practices such as conservation tillage can be profitable, but only under some (soil type, climatic, crop sequence, and management skill) circumstances (Stonehouse 1991). The general lack of profitability found for Canada mirrors similar findings for the U.S. (Miranowski; Norris and Batie; Crosson). Farmers in Canada responded by repeated requests for financial assistance from public sources, while the Canadian public repeatedly called for redressing of degradation problems. Governments in Canada reacted by introducing a plethora of new programs aimed at improving natural resource conservation and environmental protection throughout agricultural regions (Stonehouse and Bohl; Cressman).

In Ontario, the Ontario Soil Conservation and Environmental Protection Assistance Program (OSCEPAP) was introduced in 1984. Designed to assist farmers financially in adopting and using conservation practices that would reduce resource degradation and enhance soil productivity, this provincial government initiative was allotted a budget of $18 million (Cdn). The following year (1985) saw the introduction of a 5-year "Tillage 2000" program, aimed at evaluating and demonstrating the benefits of alternative conservation-oriented crop rotation and soil conservation practices. In 1986, the first of a series of federal-Ontario government joint agreement projects was launched; called the Soil and Water Environmental Enhancement Program (SWEEP), it focused on ways of reducing farmland-related phosphorus run-off into Lake Erie. Then in 1987, the Land Stewardship Program (LSP) was launched initially as a 3-year provincial government project, but subsequently renewed in 1990 for another 3 years as a joint federal-provincial project under the auspices of the Canada Green Plan (CGP).

The 5-year CGP represented the most comprehensive and pervasive initiative yet devised by the federal government on behalf of resource conservation, environmental protection, and sustainability (Ministry of Supply and Services Canada 1990). National in scope, CGP agreements were signed with each of Canada's 10 provinces. The CGP sought to promote sustainable development and a healthier environment through combining local conservation initiatives with coordination of conservation programs nationally across all walks of life and sectors of the economy. The LSP therefore should be viewed as one facet of the agricultural component of CGP, which is defined as a broadly-based national environmental protection program.

Intent of LSP

Building on the experience of its forerunner, the OSCEPAP, the LSP was designed to encourage the adoption of comprehensive conservation farming systems in all regions of Ontario and in all types of farm business specialization. The emphasis, as with predecessor programs, was on a voluntary comprehensive approach toward improving conservation standards, maintaining soil productivity levels, and developing a long-term commitment to a responsible stewardship ethic.

Where LSP differed from its predecessors was that it called for the formulation and implementation of an overall farming system plan for conservation, using any and all conservation practices, such as tillage, rotations, cover crops, and structures, that were deemed appropriate to the particular farm circumstances and degradation problems. In contrast, previous programs had concentrated more on specific aspects of degradation, such as soil erosion, or on specific environmental pollutants, such as sediment or phosphorus, or on particular conservation techniques, such as tillage. To ensure acceptability of farming system plans submitted by intending participants, a review by a panel of peers and government representatives was undertaken. The review process was intensive. Not all of the submitted plans were approved, and many more returned for modification before being finally accepted.

Components of the LSP

The first-round LSP, in operation from 1987 to 1990, received a public funding allocation of $40 million (Cdn). Over three-quarters of this (or $31.3 million) was designated for financial assistance to farmers with acceptable farming system plans, called Land Stewardship Inventory and Action Plans (LSIAP). Each submitted LSIAP had to detail past management practices (including land use and cropping sequences, tillage and other field practices, conservation equipment and structures in place), to itemize resource degradation problems in rank order of importance, and to outline a 3-year action plan to remedy these problems.

In order to ensure reasonably equitable geographical distribution of these farmer assistance funds, some 30 percent of the LSP funds (which represented about $12 million Cdn) were allocated to each county or district, based on mid-1980s row crop and cereal crop acreages. While every county or district received funding, the preponderance or assistance funds flowed to the southwestern counties where row and cereal crops were (and remain) concentrated. It was assumed that soil degradation problems intensified in close association with row and cereal crop acreage densities.

A further 7.5 percent of LSP funding (or $3 million Cdn) was allocated to program delivery and service provision in the field (Figure 2). This included covering the costs of organizing and operating committees (mainly comprising farmers) at each county or district level, under the auspices of the Ontario Soil and Crop Improvement Association (OSCIA). Each county/district committee was responsible for reviewing and assessing all LSIAPs submitted by intending farmer participants. The use of OSCIA staff for program delivery permitted the (overcommitted) staff of the Ontario Ministry of Agriculture, Food and Rural Affairs (OMAFRA) to concentrate on providing participating farmers with technical information and assistance, planning advice, and help with plan implementation. Funds allocated to OMAFRA staff and programs for education and extension purposes amounted to 6 percent of total funding, or $2.4 million (Cdn). The final 8.25 percent, or $3.3 million (Cdn) went to research into conservation practices and stewardship.

The second-round LSP (designated by the acronym LS2P to distinguish it) ran from September 1990 to March 1994, and received some $31.65 million (Cdn). Of this, $22.1 million (which represents 70 percent) was earmarked for financial assistance for stewardship practices and structures to farmer participants whose plans passed the peer- and government-representative committee review (Figure 3).

For the LS2P, applicants were required to present a more detailed plan than the one for LSP. The LS2P plan, referred to as the Conservation Farm Plan (CSP), furthermore allowed for less financial assistance coverage than the plan for LSP. While the emphasis under LSP was placed on land management, degradation and soil productivity concerns, that for LS2P was as much on (off-farm as well as on-farm) environmental protection as it was on soil productivity

Figure 2. Allocation of Land Stewardship Program (1987–1990) funding

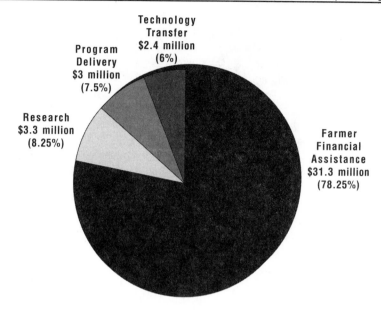

and conservation. In particular, LS2P applicants were encouraged to identify and address problems concerning animal manure handling, dairy washwater disposal, and pesticide handling. Financial assistance was capped at the lesser of $10,000 or 50 percent of total CSP expenditures (less any previous OSCEPAP or LSP funding received). This last provision was designed to encourage new applicants.

Some 20 percent of LS2P funding (or $6.25 million Cdn) was slated to cover technology transfer and administrative costs incurred by the OMAFRA staff charged with providing farmers with technical information and assistance and planning advice. The remaining 10 percent (or $3.3 million Cdn) was designated for program delivery by OSCIA staff.

Assessment of the LSP

A critical assessment of the LSP, and indeed of any soil and water conservation program, should take account of elements of both success and failure. Attempts should be made to measure program performance against specified program goals. In undertaking this assessment of the LSP, reference was made to

Figure 3. Allocation of Land Stewardship II Program (1990–1994) funding

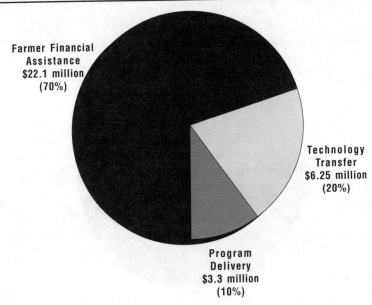

Farmer Financial Assistance $22.1 million (70%)

Technology Transfer $6.25 million (20%)

Program Delivery $3.3 million (10%)

Ministry of Supply and Services Canada publications on the principles of program evaluations (1981a; 1981b). Reference was also made to the independent evaluations of LSP (Thompson) and of LS2P (Taylor).

Pre-specified program goals. The pre-specified goals of the first-round LSP (1987-1990) were extremely general in nature. These goals were stated to be as follows:

- to respond to a farm lobby for assistance
- to protect the long-term productivity of the soil
- to provide incentive for experiments with conservation projects

As such, they were viewed as so non-specific as to be of no help in undertaking program evaluation (Thompson). The non-specific nature of the stated goals was apparently rooted in the hurried way in which the LSP was introduced during the run-up period to an Ontario provincial election. Intended to show that the then-ruling party was meeting the expressed needs of both public and farmers that the government provide support for soil and water conservation, the LSP was conceived and formulated during the summer of 1987 and implemented on August 31,1987 (Thompson).

The provincial election was held on September 10, 1987. The decision (by the OMAFRA) to stage a mid-term program evaluation prompted the staff of OMAFRA's Soil and Water Management Branch responsible for program design and implementation to develop more specific goals (Thompson). The goals follow:

- to address a long-term commitment towards the reduction of soil degradation and water management problems by responding to the expressed needs of Ontario farmers for assistance to implement acceptable conservation practices;
- to establish the OSCIA "grassroots" delivery system, allowing counties to determine their unique problems and establish their own program emphasis;
- to establish an extension team to assist in all aspects of conservation technology and information transfer;
- to encourage 5,000 participants in farm conservation programs on acres not previously under conservation by providing financial motivation for on-farm experimentation and adoption of acceptable conservation production practices on a limited basis;
- to offer a variety of options so as not to limit the management capabilities of the farmer;
- to induce a long-term commitment on the part of farmers to a conservation ethic;
- to increase awareness of agricultural soil degradation and erosion problems and alternative viable economic solutions on the part of farmers, industry and extension personnel.

With the benefit of experience gained with the first-round LSP, the following goals were specified for the second-round LS2P (Taylor):

- to help minimize agriculture's impact on soil quality, surface and ground waters by integrating technology transfer with financial incentives to modify current farm practices;
- to affect 1 million tillable acres over the life of the program (12 percent of Ontario's 8 million tillable acres);
- to promote and develop conservation farm planning and the encouragement of its use as a management tool;
- to facilitate delivery of the financial assistance component of the program by local committees of a farm organization; and
- to attract an additional 8,000 farmers to engage in conservation farm planning.

Successful elements

Perhaps the most obvious successful element to note about the LSP was its popularity with both farmers and the OMAFRA extension staff. The targeted figure of 5,000 applicants under the first-round LSP was far exceeded (at nearly 7,500) (Thompson), while the 8,000-applicant target for LS2P was almost achieved with 7,754 participants reported (Taylor). Moreover, the applicants represented mainly new participants in public fund-assisted natural resource conservation programs with each round of the LSP. This was ensured by the structuring of the financial assistance component of LSP, whereby applicants were eligible for public subsidies net of all previously received assistance under both the OSCEPAP and LSP programs. The same structuring also helped ensure that mostly "new" acres were being introduced to new or additional conservation measures with each round of the LSP. The target acreage under the LS2P was apparently exceeded with more than 1.1 million acres being affected by a wide variety of soil conservation measures (Taylor).

A second element of success lay in the requirement for participants to be comprehensive in their approach to conservation planning. This was achieved through the instruments of Land Stewardship Inventory and Action Plans (LSIAP) under the first-round LSP and Conservation Farm Plans (CFP) under LS2P. Both instruments required participants to provide a land resource inventory and cropping plan, to identify all problems associated with soil degradation and subsequent water pollution, and to delineate appropriate remedial measures. A wide range of measures, including conservation structures, tillage, and rotations, was made available for applicants to select from. The CFPs were especially comprehensive in that they required the identification of problems with and the delineation of suitable remedial practices for livestock manure, washwaters and wastewaters, and for the storage and handling of synthetic pesticides. Between them, the LSIAPs and CFPs likely encouraged farmers to adopt a more holistic approach to natural resource management and environmental protection, but there is no definitive indication that this was the case.

A third successful element was the development and emplacement of a much improved infrastructure for using public intervention programs to foster greater conservation efforts on the part of farmers. Under the first-round LSP, this was manifested by the establishment of "grassroots" delivery system based on the use of the OSCIA, a non-governmental organization (NGO) comprising farmers. By harnessing the OSCIA resources, the government both ensured a larger element of peer review of LSIAPs (OMAFRA staff also reviewed every application) and relieved the already overworked OMAFRA staff of most of the LSP delivery duties.

The advantages of peer review are that peers often provide the most critical of reviews, and, in selecting applicants for LSP participation, absolve the government of any accusations of favoritism. The OMAFRA, by divesting itself of the delivery component, was able to concentrate its efforts on provision of technical information and assistance and conservation planning guidance to LSP applicants. In particular, the OMAFRA was enabled to form a team of specialist soil and water conservation advisors.

The successful combination of an NGO for grassroots program delivery and the OMAFRA for extension advisory services was continued for the second-round LSP.

Overall, LSP certainly attracted much favorable attention in Ontario's farming community by providing a winning combination of financial incentives with conservation technology transfer. This type of voluntary compliance approach to resource conservation and environmental protection is much preferred by farmers to any kind of compulsion, penalty system, or even cross-compliance (Duff et al.; Taylor).

Elements of failure

The haste with which the first-round LSP was conceived and introduced in 1987 led to the specification of program goals that were inappropriate in terms of their generality and lack of measurability. The vagueness of such goals as "to respond to a farm lobby for assistance" leaves open several questions such as "were sufficient numbers of farmers reached by the program" or "was program assistance dispensed to those farmers most in need of it?"

The subsequent development of more specific "intended" goals to assist with a mid-term evaluation of the first-round LSP almost guaranteed an element of success (Thompson, P31). Even at that stage, several intended goals were unsuitably non-specific and non-measurable. Goals for LS2P were in some cases much more precise and measurable but in other cases were again vague, general, and immeasurable. For example, how can one reasonably measure the goal "to help minimize agriculture's impact on soil quality, surface and ground water by integrating technology transfer with financial incentives to modify current farm practices?" Against what standard should this impact to be minimized be compared? How should soil and water quality be measured? Which current farm practices ought to be modified—and how? Other questions could be posed.

Of far greater concern was the intent of the original (1987) program. The LSP was designed, for political reasons, to appeal to the farming community by responding to a farm lobby calling for governmental action on the conservation front, to assuage the fears of farmers about long-term soil productivity, and to appease the public's and farmers' concerns about the environment. Except for this last intent, which was ranked very much the lowest in the minds of LSP planners, these program designs were inappropriate. In particular, long-term soil productivity in Canada generally as well as in Ontario is simply not an issue. The long-term food production potential of this country is not in jeopardy, despite ongoing soil degradation problems (Van Kooten and Furtan; Smit et al.; Stonehouse and Bohl). What is an issue for the farmers is that degradation has been costing them (not insignificant amounts of) money, and, because most soil conservation measures are not profitable (Stonehouse 1995), preventing or remedying degradation problems was bound to cost farmers even more money in the absence of public subsidies.

A much more pressing issue, with the public at large as well as farmers, has been the far greater extent of damage from farming activities off the site than on site (Stonehouse and Bohl; Fox et al.). It is these off-farm damage costs (or externalities) that provide justification for governmental intervention and public financial assistance to farmers, rather than the much smaller on-farm costs (Stonehouse 1994; Stonehouse 1996). The LSP failed to address this crucial issue. Even where minimizing the environmental impact of agricultural activities was cited as a program goal under LS2P, there was no attempt to specify this goal in quantifiable terms. While the LSP may indeed have helped in reducing the environmental impact of agricultural activities, the extent of this was not known, the difference made by the LSP was not measured and was not compared against any standard, and therefore it is difficult to assess any measure of success.

Closely associated with the failure to link, in specific, measurable terms, the LSP with externalities, was the failure of LSP to undertake any kind of targeting. Priorities for use of LSP funding to farmers were established on the basis of (a) row crop acreages, by counties and districts, under an assumed linkage between row crops and soil degradation, and within each county, on the basis of (b) first come, first served; and (c) the acceptability and approval of conservation plans.

No attempt was made to ensure that available funding was directed to those farm situations encompassing the most severe soil productivity problems, much less to those farms causing the most environmental damage. Allocation by county row-crop acreages does not necessarily endow effective or efficient use of public funds. The ultimate payoff in targeting would have been to allocate available funds to those farms whose conservation plans would have led to the highest net social benefits (i.e., public benefits in terms of reduced externalities plus private (on-farm) benefits of conservation, if any, less public costs in terms of farmer subsidies, program administration and implementation, and private (on-farm) costs of investing in and operating conservation measures) (Stonehouse 1996).

The failure to target, and therefore to ensure efficient use of LSP subsidy money, was exacerbated by overall funding scarcities. Both parts of the LSP were well over-subscribed; many farmers applied for conservation plan subsidy money and had to be turned away disappointed. Furthermore, it was noted (Thompson; Taylor) that the majority of participants were above-average farmers (in management abilities) who were previously using conservation techniques, so that the LSP induced the adoption and use of additional conservation measures on the same farms or of the use of existing measures on "new" tracts of land on those farms.

One might question therefore the extent to which awareness of soil degradation and erosion problems was increased in the farming population in general, and in the minds of those farmers causing the worst environmental damage in particular.

It has been noted above as an element of success that the LSP encouraged the use of comprehensive and holistic approaches to conservation planning. At the same time, each participant was charged with the task of identifying the degradation problems occurring on the farm, and with the responsibility of prescribing the most appropriate remedial measures. Even among above-average farmers, the ability to correctly identify both problems and remedies can be questioned (Thompson, P109). Although greater human and monetary resource commitments would have been entailed, the LSP should have embodied a larger element of independent, objective review of each applicant's degradation problems and proposed remedies, using among other instruments on-site inspections and interviews.

Finally, the LSP, as with so many other conservation programs, failed to be integrated and coordinated with other conservation programs. For example, the OSCEPAP, which preceded LSP, and to which the LSP was supposedly linked, offered higher funding support for conservation structures (such as terraces, berms, and drop structures) than did the LSP. There was understandable reluctance therefore on the part of LSP participants to prescribe and invest in structures, even though these may have been the most appropriate conservation measures (Thompson, 1991).

Another example was the encouragement of the establishment of tree windbreaks to combat wind erosion, the trees being supplied at a subsidized rate by the Ontario Ministry of Natural Resources (OMNR). While the LSP served to increase demand for trees, the OMNR failed to maintain a commensurate increase in supply. Much worse was the failure to coordinate the LSP, and other conservation programs, with the long-extant, mainstream programs for agriculture designed to boost productivity and food output, the better to maintain competitiveness abroad and low food prices in domestic markets. Most of Ontario's agriculture output-boosting programs have long been focused on the very row crops whose degradation impacts the LSP sought to mitigate. Otherwise these same output-boosting programs have tended to encourage cultivation of marginal lands, cultivation right out to field boundaries and particularly right up to watercourse and ditch banks, removal of hedges and protective tree lines for field enlargement, more intensive tillage practices, and other practices associated with increased soil degradation.

In short, Ontario's output-boosting programs, representing the majority of governmental intervention programs on behalf of agriculture, have been largely at variance with conservation programs.

Overall, this voluntary compliance conservation program suffered from having inappropriate goals; from a failure to target those farmers with the potential to generate the highest net social benefits from conservation and therefore the most effective use of scarce public funds; from a failure to incorporate enough assurance mechanisms to be able to judge the appropriateness of individual farm conservation plans; and from failing to reconcile inconsistencies with mainstream agriculture programs and to coordinate the LSP with other conservation programs.

Ambiguous and unknown elements

Not all elements of the LSP can be categorized as definitively successful or otherwise. Some must be viewed as ambiguous, or even unknown. Ambiguity can arise from the imprecise wording or nonquantitative terms of the program goals. Examples are furnished by the first and the seventh of the first-round LSP goals, or the first of the LS2P goals.

In the absence of something specific, some measurable outcome, it is difficult to judge how well the LSP achieved "the addressing of a long-term commitment towards the reduction of soil degradation and water management problems by responding ...." For example, what length of time horizon was meant by "long-term"; was the "reduction in soil degradation" aimed at a large one or only a marginal one? Similarly, how can one evaluate the performance of the goal to induce a long-term commitment on the part of farmers to a conservation ethic?

There was no post-LSP survey of Ontario farmers to establish whether this goal might have been achieved, so this must be viewed as an unknown element. Even with the help of a farmer survey, one would have faced difficulties in assessing LSP performance because of the ambiguous wording of the goal to be achieved. For example, what was meant by "long-term commitment"; which body of farmers did the LSP planners have in mind, all Ontario farmers, only LSP participants, specifically the non-LSP participants, some other subset(s)?

Analogous criticisms can be leveled at the first and third goals specified for the LS2P. For the first one, how would one proceed to evaluate a goal "to help minimize agriculture's impact on soil quality, surface and ground waters by integrating ..."? Again no post-LSP survey was undertaken anyway, but even if there had been one, how would "helping to minimize something's impact" best be measured? For the third goal, how should one judge how well the LSP promoted and developed conservation farm planning and encouraged its use as a management tool? How many farmers' use of conservation farm planning would have constituted successful performance of the LSP, which farmers should these have been—LSP participants, non-participants, those with major environmental damage problems?

Taking all ambiguous and unknown elements into account, it can be said that too many of the prespecified or intended goals of LSP were left in vague, general and nonmeasurable terms, despite experiences gained from the mid-term review of the first-round LSP. All goals should have been specific, preferably measurable quantitatively, and therefore readily assessable. Second, it can be said that follow-up surveys and ongoing monitoring would have been necessary to provide objective assessments of LSP's performance.

Conclusions

Considering the review of all elements, successful and otherwise, the following conclusions about the LSP were drawn:

- this voluntary compliance, financial incentive conservation program was extremely well received by the farming community in Ontario, despite its more obvious drawbacks (from farmers' standpoint) of (a) limited overall funding; (b) questionable funding allocation procedures (first come, first served); (c) inadequate funding to support and therefore encourage investments in conservation structures;
- while the LSP apparently attracted "new" farmers to the overall conservation effort, these were farmers with above-average management capabilities and prior experience in conservation techniques, so that the extent to which the LSP raised conservation needs awareness and inculcated an improved long-term stewardship ethic is questionable and in any case, is difficult to gauge;
- arguably the most successful set of aspects was that the LSP engendered a highly effective program delivery mechanism, using a grassroots NGO as the instrument; the LSP contributed to the further development and enhancement of a conservation extension infrastructure, the better to assist Ontario farmers with conservation technology transfer, adoption, adaptation, and

ongoing use; and the LSP fostered greater cooperation among government, NGOs, and the farming community in matters of conservation planning and implementation;

- arguably the most notable elements of failure were first that the LSP's original focus on protecting the long-term productivity of Ontario was totally misdirected because productivity was and is not perceived to be threatened for the foreseeable future; second that when the LSP's focus was subsequently redirected to conservation of natural resources and environmental protection, there was no attempt to target the use of LSP human and financial resources on the most resource-degrading and environmentally damaging cases; and therefore third, that the LSP did not represent the best interests of the general public by providing for less than fully effective and efficient use of the public's money;
- one subsidiary element of failure was connected with the vague, nonspecific and nonmeasurable wording of a number of the LSP goals, and the resultant difficulties posed in assessing the performance of the LSP;
- another subsidiary element of failure was associated with the lack of coordination and integration between the LSP and other conservation programs, and even more so with other (nonconservation, production-boosting) programs for agriculture.

In summary, the politically motivated LSP, with its voluntary compliance, public assistance, nontarget characteristics, represented an inefficient and ineffective application of public money, and a failure to meet with the broad public's interests. This program, however, was not without its element of success. Highly popular with farmers, it was greatly oversubscribed, it contributed toward improvements in the conservation extension infrastructure serving to encourage greater conservation effort, and it probably, albeit indirectly, boosted Ontario farmers' awareness for conservation needs and a better stewardship ethic. The public's needs could have been so much better met at the same time as the farmers' if the most notable elements of failure had been obviated.

References

Anderson, M., and L. Knapik. 1984. Agricultural land degradation in western Canada: A physical and economic overview. Regional Development Branch, Agriculture Canada, Ottawa, Ontario.

Baffoe, J.K., D.P. Stonehouse, and B.D. Kay. 1987. A methodology for farm-level economic analysis of soil erosion effects under alternative crop rotational systems in Ontario. Can. J. Agr. Econ. 35(4): 55-73.

Cressman, D.R. 1994. Remedial action programs for soil and water degradation problems in Canada. In T.L. Napier, S. Camboni, and S. El-Swaify [eds.]. Adopting conservation on the farm: An international perspective on the socioeconomics of soil and water conservation. Soil and Water Conservation Society, Ankeny, Iowa.

Crosson, P.R. 1991. Cropland and soils: past performance and policy challenges. In K. Frederick and R. Sedjo [eds.]. America's renewable resources. Resources for the Future, Washington, D.C.

Duff, S.N., D.P. Stonehouse, S.G. Hilts, and D.J. Blackburn. 1991. Soil conservation behavior and attitudes among Ontario farmers toward alternative government policy responses. J. Soil and Water Cons. 46(3): 215-219.

Dumanski, J., D.R. Coote, B. Luciak, and C. Lok. 1986. Soil conservation in Canada. J. Soil and Water Cons. 41(4): 204-210.

Fox, G.C., A. Weersink, G. Sarwar, S. Duff, and B. Deen. 1991. Comparative economics of alternative agricultural production systems: a review. N.E.J. Agr. and Res. Econ. 20(1): 124-142.

Kay, B.D., and D.P. Stonehouse. 1984. The growth of intensive agriculture in Ontario. Notes on Agric. 19(2): 5-8. University of Guelph, Guelph, Ontario.

Miller, M.H. 1986. Soil degradation in eastern Canada: Its extent and impact. Can. J. Agr. Econ. 33: 7-18.

Miller, M.H., P.H. Groenevelt, and D.P. Stonehouse. 1988. Stewardship of soil and water in the food production system. Notes on Agric. 22(1): 5-12. University of Guelph, Guelph, Ontario.

Ministry of Supply and Services Canada. 1981. Principles for the evaluation of programs by federal departments and agencies. Program Evaluation Branch, Ottawa, Ontario.

Ministry of Supply and Services Canada. 1981. Guide on the program evaluation function. Program Evaluation Branch, Ottawa, Ontario.

Ministry of Supply and Services Canada. 1990. Canada's green plan for a healthy environment. Cat. No. En21-94/1990E. Government of Canada, Ottawa, Ontario.

Miranowski, J.A. 1984. Impacts of productivity loss on crop production and management in a dynamic economic model. Am. J. Agr. Econ. 66(1): 61-71.

Norris, P.E., and S.S. Batie. 1987. Virginia farmers' soil conservation decisions: an application of Tobit analysis. South. J. Agr. Econ. 19(1): 79-90.

Smit, B., M. Brklacich, R. McBride, Y. Yongyuan, and D. Bond. 1988. Assessing implications of soil erosion for future food production: a Canadian example. Geoforum. 19(2): 246-259.

Statistics Canada. 1961, 1966, 1971, 1976, 1981, 1986, 1991. Census of Agriculture. Cat. No. 96-102. Government of Canada, Ottawa, Ontario.

Statistics Canada. 1992. Agricultural Economic Statistics. Cat. No. 21-603 (occasional). Government of Canada, Ottawa, Ontario.

Stonehouse, D.P. 1991. The economics of tillage for large-scale mechanized farms. Soil and Till. Res. 20: 333-351.

Stonehouse, D.P. 1994. Canadian experiences with the adoption and use of conservation practices. In adopting conservation on the farm: An international perspective on the socioeconomics of soil and water conservation. T. L. Napier, S.M. Camboni, and S.A. El-Swaify [eds.]. Soil and Water Conservation Society, Ankeny, Iowa. pp. 369-396.

Stonehouse, D.P. 1995. Profitability of soil and water conservation in Canada. J. Soil & Water Cons. 50(2): 215-219.

Stonehouse, D.P. 1996. A targeted policy approach to inducing improved rates of conservation compliance in agriculture. Can. J. Agr. Econ. 44(2): 105-119.

Stonehouse, D.P., J.K. Baffoe, and B.D. Kay. 1987. The impacts of changes in key economic variables on crop rotational choices on Ontario cash-cropping farms. Can. J. Agr. Econ. 35(2): 403-420.

Stonehouse, D.P., B.D. Kay, J.K. Baffoe, and D.L. Johnston-Drury. 1988. Economic choices of crop sequences on cash cropping farms with alternative crop yield trends. J. Soil and Water Cons. 43(3): 266-269.

Stonehouse, D.P., and M.J. Bohl. 1990. Land degradation issues in Canadian agriculture. Can. Pub. Pol. 14(4): 418-431.

Taylor, E. P. 1993. Land stewardship II program evaluation. Ontario Ministry of Agriculture, Food and Rural Affairs, Resources and Regulations Branch, Guelph, Ontario.

Thompson, G. 1991. A mid-term evaluation of the Ontario Land Stewardship program. Unpublished M.Sc. thesis, Department of Agricultural Economics and Business, University of Guelph, Ontario.

Turvey, C.G. 1991. Environmental quality constraints and farm-level decision making. Am. J. Agr. Econ. 73(4): 1399-1404.

Van Kooten, G.C., and W.H. Furtan. 1987. A review of issues pertaining to soil deterioration in Canada. Can. J. Agr. Econ. 35(1): 33-54.

Zantinge, A.W., D.P. Stonehouse, and J.W. Ketcheson. 1986. Resource requirement yields and profits for monocultural corn with alternative tillage systems in southern Canada. Soil and Till. Res. 8: 201-209.

16

The Permanent Cover Program: Is Twice Enough?

Jill S. Vaisey, Ted W. Weins, and Robert J. Wettlaufer

The Canadian prairie provinces (Figure 1) contain about 82 percent of Canada's agricultural land (Statistics Canada 1992). The agricultural portion of these provinces is primarily a Central Plains ecoregion comprised of two ecozones: prairie and forested boreal plains. The prairie ecozone to the south is composed mainly of agricultural cropland and remaining grasslands while the agricultural and forested Boreal Plains Ecozone lies to the north. Agriculture occupies about 56 million hectares of the 121 million hectare Central Plains.

Agricultural operations are relatively recent, being less than 100 years old in most of the prairie provinces. Much of the land settlement occurred between 1900 and 1913 although some farming had been initiated along Manitoba watercourses almost 100 years earlier (Acton 1995). By 1931, 60 percent of the vast Canadian grasslands (originally about 500,000 square km) were under annual cultivation. Use of farm practices more suited to the moister conditions found in eastern Canada and Europe, along with the prolonged drought of the 1930s, and settlement patterns based on location rather than suitability for agriculture resulted in almost a decade (1930s) of continuous soil erosion events. Thousands of farms were abandoned across the prairies even in the moister dark brown soil zone that overlaps much of the moist mixed-grassland ecoregion of the prairie ecozone.

In response to the drought, Canada's parliament created the Prairie Farm Rehabilitation Administration (PFRA) in 1935 "to promote within the soil drifting areas of Manitoba, Saskatchewan, and Alberta, systems of farm practice, tree culture and water supply that would afford greater economic security." This sustainable development initiative was complemented by intensive research efforts at federal experimental farms and other research institutions to integrate soil conservation into production practices.

Although more suitable methods for dryland farming were developed from the 1940s to the 1960s, the cycle of droughts continued and soil health continued to deteriorate, especially in areas where much of the land was summer-

Figure 1. The Canadian Prairie Provinces

fallowed. Summerfallow, in the prairie provinces context, is the practice of not cropping for the spring to fall growing season to conserve moisture and regenerate soil fertility. It usually involves mechanical cultivation of the soil to control weeds. The commonly practiced wheat-summerfallow rotation is now believed a major contributing factor for decline in organic matter, increased soil erosion, and increased risk of salinity (Standing Senate Committee 1984). As shown in Table 1, the practice increased between 1941 and 1961 and has declined steadily since then.

Another change has been an intensification of prairie agriculture. Since the 1950s there have been increased numbers of crops, increased production, increased use of inputs, and larger more mechanized farms. At the same time, the number of farms has decreased dramatically.

1980-1990 - A decade of program response

The 1980s saw not only a new cycle of prairie drought but also a new environmental thrust with enhanced focus on soil conservation and sustainability. In 1984, Canada's Senate Committee on Agriculture, Fisheries, and Forestry released "Soil at Risk - Canada's Eroding Future." This document, combined with intensive communication effort by Senator H. Sparrow, then Committee Chair, resulted in increased public awareness of soil quality issues. Soil degradation has been estimated to cost Canadian agriculture $500-700 million[1] (Dumanski et al. 1986 reported in Government of Canada 1991) per year in reduced yields and increased input costs. This risk to prairie soils is primarily caused by moderate to severe wind erosion, water erosion, and salinization.

The 1980s was a period of increasing federal-provincial cooperation in an effort to address environmental issues, particularly soil degradation. Both partners contributed resources to conservation programming. Examples include the following:

- *ERDAs*: In the mid-1980s, the Government of Canada signed Economic and Regional Development Agreements (ERDAs) with provincial governments. The ERDAs were a joint federal-provincial approach to economic strategies that coordinated financial action in specific areas. In Manitoba and Saskatchewan these 5-year agreements (1984/85–1988/89) included a soil conservation component, amounting to $4.5 million and $8.0 million, respectively, over the period. PFRA also assigned some resources—$150,000 annually—to Alberta for complementary programming in that Province.
- *NSCP*: Canada initiated the National Soil Conservation Program (NSCP) in 1989. Three-year agreements were signed between the federal and provincial governments under Accords on Soil and Water Conservation and Development. NSCP provided assistance to producers to help address specific soil degradation problems in each province. NSCP also encouraged new partnerships among government and non-government organizations. In the Prairie Provinces, assistance amounted to a total of $7.5 million, cost-shared between the federal and provincial governments.

In Manitoba, Saskatchewan, and Alberta one component of the NSCP was the Permanent Cover Program (PCP), which is discussed more thoroughly in the latter sections of this chapter.

In both the above programs, conservation activities were targeted to soil degradation issues associated with annual crop production. Most resources in-

Table 1. Summerfallow area in the prairie provinces, 1941–1991 (thousands of hectares)

	1941	1961	1971	1986	1991	1961–91 % change
Manitoba	1,118	1,307	1,074	509	297	−77%
Saskatchewan	5,586	6,952	6,701	5,658	5,713	−18%
Alberta	2,649	3,015	2,836	2,126	1,771	−41%
Prairies	9,353	11,274	10,611	8,293	7,781	−31%

Source: Statistics Canada, Census of Canadian Agriculture 1941–1991

volved the demonstration and piloting of soil conservation activities and were delivered with and through local agricultural organizations.

The experience gained in these programs contributed to the recognition that conservation involved more than just soil and more than just agriculture. The next round of federal-provincial programming involved a wider range of partners and a wide range of activities associated with primary agricultural production.

In 1990, Canada released a $3 billion environmental action plan, "Canada's Green Plan," which included $170 million to expand environmental sustainability in the agricultural sector. Eight key issue areas were identified: soil conservation, water quality, water quantity, habitat conservation, air quality and climate change, energy use, pollution/waste management, and conservation of genetic resources. Approximately three quarters of the Agricultural Green Plan resources were delivered through federal-provincial agreements, which expired in March 1997. In the prairie provinces, the targeted issues were soils, water quality and quantity, and wildlife with a total resource allocation of $60 million.

The Permanent Cover Program

The Permanent Cover Program (PCP) is an example of a program implemented primarily for soil conservation reasons, that has significant benefits for other environmental issues. It was delivered by the Government of Canada, through PFRA, within four provinces — Manitoba, Saskatchewan, Alberta, and British Columbia.

Description

The PCP was first announced in 1989 as a 3-year program in Manitoba, Saskatchewan, and part of Alberta. The primary objective was to reduce soil degradation on marginal lands that had high erosion risk under annual cultivation. Marginal lands, classes 4, 5, and 6 under the Canada Land Inventory, were targeted for conversion to alternative sustainable uses under permanent cover.

The program provided a seeding payment of $50/ha ($20/acre) to convert eligible lands from annual crops to perennial forage or tree cover. Farmers who chose to sign long-term contracts subsequently received a one-time payment based on a bid price (based on the market value of similar land) after cover was established and after signing a 10- or 21-year land use agreement. No further payments were made. A caveat was registered on the PCP land to safeguard Canada's interests over the contract years. Producers continue to use the land, primarily for cattle grazing or forage production. The program was very popu-

214

lar and was fully subscribed within the first few months. It converted 168,000 hectares (416,000 acres) from annual crop to forage.

An extension to the first program was announced in July 1991. Bid prices were replaced with fixed prices based on the norm of acceptable bids from the first program, and all participants were obliged to sign a long-term contract. More opportunities for partnerships with environmental organizations were included. PCP II was expanded to include eligible land in the Peace River Region of British Columbia and Alberta. Initial payments of $50/ha ($20/acre) for seeding were followed by a one-time payment after the cover was established. The contract payments were as follows: $50/ha ($20/acre) for a 10-year contract, ($75/ha ($30/acre) in Alberta and British Columbia); $125/ha ($50/acre) for a 21-year contract ($160/ha ($65/acre) in Alberta and British Columbia). Applications were accepted until the summer of 1992, resulting in long-term contracts for the conversion of an additional 354,000 hectares (874,000 acres).

Table 2 shows that, together, PCP I and II included some 15,000 contracts and 522,000 hectares of marginal land. Most of the land was converted to alternative productive uses under long-term contracts of which 64 percent were for 21 years. The two programs cost Canada $74 million in payments.

Although there were many reasons for a producer to take the longer-term contract, the decision partly reflected a desire for good stewardship and a belief that forage production would be a more sustainable use than annual crops.

Continuing obligations and issues

The PCP was a relatively short-term program (1989 to 1992 sign-up) with long-term implications. There is an ongoing commitment to monitor land under PCP contracts. Procedures involve the annual inspection of a sample of PCP sites to ensure that the land continues under permanent cover. The number selected varies according to the total sites in a district and the perceived risk.

There is also an obligation to manage any early withdrawals from the program. Recognizing that circumstances change, all contracts contain 'buy-out' provisions. The amount of liquidated damage depends on the number of months remaining in the contract, and declines as the contract gets closer to the end. Withdrawal within a short period of contract signing results in liquidated damages exceeding the original contract payment. Towards the end of the contract, the liquidated damages are much less than the contract payment. The formulas used are described as follows:

- Ten-year contract: original payment x 1.25% x number months remaining = liquidated damage
- Twenty-one-year contract: original payment x 0.60% x number months remaining = liquidated damage

Table 2. Permanent Cover Program: Contracts and area

Program	British Columbia		Alberta		Saskatchewan		Manitoba		Totals	
	Contracts	ha	Contracts	ha	Contracts	ha	Contracts	ha	Contracts	ha
PCP I										
Seeding only	0	0	1,503	54,039	0	0	789	20,877	2,292	74,916
10 yr	0	0	311	10,972	721	23,425	0	0	1,032	34,397
21 yr	0	0	468	18,707	1,097	40,340	0	0	1,565	59,047
PCP II										
10 yr	36	1,265	1,645	54,990	1,091	34,949	792	22,519	3,564	113,723
21 yr	61	3,567	2,179	81,997	2,928	110,361	1,388	43,990	6,556	239,915
Totals	**97**	**4,832**	**6,106**	**220,705**	**5,837**	**209,075**	**2,969**	**87,386**	**15,009**	**521,998**

Source: PFRA, PCP program records.

The first of the 10-year contracts will begin to expire in 1998. This generates an obligation to manage Canada's withdrawal. It also opens the question of whether farmers will maintain their forage stands of their own volition. In land under 21-year contracts, forage rejuvenation will be required to keep the stands productive. This process also needs to be managed.

Benefits of PCP

PCP resulted in a number of benefits to Canada and to landowners.

Benefits to Canada. PCP resulted in a wide variety of benefits, some of which accrued to government and some to society:

- Reduced federal government payments under former acreage-based programs targeted at annual crop production. Total savings to Canada from these programs (Guaranteed Revenue Insurance Program, Net Income Stabilization Account and Crop Insurance) were estimated to be $11.8 million in 1993, from the 522,000 hectares enrolled in PCP (PFRA 1993).
- Reduced soil degradation. In 1994, PFRA conducted an analysis of the economic value of environmental benefits gained from wind erosion reduction (PFRA 1994). Using models for residue and erosion (CanHelp), a model on productivity loss (Greer et al. 1992), the wind erosion equation, and crop yield data from provincial Crop Insurance, PFRA estimated between $2 million and $5 million dollars of soil productivity were saved by putting 320,000 hectares of land into permanent cover. This saving will accrue to the recipients of the soil in the future.
- Improved water quality in nearby surface water due to reduced soil sedimentation and associated chemical residues.

- Enhanced wildlife habitat.
- Increased carbon sequestration. In 1994, the Working Group on Agriculture and Greenhouse Gases estimated economical opportunities to reduce net greenhouse gas emissions in Canada. They project that an increase of 1.5 million to 2 million hectares in perennial forages in western Canada would equate to between 0.9 million and 1.2 million tonnes of carbon sequestered per year.
- Reduced expense by local governments for removing wind and waterborne sediments from road ditches and drains.

Benefits to Participants. In 1994, PFRA commissioned an "after establishment" study to determine landowner attitudes to the Permanent Cover Program (Western Opinion 1994). A sample of 500 clients who had seeded PCP land were interviewed. Impacts identified were as follows:

- decreased soil erosion (74 percent of respondents),
- decreased operating costs (70 percent),
- increased wildlife habitat (65 percent),
- increased size of livestock herd (64 percent),
- decreased need for purchased feed (60 percent), and
- increased net farm income (56 percent).

With regard to operating costs, the survey indicated both savings and increases. There were fewer costs on annual cropland resulting from reduced need for gully repair and rock picking, and reduced fertilizer costs. There were also extra costs related to fencing, moving livestock or forage, water supply development, labor, wildlife damage, and travel time. However, a majority of respondents reported increased net farm income.

A significant benefit of PCP, from the participants' perspective, is their ability to continue using their PCP land and thereby continue generating income from their resource. The uses were as follows:

- Haying—79 percent of respondents said that their farm operation involved haying on lands enrolled. The estimated average annual hay yield from the PCP land was 4.0 tonnes per hectare (1.8 tons per acre), and those who hayed PCP land did so on an average of 60 hectares (152 acres).
- Feeding—85 percent fed their PCP hay to their own livestock, while 12 percent sold the hay to other livestock producers or dehydrated forage operators (1 percent).
- Grazing—64 percent indicated that livestock graze on their PCP land, most commonly beef cattle (96 percent).

For the future, most respondents (93 percent) planned to keep the forage stand as long as possible. Half the respondents said they planned to increase livestock production (54 percent), seed more annual crop land to forage (52 percent), or shift PCP hay land to pasture use (48 percent). Just 18 percent indicated they planned to return PCP land to annual crop after their contract expires.

In summary, most respondents felt that the PCP was a good program.

Other conservation efforts

Agricultural sustainability, including resource conservation, is an issue in many other countries. Some countries have taken significant initiatives, as discussed elsewhere in these proceedings. Some, perhaps many, of the programs, such as the American Conservation Reserve Program and the Canadian Permanent Cover Program, have significant wildlife benefits.

The linkage between agriculture and the environment is also recognized by environmental interests. Some private agencies such as Ducks Unlimited and the Nature Conservancy make arrangements with private landowners and operators to influence their land use decisions in favor of wildlife. Ducks Unlimited, for example, offers annual payments to farmers for leases and delayed hay cut, to increase cover on prairie cropland.

Environmental programs such as the North American Waterfowl Management Plan (NAWMP) have similar targets. This 15-year agreement between Canada and the United States, and most recently Mexico, targets the protection and management of lands for waterfowl benefits. A significant target area is the prairie pothole region of Canada, most of which is private agricultural land. The Prairie Habitat Joint Venture (PHJV), the largest program under NAWMP, has secured 366,000 ha (906,000 acres) via lease, land management agreement, flood easement, and purchase for habitat—25 percent of the original objective— between 1986 and 1996. Total PHJV contributions for that period equal $186 million (Knutson 1996).

The NAWMP was one of the first programs to recognize that prairie landowners have few if any methods of capturing economic benefit from the wildlife or the habitat produced on their land. In contrast to some other nations, farmers in the prairie provinces cannot sell hunting rights or access, or otherwise earn a return from habitat land. While the habitat is owned by the landowner, the wildlife is owned and managed by provincial and federal governments. The wildlife sector of the prairie economy is highly regulated and many landowners view this as an obstacle to economic development. Because there are few economic gains to be made from conserving habitat, it is often viewed as a liability. Given the availability of alternative economic opportunities, the "public good at private cost" issue will continue as an impediment to habitat conservation.

Acceptability of PCP today

Would PCP be accepted today? We do not know. Some factors suggest no; others suggest yes.

International factors

The General Agreement on Tariffs and Trade (GATT) was completed in 1994 and is being implemented by the World Trade Organization (WTO). Reduction of commodity-based support mechanisms, required through the agreement, was expected to dampen the propensity of prairie farmers to grow wheat and provide farmers some additional inclination to diversify to other crops or raise cattle. The grazing benefits of PCP would be expected to remain attractive. However, current high wheat prices may mask the expected impact.

International environmental responses, such as the Convention on Biological Diversity, were expected to generate support for initiatives such as PCP. Canada ratified this Convention in December 1992, and responded with the 1995 "Canadian Biodiversity Strategy." The strategy recognizes that "optimizing the use of agricultural lands is not only an essential element of agricultural sustainability, but also can significantly contribute to the conservation of biodiversity...." As PCP supports biodiversity, it would be expected to be favorably received under this convention. However, fiscal constraints limit economic options of governments, while producers continue to need a return from their land base.

Domestic factors

In 1996, many governments faced significant fiscal constraints. Canada was no exception. One result is a significant decrease in resources available for incentive payments for conservation activities, including the conversion of land that is marginal under annual crop production to other uses.

In 1995, Canada eliminated the Western Grain Transportation Act (WGTA), partially in response to international trade requirements to reduce commodity based support, thereby eliminating the transportation assistance in moving grain to export position. The anticipated results were an incentive to use grain locally, primarily in the livestock industry, and consequently an incentive to keep PCP lands producing hay and forage. Current higher grain prices appear to have reduced or offset this expected impact, at least in the short term.

Other policy changes are also reducing an implicit bias in favor of grain. The Canadian Wheat Board quota system has been dismantled in favor of contract deliveries based on the ability to produce crops. A federal-provincial Guaranteed Revenue Insurance Program (GRIP) is being discontinued across the prairie provinces. The Net Income Stabilization Account (NISA) is being broadened from a commodity-specific approach to one that includes all farm income. Crop Insurance, which is alleged to encourage the cultivation of marginal land, is currently under review.

Public opinion surveys suggest that environmental issues, while still relevant, are no longer at the top of the public agenda. Instead, economic issues—jobs and growth—are receiving the most attention. This suggests that programs directed to adaptation or economic growth will be more accepted than those targeted to strictly environmental issues.

Local factors

PCP was implemented to protect parts of the soil resource in western Canada that are at risk of degradation under annual cultivation. The cultivation of marginal land is a function of several factors including the relative profitability of grains, livestock or forage, costs of machinery operation, and policy and program emphasis of the day. However, the main cause for the cultivation of Class 4, 5, and 6 marginal land over the last 20 to 30 years has been the world price of grain.

In the late 1980s, wheat prices were relatively low and prices for beef calves were relatively high (Figure 2). In the mid-1990s the situation had reversed itself. Economists speculate that even with higher costs of only $5.50 per tonne (15 cents a bushel) from 1975–80, wheat and beef prices of $165 /tonne ($4.50 /bu) and $1.10 /kg ($0.50 /lb) would likely have resulted in conversion of forage to grain (Ward 1996). Therefore, today's higher grain prices would likely make PCP unattractive.

Drought factor

Initial funding for PCP was secured during the 1988 drought. The droughts of the 1980s, associated dust storms, and strong awareness campaigns by agriculture agencies and the news media helped create an environment that made farmers very receptive to soil conservation programs. In the 1990s, precipitation improved (Figure 3). The need to convert marginal lands to permanent cover, therefore, would likely seem less immediate.

In the 1994 landowner attitude survey, PCP clients indicated their main reasons for enrolling in the program were that the land entered was marginal and would not support grain. This was exacerbated by low grain prices. The program provided a return from the land which had no other productive use.

Future directions in Canada

The gains made in conserving soil, water, and wildlife habitat have occurred against a backdrop of public expenditures for environmental benefits; a willingness by farmers to supplement their income through direct payments for soil and water conservation; and low commodity prices and poor demand for grain.

Farmers will always perceive the need for the land to produce a positive cash flow, regardless of whether the product is grain, livestock, forage, or recreation. However, on-farm costs resulting from conservation practices outweigh on-farm benefits. Benefits to society from conservation are likely greater than benefits accruing to individual farmers (Perlich 1992).

We have learned that we can't buy the factory—it's just too expensive, and it's not socially acceptable. Some, but not all, have learned that you can't force farmers to act in the public interest at their own cost. Some policy studies (Gray et al 1995; Sopuck 1993) concluded that governments would save money if agricultural subsidies were directed to more positive land use programs. Governments chose another course of action. Subsidies were not only decoupled from commodities, but were eliminated all together.

Today, public interest in environmental issues has given way to a preoccupation with job creation, growth of the economy, and deficit reduction. In Canada, the public purse continues to shrink. Funds for environmental enhancement are being cut and will probably continue to be cut.

Under PCP in Canada, NAWMP, and CRP in the U.S., land cover in North America has increased with benefits for society, the environment, and for the

Figure 2. Prices of #1 CWRS Wheat and Calves

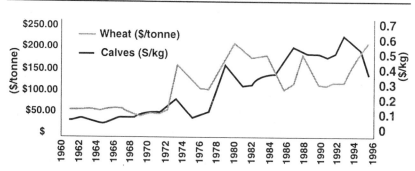

Source: Adapted from Sask Ag & Food and Canadian Wheat Board data (1996).

Figure 3. Annual Precipitation, Saskatoon, Sask., 1911-1992 (Sept.-Aug.)

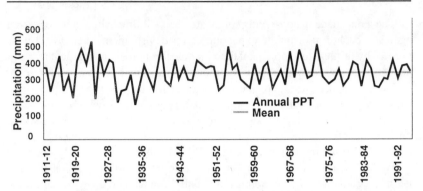

landowner. However, our environmental gains over the past decade could be wiped out by 1 or 2 years of high grain prices and an ability to sell those commodities.

We face new challenges to the sustainability of soil, water, and biodiversity. Farming technology has improved. Prices and demand for grain are rising—even soaring. China is moving up the food chain—from a traditional menu where rice provided 70 percent of calories, to a more diverse fare including meat, milk, butter, eggs, and cheese.

Despite these changes, sustainable development is increasingly recognized as being the objective for society. Since the Brundtland report, "Our Common Future," was released by the World Commission on Environment and Development in 1987, the concept has received increasing acceptance.

In Canada, all federal departments including Agriculture and Agri-Food Canada work on Sustainable Development Strategies. The objective is to integrate environmental sustainability objectives into policies, programs, and activities. The private sector looks at environmental issues from much broader perspectives. Environmental issues are relevant for the primary production, processing, and retail components of the agriculture and agri-food sector. In the National Environment Strategy for Agriculture and Agri-Food (1995), environmental issues and opportunities are identified in production, management, and marketing and trade.

While there is no single solution, the key to sustainable land use lies in the three pillars of environmental, social, and economic sustainability (Figure 4). We must learn how to integrate environmental sustainability with economic viability and social acceptability for truly sustainable development. If any of

Figure 4. Sustainable Development

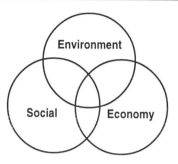

Source: Adapted from the State of the Environment reporting group of Environment Canada, June 1990.

these three pillars are missing, resource management will not be sustainable.

The challenge is to find ways for landowners to capture the benefits from the products and uses of the land, whether they are canola or corn, wheat or ducks, moose, photographs, tourism, campers, trees, native fruits. The challenge for the conservation community, and for policy-makers, is to turn our attention from agricultural policy, which has largely been changed, to other, more integrated land use policies. We simply have to find new ways for farmers to make a living from all products of the land and water.

Notes

1. All dollar values throughout this paper are in Canadian dollars.

References

Acton, D.F., and L.J. Gregorich (eds.). 1995. The health of our soils—Toward sustainable agriculture in Canada. Centre for Land and Biological Resources Research, Research Branch, Agriculture and Agri-Food Canada. Publication 1906/E.

Agriculture and Agri-Food Canada. 1995. National environment strategy for Agriculture and Agri-Food. Prepared for the Federal and Provincial Ministers of Agriculture.

Dumanski, J., D.R. Coote, G. Luciuk, and C. Lok. 1986. Soil conservation in Canada. Journal of Soil & Water Conservation 41:204-210.

Government of Canada. 1991. The state of Canada's environment—1991. Department of the Environment, Government of Canada, Ottawa.

Government of Canada. 1995. Canadian biodiversity strategy. Canada's response to the convention on biological diversity. Minister of Supply and Services Canada.

Gray, R., G. Conacher, and D. Burden. 1994. Decoupled payments for habitat conservation: A preliminary assessment of cost. University of Saskatchewan.

Greer, K.J., J.J. Schoneau, and D.W. Anderson. 1992. Assessment of economic value of topsoil. University of Saskatchewan, Canada.

Knutson, D. 1996. Personal communications. Prairie Habitat Joint Venture Coordinator, Edmonton, Alberta.

Perlich, K. 1992. The economic and environmental benefits and costs of the permanent cover program. Master's thesis, University of Saskatchewan, Canada.

Prairie Farm Rehabilitation Administration, AAFC. 1993. PCP impact on federal government programming : Update. Unpublished, Regina, Saskatchewan.

Prairie Farm Rehabilitation Administration, AAFC. 1994. Environmental benefits of permanent cover programs: An assessment of the economics of wind erosion reduction. P. Brand and M. Bonneau. Unpublished, Regina, Saskatchewan.

Prairie Habitat Joint Venture. 1989. North American Waterfowl Management Plan—Saskatchewan Implementation Strategy.

Sopuck, R.D. 1993. Canada's agricultural and trade policies: implications for rural renewal and biodiversity. Working paper number 19. National Round Table on the Environment and the Economy.

Standing Senate Committee on Agriculture, Fisheries and Forestry. 1984. Hon. H.O. Sparrow, Chair. Soil at risk—Canada's eroding future. Hon. H.O. Sparrow, Chair. Senate of Canada. Ottawa, Ontario.

Statistics Canada. 1992. Agricultural profile of Canada—Part 1. Minister of Industry Service and Technology. Catalogue No. 93-350.

Ward, B. 1996. Prairie Farm Rehabilitation Administration, Analytical division. Regina, Saskatchewan. Personal Communication.

Western Opinion Research, Inc. 1994. PFRA — Permanent Cover Program Study—Final report. Prepared for PFRA—Regina, Saskatchewan.

Working Group on Agriculture and Greenhouse Gases. 1994. Canadian Agriculture and Net Greenhouse Gas Emissions. Ottawa, Canada.

World Commission on Environmental & Development. 1987. Our common future. Gro Brundtland, Chair. Oxford University Press. New York.

17

Evolving Perspectives, Policies, and Recommendations on Soil Erosion in the United Kingdom

Michael A. Fullen

Theoretically, temperate northern Europe should experience little erosion, as moderately high effective precipitation levels occurring throughout the year encourage full vegetative protection (Langbein and Schumm 1958; Fournier 1972). Evidence of erosion in historical times is present in European landscapes, particularly in colluvial deposits (Bell and Boardman 1992) and is largely related to medieval clearances, e.g., in Luxembourg (Kwaad and Mücher 1979), Germany (Bork 1989), and England and Wales (Evans 1990a; Catt 1992). However, in the early to mid-twentieth century, there were few reports of soil erosion. According to Jacks and Whyte (1939 pp. 43, 44) "in northern and central Europe erosion is not a major agricultural factor" and "there are few reports of erosion on agricultural land in northern and central Europe." However, in several papers, mainly published after the war, there were sporadic reports of soil erosion in pre-war Britain. Brade-Birks (1944) noted wind erosion in the Fens of East Anglia and in North Yorkshire, sheet erosion in the Weald of Kent and Sussex and rill erosion in Lincolnshire. Rill erosion was observed by Oakley (1945) in Buckinghamshire and Rogers and Greenham (1948) on sandy soils in Kent.

Wartime food shortages nurtured a new post-war ethos of national self-sufficiency amongst European agricultural ministries. During the war, much land was converted to arable use and policies placed increasing reliance on domestically produced food, rather than imported, food. There were reports of erosion on agricultural soils (e.g., Morris 1942) and evidence of direct correlations between the 'plowing up' campaign of the 1940s and increased rates of colluviation (Smith et al. 1990). Accompanying technological changes assisted in the transformation of the agricultural landscape. Increased mechanization and advances in farm machinery technology encouraged more intensive agriculture. Both the widespread use and improvements in NPK fertilizers enabled farmers to avoid standard rotation practices. The increasing use of herbicides removed weeds more efficiently, resulting in bare soil in winter and spring.

A number of agricultural management trends can be identified which both extended the areas susceptible to erosion and magnified the problem (Chisci 1994). For example, the large scale removal of hedgerows and windbreaks to enable more efficient use of agricultural machinery over larger fields (Baird and Tarrant 1972; Countryside Commission 1977; Pollard et al. 1974) enhanced the erosivity of both running water (Evans and Nortcliff 1978; Reed 1983) and wind (Spence 1955; Caborn 1957; Radley and Simms 1967; Pollard and Millar 1968; Robinson 1969; Wilkinson et al. 1969; Davies and Harrod 1970; Rickard 1979).

Post-war agronomic changes encouraged the continual cultivation of arable soils, causing the oxidation and depletion of soil organic matter (Eagle 1975; Voroney et al. 1981). Generally, soils with less than 2 percent organic content by weight are considered erodible (De Ploey and Poesen 1985). Russell (1977) stated "many rotations used in arable areas of England tend to maintain a level below 1.5 percent" (organic carbon, equivalent to 2.6 percent organic matter). Concern has been expressed over falling soil organic contents in British arable soils (Arden-Clarke and Hodges 1987; Fullen 1991a; Evans 1996). Depletion of soil organic matter decreases aggregate resistance to dispersion (Low 1972; De Meester and Jungerius 1978), allowing slaking and sealing of topsoils with dispersed fines (capping or crusting). De Ploey and Mücher (1981) suggested that crusting is contributing to soil erosion problems and stated "soil crusting is one of the most important processes of temporary or permanent soil degradation in the farmlands of north western Europe."

The more frequent use of increasingly heavy farm machinery is also contributing to soil degradation problems. Compaction by agricultural machinery tends to decrease infiltration rates (Trouse 1966; Davies et al. 1973; Gaheen and Njøs 1977; Fullen 1985a; Robinson and Naghizadeh 1992), thus increasing surface runoff and the likelihood of water erosion. To some extent, recent increases in runoff and erosion on north European soils may reflect the effects of soil compaction in lowering the threshold at which rainfall becomes erosive. On saturated, highly compacted sandy soils, precipitation intensities as low as 1 mm/h caused erosive overland flow in Shropshire (Reed 1979; Fullen 1985a). Intensities of only 2 mm/h falling on loamy soils at field capacity caused erosion in Cornwall (Harrod 1994).

Land use trends further contribute to increased erosion. Cultivation of potatoes and sugar beet often exposes soils to early summer convectional rains (Fullen and Reed 1986). The threefold increase in the areas under winter cereals in Britain between 1969 and 1983 is a further contributory factor (Boardman

1991a). This practice exposes soils to voluminous and prolonged autumn and early winter rains, which are often erosive. Moreover, the use of very smooth rolled seedbeds decreases depression storage and thus contributes to increased volumes of potentially erosive overland flow. Current trends of increasing grazing densities on lowland pastures may contribute to soil compaction and erosion (Foster et al. 1990; Mulholland and Fullen 1991; Mitchell 1994).

From the 1940s to 1970s, much of the literature maintained the view that soil erosion in Britain was slight. According to Stamp (1955) "despite the overwhelming evidence of the universal dangers of soil erosion, it is still the generally held belief that it is unimportant in Britain." The perspectives of U.S. conservationists tended to dominate the literature (Mather 1982). The dual conditions of a mature vegetation cover and low rainfall erosivity regime were stressed by Bennett (1960): "England has numerous small parcels of land, but the gentleness of the rains and the numerous substantial hedges between fields tend to hold erosion to a minimum." Kohnke and Bertrand (1959, p. 17, 18) commented "in certain other areas there is much moisture in the air, but the lack of cold air masses to cool down the moist air is the reason rainfall is only gentle. England is a prime example of this condition. In spite of over a thousand years of cultivation there is little evidence of severe erosion. Much of northern Europe is in a similar condition" and "the gentle rainfall of England has a very low kinetic energy" (p. 102). At this time much European soil erosion research effort was expended on overseas colonial territories, e.g. Hudson and Jackson (1959) in Southern Rhodesia (Zimbabwe) and De Ploey (1967) in the Belgian Congo (Zaire).

Some of the U.S. perspectives are echoed in the British literature. For instance, Davies et al. (1972, p. 240) considered "surges of runoff water, which in the U.S. can gully a field to depths of more than 250–500 mm in a single storm, are practically unknown as there is no serious risk of erosion except on steep slopes." Furthermore, "there is no reason to believe the British climate is likely to change radically, and it can be concluded that the risk of water erosion is not likely to get worse in the future, as at present only very infrequent gully erosion is likely to occur on sloping land on sandy areas." In a paper entitled "Why don't we have soil erosion in England?" Hudson (1967) argued "approximately 5 percent of annual rainfall would fall at intensities greater than one inch per hour below which kinetic energy is insufficient for splash erosion" and later "probably the most important application of the measurement of erosivity is that it allows a simple explanation of why soil erosion is a serious problem in tropical and sub-tropical countries, but barely noteworthy in temperate climates" (Hudson 1971, p. 74).

In a review of soil erosion in Europe, Fournier (1972, pp. 91–92) stated "the climatic conditions of the maritime fringe of moderate temperate Europe encourage the vegetation. They allow it to form an effective protective covering, even when it has deteriorated...The traditional agricultural methods favor conservation, especially the separation of fields by hedges." However, Fournier added a cautionary note: "The systematic elimination of the hedges, favored by certain supporters of regrouping, may give rise to a dangerous situation...it is necessary to act with caution and avoid changing the traditional systems without careful consideration."

A picture of soil erosion has gradually emerged in post-war northern Europe. Generally, national agricultural ministries expressed little interest or concern over soil erosion. Rather, individual research teams from various academic institutions began to survey local erosion patterns, rates, and processes. These early surveys were often conducted by Quaternary geologists and geomorphologists who increasingly saw evidence of agriculturally induced erosion in their landscapes. A particularly valuable contribution has been made by researchers at the Laboratory for Experimental Geomorphology at the Catholic University of Leuven. There, under the leadership of the late Professor Jan De Ploey, fieldwork on the erosion of the loess soils of central Belgium has been combined with advanced research on laboratory simulations of erosion processes. Evidence of accelerated erosion on arable soils was presented in local soil survey reports. While not their primary concern, the types and severity of erosion were often reported. Some of the noteworthy U.K. Reports are reviewed by Reed (1979) and Fullen (1985b).

Douglas (1970) and Evans (1971) drew attention to the evidence of increased erosion. Evans (1971) suggested "although erosion may not be of great importance at present, surely now is the time to investigate the problem." The rather patchy and confusing patterns of soil erosion in the early 1970s were well expressed by Evans and Morgan (1974): "little work has been done in this country on the distribution and causes of soil erosion. Instances of erosion cited in arable areas are mainly of more spectacular wind erosion and, except in some Soil Survey Records, there is little in the literature on water erosion. Indeed, erosion is generally considered to be insignificant because of the low intensity of the rainfall." Since these three keynote papers, many studies of erosion have been completed. Some of the main reviews and local case studies are identified in the list at the end of this chapter, along with papers on specific sediment source areas in the U.K.

The publication of soil erosion maps has provided useful summaries of the state of soil erosion. A pioneering venture was the publication of the soil erosion map of the Federal Republic of Germany (Richter 1965). Later maps have been published and of particular note is the 'Soil Erosion Map of Western Europe,' which shows areas vulnerable to both water and wind erosion, mass movements and flash floods (De Ploey et al. 1989). In addition, Sheet 2 of the 1:10 million GLASOD (Global Assessment of Soil Degradation) Map summarizes soil degradation in Europe, Africa and west/central Asia (Oldeman et al. 1990).

In the U.K., erosion hazard assessment was incorporated in the 1:250,000 Soil Association Map of England and Wales (Soil Survey of England and Wales 1983). Soils are classified as being "at risk" or "slight risk" of erosion and their erosion type as liable to water or wind erosion or both. Of 296 Soil Associations, 50, occupying 23,186 km^2 or 15.3 percent of England and Wales, are considered liable to erosion. Combining data on rainfall erosivity, soil erodibility and land capability in England and Wales, Morgan (1985) estimated that about 20,500 km^2, or 37 percent of the arable area, is at risk of erosion. Morgan (1992) suggested between 37 and 45 percent of the arable area of England and Wales was at risk of soil erosion. Evans (1990b) estimated that 79 Soil Associations, covering 23.9 percent (33,666 km^2) of the surveyed area of England and Wales, were at moderate to very high risk of accelerated erosion. The Soil Survey of England and Wales (renamed the Soil Survey and Land Research Centre in 1991) has produced a 1:625,000 map of risk of soil erosion by water on land under winter cereal cropping (Soil Survey and Land Research Centre 1993).

In 1982, the Soil Survey of England and Wales (SSEW) and the Ministry of Agriculture, Fisheries and Food (MAFF) established an erosion survey covering about 700 km^2 in 17 districts of England and Wales. Data were acquired using aerial photography, ground-truthing, and field survey (Evans and Cook 1986; Evans 1988; Evans 1990c; Evans 1993a; Boardman and Evans 1994). Largely due to financial constraints, the survey was scaled down in 1989. The environmental group "Friends of the Earth" coordinated a soil erosion monitoring scheme, carried out by volunteers (Evans and McLaren 1994).

Integrating the available information suggests 7 areas of lowland arable farming are at risk of accelerated erosion (Boardman and Evans 1994). They are as follows:

- Lower Greensand soils of southern England (including the Isle of Wight),
- Sandy and loamy soils in the West Midlands and Nottinghamshire,
- Sandy and loamy soils in Somerset and Dorset,
- Sandy and loamy soils in parts of East Anglia,
- Chalk soils of the South Downs, Cambridgeshire, Yorkshire and Lincolnshire Wolds, Hampshire and Wiltshire,

- Sandy soils in South Devon,
- Loamy soils in eastern Scotland.

However, the notion of a moderate to severe erosion risk on U.K. soils has been challenged by Frost and Speirs (1996). A field survey of erosion in east central Scotland, after a prolonged rainfall event with a recurrence interval of about 20 years, showed relatively little localized erosion. Based on this evidence, Frost and Speirs have cast doubt on the severity of erosion in the U.K. However, these findings contradict other studies in east central Scotland, which show significant erosion on arable soils (Duck and McManus 1987; Kirkbride and Reeves 1993; Davidson and Harrison 1995; Wade 1996).

Current trends of climatic change and global warming do not provide optimistic scenarios for future soil erosion in the U.K. Most General Circulation Models (GCMs) predict that Mediterranean climatic conditions will extend further north, possibly covering much of Europe south of ~50° N. More northern areas are likely to experience greater amounts of precipitation. Increases in precipitation totals and erosivity, especially that of summer convectional storms, are likely to accelerate erosion rates (Boardman et al. 1990a; Boardman and Favis-Mortlock 1993; Evans 1990c). Furthermore, increased temperatures are likely to contribute to declining soil organic matter contents and concomitant increases in soil erodibility (Evans 1996; Royal Commission 1996).

The uplands of the U.K. are vulnerable to accelerated soil erosion, partly due to their high erodibility. Many upland soils consist of peat, which possesses a low specific gravity and thus low resistance to detachment and transport. Moreover, many upland areas have erodible humose and mineral soils. The erosivity of wind and water is enhanced by vegetation removal, which has increased due to several land management practices and amenity pressures. The burning of upland heather *(Calluna vulgaris)* moorland, whether deliberately as part of moorland management or by accidental fires, exposes bare soil. Overgrazing by sheep, afforestation, industrial pollution, and tourist pressures also disturb vegetation and thus expose soil to accelerated erosion. Keynote papers on the erosional significance of these factors are reviewed in Table 1.

Policy successes, failures, and directions for the future

> The dust is gold that bears the harvest;
> Save the soil that grows our bread;
> Let not wind and rain remove it;
> Guard with care for years
> *S. G. Brade-Birks (1944, p. 189)*

Although there is increasing evidence in the U.K. that soil erosion at least poses a moderate to severe problem at local to regional scales (Boardman and Evans 1994), government interest remains moderate. The Agricultural Development and Advisory Service (ADAS), a constituent part of the Ministry of Agriculture, Fisheries and Food (MAFF), has published two advisory bulletins to help farmers minimize erosion (ADAS 1984, 1985). A 'Code of Good Agricultural Practice for the Protection of Soil' has also been published (MAFF 1993). MAFF recognizes that 'set-aside' schemes may contribute to soil conservation (MAFF 1991, 1994; Margach 1993). An independent government-sponsored report has advised the government to devise and implement a soil protection policy (Royal Commission 1996).

The development of informed debate is critical to the future development of European soil conservation and several organizational developments are assisting with this. The European Society for Soil Conservation (ESSC) was founded in November 1989, with the mission of developing an integrated European approach to the issues of soil erosion and conservation. The ESSC consists of a group of scientists attempting to influence governmental policies and public attitudes towards erosion problems. It consists of 640 members from 46 countries, including 36 European countries (ESSC 1994). The first major meeting of the newly-launched organization at the Coventry Conference in 1990 resulted in a valuable overview of European research (Boardman et al. 1990b). The first ESSC Congress, held at Silsoe College, Bedfordshire, U.K., resulted in a further review (Rickson 1994). The second ESSC Congress was held at Munich in September 1996. Furthermore, the e-mail based 'soil erosion discussion group,' coordinated by the University of Trier (e-mail address: se-list@uni-trier.de), is proving a valuable means of discussion and communication, linking over 360 subscribers worldwide (Bernsdorf and Favis-Mortlock 1995).

The increased activity of the International Erosion Control Association (IECA) in Europe is another welcome trend. The IECA was founded in 1972 and has its headquarters in Steamboat Springs, Colorado. Its activities are mainly in the U.S., but the first European IECA Conference was held in Barcelona in May 1996. The very practical outlook of the IECA, with strong emphasis on technical, engineering, and industrial solutions to erosion and sediment control, should complement and enhance the activities of the more academically orientated European researchers.

In North America, soil erosion research and conservation have a history of more than 60 years (Sanders 1990). Since the 1930s, a great deal of experience has accumulated and consequently many North American policies, lessons, and perspectives have direct relevance to the formulation of European soil conservation policies (Boardman 1991b). In the European context, it should be noted that the U.S. system is well structured to deal with erosion policies at national, regional, and local levels. Policies are in place which are largely advisory, but do incorporate elements of coercion.

The North American soil conservation movement was galvanized by the severe erosion of the 'Dust Bowl' of the U.S. Great Plains States during the 'dirty thirties' and the dynamic leadership of Hugh Hammond Bennett (Hurt 1981). The U.S. Soil Erosion Service was founded in 1933 and became the Soil Conservation Service (SCS) in 1935, a permanent branch of the U.S. Department of Agriculture (USDA) (Soil Conservation Society of America 1980). The SCS (renamed the Natural Resources Conservation Service [NRCS] in 1994), remains the main body responsible for soil conservation, with more than 13,000 employees who are directly under the jurisdiction of the Secretary of Agriculture. The Soil Survey Division of NRCS provides local and regional soil maps, essential for the formulation of soil conservation strategies. At the NRCS National Soil Survey Laboratory in Lincoln, Nebraska, the Soil Survey Division also analyzes samples collected by the staff of about 1,000 scientists, thus ensuring accurate and comparable data.

To administer NRCS policies, soil conservation districts are established, usually adopting county boundaries (the administrative sub-unit of the state). Each district has a conservationist, who is responsible for providing local soil conservation advice. Every district also possesses a Board of Supervisors, consisting of unpaid citizens who share an interest in soil conservation. There are 2,950 conservation districts, covering most of the 171 million ha of U.S. cropland (Steiner 1990). Each state appoints a state conservationist, responsible to the NRCS headquarters in Washington, D.C.

The NRCS adopts a voluntary land-use planning approach to soil conservation. Upon request by farmers, a conservationist will recommend appropriate strategies for reducing soil erosion to acceptable levels. The NRCS does not possess mandatory powers of land control, but non-implementation of conservation policies can exclude farmers from federal and state grant aided programs. The Conservation Title of the Food Security Act (1985) and the Food, Agriculture, Conservation and Trade Act (1990) introduced elements of coercion into soil conservation policies (Napier 1990; Esseks and Kraft 1991). Sodbuster provisions deny farmers access to farm program benefits if they crop highly erodible land without an approved soil conservation plan. Swampbuster provisions aim at preventing conversion of wetland to crop production, while the Conservation Reserve Program (CRP) attempts to retire highly erodible land from agricultural production. The Conservation Compliance Provisions have the greatest potential for long-term reductions in erosion. According to this legislation, operators of highly erodible land were required to have an officially

232

approved plan by January 1, 1990, with full implementation by January 1, 1995. Non-compliance resulted in the loss of all USDA farm benefits until an approved plan was implemented (Napier 1990).

Much can be learned from the North American strategies and policies, which need to be modified and adapted to European conditions. For instance, set-aside policies were borrowed from North America and established to reduce European grain surpluses (Marsh 1987; Clarke 1992). However, there is a great deal of potential for set-aside to be targeted on steep and erodible land (Boardman 1988; Fullen 1991b) and preliminary evidence from the South Downs suggests it is very effective for soil and water conservation (Evans and Boardman, 1996, unpublished data). A "more directed policy" has been proposed by the Royal Commission, which recommended "the government make maximum use of national discretion to adopt environmental and soil protection criteria in the selection of land for set-aside, and encourage this approach to set-aside at EU level" (Royal Commission 1996, Recommendation 24).

There has been some recent limited European soil conservation legislation. For instance, some soil conservation requirements have been incorporated into state law in 1991 in the German Lander of Baden-Württemberg (Jäger 1994) and in the Netherlands by a ruling from the Dutch Agricultural Board (*Landbouwschap*) in 1992 (Boardman et al. 1994). In a review of more than 50 years of U.S. soil conservation experience, Sanders (1990) argued that policies must be developed that are based on thorough analysis and understanding of the problems and should remedy the causes of erosion, rather than simply treating the symptoms.

Rational policies must be designed on national, regional, and local scales. Governments have a crucial role to play, especially at a national level. As an item of policy, governments should state their concern for the status of their national soils and their commitment to proper soil use and conservation. To some extent, the recent Royal Commission (1996) report addresses these issues. Government involvement should not be authoritarian or punitive, but should aim at facilitating conservation by assisting with the identification of problems, the tackling of underlying causes of soil misuse and the encouragement of the necessary action. A senior administrative body or commission would be necessary for such a task. The body should, in consultation with interested parties, establish, promote, and finance research priorities.

The availability of accurate, high quality soil data is pivotal to a successful policy. A national inventory of land resources is necessary; gaps in knowledge could be identified and, where necessary, studies commissioned. National soil survey organizations should play a vital role in providing information (Young 1991). In the U.K., soil survey responsibility in England and Wales is with the

Soil Survey and Land Research Centre (SSLRC), which has its headquarters at Silsoe. The responsible body in Scotland is the Soil Survey of Scotland, based at the Macaulay Land Use Research Institute (MLURI), Aberdeen. In Northern Ireland, soil survey is undertaken by the Department of Agriculture for Northern Ireland (DANI), based at Hillsborough.

European governments have tended to regard national soil survey organizations as rather esoteric entities, their finances often vulnerable to the whims of finance ministries. Much can be learned from the U.S. experience, where the Soil Survey is a respected, properly funded organization, with a relatively high profile in public awareness. European Soil Surveys should consider adopting the policy of U.S. county soil surveys of incorporating Universal Soil Loss Equation (USLE) soil erodibility (K) value assessments into their mapping at the series scale. This approach has proved useful for regional erodibility mapping in Belgium (Pauwels et al. 1980), Denmark (Madsen et al. 1986), and Germany (Becher et al. 1980; Jäger 1994). In addition, temporal trends in the organic content of arable soils should be closely monitored (Fullen 1991a). The SSLRC is evaluating changes in soil organic matter between the early 1980s and 1995. Concentrations are decreasing on arable topsoils and many former peats are now classified as humose mineral soils (Bullock and Burton 1996).

Paucity of information on the costs of erosion impede full evaluation of its effects. Collection of data and evaluation of both on-site and off-site costs are problematic. Many costs are difficult to quantify, borne by various groups (e.g., local councils, water authorities, insurance companies and householders), inherently difficult to collate, and not necessarily directly due to soil erosion. Costing of erosion and related flooding episodes on the South Downs placed off-site costs between $100,000 and $400,000 and up to almost $2 million at Rottingdean in 1987 (Boardman et al. 1994). A tentative costing of both on-site and off-site erosion in England and Wales produced a total of £980 million per year (Evans 1995). Comprehensive and accurate costing of soil erosion would be helpful, both in evaluating the problem and planning policy responses. Adoption of the "polluter pays" principle would promote more effective conservation. The Royal Commission advised that, where it can be shown that failure to follow good management practice caused soil erosion damage to roads or property, then the costs of remedial action should be recovered from the responsible land owner or manager. Royal Commission (1996), Recommendation 19, stated "we recommend local authorities make more effective use of their existing powers to recover the costs of remedying such damage."

On a regional scale, skilled personnel are necessary for consultative duties. With reference to the U.K., Boardman (1988, 1991b) suggested the establishment of a small soil conservation unit within MAFF. Morgan and Rickson (1990) argued that the Danish Land Development Service (*Hedeselskabet*), or Europe's only designated soil conservation service, the *Landsgraedsla Rikisins* of Iceland, could act as organizational models generally within Europe. The establishment of European soil conservation services, whether as distinct entities or as subdivisions of agricultural advisory services, merits discussion. Essential components of any soil conservation service should be free information and advice to agriculturists and interested bodies. Advisory services should also freely disseminate information to the public, particularly educational establishments. The current U.K. policy of charging fees for ADAS advice is counterproductive. It is also imperative that soil conservation field demonstrations are organized, so that farmers can see tangible evidence of the benefits of conservation. In the U.K., the Environment Agency was established by law on April 1, 1996, with a remit to protect the U.K. environment, including its soils. It is likely the Environment Agency will play a significant role in soil conservation (Evans 1996).

As erosion occurs on a field scale, local conservation policies are essential (Evans 1990c). The agriculturist must be able to freely call upon the advice and expertise of soil scientists. In this respect, Europeans have much to learn from the U.S. NRCS. The advisor should assist the farmer in identifying the causes of erosion and selecting the appropriate technologies. The evolution of a conservation plan should be an interactive process between advisor and agriculturist, leading to the development of a range of possible strategies. The costs of remedial or preventative measures may be prohibitive and so the U.S. cost-share system seems appropriate. The NRCS can meet up to 75 percent of the cost of conservation measures. Sanders (1990) stressed the need for local voluntary organizations to discuss erosion problems. The U.S. county soil conservation district could provide a useful model. In such a forum, interested parties meet and discuss local erosion problems and potential solutions.

The Australian Landcare system offers a possible model for group participation in soil conservation. Landcare started in Victoria in 1986 and has grown to encompass 25 percent of the farming community (Campbell 1995). Landcare adopts an integrated and holistic approach to resource sustainability and is a cooperative venture between federal and state government, extension services, consultants, and farmers (Curtis and DeLacy 1995). More than 2,000 Landcare groups cooperate on local land degradation issues, which are usually managed at the catchment scale. The issues addressed include the identification of land degradation problems, the implementation of solutions, the development of demonstration sites and education, in particular encouraging "land literacy" among the participants (Campbell 1995).

Recommendations of the Royal Commission (1996) to the U.K. government embrace many of the proposed strategies, including:

- Recommendation 7: "To complement the monitoring of air and water quality, we recommend the setting-up of a national soil quality monitoring scheme, for which responsibility should lie with central government."
- Recommendation 16: "We recommend that the free advisory visits on pollution prevention and conservation should continue, and should also include advice on appropriate measures for the more sustainable use of soils, including the control of soil erosion."
- Recommendation 18: "We recommend the government consider ways of encouraging farmers to seek advice on erosion control and to draw up farm management plans aimed at minimizing the loss of soil from their land."
- Recommendation 27: "We recommend the Agriculture Departments put research in hand to explore how far erosion of upland peat could be halted or reversed by changes in land use practices."

To some extent, suggested soil conservation organizational structures are developing in South Limburg (the Netherlands), where a coordinated soil conservation project has been conducted since 1991. This involves active collaboration between government (provincial and municipality), agricultural advisory services, three university research institutes and local farmers. Demonstration projects and information dissemination are important components of the program (Boardman et al. 1994). The Limburg *Erosienormeringsprojekt* may well act as a future model for soil conservation policy in northern Europe.

Conclusions

The extent and severity of erosion on U.K. soils have markedly increased over the last 50 years, particularly on arable land. Collation of information from disparate local-scale studies has contributed to increased knowledge of the extent, rates, patterns, and impacts of soil erosion. However, government action and advice on soil conservation have been limited. Policies at national, regional, and local scales should include the initiation of a soil conservation service, full mapping, monitoring, and costing of erosion risk by the three national soil survey organizations and targeting 'set aside' on steep and erodible land.

Selected keynote references on soil erosion in the U.K.

Reviews of U.K. soil erosion
Morgan (1980) Collated evidence of soil erosion in Britain
Gardiner and Burke (1983) Reviews soil erosion in all Ireland, including the drumlin belt of Northern Ireland
Fullen (1985b) Review of soil erosion in Britain
Evans and Cook (1986) Review of soil erosion in Britain
Bullock (1987) Review of causes and possible solutions of soil erosion in the U.K.
Speirs and Frost (1987) Review of causes and extent of soil erosion in the U.K.
Institute of Biology (1991) Argues soil erosion is a local problem and proper evaluation requires the development of a rigorous assessment methodology
Arden-Clarke and Evans (1993) Analysis of agronomic and economic factors promoting soil erosion in the U.K.
Boardman and Evans (1994) Review of results of SSEW/MAFF monitoring scheme
Evans (1993b) Comprehensive review of the causes and severity of soil erosion in upland Britain

Reviews of soil erosion in England and Wales
Evans (1980) Examined factors promoting soil erosion in lowland areas in a survey incorporating 16 English Counties
Morgan (1985) Assessment of the scale of erosion on arable soils, rainfall erosivity patterns, and possible future effects of erosion on crop productivity
Evans and Cook (1986) Preliminary review of results of SSEW/MAFF monitoring scheme
Evans (1990b) Classifies soil associations in terms of erosion risk
Chambers et al. (1992) Results of erosion survey at 13 sites
Evans (1992a) Historical review of soil erosion in England and Wales

Reviews of soil erosion in Scotland
Ragg (1973) Anecdotal account of several soil erosion episodes in different parts of Scotland
Speirs and Frost (1985) Identifies factors promoting high soil erosion rates in eastern Scotland
Duck and McManus (1990) Evidence of increased erosion rates in the Midland Valley, based on reservoir sedimentation

Grieve and Hipkin (1996) Review of erosion in both upland and lowland Scotland, which concludes that accelerated erosion is a significant problem

Sediment source areas in the U.K.

Arable soils

Spence (1955) Reports 'blow' in East Anglian Fens in May 1955

Pollard and Millar (1968) Wind erosion in the East Anglian Fens

Al-Ansari et al. (1977) Increased river sediment concentrations in Glen Almond, Perthshire, due to turnip lifting

McCabe and Collins (1977) Report of colluvial deposits on arable fields on drumlins in County Cavan (Republic of Ireland), considered typical of features in the drumlin belt of Northern Ireland

Evans and Nortcliff (1978) Gully erosion on the Cromer Ridge of Norfolk

Foster (1978) A case study of gully erosion on the Yorkshire Wolds

Pidgeon and Ragg (1979) Wind erosion and crop damage in Morayshire

Reed (1979) Soil erosion in the West Midlands of England

Frost and Speirs (1984) Field survey of erosion near Kelso, Roxburghshire

Colborne and Staines (1985) Water erosion in south Somerset

Fullen (1985a) Water erosion in east Shropshire

Fullen (1985c) Wind erosion in east Shropshire

McManus and Duck (1985) Erosion estimates based on reservoir sedimentation in the Ochil Hills, Scotland

Boardman (1986) Review of soil erosion in south-east England

Boardman and Spivey (1987) Case study of soil erosion in Derbyshire

Duck and McManus (1987) Field study in Angus, Scotland, demonstrating that significant erosion can occur due to prolonged low-intensity rain

Boardman (1991a) Reports relationships between land use and water erosion risk on the South Downs

Watson and Evans (1991) Field survey of rilling around Stonehaven, north-east Scotland

Evans (1992b) Comparative analysis of erosion in Bedfordshire and Lincolnshire

Boardman (1993) A geomorphological analysis of soil erosion on the South Downs

Kirkbride and Reeves (1993) Field study in Angus and Fife, east central Scotland, suggesting runoff produced by prolonged low-intensity rain can be erosive

Boardman (1995) Reviews 15-year survey of flooding and erosion on the South Downs

Davidson and Harrison (1995) Water erosion in central Scotland

Wade (1996) Case study of gully erosion in Fife, demonstrating that snow melt can be highly erosive on sandy soils

Industrial sites

Bridges and Harding (1971) Report on the eroded state of soils due to industrial activities in the lower Swansea Valley

Burnt moorland

Fairbairn (1967) Severe gully erosion along the River Findhorn in the Highland Region of Scotland, due to drainage realignment and moorland burning

Imeson (1971) Effects of heather moorland burning on soil erosion on the North York Moors

Curtis (1975) Evidence of heather moorland burning promoting gully erosion on the North York Moors

Arnett (1979) Runoff plot study showing erosion rates on burnt moorland at 20 times that of *Calluna* moor on the North York Moors

Arnett (1980) Effects on the 1976 Glaisdale 'wildfire' on gully erosion on the North York Moors

Kinako and Gimingham (1980) Effects of heather burning on soil erosion in north-east Scotland

Alam and Harris (1987) Survey of severely burnt moorland on the North York Moors

Overgrazed uplands

Harvey (1974) Field survey and measurement of extreme gully erosion on the Howgill Fells of Cumbria.

Evans (1977) Effects of sheep overgrazing on soil erosion in the Peak District

Evans (1990b) Classifies upland soil associations in terms of erosion risk

Evans (1990d) Review of water and wind erosion in upland Britain with particular reference to the Peak District. Views erosion as the product of several interacting factors, particularly overgrazing

Evans (1992c) Review of factors promoting soil erosion in the Peak District, emphasizing the importance of overgrazing

Royal Commission (1996) Expresses concern over severity of erosion on upland peats in the U.K, which it attributed mainly to overgrazing by sheep

Upland peats

Bower (1962) Review of causes of erosion in upland peat bogs

Radley (1962) Review of causes of peat erosion in the Peak District

Tomlinson (1981) Nature of peat erosion in uplands of Northern Ireland
Carling (1986) Causes of peat slides in the Northern Pennines
Tallis (1987) Study of the varied types, forms, and periods of peat erosion in the
    Peak District
Burt and Labadz (1990) Erosion of the blanket peats of the Southern Pennines,
    largely attributed to industrial pollution damaging vegetation
Tallis and Anderson (1992) Nature of peat erosion in the northern Peak District
Birnie (1993) An erosion pin survey on the Shetland Islands which showed
    erosion rates on bare peat to be 1-4 cm/y

Upland forests
Newson (1980) Effects of forest drainage ditches on accelerated erosion in up-
    land Wales
Battarbee et al. (1985) Forest ditches as sediment source areas in south-west
    Scotland
Moffat (1988) Review of complex erosional responses to forest management
Soutar (1989) Study which presents thesis that upland afforested areas are im-
    portant sediment source areas in the U.K.

Recreational footpaths
Bayfield (1973) Gully erosion associated with footpaths on Cairn Gorm,
    Scotland
Coleman (1981) Quantitative analysis of factors promoting footpath erosion in
    the English Lake District
Watson (1985) Tourist footpaths as sediment source areas on Cairn Gorm
Ferris et al. (1993) Evidence of increased erosion of recreational footpaths in
    the Mourne Mountains of Northern Ireland

References

Agricultural Development and Advisory Service. 1984. Soil Erosion by Water.
    Ministry of Agriculture, Fisheries and Food Advisory Leaflet 890.
Agricultural Development and Advisory Service. 1985. Soil Erosion by Wind.
    Ministry of Agriculture, Fisheries and Food Advisory Leaflet 891.
Alam, M.S., and Harris, R. 1987. Moorland soil erosion and spectral reflec-
    tance. International Journal of Remote Sensing 8:593-608.
Al-Ansari, N.A., M. Al-Jabbari, and J. McManus. 1977. The effects of farming
    upon solid transport in the River Almond, Scotland. International Associa-
    tion of Scientific Hydrology Publications 122:118-125.

Arden-Clarke, C., and D. Hodges. 1987. Soil erosion: the answer lies in organic farming. New Scientist 12 February: 42-43.

Arden-Clarke, C., and R. Evans. 1993. Soil erosion and conservation in the United Kingdom. In D. Pimental (ed.). World Soil Erosion and Conservation. Cambridge University Press, Cambridge, U.K. pp. 193-215.

Arnett, R.R. 1979. The use of differing scales to identify factors controlling denudation rates. In A. F. Pitty (ed.). Geographical Approaches to Fluvial Processes. Geoabstracts, Norwich, U.K. pp. 127-147.

Arnett, R.R. 1980. Soil erosion and heather burning on the North York Moors. In J.C. Doornkamp, K.J. Gregory and A.S. Burn (eds.). Atlas of Drought in Britain. I.B.G. Publications, London, U.K. pp. 45.

Baird, W.W., and J.R. Tarrant. 1972. Vanishing hedgerows. Geographical Magazine 44:545-551.

Battarbee, R.W., P.G. Appleby, K. Odell, and R.J. Flower. 1985. ^{210}Pb dating of Scottish lake sediments, afforestation and accelerated soil erosion. Earth Surface Processes and Landforms 10:137-142.

Bayfield, N.G. 1973. Use and deterioration of some Scottish hill paths. Journal of Applied Ecology 10:635-644.

Becher, H.H., R. Schafer, U. Schwertmann, O. Wittmann, and F. Schmidt. 1980. Experiences in determining the erodibility of soils following Wischmeier in some areas in Bavaria. In M. De Boodt and D. Gabriels (eds.). Assessment of Erosion. Wiley, Chichester, U.K. pp. 203-206.

Bell, M., and J. Boardman (eds.). 1992. Past and Present Soil Erosion. Archaeological and Geographical Perspectives. Oxbow, Oxford.

Bennett, H.H. 1960. Soil erosion in Spain. Geographical Review 50:59-72.

Bernsdorf, B., and D. Favis-Mortlock. 1995. An e-mail based soil erosion discussion list. Letter to the editor. Soil Use and Management 11(1).

Birnie, R.V. 1993. Erosion rates on bare peat in Shetland. Scottish Geographical Magazine 109(1):12-17.

Boardman, J. 1986. The context of soil erosion. Seesoil 3:2-13.

Boardman, J. 1988. Public policy and soil erosion in Britain. In J.M. Hooke (ed.). Geomorphology in Environmental Planning. J. Wiley, Chichester, U.K. pp. 33-50.

Boardman, J. 1991a. Land use, rainfall and erosion risk on the South Downs. Soil Use and Management 7:34-38.

Boardman, J. 1991b. The Canadian experience of soil conservation: a way forward for Britain? International Journal of Environmental Studies 37:263-269.

Boardman, J. 1993. The sensitivity of downland arable land to erosion by water. In D.S.G. Thomas and R.J. Allison (eds.). Landscape Sensitivity. J. Wiley, Chichester, U.K. pp. 211-228.

Boardman, J. 1995. Damage to property by runoff from agricultural land, South Downs, southern England, 1976-93. Geographical Journal 161(2):177-191.

Boardman, J., and D. Spivey. 1987. Flooding and erosion in west Derbyshire, April 1983. East Midland Geographer 10(2):36-44.

Boardman, J., R. Evans, D.T. Favis-Mortlock, and T.M. Harris. 1990a. Climatic change and soil erosion on agricultural land in England and Wales. Land Degradation and Rehabilitation 2:95-106.

Boardman, J., I.D.L. Foster, and J.A. Dearing (eds.). 1990b. Soil Erosion on Agricultural Land. J. Wiley, Chichester, U.K.

Boardman, J., and D.T. Favis-Mortlock. 1993. Climatic change and soil erosion in Britain. Geographical Journal 159(2):179-183.

Boardman, J., and R. Evans. 1994. Soil erosion in Britain: a review. In R. J. Rickson (ed.). Conserving Soil Resources: European Perspectives. CAB International, Wallingford, U.K. pp. 3-12.

Boardman, J., L. Ligneau., A. de Roo, and K. Vandaele. 1994. Flooding of property by runoff from agricultural land in northwestern Europe. Geomorphology 10:183-196.

Bork, H-R. 1989. Soil erosion during the past millennium in central Europe and its significance within the geomorphodynamics of the Holocene. In F. Ahnert (ed.). Landforms and Landform Evolution in West Germany. Catena Supplement 15, Cremlingen-Destedt. pp. 121-131.

Bower, M.M. 1962. The cause of erosion in blanket peat bogs. Scottish Geographical Magazine 78:33-43.

Brade-Birks, S. G. 1944. Good Soil. English University Press, London.

Bridges, E.M., and D.M. Harding. 1971. Micro-erosion processes and factors affecting slope development in the Lower Swansea Valley. In D. Brunsden (ed.). Slopes: Form and Process. I. B. G. Special Publication No. 3, Alden & Mowbray, Oxford, U.K. pp. 65-79.

Bullock, P. 1987. Soil erosion in the U.K. — an appraisal. Journal of the Royal Agricultural Society of England 148:144-157.

Bullock, P., and R.G.O. Burton. 1996. Organic matter levels and trends in the soils of England and Wales. Soil Use and Management 12(2):103-104.

Burt, T., and J. Labadz. 1990. Blanket peat erosion in the Southern Pennines. Geographical Review 3(4):31-35.

Caborn, J.M. 1957. Shelterbelts and Microclimate. Forestry Commission Bulletin No. 29. H.M.S.O., Edinburgh, U.K.

Campbell, C.A. 1995. Landcare: participative Australian approaches to inquiry and learning for sustainability. Journal of Soil and Water Conservation 50(2):125-131.

Carling, P.A. 1986. Peat slides in Teesdale and Weardale, northern Pennines,

July 1983: description and failure mechanisms. Earth Surface Processes and Landforms 11: 193-206.

Catt, J.A. 1992. Soil erosion on the Lower Greensand at Woburn Experimental Farm, Bedfordshire - evidence, history and causes. In M. Bell and J. Boardman (eds.). Past and Present Soil Erosion. Archaeological and Geographical Perspectives. Oxbow, Oxford, U.K. pp. 67-76.

Chambers, B.J., D.B. Davies, and S. Holmes. 1992. Monitoring of water erosion on arable farms in England and Wales. Soil Use and Management 8(4):163-170.

Chisci, G. 1994. Perspectives on soil protection measures in Europe. In R. J. Rickson (ed.) Conserving Soil Resources: European Perspectives. CAB International, Wallingford. pp. 339-353.

Clarke, J. (ed.) 1992. Set-Aside. British Crop Protection Council Monograph No. 50, Farnham.

Colbourne, G.J.N., and S.J. Staines. 1985. Soil erosion in south Somerset. Journal of Agricultural Science, Cambridge 104:107-112.

Coleman, R. 1981. Footpath erosion in the English Lake District. Applied Geography 1:121-131.

Countryside Commission. 1977. New Agricultural Landscapes: Issues, Objectives and Action. Cheltenham, U.K.

Curtis, A., and T. DeLacy. 1995. Evaluating landcare groups in Australia: How they facilitate partnerships between agencies, community groups and researchers. Journal of Soil and Water Conservation 50(1):15-20.

Curtis, L. 1975. Landscape periodicity and soil development. In R. Peel, M. Chisholm and P. Haggett (eds.). Processes in Physical and Human Geography. Heinemann Educational, London, U.K. pp. 249-265.

Davidson, D.A., and D.J. Harrison. 1995. The nature, causes and implications of water erosion on arable land in Scotland. Soil Use and Management 11:63-68.

Davies, D.B., and M.F. Harrod. 1970. The processes and control of wind erosion. National Agricultural Advisory Service Quarterly Review 88:139-150.

Davies, D.B., D.J. Eagle, and J.P. Finney. 1972. Soil Management. Farming Press, Ipswich, U.K.

Davies, D.B., J.B. Finney, and S.J. Richardson. 1973. Relative effects of tractor weight and wheel-slip in causing soil compaction. Journal of Soil Science 24:399-409.

De Meester, T., and P.D. Jungerius. 1978. The relationship between the soil erodibility factor K (Universal Soil Loss Equation), aggregate stability and micromorphological properties of soils in the Hornos area, S. Spain. Earth Surface Processes 3:379-391.

De Ploey, J. 1967. Erosion pluviale au Congo Occidental. In Isotopes in Hydrology. International Atomic Energy Authority, Vienna, Austria. pp. 291-301.

De Ploey, J., and H.J. Mücher. 1981. A Consistency Index and rainwash mechanisms on Belgian loamy soils. Earth Surface Processes and Landforms 6:319-330.

De Ploey, J., and J. Poesen. 1985. Aggregate stability, runoff generation and interrill erosion. In K.S. Richards, R.R. Arnett, and S. Ellis (eds.) Geomorphology and Soils. Allen & Unwin, London, U.K. pp. 99-120.

De Ploey, J., A.V. Auzet., H-R. Bork, N. Misopolinos, G. Rodolfi, M. Sala, and N.G. Silleos. 1989. Soil Erosion Map of Western Europe. Catena Verlag, Cremlingen-Destedt, Germany.

Douglas, I. 1970. Sediment yields from forested and agricultural lands. In J. A. Taylor (ed.) The Role of Water in Agriculture. University College of Wales, Aberystwyth Memorandum No. 12. Permagon Press, Oxford, U.K. pp. 57-88.

Duck, R.W., and J. McManus. 1987. Soil erosion near Barry, Angus. Scottish Geographical Magazine 103(1):44-46.

Duck, R.W., and J. McManus. 1990. Relationships between catchment characteristics, land use and sediment yield in the Midland valley of Scotland. In J. Boardman, I.D.L. Foster, and J.A. Dearing (eds.). Soil Erosion on Agricultural Land. J. Wiley, Chichester, U.K. pp. 285-299.

Eagle, D.J. 1975. A.D.A.S. ley fertility experiments. In Soil Physical Conditions and Crop Production. Ministry of Agriculture, Fisheries and Food. H.M.S.O., London, U.K. pp. 344-359.

Esseks, J.D., and S.E. Kraft. 1991. Land user attitudes toward implementation of conservation compliance farm plans. Journal of Soil and Water Conservation 46(5): 365-370.

European Society for Soil Conservation. 1994. Newsletter 4. Trier, Germany.

Evans, R. 1971. The need for soil conservation. Area 3:20-23.

Evans, R. 1977. Overgrazing and soil erosion on hill pastures with particular reference to the Peak District. Journal of the British Grassland Society 32:65-76.

Evans, R. 1980. Characteristics of water-eroded fields in lowland England. In M. De Boodt, and D. Gabriels (eds.) Assessment of Erosion. Wiley, Chichester, U.K. pp. 77-87.

Evans, R. 1988. Water erosion in England and Wales 1982-84. Report to the Soil Survey and Land Research Centre, Silsoe.

Evans, R. 1990a. Soil erosion: its impact on the English and Welsh landscape since woodland clearance. In J. Boardman, I.D.L. Foster, and J.A.

Dearing (eds.). Soil Erosion on Agricultural Land. J. Wiley, Chichester, U.K. pp. 231-254.

Evans, R. 1990b. Soils at risk of accelerated erosion in England and Wales. Soil Use and Management 6:125-131.

Evans, R. 1990c. Water erosion in British farmers' fields - some causes, impacts, predictions. Progress in Physical Geography 14:199-219.

Evans, R. 1990d. Erosion studies in the Dark Peak. North of England Soils Discussion Group Proceedings 24:39-61.

Evans, R. 1992a. Erosion in England and Wales - the present the key to the past. In M. Bell and J. Boardman (eds.). Past and Present Soil Erosion. Archaeological and Geographical Perspectives. Oxbow, Oxford, U.K. pp. 53-66.

Evans, R. 1992b. Rill erosion in contrasting landscapes. Soil Use and Management. 8(4):170-175.

Evans, R. 1992c. Erosion of rough grazings in Britain. In: J. Boardman (ed.). 1st International European Society for Soil Conservation Congress Post-Congress Tour Guide. pp. 18-22.

Evans, R. 1993a. Extent, frequency and rates of rilling of arable land in localities in England and Wales. In S. Wicherek (ed.). Farm Land Erosion in Temperate Plains, Environments and Hills. Elsevier, Amsterdam, the Netherlands. pp. 177-190.

Evans, R. 1993b. Sensitivity of the British landscape to erosion. In D.S.G. Thomas and R.J. Allison (eds.) Landscape Sensitivity. Wiley, Chichester, U.K. pp. 189-210.

Evans, R. 1995. Soil erosion and land use: towards a sustainable policy. Cambridge Environmental Initiative. Professional Seminar Series No. 7:15-26.

Evans, R. 1996. Some factors influencing accelerated water erosion of arable land. Progress in Physical Geography 20(2):211-221.

Evans, R., and R.P.C. Morgan. 1974. Water erosion of arable land. Area 6:221-225.

Evans, R., and S. Nortcliff. 1978. Soil erosion in north Norfolk. Journal of Agricultural Science, Cambridge 90:185-192.

Evans, R., and S. Cook. 1986. Soil erosion in Britain. Seesoil 3:28-58.

Evans, R., and D. McLaren. 1994. Monitoring water erosion on arable land. Friends of the Earth briefing document.

Fairbairn, W.A. 1967. Erosion in the River Findhorn Valley. Scottish Geographical Magazine 83:46-52.

Ferris, T.M.C., K.A. Lowther, and B.J. Smith. 1993. Changes in footpath degradation 1983-1992: a study of the Brandy Pad, Mourne Mountains. Irish Geography 26(2):133-140.

Foster, I.D.L., R. Grew, and J.A. Dearing. 1990. Magnitude and frequency of sediment transport in agricultural catchments: a paired lake-catchment study in Midland England. In J. Boardman, I.D.L. Foster, and J.A. Dearing (eds.). Soil Erosion on Agricultural Land. J. Wiley, Chichester. U.K. pp. 153-171.

Foster, S. 1978. An example of gullying on arable land in the Yorkshire Wolds. The Naturalist 103:157-161.

Fournier, F. 1972. Soil Conservation. Aspects of Soil Conservation in the Different Climatic and Pedologic Regions of Europe. Nature and Environment Series, Council of Europe, France.

Frost, C.A., and R.B. Speirs. 1984. Water erosion of soils in south-east Scotland - a case study. Research and Development in Agriculture 1(3):145-152.

Frost, C.A., and R.B. Speirs. 1996. Soil erosion from a single rainstorm over an area in East Lothian, Scotland. Soil Use and Management 12(1):8-12.

Fullen, M.A. 1985a. Compaction, hydrological processes and soil erosion on loamy sands in east Shropshire, England. Soil and Tillage Research 6:17-29.

Fullen, M.A. 1985b. Erosion of arable soils in Britain. International Journal of Environmental Studies 26:55-69.

Fullen, M.A. 1985c. Wind erosion of arable soils in east Shropshire (England) during spring 1983. Catena 12:111-120.

Fullen, M.A. 1991a. Soil organic matter and erosion processes on arable loamy sand soils in the West Midlands of England. Soil Technology 4:19-31.

Fullen, M.A. 1991b. A comparison of runoff and erosion rates on bare and grassed loamy sand soils. Soil Use and Management 7(3):136-139.

Fullen, M.A., and A.H. Reed. 1986. Rainfall, runoff and erosion on bare arable soils in east Shropshire, England. Earth Surface Processes and Landforms 11:413-425.

Gaheen, S.A., and A. Njøs, A. 1977. Long term effects of tractor traffic on infiltration rate in an experiment on a loam soil. Agricultural University of Norway, Department of Soil Fertility Management Report 87.

Gardiner, M.J., and W. Burke. 1983. Soil erosion and conservation measures in Ireland. In A.G. Prendergast (ed.). Soil Erosion. Abridged Workshop Proceedings, Florence, October 1982. Commission of the European Communities, Luxembourg. pp. 53-57.

Grieve, I.C., and J.A. Hipkin. 1996. Soil erosion and sustainability. In: A.G. Taylor, J.E. Gordon and M.B. Usher (eds.). Soils, Sustainability and the Natural Heritage. H.M.S.O., Edinburgh. pp. 236-248.

Harrod, T.R. 1994. Runoff, soil erosion and pesticide pollution in Cornwall. In R.J. Rickson (ed.). Conserving Soil Resources: European Perspectives. CAB International, Wallingford, U.K. pp. 105-115.

Harvey, A. 1974. Gully erosion and sediment yield in the Howgill Fells, Westmorland. In: K.J. Gregory and D.E. Walling (eds.). Fluvial Processes in Instrumented Watersheds. I.B.G. Special Publication No. 6, Alden & Mowbray, Oxford. pp. 45-58.

Hudson, N.W. 1967. Why don't we have soil erosion in England? In J.A.C. Gibb (ed.). Proceedings of the Agricultural Engineering Symposium, Paper 5/B/42, National College of Agricultural Engineering, Silsoe, U.K.

Hudson, N.W. 1971. Soil Conservation. Batsford, London.

Hudson, N.W., and D.C. Jackson. 1959. Results achieved in the measurement of erosion and runoff in Southern Rhodesia. In Proceedings of the Third Inter-African Soils Conference, Dalaba, pp. 575-583.

Hurt, R.D. 1981. The Dust Bowl: An Agricultural and Social History. Nelson-Hall, Chicago.

Imeson, A.C. 1971. Heather burning and soil erosion on the North Yorkshire Moors. Journal of Applied Ecology 8:537-542.

Institute of Biology. 1991. Soil erosion in the U.K. Institute of Biology Policy Studies No. 4, London.

Jacks, G.V., and R.O. Whyte. 1939. The Rape of the Earth. A World Survey of Soil Erosion. Faber & Faber, London.

Jäger, S. 1994. Modeling regional soil erosion susceptibility using the Universal Soil Loss Equation and GIS. In R.J. Rickson (ed.). Conserving Soil Resources: European Perspectives. CAB International, Wallingford, U.K. pp. 161-177.

Kinako, P.D.S., and C.H. Gimingham. 1980. Heather burning and soil erosion on upland heaths in Scotland. Journal of Environmental Management 10:277-284.

Kirkbridge, M.P., and Reeves, A.D. 1993. Soil erosion caused by low-intensity rainfall in Angus, Scotland. Applied Geography 13:299-311.

Kohnke, H., and A.R. Bertrand. 1959. Soil Conservation. McGraw-Hill, New York.

Kwaad, F.J.P.M., and H.J. Mücher. 1979. The formation and evolution of colluvium on arable land in northern Luxembourg. Geoderma 22:173-192.

Langbein W.B., and S.A. Schumm. 1958. Yield of sediment in relation to mean annual precipitation. Transactions of the American Geophysical Union 39:257-266.

Low, A.J. 1972. The effect of cultivation on the structure and other physical characteristics of grassland and arable soils (1945-70). Journal of Soil Science 23:363-380.

Madsen, H.B., B. Hasholt, and S.W. Platou. 1986. The development of a computerized erodibility map covering Denmark. In G. Chisci and R.P.C. Morgan (eds.). Soil Erosion in the European Community: Impact of Changing Agriculture. A.A. Balkema, Rotterdam, the Netherlands. pp. 143-154.

Margach, L. (ed.). 1993. Set-Aside Roundup. The Way Ahead. Farming News, London, U.K.

Marsh, J. 1987. The case for 'set-aside.' Span 30(2):50-52.

Mather, A.S. 1982. The changing perception of soil erosion in New Zealand. Geographical Journal 148:207-218.

McCabe, F., and J.F. Collins. 1977. Soil type, soil slope and topsoil depth relationships on a Co. Cavan drumlin. Irish Geography 10:19-27.

McManus, J., and R.W. Duck. 1985. Sediment yield estimates from reservoir siltation in the Ochil Hills, Scotland. Earth Surface Processes and Landforms 10:193-200.

Ministry of Agriculture, Fisheries and Food. 1991. Set Aside. Advisory Leaflet PB0299.

Ministry of Agriculture, Fisheries and Food. 1993. Code of Good Agricultural Practice for the Protection of Soil. MAFF Publication PB0617, London.

Ministry of Agriculture, Fisheries and Food. 1994. Arable Area Payments 1994/95. Explanatory Guide: Part II. MAFF Publication PB1872.

Mitchell, D. J. 1994. Relationships between soil and agronomic parameters and sediment yields. In R.J. Rickson (ed.). Conserving Soil Resources: European Perspectives. CAB International, Wallingford, U.K. pp. 215-231.

Moffatt, A.J. 1988. Forestry and soil erosion in Britain - a review. Soil Use and Management 4(2):41-44.

Morgan, R.P.C. 1980. Soil erosion and conservation in Britain. Progress in Physical Geography 4(1):24-47.

Morgan, R. P. C. 1985. Assessment of soil erosion risk in England and Wales. Soil Use and Management 1:127-131.

Morgan, R.P.C. 1992. Soil conservation options in the U.K. Soil Use and Management 8(4):176-180.

Morgan, R.P.C., and R.J. Rickson. 1990. Issues on soil erosion in Europe: the need for a soil conservation policy. In J. Boardman, I. D. L. Foster, and J. A. Dearing (eds.). Soil Erosion on Agricultural Land. J. Wiley, Chichester, U.K. pp. 591-603.

Morris, F.G. 1942. Severe erosion near Blaydon, County Durham. Geographical Journal 100:257-261.

Mulholland, B., and M.A. Fullen. 1991. Cattle trampling and soil compaction on loamy sands. Soil Use and Management 7:189-193.

Napier, T.L. 1990. The evolution of U.S. soil-conservation policy: from voluntary adoption to coercion. In J. Boardman, I.D.L. Foster and J.A. Dearing (eds.). Soil Erosion on Agricultural Land. J. Wiley, Chichester, U.K. pp. 627-644.

Newson, M. 1980. The erosion of drainage ditches and its effects on bedload yields in mid-Wales. Reconnaissance case studies. Earth Surface Processes 5:275-290.

Oakley, K.P. 1945. Some geological effects of a "cloud-burst" in the Chilterns. Records of Buckinghamshire 15:265-280.

Oldeman, L.R., R.T.A. Hakkeling, and W.G. Sombroek. 1990. World Map of the Status of Human-Induced Soil Degradation. ISRIC/UNEP, Wageningen.

Pauwels, J.M., J. Aelterman, D. Gabriels, A. Bollinne, and P. Rosseau. 1980. Soil erodibility map of Belgium. In M. De Boodt and D. Gabriels (eds.). Assessment of Erosion. J. Wiley, Chichester, U.K. pp. 193-201.

Pidgeon, J.D., and J.M. Ragg. 1979. Soil, climatic and management options for direct drilling cereals in Scotland. Outlook on Agriculture 10:49-55.

Pollard, E., and A. Millar. 1968. Wind erosion in the East Anglian Fens. Weather 23: 415-417.

Pollard, E., M.D. Hooper, and N.W. Moore. 1974. The present position. In E. Pollard (ed.). Hedges. Collins, London, U.K. pp. 59-68.

Radley, J. 1962. Peat erosion on the High Moors of Derbyshire and West Yorkshire. East Midland Geographer 3:40-50.

Radley, J., and C. Simms. 1967. Wind erosion in East Yorkshire. Nature 216:20-22.

Ragg, J.M. 1973. Factors in soil formation. In J. Tivy (ed.). The Organic Resources of Scotland. Oliver & Boyd, Edinburgh, U.K. pp. 38-50.

Reed, A.H. 1979. Accelerated erosion of arable soils in the United Kingdom by rainfall and runoff. Outlook on Agriculture 10:41-48.

Reed, A.H. 1983. The erosion risk of compaction. Soil and Water 11:29,31,33.

Richter, G. 1965. Bodenerosion und Raumforschung. Bundesanstalt für Landeskunde und Raumforschung, Bad Godesberg, Germany.

Rickard, P. 1979. Blowing - what it costs and ways to prevent it. Arable Farming 6:79-83.

Rickson, R.J. (ed.). 1994. Conserving Soil Resources: European Perspectives. CAB International, Wallingford, U.K.

Robinson, D.A., and R. Naghizadah. 1992. The impact of cultivation practice and wheelings on runoff generation and soil erosion on the South Downs: some experimental results using simulated rainfall. Soil Use and Management 8(4):151-156.

Robinson, D.N. 1969. Soil erosion by wind in Lincolnshire, March 1968. East Midland Geographer 4:351-362.

Rogers, W.S., and D.W. Greenham. 1948. Soil management, with special reference to fruit plantations. Journal of the Royal Agricultural Society of England 109:194-211.

Royal Commission 1996. Report of the Royal Commission on Environmental Protection. Nineteenth Report. Sustainable Use of Soil. Report Cm 3165. Sir John Houghton (Chairman). H.M.S.O., London.

Russell, E.W. 1977. The role of organic matter in soil fertility. Philosophical Transactions of the Royal Society, London B281:209-219.

Sanders, D.W. 1990. New strategies for soil conservation. Journal of Soil and Water Conservation 45(5):511-516.

Smith, J.P., M.A. Fullen, and S. Tavner. 1990. Some magnetic and geochemical properties of soils developed on Triassic substrates and their use in the characterization of colluvium. In J. Boardman, I.D.L. Foster and J.A. Dearing (eds.). Soil Erosion on Agricultural Land. J. Wiley, Chichester, U.K. pp. 255-271.

Soil Conservation Society of America. 1980. Soil Conservation Policies: An Assessment. Soil Conservation Society of America, Ankeny.

Soil Survey and Land Research Centre. 1993. 1:625,000 Map of risk of soil erosion in England and Wales, by water on land under winter cereal cropping. Soil Survey and Land Research Centre, Silsoe.

Soil Survey of England and Wales. 1983. 1:250,000 Soil Association Map of England and Wales. Lawes Agricultural Trust, Harpenden.

Soutar, R.G. 1989. Afforestation and sediment yields in British fresh waters. Soil Use and Management 5(2):82-86.

Speirs, R.B., and Frost, C.A. 1985. The increasing incidence of accelerated soil water erosion on arable land in the east of Scotland. Research and Development in Agriculture 2:161-167.

Speirs, R.B., and C.A. Frost. 1987. Soil water erosion on arable land in the United Kingdom. Outlook on Agriculture 4:1-11.

Spence, M.T. 1955. Wind erosion in the Fens. Meteorological Magazine 34:304-307.

Stamp, D. 1955. Man on the Land. Collins, London.

Steiner, F.R. 1990. Soil Conservation in the United States. Policy and Planning. John Hopkins University Press, Baltimore.

Tallis, J.H. 1987. Fire and flood at Holme Moss: erosion processes in an upland blanket mire. Journal of Ecology 75:1099-1129.

Tallis, J., and P. Anderson. 1992. Excursion to Holme Moss. In: J. Boardman (ed.). 1st. International European Society for Soil Conservation Congress Post-Congress Tour Guide. pp. 24-28.

Tomlinson, R.W. 1981. The erosion of peat in the uplands of Northern Ireland. Irish Geography 14:51-64.

Trouse, A.C. 1966. Alteration of the infiltration permeability capacity of tropical soils by vehicular traffic. In Proceedings of the First Pan-American Soil Conservation Congress. São Paulo, Brazil. pp. 1103-1109.

Voroney, R.P., J.A. Van Veen, and E.A. Paul. 1981. Organic C dynamics in grassland soils 2. Model validation and simulation of the long-term effects of cultivation and rainfall erosion. Canadian Journal of Soil Science 61:211-224.

Wade, R. 1996. A snowmelt generated soil erosion event in Fife, Scotland. In: A.G. Taylor, J.E. Gordon, and M. B. Usher (eds.). Soils, Sustainability and the Natural Heritage. H.M.S.O., Edinburgh. pp. 264-266.

Watson, A. 1985. Soil erosion and vegetation damage near ski lifts at Cairn Gorm, Scotland. Biological Conservation 33:363-381.

Watson, A., and R. Evans. 1991. A comparison of estimates of soil erosion made in the field and from photographs. Soil and Tillage Research 19:17-27.

Wilkinson, B., W. Broughton, and J. Parker-Sutton. 1969. Survey of wind erosion on sandy soils in the East Midlands. Experimental Husbandry 18:53-59.

Young, A. 1991. Soil monitoring: a basic task for soil survey organizations. Soil Use and Management 7(3):126-130.

18

Landscape and Nature Conservation Policies in Denmark

Alex Dubgaard

With no mountains and few other natural barriers to cultivation and development, most of Denmark's land area is in some form of intensive use. At present, 17 percent of the country is built upon or used for infrastructure. However, it is not urban development that has changed the Danish landscape most profoundly. A strong tradition in land-use regulation has confined urban development and maintained the impression of rural landscapes in most of the country. It is agricultural expansion and intensification that have exerted the most penetrating influence on the visual and ecological characteristics of the Danish countryside. At the beginning of the nineteenth century about half the country's area was semi-natural environments like meadows, heathlands, and dry pastures. At present, the share is less than 5 percent. Some of the former semi-natural areas—especially heathlands—have been afforested, but most of them are now used for arable cultivation.

Until quite recently, few in Denmark disputed the social desirability of agricultural expansion and intensification. The loss of traditional landscape amenities was widely accepted as a necessary sacrifice to ensure economic growth and prosperity. Also, in contrast to industry, agriculture was generally viewed as a non-polluting activity. However, at the beginning of the 1980s, it became clear that modern agricultural practices have serious detrimental effects on the environment. In addition to this, agricultural surplus problems and increasing support payments gradually eroded the social legitimacy of continued agricultural expansion. As a result, a considerable body of legislation and policy programs have been implemented since the mid-1980s to control agricultural pollution and preserve and enhance landscape amenities and biodiversity.

The main purpose of this report is to review the policy programs and legislation that affect the visual, ecological, and recreational qualities of the Danish landscape. The policy measures applied will be assessed with emphasis on the economic and property rights implications, rather than the legal and regulatory features. A more detailed inquiry into the legal and administrative aspects of Danish landscape policies can be found in Primdahl (1996). To provide background information on the issues relating to nature and landscape protection in Denmark, the report will also summarize the structural development in Danish agriculture and the policy programs introduced to control nitrogen and pesticide pollution.

Transformation of the Danish landscape

Land reclamation

Human activity has shaped the Danish landscape during the last 5,000 years—with increasing intensity during the last few centuries. A memorandum from the Ministry of the Environment (1995) summarizes the development during the past 200 years as follows: "In the beginning of the nineteenth century almost half the country's area was semi-natural landscapes like meadows, heathlands, and dry pastures. At present, the share is less than 5 percent. In the same 200-year period the total surface area of the Danish lakes has been more than halved and at least 90 percent of the water courses have been dredged, deepened, straightened or culverted. In this century 75 percent of the ponds have been drained and reclaimed and 60 percent of the hedgerows and field banks have been abolished."

The reclamation of the vast Danish heathlands and wetlands gained pace in the second half of the nineteenth century—inspired and supported by The Danish Heath Society (founded in 1866). After the loss of Schleswig-Holstein in the war in 1864, land reclamation was viewed as a national cause, disputed only by a few poets and aesthetes praising the heath for its beauty and solitude. The efforts to reclaim heartlands and wetlands were largely completed by the middle of the twentieth century. At the beginning of the 1950s, semi-natural grasslands, heaths, and bogs had been reduced to about 10 percent of the country's area (Statistical Yearbook 1958).

Although extensive land reclamation was coming to an end by the middle of this century, Table 1 shows that the trend toward a reduction in natural and semi-natural areas has continued during the last four to five decades. A review of the structural changes in Danish agriculture will probably be useful as background information on both the causes of this development and the policy measures adopted to halt and reverse the trend toward an impoverishment of the visual, ecological, and recreational qualities of the Danish landscape.

Structural changes in Danish agriculture

Table 2 shows that agricultural production in Denmark increased by about 60 percent from 1960 to 1993. Compared with several other European coun-

Table 1. Land-use scenario, 1951–1995

	—1951—		—1965—		—1995—	
	%	1000 ha	%	1000 ha	%	1000 ha
Arable land						
(incl. horticulture)	64	2,752	62	2,690	58	2,510
Permanent grasslands*	10	410	8	330	4	200
Forests, woodlands	10	440	11	470	12	500
Heaths, bogs, dunes	6	280	5	220	4	200
Lakes, watercourses	2	60	2	70	2	65
Hedgerows, ditches,						
field lanes	NA	NA	3	140	3	120
Urban, summerhouses,						
infrastructure, etc.	8	350	9	390	17	713
Total area of Denmark	100	4,293	100	4,307	100	4,308

* Including culture grass, semi-natural dry grassland, meadows, marshlands, etc.

Sources: Statistical Yearbooks, and Ministry of the Environment (1995).

Table 2. Production and factor use in Danish agriculture, 1960–1993

	1965	1975	1985	1990	1993
			(1960 = 100)		
Number of holdings	88	68	47	40	35
Gross production					
(in fixed prices)	105	103	143	156	161
Intermediate inputs					
(in fixed prices)	103	110	140	141	148
Fertilizer nitrogen per ha*	160	290	338	355	323
Pesticides** (1975 = 100)	NA	100	240	270	200
Total agricultural area	97	95	93	91	89
Labor (man-years)	76	48	36	30	29

* Pure nitrogen in chemical fertilizer. Set-aside land excluded.
** Number of treatments per ha.

Sources: Agricultural Statistics. Danmarks Statistik: Dubgaard (1987).

tries, this is a rather modest growth. It has meant that in most of the country livestock densities do not exceed tolerable levels from a waste viewpoint. Still, structural change and specialization have changed Danish agriculture profoundly during the last three to four decades. This in turn has affected the visual and biological characteristics of the landscape.

The number of holdings has been reduced by almost two-thirds since 1960, while the agricultural labor force has decreased by more than 70 percent. The main determinants of this trend are technological and economic changes beyond the control of the individual farmer. Deteriorating sector terms of trade—especially when measured in relation to the wage rate—have provided strong economic incentives to implement labor-saving technologies and increase farm size.

The search for economies of scale has led to increasing specialization in Danish agriculture. Mixed farming was still predominant up to the mid-1960s—with cattle on 80 percent of the farms and three out of four farms having both cattle and pigs. In the mid-1990s the share of farms with cattle had dropped to about 45 percent and only 15 percent of Danish farms had a mixed livestock production. In addition, regional specialization has led to a concentration of the dairy herd in the western parts of the country (with predominantly sandy soils).

Another feature which has affected the environment is the growing intensity of fertilizer and pesticides. Table 2 shows that the input of nitrogen fertilizer per ha has more than tripled since 1960 while pesticide use has doubled since the mid-1970s. There is no direct link between the use of chemical inputs and environmental damage. Still, the steep increase in the intensity of fertilizer and pesticides has been accompanied by increasing pollution problems and a loss of biodiversity in the farmed countryside (Ministry of the Environment 1995).

Specialization and intensification have affected land-use patterns leading to a shift from grass and root crops to grain production and a conversion of permanent grass to arable land. Table 3 shows that the area under grass and root crops has dropped from 50 to 28 percent of the corresponding increase. The area under permanent grass was reduced from 11 to only 7 percent of the agricultural area.

Table 3. Utilization of the agricultural area* in Denmark, 1960–1993

	1960	1975	1985	1990	1993
			%		
Cereals, pulses, and (oil) seeds	49	64	70	71	65
Root crops	18	10	8	8	7
Temporary grass	21	16	13	12	14
Permanent grass	11	9	8	8	7
Set-aside	1	1	1	1	7
Total	**100**	**100**	**100**	**100**	**100**

* The agricultural area equals 63 percent of the total area (cf. Table 1).

Source: Landbrugsstatistik (Agricultural Statistics), Danmarks Statistik.

Landscape effects of agricultural change

Specialization and intensification have reduced the amenities associated with mixed agriculture and probably contributed to the reduction in biodiversity seen in the agricultural landscape during the last few decades. Fields have been amalgamated into larger units causing hedgerows, dikes, ditches, and other field boundaries to be abolished. Efforts to create more uniform fields led to the removal of a large number of ponds. Extensive grasslands have become economically marginal and to a large extent reclaimed for arable cultivation or abandoned by agriculture. Many studies signify that the principal factors causing the impoverishment of wildlife in agricultural landscapes are habitat loss and growing vegetational uniformity. Recent research indicates that increasing pesticide intensity has contributed significantly to the decline in farmland bird species (Danish Environmental Protection Agency 1989).

Review of Danish agri-environmental policies

The Danish economy experienced its industrial take-off in the 1890s (Hansen 1972). However, agriculture retained its position as the main export sector until the beginning of the 1960s when agricultural exports were surpassed by industrial products. Agriculture was seen by a great deal of the public as the backbone of the Danish economy for much longer. At present, agricultural products account for about one-fifth of the total Danish exports of goods and services, and the farm sector's share of total employment has dropped to 4 percent— compared to more than 20 percent at the beginning of the 1950s.

In the 1980s, the public perception of the social role of agriculture changed rather abruptly.[1] A growing amount of scientific evidence indicated that modern agricultural practices had significant detrimental effects on the environment, and mounting agricultural surpluses made it increasingly difficult to justify these practices. As a result, land-use policies started to change, shifting the focus from the enhancement of agricultural productivity to the preservation of environmental values and landscape amenities.

Figure 1 provides an overview of the Danish agri-environmental policy measures implemented since the mid-1980s.

Nitrogen pollution programs

From the outset—at the beginning of the 1980s—the Danish nitrogen pollution debate has focused attention on the nitrate contamination of drinking water as well as nitrogen pollution of surface water, especially eutrophication and degradation of the marine environment. Excess nutrients over-fertilize water bodies, causing algal blooms that deplete oxygen as they decay. This may result in a loss of amenity and recreational opportunities—and in severe cases of fish

Figure 1. Danish agri-environmental policy measures

Nitrogen and Phosphorus Pollution Programs
Manure storage requirements
- Minimum storage capacity approximately nine months
- Liquid manure must be stored in tanks of impermeable material
- Solid manure and silage must be stored on basis of impermeable material with run-off to storage tank

Manure application standards
- Ban on autumn and winter spreading of liquid manure
- Minimum utilization rates for nitrogen content in manure (from 1997):
  - Pig slurry = 50 percent
  - Cattle slurry = 45 percent
  - Other animal waste = 40 percent

Fertilizer intensity
- Mandatory nutrient bookkeeping and control
- Total nitrogen use must not exceed economically optimal N-rates
- Amount of nitrogen produced in manure must not exceed
  - 180 kg N per ha on pig farms
  - 240 kg N per ha on cattle farms
- Support to farmers reducing N-fertilizer levels (EU Regulation 2078/92)

Other nutrient pollution provisions
- Autumn crop cover on at least 65 percent of cultivated area
- Two-meter non-cultivated zones along water courses and lakes

Pesticide Reduction Program
Pesticide application provisions
- Mandatory education of farmers using spraying equipment
- Mandatory pesticide application bookkeeping
- Support to ray grass as a nitrogen catch crop in cereals (EU Regulation 2078/92)

Reregistration of pesticides
- Removal of chemicals that are significantly toxic, carcinogenic, highly mobile in soil, slowly decaying, or liable to cause embryonic or genetic changes

Eco-tax (ad valorem) on pesticides
- 37 percent on insecticides
- 15 percent on herbicides, fungicides, and growth regulation products

Organic farming
- One-time conversion grant = 850 DDK per ha
- Current support = 850 DDK per ha per year

---

kills. Nitrogen is considered the limiting factor on marine plant growth, while phosphorus limits freshwater plant growth (Committee on Environmental Research 1991). Ammonia emissions, mainly from livestock operations and manure handling, contribute to the eutrophication of the marine environment. Acidification resulting from ammonia volatilization is a threat to the unique flora found on heathland and elevated bogs (Ministry of the Environment 1995).

In 1987, the Danish Parliament approved an extensive *Aquatic Environment Program* to protect the environment from nutrient pollution. The program called for an 80 percent reduction in phosphorus emissions, mainly through the improvement of urban waste water treatment[2] and a 50 percent reduction in nitrogen losses from agriculture. In 1991, the Aquatic Environment Program was supplemented by the *Action Plan for a Sustainable Agriculture*. The most important provisions are listed in Figure 1.

The *EU (European Union) Nitrate Directive* from 1991 requires the designation of *Nitrate Sensitive Areas*. In these areas provisions for good agricultural practices must be implemented to protect groundwater resources. Denmark has designated the whole agricultural area as nitrate sensitive (Ministry of Agriculture and Fisheries 1996). Consequently, the nitrogen application standards listed above apply everywhere in the country. The provisions of the Nitrate Directive are more or less in line with the Danish nitrate application standards that had already been implemented through national legislation.

The reduction targets for urban nutrient pollution are being realized as planned, but as yet the agri-environmental measures have failed to solve the nonpoint nitrogen pollution problems (Ministry of the Environment 1995). New measures are being implemented to protect groundwater from nitrate contamination. The most important are the designation of groundwater protection areas and afforestation (addressed below).

From a land economics perspective it is worth mentioning that agri-environmental policies have (re)linked livestock production to arable land. The new manure application standards require that there is sufficient agricultural land available to facilitate adequate plant uptake of the nitrogen produced on a farm in the form of animal waste. Until now, this has not caused serious problems for livestock farmers in Denmark. However, an expansion of livestock productions requires a proportionate increase in the holding's agricultural area. In areas with high stocking rates this has led to increasing competition for land and rising land prices.

Pesticide reduction program

As in many other European countries, pesticide application has led to groundwater contamination in Denmark. Water pollution is not, however, the only environmental hazard associated with the use of pesticides. In Denmark considerable emphasis is placed on the side effects of pesticides on flora and fauna. The rapidly increasing intensity of pesticide use is probably one of the causes of the decline in a number of farmland bird populations. In 1986 an action plan was introduced to reduce pesticide application. The goal of reduced pesticide application is interpreted as an equivalent percentage reduction in the *amount of active ingredients* as well as the *number of treatments applied* (the number of treatments being measured as the number of times the whole agricultural area can be treated with the purchased amount of pesticides assuming that *labeled* dosages are applied). The use of the number of treatments as a measure of application intensity is based on the assumption that the frequency of spraying is an important indicator of the side effects on flora and fauna from pesticide application and the danger of development of resistance in pest populations.

The provisions introduced to realize the program's targets are stricter pesticide registration standards; mandatory education of farmers using spraying equipment; and mandatory pesticide application bookkeeping. Pesticides licensed under the old legislation must be reregistered. The aim is to prevent the continued use of chemicals which are significantly toxic, carcinogenic, highly mobile in soil, slowly decaying or liable to cause embryogenic or genetic changes. This part of the program has recently been bolstered by new legislation limiting pesticide producers' possibilities to inhibit the reregistration process through appeals. The ongoing efforts to reregister "old" pesticides in Denmark are in line with the EU pesticides directive (from 1991) demanding a reregistration of all pesticides licensed before 1993.

Economic instruments in Danish agri-environmental policy

Eco-tax on pesticides

The regulatory measures introduced to curtail pesticide use have failed to attain the targeted reductions. In January 1996, an eco-tax was imposed on pesticides to strengthen the incentives for farmers to reduce pesticide intensity. The new levy is an *ad valorem* tax at the retail level. The tax rate is 37 percent for insecticides and 15 percent for herbicides, fungicides, and growth regulation products (Danish Environmental Protection Agency 1995). The tax revenue will be remitted to the farm sector in the form of a reduction in land taxes.

It is still too early to tell what the impact of the levy will be in terms of reduced pesticide utilization, but model simulations indicate that the price induced reduction in pesticide use will probably be less than 10 percent (Dubgaard 1987; Danish Environmental Protection Agency 1995). Even so, introducing an eco-tax rather than strengthening the bureaucratic efforts to control farmers may signal an increasing political willingness to use the Polluter-Pays-Principle in relation to agriculture.

Elimination of support to drainage and irrigation

In fact, economic policy instruments, in the form of the *removal* of economic support to environmentally detrimental activities in agriculture, have previously been employed in Denmark to protect landscape amenities. To reduce the economic incentives to reclaim marginal land and habitats, subsidies to drainage were abolished in 1987. Irrigation can be environmentally damaging by drawing down the water table and drying out wetlands and watercourses. Subsidies for irrigation investments were abolished in 1983.

Support to organic farming

The standards for organic farming prohibit the use of chemical fertilizers and pesticides, and it is the general perception that organic practices are environmentally more benign than conventional agriculture (Ministry of Agriculture and Fisheries 1996). In 1987, the first act was passed to promote organic farming in Denmark. After a number of revisions, the program now offers farmers an 850 DKK[3] one-time conversion grant per ha and a current subsidy of the same magnitude, i.e., 850 DKK per ha per year (Ministry of Agriculture 1994). In 1996, about 2 percent of the agricultural area in Denmark had been converted to organic farming or was in a conversion program. Dairy production was the principal enterprise on most of the organic farms, and two-thirds of the area was used for grass and green fodder (Ministry of Agriculture and Fisheries 1996).

Policies to preserve and restore landscape qualities

There is not a tradition in Denmark for the designation of areas as national parks. Neither is there a strong tradition for public land acquisitions. Landscape and nature preservation policies have been based primarily on regulation limiting private property rights. In 1805, an ordinance was implemented protecting forests and woodlands from conversion to other uses. The first nature conservation act was passed in 1917. Under this legislation, conservation orders could be imposed on specifically designated areas with a one-time compensation paid to the owners for economic losses, i.e., the discounted value of rent foregone.

In the 1960s, general planning regulation was introduced, zoning all land as either urban or rural—with land in the rural zone being protected from urban development. The protection of the countryside from development was further strengthened by the imposition of forest and coast line preservation zones. From an economic perspective, general regulation entails a redesignation of property rights (without compensation) deferring the opportunity to develop the land from the individual owner to society.

Since the mid-1980s, Danish landscape policies have taken a more pro-active approach. New legislation allows state acquisition of land for afforestation and landscape management. During the last few years the national landscape improvement efforts have been supplemented with EU programs supporting afforestation, extensification of agricultural land and landscape management.

Figure 2 provides an overview of the National and EU landscape policy programs implemented in Denmark.

Conservation orders

Conservation orders and management directions can be imposed on specifically designated areas of particular scenic or ecological importance and on species of plants and animals. Landowners are compensated for losses incurred as a result of conservation restrictions and public access rights imposed on their property. Since the first nature conservation act was passed in 1917, 5 percent of the Danish land area has come under a conservation order.

General nature protection regulation

In 1983, a Nature Protection Act was passed introducing general provisions to protect nature areas and habitats in the agricultural landscape from intensification or elimination. After subsequent revisions of this legislation, even quite small nature areas and habitats are protected.

- Heaths, bogs, salt marshes, meadows, and semi-natural pastures larger than 0.25 ha;
- Ponds and lakes larger than 0.01 ha.

The major part of these areas is private property. In contrast to losses incurred from restrictions under conservation orders, landowners are *not* eligible for compensation when limitations are imposed through general regulation. It has been estimated that 250,000 ha of heaths, bogs, salt marshes, meadows, and semi-natural dry pastures are protected under the general provisions of this legislation (Ministry of the Environment 1995). However, it is becoming increasingly difficult to sustain the extensive agricultural management (grazing or hay making) required to preserve the visual and ecological characteristics of semi-natural grasslands. These landscape features were previously an unintended result of commercial agricultural activities. During the last few decades extensive grasslands have become economically marginal. In principle, owners have an obligation to "maintain" the landscape through (some form of) agricultural use. Economically this could transform marginal land from an asset to a liability for the owners. In practice, however, there is no effective enforcement of landowners' obligations to use land which is no longer rent-bearing. Instead, public support is provided to extensive grazing.

Bird protection and habitat directives

Denmark has designated 9,600 km^2 as bird protection areas under the EU Bird Protection Directive from 1979. Salt and brackish waters comprise 75 percent of the designated area, but considerable agricultural areas are also included (Ministry of Agriculture and Fisheries 1996). The designated areas are protected under conservation orders and the Nature Protection Act. One hundred and seventy-five large areas have recently been designated as protected habitats for flora and fauna under the EU Habitat Directive from 1992. Most of these areas are already protected under the Bird Protection Directive or conservation orders.

Other environmental measures

Figure 2. National and EU landscape policy measures

General regulations without compensation
- Planning regulation zoning land as either urban or rural
- Forest and coastline preservation zones
- Preservation of heaths, bogs, salt marshes, meadows, and semi-natural pastures larger than 0.25 ha
- Preservation of most water courses, and ponds and lakes larger than 0.01 ha
- Two-meter-wide uncultivated zones along watercourses

Selective conservation with compensation
- Conservation orders and management directions
- Ban on re-drainage of land with a high ochre pollution potential

Support programs
- Afforestation program
- Nature restoration (lakes, watercourses, heathland, bogs, etc.)
- Grants for the planting of shelterbelts
- Support to landscape management
- Extensification of permanent grassland (EU Regulation 2078/92)
- Set-aside agreements of 20 years duration (EU Regulation 2078/92)

In the western parts of the country many soils have a large content of fer-ruginous sediments and drainage has led to serious ochre pollution of water bodies. In 1985 legislation was passed prohibiting (new) drainage of land with a high ochre pollution potential. Up to 10 percent of the agricultural area in Denmark is affected by this legislation. Implicitly the ochre pollution legisla-tion acknowledges that land ownership entails a right to land improvement through drainage. Should a drainage permission application be turned down due to high ochre pollution potential, the landowner is eligible for compensation.

It is also worth mentioning that there has been a ban on straw burning in fields since 1990 to reduce air pollution and prevent damage to hedgerows, woods, and other habitats.

Nature restoration and landscape management

In the mid-1980s, it was anticipated that agricultural surpluses would re-quire changes in the EC's Common Agricultural Policy that in turn would lead to widespread marginalization and abandonment of agricultural land. The de-liberations about marginalization and surplus land led to the formulation of new landscape policies in Denmark. With the Nature Management Act of 1989 a strategy was introduced aimed at actively promoting alternative land use op-tions such as afforestation and reestablishment of lakes and wetlands. The Na-ture Management Act (since 1992 incorporated into the Nature Protection Act) provides the following policy measures: land acquisitions; afforestation; resto-ration of important habitats and landscape amenities; and management agree-ments with private landowners. Under the terms of the act, expropriation is pos-sible, but as yet all land acquisitions have been based on voluntary agreements.

Nature restoration

Typically nature restoration projects are undertaken to re-establish lakes, which have been drained and reclaimed, and to return streams and rivers, which have been straightened, to their original meandering courses. Nature restora-tion takes place mainly on land which has been purchased by the state. To a lesser extent individual agreements are established with private landowners who receive support to create or restore small habitats, trails, etc.

Afforestation programs. The Danish afforestation program (initiated in 1989) is the most conspicuous result of the marginalization discussion mentioned above. It aims at doubling the forest area (from 12 to 24 percent of the country's terri-tory)—over the next 100 years, though. Of course, with a 100-year time hori-zon the target of doubling the forest area is rather elusive. The more specific objective is annual afforestation of 5,000 ha of agricultural land. About half of the target area is intended to be afforested by the state.

At the outset, afforestation was seen primarily as a residual land use option in marginal areas. However, the marginalization predictions of the late 1980s have proved wrong, partly because the CAP reform has linked support to land use rather than yield. Still, it is probably of greater significance that agri-environmental legislation has linked livestock production to arable land (cf. Section 3.1). In areas with high stocking rates, demand for land is increasing and land prices are rising, irrespective of soil quality. Now the emphasis of the state afforestation scheme is placed on multipurpose afforestation projects in urban fringe areas where landscape amenity and recreation play a major role (Forestry and Nature Protection Agency 1994A).

Support to private afforestation is differentiated with the highest rate obtainable in designated afforestation zones. The afforestation zones (approximately 180,000 ha) are located primarily on poor soils, in urban fringe areas, adjacent to existing forests, and in groundwater protection areas. From 1994, support to private as well as public afforestation projects qualifies for EU cofinancing.

Sustainable forestry program. The Forest Declaration of the Rio Conference (1992) and the Helsinki Conference on Forest Protection (1993) demand that initiatives be taken by the national governments to ensure the sustainable use of forests. In a report from 1994, the Danish Forestry & Nature Protection Agency (1994B) claims that Danish forestry—with the afforestation program—*is* sustainable to a considerable extent. However, initiatives have to be taken to conserve and enhance biodiversity and amenities, through the designation of uncultivated forest areas and "traditionally" managed forests. Investigations seeking to establish criteria for sustainable forest management are ongoing.

Accompanying measures (EU Regulation 2078/92). The CAP Reform introduced a set of *Accompanying Measures* to support environmentally benign practices in agriculture. The EU programs do not differ significantly in scope from the national agri-environmental policies already implemented in Denmark, but the approach is different. As outlined above, the principal approach to land use policies in Denmark is to suspend the right to use land in ways which are socially undesirable. If the economic burdens are unequally distributed, then the affected landowners are compensated for the losses incurred. The EU programs, on the other hand, are based exclusively on voluntary participation and economic incentives (support). In 1994, the following schemes were implemented in Denmark[4]:

- extensification of permanent grassland;
- support to pesticide-free margins along streams and lakes;
- support in reducing N-fertilizer levels;
- support in the establishment of ray grass as a nitrogen catch crop in cereals;
- set-aside agreements of 20 years duration; and

- support to organic farming.

These programs are targeted towards Environmentally Sensitive Areas (ESAs). Approximately 350,000 ha have been designated as environmentally sensitive in Denmark.

Assessment of landscape restoration and afforestation programs. The previous trend toward a reduction in semi-natural landscapes and habitats seems to have been reversed during the last few years (Ministry of the Environment 1995). In the period 1989–1994, the implementation of the Nature Protection Act has resulted in the reestablishment of 3,000 ha lakes, the restoration and management of 13,000 ha meadows and other semi-natural grazing areas, the establishment of 1,600 new ponds, and the restoration of 60 km of watercourses (Ministry of the Environment 1995).

Under the Accompanying Measures (EU Regulation 2078/92) only the permanent grassland scheme has acquired some success, with 40,000 ha signed up under a management agreement (Ministry of Agriculture and Fisheries 1996). However, investigations indicate that a majority of the farmers entering into management agreements for permanent grassland may be subsidized for practices they would have applied anyway (Primdahl 1996). The areas under the other schemes are negligible (apart from the scheme supporting organic farming, where national support programs were implemented before the introduction of the Accompanying Measures). To improve the effectiveness of the program it has now been decided to increase support payments and transfer the administration of the program from the Ministry to the county level.

In the period 1989–1994, state afforestation amounted to a total of about 4,000 ha (Ministry of the Environment 1995). This is only one-third of the targeted state afforestation level of 2,500 per year. State afforestation has fallen short of the target partly because a large share of the available funds (60 percent) has been allocated to multiple use afforestation projects in urban fringe areas. This has increased the costs of land acquisition considerably and thus reduced the area that could be afforested with the funding available for this scheme (Forestry and Nature Protection Agency 1994A). *Supported* afforestation by private landowners has been negligible (Forestry and Nature Protection Agency 1994A). However, the Forestry and Nature Protection Agency estimates that unsupported private afforestation totals some 1,600 ha annually. It is not clear why landowners prefer to plant forests without support, but it probably plays an important role that conditions attached to the grant limit the right to sell afforested land. It seems that most private afforestation is undertaken, not by farmers, but by entrepreneurs with a preference for having the disposal of the land. As far as farmers are concerned, they apparently hesitate to engage in supported afforestation leading to irreversible changes in land use.

Landscape change perspectives

In a memorandum from 1995, the Danish Ministry of the Environment presented a tentative land-use scenario shown in Table 4. The scenario, covering the period 1995–2025, is based on the following presumptions:

- present set-aside schemes are changed or abolished;
- afforestation: 5,000 ha annually from the year 2000;
- nature restoration: 1,000–2,000 ha per year;
- land use for urban development and infrastructure grows at the present (low) rate.

The most significant results would be a 125 percent expansion of permanent grasslands, a 50 percent increase in the area covered by lakes and water courses, and a 25 percent increase in the forest area. Considering the deficiencies of the present programs this scenario may seem overly "optimistic." On the other hand, the annual required changes in land-use are relatively modest. If the present programs to change land use in Denmark are intensified and sustained, a significant enhancement of landscape amenities and biodiversity will be the long-term result.

Conclusions

With 75 percent of the countryside in agricultural use, it is obvious that farming practices have a decisive influence on the aesthetic, ecological, and recreational qualities of the Danish landscape. It is important to note that the interrelationship between agriculture and landscape amenity often has two sides: intensification generally reduces amenities and biological diversity; some

Table 4. Land-use scenario, 1995–2025

| | ——1995—— | | ——2025—— | |
	%	1000 ha	%	1000 ha
Arable land	58	2,510	48	2,050
Permanent grassland*	4	200	10	450
Forest	12	500	15	635
Heaths, bogs, dunes	4	200	5	210
Lakes, watercourses	2	65	2	95
Hedgerows, ditches, lanes	3	120	3	130
Urban and infrastructure	17	713	17	738
Whole Danish area	100	4,308	100	4,308

* Including dry semi-natural pastures, meadows, marshlands, salt meadows, etc.

Source: Ministry of the Environment, 1995b.

unintensive agricultural use is usually required to preserve the traditional qualities of a cultural landscape. In Denmark, it is the underlying assumption of most landscape preservation policies that the socially desired countryside is not the "natural" environment, but a cultural or semi-natural landscape. In other words, some form of management, in the form of (nonintensive) agriculture or forestry, is required to maintain the desired landscape qualities. During the last three to four decades economic and technological changes have led to an (over)intensification of most of the agricultural area in Denmark. Concurrently a relatively small, but aesthetically and biologically important, share of the agricultural area has become economically marginal and gradually abandoned by agriculture—leading to a loss of amenities and biodiversity.

A large body of national and EU programs has been implemented to change land use in Denmark toward environmentally and aesthetically more benign practices. The policy measures applied are an amalgam of traditional, mostly uncompensated limitations of property rights and voluntary approaches offering farmers economic incentives for the provision of landscape amenities. So far this policy package has been only partly successful in achieving the specified targets. But the overall trend toward an impoverishment of the amenities of agricultural landscapes seems to have been reversed. Still, the experiences gained so far indicate that considerably stronger economic incentives will be required to significantly boost farmers' willingness to engage in nature restoration and landscape management programs.

Notes

1. The evolution of the Danish agri-environmental debate and policy process is reviewed in Dubgaard (1991a, 1991b, and 1994).
2. In the mid-1980s, nitrogen pollution was ascribed primarily to agricultural activities while phosphorus emissions were attributed mainly to urban sources (Committee on Environmental Research 1991).
3. 1 ECU = 7.3 DKK
4. Primdahl (1996) provides a detailed description of the implementation of the Accompanying Measures in Denmark.

References

Committee on Environmental Research (1991). Kvaelstof, Fosfor og Organisk Stof I Jord og Vandmiljoet (Nitrogen, Phosphorus, and Organic Matter in the Soil and the Aquatic Environment), Rapport fra Konsensuskonference Jan - Febr 1991 (Report from a Consensus Conference January-February 1991). Undervisningsministeriets Forskningsafdeling (Research Department, Danish Ministry of Education). Copenhagen.

Danish Environmental Protection Agency (1989). Fugle Foretraekker usprojtede marker (Birds Prefer Unsprayed Fields). Miljo Danmark (Danish Environment), No. 1-2, Copenhagen.

Danish Environmental Protection Agency (1995). Danish Tax on Pesticides, Act No. 416 of June 14, 1995, Ministry of Environment and Energy.

Dubgaard, A. (1987). Avendelse af afgifter til regulering af pesticidforbruget (Taxation as a Means to control Pesticide Use), Report No. 35, Statens Jordbrugsokonomiske Institut (Institute of Agricultural Economics), Copenhagen.

Dubgaard, A. (1991a). Agriculture and Polluter Pays Principle — Denmark, in D. Baldock and G. Bennet (eds.). Agriculture and the Polluter Pays Principle—a Study of Six EC Countries, Institute for European Environmental Policy. London.

Dubgaard, A. (1991b). The Danish Nitrate Policy in the 1980s, Report No. 59, Statens Jordbrugsokonomiske Institut (Institute of Agricultural Economics), Copenhagen, Denmark.

Dubgaard, A. (1994). The Danish Aquatic Environment Programs: An Assessment of Policy Instruments and Results, in T.L. Napier, S.M. Camboni, S.A. El-Swaify (eds.). Adopting Conservation on the Farm. An International Perspective on the Socioeconomics of Soil and Water Conservation, Soil and Water Conservation Society, Ankeny, Iowa.

Forestry and Nature Protection Agency (1994a). Evaluering af naturforvaltningsprojekterne 1989-1992 (Evaluation of Nature Management Projects 1989-1992). Copenhagen.

Forestry and Nature Protection Agency (1994b). Strategi for Baeredygtig Skovdrift (Strategy for Sustainable Forestry), Betaenkning nr. 1267 (Report no. 1267), Copenhagen.

Ministry of Agriculture (1994). Miljovejledning. Vejledning om tilskud til miljovenlige jordbrugsforanstaltninger. (Guidelines for Agri-Environmental Support Schemes), Landbrugsministeriet, Jordbrugsdirektoratet, June 1994, Copenhagen.

Ministry of Agriculture and Fisheries (1996). Betaenkning fra udvalget om Natur, miljo og Eu's landbrugspolitik (Nature, Environment and the European Union's Agricultural Policy), Betaenkning nr. 1309 (Report no. 1309), Copenhagen.

Ministry of the Environment (1995). Natur-og Miljopolitisk Redegorelse (Nature and Environmental Policy Memorandum), Miljo-og Energiministeriet, Copenhagen.

Primdahl, J. (1996). Response of Member States: Denmark, in M. Whitby (ed.). The European Environment and Cap Reform. Policies and Prospects for Conservation, CAB International, U.K.

19

The CAP Reform and Danish Agriculture:
An Integrated Environmental and Economic Analysis
Jesper S. Schou

The interactions between agricultural production and the environment are complex and multi-faceted. On the one hand, environmental conditions together with the climate set out the natural conditions for agricultural activities, and on the other hand agricultural activities interfere with and change the state of the environment. In this chapter the second relationship is brought into focus, that is how agricultural production affects the state of the environment, and furthermore, it is investigated how changes in agri-environmental policies affect the environmental stress caused by agricultural production.

A large number of agri-environmental indicators are relevant when describing the interactions between agricultural production and the environment. Examples of indicators are production inputs, land use, the number of livestock, types of farm management, energy and nutrient balances, emissions, effect processes, animal welfare, and human perceptions of landscape. In Denmark the contamination of aquatic recipients (ground water, lakes, streams, and coastal waters) with nitrates and pesticide residues has received a great deal of attention. This is mainly due to the fact that more than 95 percent of all drinking water in Denmark is derived from ground water resources. Furthermore, the emergence of eutrophication problems in coastal waters in the early 1980s (see, e.g., Dubgaard 1991 and Rude and Frederiksen 1994) contributed to attention being directed toward the use of nitrogen fertilizer in agriculture. Therefore this chapter mainly focuses on contamination effects such as nitrate leaching, but a qualitative analysis of nature and landscape effects is also included.

Of course, other industries are also responsible, besides agriculture, for the development in the problems with nitrogen pollution. Concerning nitrogen (and phosphorus) some large point-sources in the industry have been identified in Denmark, but their emissions are now considered to be at an acceptable level (Danish Environmental Protection Agency 1995). This leaves nitrate pollution from agriculture as the remaining primary source. This is also the case with respect to pesticide use, where agriculture is responsible for more than two-thirds of total consumption, while the wood and forest industry accounted for about 25 percent in 1994 (Statistics Denmark 1995a).

Agri-environmental policies in Denmark

The environmental problems caused by agriculture have not passed unnoticed. Within the last decade various new environmental regulations have been implemented. These regulations have primarily been of the command-and-control type; i.e., they are designed as rules which farmers should comply with at the farm level and which subsequently are controlled by the authorities. Table 1 shows the main environmental regulations regarding fertilizer, manure and pesticide use, which are regarded as the main contributors to the agri-environmental problems in Denmark. As seen, the command-and-control measures directed toward pesticides were supplemented with a graduated tax on pesticide sales in 1996. Earlier a tax of 3 percent was implemented in order to finance the increases in research and advisory services (see, for example, Dubgaard 1996 and Schou 1996 for further details).

The overall goals of the environmental policies are 50 percent reductions in both the nitrate losses to the aquatic environment and in pesticide use. The first goal relates to the obligations to which Denmark is committed by the Oslo and Paris Convention, while the second is a national goal related to the agenda 21 on biodiversity and wildlife preservation.

Cross achievement effects

Besides the national initiated agri-environmental policies, Danish agriculture since 1972 has been subject to regulation under the Common Agricultural Policy (CAP) of the European Union (EU). Initially the aim of the CAP was to improve income and productivity in the agricultural sector, but with the 1992 CAP reform environmental concerns have also been included in the policy-statement (EEC Directive 2078/92). The most important elements of the CAP reform are the changes in the agricultural income support scheme. Prior to the reform, income support was given through price support, but the reform has reduced intervention prices substantially and introduced a factor-based income support system instead.

The price support scheme enabled EU producers to sell their agricultural outputs at prices substantially above world prices. Due to the higher prices, producers had an incentive to produce more and thus also to use more inputs, such as fertilizers and pesticides. Changes in the relative prices between crops also induced changes in land use and cropping patterns. In addition, the high

Table 1. Danish nitrate and pesticide policies toward agriculture

Nitrate policies

Aim: Nitrate leaching from Danish agriculture should be reduced by half before the year 2000 compared to the level in 1984. Point sources should be eliminated.

Measures

- Maximum application of manure equivalent to 2.3 lu* per ha on farms with cattle production; 1.7 lu per ha on farms with pig production; and 2.0 lu per ha on other farms.
- Application of manure is in general prohibited from harvest to February 1.
- Minimum nine-month storage capacity for animal manure.
- Minimum of 65 percent winter crops on all farms.
- Minimum utilization of nitrogen in manure of 45 percent on farms with cattle production; 50 percent on farms with pig production; and 40 percent on other farms. On farms with deep bedding stables the minimum utilization is 15 percent.
- Prepare a plan of crop rotation (land-use), crop yields, and nutrient account on farms larger than 10 ha.
- Spot check of key figures concerning land use, crop yields, and nitrogen application.

Pesticide policies

Aim: The use of pesticides in Danish agriculture should be reduced by half before the end of 1997 compared to the average level in 1981–1985.

Measures

- Increased research and advising in the use of pesticides.
- All persons who use pesticides professionally should pass a test in handling and applying pesticides.
- A plan of pesticide use should be made on all farms larger than 10 ha.
- Pesticide use is prohibited on environmentally sensitive areas (e.g., meadows and water supply plants) and up tills a distance of 2 meters from barrows, streams, and lakes.
- All pesticides traded in Denmark should be evaluated for their toxic effects on humans and the environment before approval.
- A graduated tax on pesticide sale (introduced in January 1996): 27 percent for insecticides, 13 percent for herbicides and fungicides, and 3 percent for other pesticides.

* Lu: Danish livestock units; the nitrate production from one lu is defined as the yearly nitrogen content in manure from one milk cow of a larger breed.

prices made it profitable to cultivate previously marginalized land. From an environmental point of view, the price support scheme therefore to some extent has contributed to increasing the environmental problem caused by agricultural production in the EU. However, it should be recognized that technological progress, changes in farm structure, and introduction of new crop types probably have been the main reasons for the intensification of agricultural production. In Denmark, increases in the use of fertilizers and pesticides took place from the 1950s to the early 1970s—that is, before the Danish EU membership and in a period when Danish prices on sales crops were falling in real terms (DIAFE 1968; Larsen et al. 1996).

With the 1992 CAP reform, intervention prices on cereals have been reduced by approximately one-third, whereas oilseed and pulses are produced at world prices. Prices on beef have also been reduced, but only by about 10 percent.[1] Even though prices have been reduced, the level of support is unchanged due to the introduction of hectare, set-aside, and animal premiums. The price changes and factor support premiums for Danish agriculture are shown in Table 2 together with the quantitative restrictions of the reform.

Price reductions affected the environmental stress from agriculture in two ways; first, the reduction in sales prices led to reduced use of variable inputs such as fertilizers and pesticides, and second, changes in relative prices between crops led to changes in land use. Furthermore, the price reduction led to a decrease in total cropped area (marginalization). However, as farmers were obliged to set-aside a certain percentage of their cropped land in order to receive hectare premiums, it is not possible statistically to detect the marginalization effect, as it is contained within the set-aside. In the next section, the results from the analysis of effects of the CAP reform on the structure of production and the derived environmental effects are presented.

Besides the changes in the income support-scheme, a small part of the CAP budget was designated for subsidizing environmentally beneficial production methods in agriculture. In the analysis the effects of these "accompanying measures" have not been included, since they only have had very little importance in Denmark until now.[2]

Table 2. Changes in agricultural support schemes due to the CAP reform

	Price change	Factor support premiums	Quantitative restrictions
Cereals	−28%	2.196 Dkr/ha	—
Oilseeds	−44%	3.836 Dkr/ha	—
Pulses	−44%	3.172 Dkr/ha	—
Set-aside	—	2.781 Dkr/ha	—
Area eligible for premiums	—	—	Max. 2,018,000 ha
Milk	+2.4%	—	Max. 4,430 x 10^6 kilo
Beef	+2.4%	—	—
Bulls and bullocks	—	841 Dkr/animal	Max. 325,000 animals
Nursing cows	—	1,122 Dkr/animal	Max. 136,000 animals

Source: DIAFE 1995.

The effects of the CAP reform

As pointed out, the most important elements of the CAP reform are the changes in the agricultural support scheme. These changes have resulted in substantially lower prices on cash crops, but have been replaced by factor support premiums. If farmers are profit maximizing, they will respond to these economic changes by changing their structure of production. These changes will be quantified using a Sectorial Model developed at the Danish Institute of Agricultural and Fisheries Economics (Jensen 1996). The Sector Model is a national static-comparative econometric model, which for given economic conditions shows the land use, the quantity of production inputs used, and the yields per hectare, assuming that farmers maximize profits. Once the economic conditions change, the model thus determines the changes in land use, the number of livestock, and input use, etc.

In the analysis, we compare two situations. The first is the 1989 situation prior to the implementation of the CAP reform, while the second is the situation after the CAP reform has been fully implemented (partial equilibrium), but assuming that no changes other than the reform have taken place in the period in between. Of course it is not realistic to assume that there will be no other changes, as there have been additional environmental regulations and technological developments, which would influence the outcome of the analysis if they were taken into account. However, if one accepts the static-comparative nature of the analysis and just compares the 1989 situation with the situation after farmers have made full adjustment to the CAP reform, one gets a good impression of the direction and the magnitude of the changes caused by the CAP reform in itself. It is therefore important to have in mind when interpreting the results that the analysis is a static-comparative *ceteris paribus* type of analysis, showing the isolated effects of the CAP reform, and *not* a projection of the actual changes in Danish agriculture and the environmental stress.

Having analyzed the consequences of the CAP reform for the structure of production, we proceed to estimate the derived effects on the aquatic environment and we perform a qualitative analysis of the effects on nature and landscape. The points of reference for these estimates are changes in the structure of production estimated with the Sector Model. These changes are used as an input to a national, regional distributed model which is developed at the Danish Environment Research Institute (The NP-model). The model quantifies the interaction between the geographical distribution of land use, animal husbandry and fertilization practices, and nitrate losses and contamination of the aquatic environment divided into 12 regions on basis of counties (Paaby et al. 1996). The nitrogen load to marine waters is composed of two elements, namely nitrogen leaching from the root zone and ammonia evaporation. Nitrogen leaching has been estimated using data from empirically estimated leaching functions for different types of crops (Simmelsgaard 1991), whereas ammonia evaporation has been estimated using emission coefficients for animal husbandry.

As seen from Figure 1, the concept of model integration is one way, since there is no feedback mechanism from the environmental model to the econometric production model. The analysis originates by defining the initial (exogenous) conditions for the Sectoral Model on the basis of which changes in the structure of production are estimated on a national level. It should be noted that the Sector Model is not a regional model. Therefore, the results from the Sector Model are distributed geographically using a rather simple disaggregation procedure assuming that the relative changes in all variables are the same in all regions. Afterwards, the results are fed into the NP-model, which estimates the nitrate, phosphorus, and ammonia emissions and the resulting loads to the aquatic recipients.

Changes in the structure of production

The effects on the structure of production will be illustrated by the changes in land use, number of livestock, and the use of nitrogen fertilizers and pesti-

Figure 1. The used concept of integration

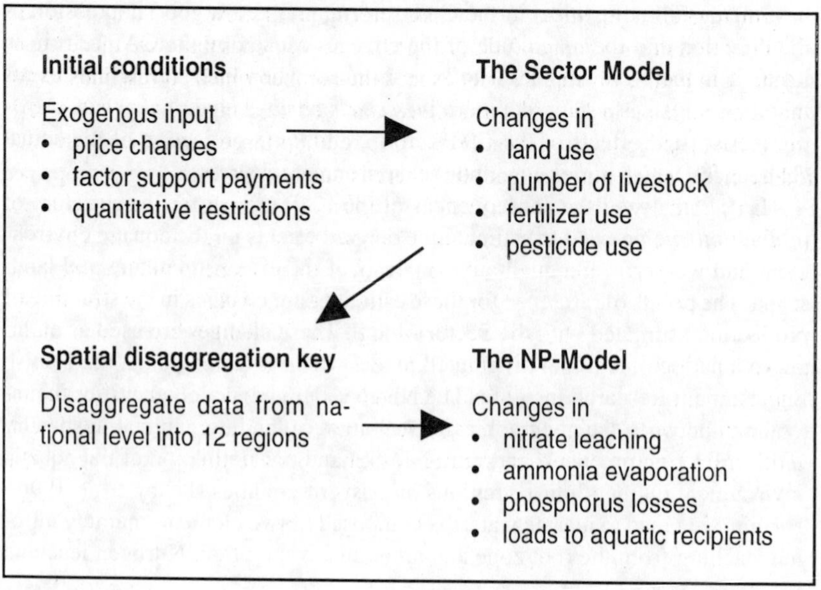

Initial conditions

Exogenous input
- price changes
- factor support payments
- quantitative restrictions

The Sector Model

Changes in
- land use
- number of livestock
- fertilizer use
- pesticide use

Spatial disaggregation key

Disaggregate data from national level into 12 regions

The NP-Model

Changes in
- nitrate leaching
- ammonia evaporation
- phosphorus losses
- loads to aquatic recipients

cides. These changes will be modeled, assuming that the changes in the support scheme are as indicated in Table 2. Furthermore, it is assumed that milk production cannot exceed the milk quota, and that the number of nursing cows, bulls, and bullocks cannot exceed the respective quota. It is also assumed that there can be no more than two dairy cow equivalents per hectare with fodder crops, which is a requirement for receiving animal premiums. Concerning pig production it is assumed that relative prices in the pig sector do not change and, therefore, that pig production is unchanged after the CAP reform. Finally, set-aside is set exogenously at 214.000 hectares or about 14 percent of the area with reform crops corresponding to the 1994 set-aside in Denmark (Statistics Denmark 1995b).

The estimates, which are listed in Table 3, show that for most crops the cropped area is reduced due to the CAP reform. This reduction can be attributed to the set-aside requirement in the CAP reform. However, the reduction in the cropped area is not the same for all crops. If land use is set at 100 prior to the reform (basis period: 1989) it remains at a 100 for potatoes and sugar beets, whereas it is reduced to 85 for winter wheat, 92 for spring barley and 86 for pulses. The differences in the reduction in land-use can be attributed to the changes in the income support system, as crops with low yields and production intensities (e.g., barley) are favored at the expense of higher yielding crops (e.g., wheat). Also, the difference in the change of the intervention prices between crops influences the land use.

With regard to the number of livestock, the estimates show that the CAP reform will not result in any changes. This is due to the assumptions that were made about the stock of cattle, which could not exceed the number which corresponds to the quotas on milk, bullocks, and nursing cows, and the unchanged prices in the pork sector. The production in the cattle sector is of importance for the land used for fodder crops, since it is assumed, that there is one hectare with fodder crops per two cattle. It follows that the area used for production of roughage does not change.

The use of nitrogen fertilizer and pesticides falls as a result of the CAP reform. The average reduction in fertilization is 25 percent, while the average reduction in pesticide use is 35 percent. These reductions can be attributed both to a lower average per hectare application on cultivated lands, but also to the set-aside, which accounts for about 10 percent of the total agricultural area. The reduction in total fertilization includes both commercial fertilizer and animal manure. Since the number of livestock is unchanged, the entire reduction in fertilization has to come from the use of commercial fertilizers corresponding to more than a 30 percent reduction.

Environmental consequences

In order to estimate the environmental consequences of the CAP reform, the environmental effects of the changes in the structure of production are analyzed. The change in land use and the change in the use of fertilizers and pesticides result in two types of environmental effects: (1) effects on nature and landscape (due to changes in land use, flora and fauna) and (2) contamination effects (due to emissions of nitrates and pesticides).

In the analysis we try to investigate both types of effects as we look at a number of qualitative and quantitative indicators. Concerning the effects on nature and landscape, we have made no attempt to quantify the effects. This is primarily due to the highly site-specific features of the effects, which cannot be comprised in this rather highly aggregated level of analysis. Instead we use four qualitative indicators for nature and landscape: the cultivated area, the area with set-aside and permanent grassland, the average nitrogen application per hectare, and the average expenses for pesticides per hectare. The contamination effects are quantified in terms of both nitrogen emissions and the resulting nitrogen load to marine waters by use of the NP-model.

In Table 4 the effects of the CAP reform on the four indicators for nature and landscape are listed. The changes are generally environmentally positive, although attention to some modifications in respect to the interpretation should be given. The change in the cultivated area is caused by the set-aside requirement alone, and is therefore not to be interpreted as a long-run marginalization

Table 3. Effects of the CAP reform on the agricultural structure of production in Denmark; calculations with The DIAFE Agricultural Sector Model

| Index: 1989 = 100 | Area | Per hectare | | |
		Yield	Fertilization	Pesticides
Spring barley	92	91	75	68
Winter wheat	85	95	93	71
Rye	90	92	77	85
Oats	92	95	82	71
Pulses	86	87	43	60
Rape	88	85	61	60
Root crops	100	100	100	100
Roughage and grassland	100	100	100	100
Set-aside	*	—	—	—

*Set-aside is set exogenously on basis of Statistics Denmark.

Source: Schou et al, 1996.

278

of farmland. With respect to the change in permanent grassland and set-aside, the changes are also primarily caused by the set-aside requirements, although the CAP reform tends to promote permanent grassland compared to grassland in rotation.

The positive effects of the changes in the cultivated area and permanent grassland are therefore dependent on the benefits of the set-aside regime and the set-aside requirements. One key question is whether set-aside is permanent or rotational, since the diversity of the flora and fauna is higher if land is set-aside for a period of 2 years or more. It is also important whether the set-aside land is cut in order to prevent spreading of weeds and at what time of the year this is done. Both the set-aside areas and how the set-aside areas should be taken care of are decided upon in the EU, and therefore the consequences for nature and landscape of set-aside are highly dependent on the bureaucratic year-to-year decisions. For example, the set-aside requirement was originally 18 percent of the area with sales crops, but with the implementation of the CAP reform it was reduced to 10 percent in 1995-1996 and to 5 percent in 1996-1997. With this tendency, which is caused by falling production of sales crops within the EU due to the cuts in intervention prices, the role of the set-aside policy is likely to be ruled out within near future, reducing the beneficial effects on nature and landscape.

Turning to the contamination effects, the CAP reform will reduce nitrogen leaching from the root zone by about 20 percent in Denmark, while it will reduce the nitrogen load to marine waters by a bit less. The reduction is mainly due to the reduction in fertilizer application, but the set-aside requirements and the shift from wheat cropping to the less intense cropping of winter barley also contribute to the reduction. Concerning the beneficial effect of the set-aside, as mentioned before, the requirement has been reduced successively from the original level of 18 percent in 1993-1994 to 5 percent in 1996-1997; the decrease in set-aside will of course reduce the environmental benefits, if the land, which was previously set aside, is turned into cropland again.

Table 4. Indicators for the effects of the CAP reform on nature and landscape in Denmark

	Change	Resulting effect
Cultivated area	−10%	+ / 0
Fertilizer application per ha	−20%	+
Costs of pesticides per ha	−30%	+
Permanent grassland and set-aside	210%	+

Source: Schou et al, 1996.

279

Conclusion

Once the CAP reform is fully implemented, it is estimated it will reduce some of the environmental problems caused by agricultural production in Denmark, due to changes in land use and the use of fertilizers and pesticides. Using nitrogen emissions as an example, the CAP reform is expected to reduce leaching from the root zone and loading to marine waters by about 20 percent. These reductions are substantial, but it is safe to say that they are not sufficient to reach the goals in the Danish Action Plan on the Aquatic Environment, which is a 50 percent cut in nitrogen emissions and in the loss to marine and surface waters. Neither is the aim of the Pesticide Action Plan of a 50 percent cut in pesticide use likely to be fulfilled due to the reform. Concerning the effects on nature and landscape, the effects of the CAP reform are considered to be beneficial primarily because of an increase in the area with permanent grassland and set-aside, but also the reduction in fertilizer and pesticide use due to the price cuts on sales crops makes a positive contribution.

Although this was not the main purpose of the 1992 reform, i.e., the shift in policy measures from price support to factor-based support and the set-aside requirements, it will result in an overall reduction in the environmental problems caused by agricultural production in Denmark. This accentuates the need for taking cross-achievement effects into account when designing agricultural policies (and vice versa). Also, it has led to a discussion of whether the CAP should be made into a cross-compliance policy.[3] Traditionally, the idea of cross-compliance has been that farmers should deliver some "environmental services" in return for the income support they receive (see, for example, Baldock 1993; Russell and Fraser 1995). This version of cross-compliance is labeled the red-ticket or the "stick" version, where the farmer has to comply with certain environmental standards in order to keep the current level of support in the future. Another form of cross-compliance is the "carrot" or green-ticket version, where farmers receive additional support in return for changing to environmentally friendly farming practices.

The green-ticket cross-compliance has already been implemented within the CAP in connection with the 1992 reform, as environmental concerns were integrated into the CAP with the "accompanying measures." These include subsidies to structural and production-related measures that have beneficial effects on the environment. With the accompanying measures, it was decided to open up for using a small part of the CAP budget for subsidizing environmentally beneficial initiatives in agriculture. In Denmark, this has led to the development of a support scheme where farmers voluntarily can enter an agreement with the authorities of changing agricultural practice in return for fixed subsidies. The support scheme includes reduction of fertilization by 40 percent compared with the recommended level, keeping grassland as permanent pastures, abandoning pesticide use on field margins, set-aside for a 20-year period, public access to farmland, under sowing of rye grass in cereals, or combinations of these. In the analysis, the effects of the accompanying measures have not—as mentioned earlier—been included, since they only have very little importance.

One important feature of both the stick and carrot type of cross-compliance is that the environmental benefits are highly politically dependent. If agri-environmental policies are to lead to beneficial changes in the environmental strain from agriculture, it is important to take into account that the time horizon for changes in agricultural production to result in measurable environmental changes is typically 5 to 10 years. The CAP seems to have been entering a volatile period after the 1992 reform with the changes in set-aside requirements and the current discussion of a reduction in hectare premiums caused by the high prices on world markets. Furthermore, the expansion of the EU to include some of the Central-Eastern European countries may in the long run lead to radical changes in the CAP.

A red-ticket cross-compliance version of the CAP, where both objectives are included, seems to hold strong elements of contrast for both the long-term perspective in the environmental policy and the need for short-term adjustments in support policy. If one still wants to hold on to both the income support objectives and the environmental objectives in the CAP, it therefore seems to offer a more efficient possibility to follow a two-winged strategy, where the environmental objectives are pursued through green-ticket cross-compliance like the accompanying measures, whereas the income support objectives are pursued through factor support payments or even more decoupled support schemes.

Notes

1. The CAP reform only affects the plant and cattle sectors, whereas the pig sector is virtually unaffected. This is because the pig sector primarily has been subsidized in order to counterbalance the increased costs to fodder due to the price support to cereals.
2. In 1994 only about 48,000 hectares out of a total agricultural area in Denmark of 2,691,000 hectares received subsidies under the accompanying measures.
3. In Denmark this intention is stated in a report from January 1996 conducted by *The Committee on Nature, the Environment and the CAP*, including amongst others: The Danish Ministry of Agriculture and Fisheries, The Danish Ministry of Energy and The Environment, and The Danish Farmers Association.

References

Baldock, D. 1993. The concept of "Cross-Compliance" and its possible application within the CAP, Institute for European Environmental Policy, London.

Danish Environmental Protection Agency. 1995. Vandmiljø-94 (The Aquatic Environment-94), Redegørelse fra Miljøstyrelsen, nr. 2 (in Danish with English summary).

DIAFE. 1968. The Danish Agricultural Economy in 50 years: 1918 to 1968, Danish Institute of Agricultural and Fisheries Economics (in Danish).

DIAFE. 1995. Paper on the consequences of the CAP reform for agricultural producer prices in Denmark (in Danish), Danish Institute of Agricultural and Fisheries Economics, Unpublished.

Dubgaard, A. 1991. The Danish Nitrate Policy in the 1980s, report no. 59, Danish Institute of Agricultural and Fisheries Economics.

Dubgaard, A. 1996. Water conservation policies in Denmark. Paper presented at the symposium on "Soil and Water Conservation Policies: Successes and Failures." Prague Agricultural University, September 17 to 19, 1996.

Jensen, J.D. 1996. An applied econometric model of Danish agricultural production and input demand. Paper at the VIII EAAE Congress in Edinburgh, September 5, 1996.

Larsen, I. O. Olesen, and S. Sørensen. 1996. Agricultural account statistics 1972/72 to 1993/94—Structural and economic development since EU asseccation, Report no. 86. Danish Institute of Agricultural and Fisheries Economics (in Danish with English summary).

Paaby, H., F. Møller, E. Skop, J.J. Jensen, B. Hasler, H. Bruun, and W.A.H. Asman. 1996. The costs of reducing nitrogen loads to aquatic recipients—Method, model and analysis, Report no. 165, Danish Environmental Research Institute (in Danish with English summary).

Rude, S. and B.S. Frederiksen. 1994. National and EC Nitrate Policies - Agricultural aspects for 7 EC countries, report no. 77, Danish Institute of Agricultural and Fisheries Economics.

Russell, N.P. and I.M. Fraser. 1995. The potential impact of environmental cross-compliance on arable farming, Journal of Agricultural Economics 46(1) (1995), pp. 70-79.

Schou, J.S., H. Paaby, J.D. Jensen, and H. Vetter. 1996. Agricultural policies and environmental regulation — part two (in Danish), Miljøprojekt nr. 321, Miljø- og Energiministeriet, Miljøstyrelsen 1996.

Simmelsgaard, S.E. 1991. Estimating functions for Nitrogen Leaching—Nitrogen leaching as a function of fertilizer application for different crops on loam and sandy soils. In Rude, S., Nitrogen Fertilizers in Danish Agriculture—present and future application and leaching, Report no. 62, Danish Institute of Agricultural and Fisheries Economics.

Statistics Denmark. 1995a. Pesticidsalget 1994 (Pesticide sales 1994), Statistiske efterretninger MILJØ 1995:15 (in Danish with English summary).

Statistics Denmark. 1995b. Agricultural Statistics 1994.

20

Rationality Is in the Eye of the Actor

Paul Vedeld and Erling Krogh

The authors share responsibility completely equally—for good and for bad. We would, however, like to thank associate professor Terje Kvilhaug, IES, associate professor Arild Vatn, IES, and professor Lars Bakken, DSS, all AUN, for valuable comments.

After 1945, the Norwegian agrarian structure underwent significant changes that directly and indirectly led to increased environmental problems. The last 35 to 40 years are ones of unprecedented production increases; grain yields per ha and the yield per cow have approximately doubled, contributing to economic growth and welfare increase of the Norwegian society. The intensification of farming[1] is part of a general trend of increased capital input in agriculture. There is also a general increase in the level of capital demanding input use and/or labor saving input use, including the use of nitrogen fertilizers and nitrogen rich fodder.

The increased use of chemical fertilizer per unit of land coupled with a higher percentage of open-field acreage for production has caused a substantial growth in diffuse drainage of excess nutrients into watersheds and groundwater reservoirs making pollution problems more serious and imposing social costs to the society. Regional and on-farm specialization as well as a general intensification of farming engineered by both market forces and public policies are usually stressed as the most important driving forces behind increased surface runoff of nitrogen (Vatn 1989; Simonsen 1989; Sødal and Aanestad 1990; Simonsen et al. 1992).

Scientists and politicians in Norway have increasingly emphasized the need for sound management of our environment. The World Commission on Environment and Development submitted its report in 1987 (WCED 1987) and this helped initiate a process in Norway where each sector in the society was made responsible for establishing plans to reduce its contribution to the environmental problems (St.m.46.1988-89).

In 1987, the Norwegian government signed the North Sea Treaty, which specified aims for reducing the runoff of nutrients to the North Sea. In 1992, the

National Plan for the fulfillment of the treaty passed the Norwegian Parliament (Min. of Environment, MoE 1992).[2]

The public awareness of problems related to pollution has also increased and the present focus on these problems must also be seen and interpreted in this light. According to the Norwegian Pollution Control Authority (NPCA) (1989), agriculture is the main source of nitrogen runoffs to watersheds in Norway with an estimated 36 percent of the quantities (diffuse runoffs mainly).

This has also led to many different suggestions on how to solve these problems. In the last 10 years there has been a lively debate in Norway around these issues. This debate has highlighted not only disagreements on what is the problem, but also on what causes the problem and how to reduce the problem in the best possible ways.

The aim of this chapter is to give insight in the scientific debate over the problem of nitrogen surface runoff and its dependency on agronomic practices and the choice of policy instruments to reduce the diffuse runoff of nutrients from agriculture. In particular, we analyze the debate around the use of ambient taxes on nitrogen fertilizers raised by economists versus information and "command and control" measures raised by agronomists in order to reduce runoff.

From this, the question of rationality between different sciences will be discussed. We also debate the role of research in society and the contact between researchers and the decision-makers. If rationality is relational and contextual, socially constructed, what implications does it have for "good governance"? Can policy-makers choose "rationally" between different scientific advice for the "same" problem in question? And do they?

A presentation of two world views and the case

We address agronomists' and economists' differing world views concerning diffuse runoff of nutrients from Norwegian agriculture over the last 10 years by describing the historical debate over how to mitigate these problems.

An agronomic world view

Diffuse nitrogen runoff was first stated as a serious pollution problem in two committee reports from the Ministry of Agriculture (MoA) and Ministry of Environment (MoE) in 1984 and 1986 (MoA 1984, 1986). Responding to these reports the ministries developed an action program against pollution from agriculture, "Handlingsplan mot landbruksforurensinger," lasting from 1985 to 1988. The ministries engaged GEFO[3] to realize the program.

The research director of GEFO expresses the relationship between nitrogen fertilizing and runoff causing pollution in the following way: "Correct fertilization, that means good utilization of fertilizer, will minimize pollution" (Rognerud 1989).[4]

Rognerud discusses the design of a good fertilization program and stresses that "with a basis in soil and plant conditions such as the plant nutrient needs, how the plants grow, the quality of the soils, and climatic conditions, the program tries to adapt to "correct fertilizer amount." The idea behind the program is that the plants are supplied with optimal nutrient amounts at the appropriate time and well adapted to local conditions. A side effect of such optimal plant nutrition is a minimization of nutrient and soil losses through surface runoff and leaching. Fertilizer planning, split application of fertilizer, reduced tillage, and increased use of catch crops are presented as important measures to secure correct fertilizer use.

The agronomic understanding: When the fertilizer use is adapted to plants and soils in an optimal way based on the "needs of the plants and the soils" and according to a logic shared by farmers and agronomist researchers, the runoff is minimized. What is optimal is strongly dependent upon micro-level variations in soil qualities, climate, etc., and is difficult to model from outside. "The farmer knows his soils and better than any outsider."

In Norway, this agronomic understanding has been and is continuously developed in an interplay between researchers and the extension service. The extension service puts strong emphasis on disseminating information and knowledge to farmers through practical down-to-earth extension service advice. The action plan started by GEFO reflects this view. On "farm days" in different farms over the entire country, practical and physical measures were demonstrated and discussed in active cooperation and dialogue among farmers, extension service personnel, and researchers. In the evaluation of the Action Plan the positive side effects of this cooperation are highlighted: "Close contact between the consultant and the farmer in each farm in the trial areas has increased the interest in different measures. Positive and scientifically well-founded information is well received" (Rognerud 1990). The fact-to-face contact between extension officer and the farmer is seen as good both for the transfer of concrete knowledge and skills related to environmental measures and to enhance and promote environmental awareness among farmers.

At the same time as farmers' environmental awareness increases, researchers and extension officers are also influenced by the farmers' view that the environmental measures should not threaten their economic base. In the North Sea Treaty of 1987, Norwegian authorities committed themselves to reduce runoff by 50 percent. The research director for GEFO writes in a Farm Handbook: "We must anticipate that strong measures will be enforced to reach this level of runoff and the more we are able to manage voluntarily, the better. With as correct fertilizer use as possible, there is a possibility to avoid the fertilizer tax increases" (Rognerud 1989:41-42). The interaction among farmers, extension service officials, and the researchers leads to the development of a common horizon of understanding.

An economic world view

During the summer of 1988 an algae flowering along the coast of Norway resulted in the death of fish and other "visible sea living creatures." In the media, agriculture was the number one suspect, and both the political and the administrative leaders in the Ministry of Environment asked for "efficient policy instruments to reduce nutrient runoff to the North Sea." Concomitantly, the ministries started the work to follow up Norway's commitments relative to the North Sea Treaty.

In the winter of 1989, a report was published from the Department of Agricultural Economics at AUN giving a general assessment of different types of environmental policy instruments in agriculture (Holm et al. 1989). It emphasized environmental ambient taxes as the most cost-efficient instrument to reduce nitrogen runoff. Based on this, in March 1989 NPCA raised a proposal to substantially increase taxes on chemical fertilizers in order to reduce runoff.

At the same time, the Department of Agricultural Economics was also given the assignment by MoA to produce a report discussing the use of environmental taxes in agriculture, following a parliamentary decree of October 1988. In the so-called Simonsen report (1989) a resource economic perspective on the use of ambient taxes was developed.

The report combines a private and a welfare theoretical economic perspective on the use of taxes; it is in line with what one may call a rational comprehensive planning approach (Faludi 1983), where the market is to solve the problems, but by means of public control and guidance—what Randall (1988) calls a "market failure, government fix tradition."

Nutrient runoff is seen as a negative external effect of crop and livestock production. Following the polluter pays principle, farmers are charged a tax on nitrogen in fertilizers to make nitrogen use less profitable. Assuming that farmers are profit maximizers, it is assumed in the report that farmers will adapt to what is profit maximization fertilizer use, given different levels of the tax. Profit maximization fertilizer use is calculated by means of partial models, using continuous (and differentiable) production functions between yields and nitrogen use for different crops and geographical areas with corresponding runoff functions. The models are used to predict both yield reductions and runoff reductions with different tax levels. These functions are based on average figures for yields and runoffs in different productions in different regions (see Figure 1).

288

The main social cost following a tax is the loss in yields following reduced nitrogen use. For the farmers the tax becomes as an extra cost. Simonsen (1989) states that the tax revenue may be transferred back to the farmers, through production independent transfers, given that there is political will for such transfers.

The economic understanding: Farmers adapt their nitrogen fertilizer use according to what is profitable for them as individuals and according to where the marginal returns equal marginal costs. There is local variation in various factors impacting the production function, but on average and on aggregate, variation is not so important for the economists. Since farmers are profit maximizers, they will, with a price change following an optimal tax, change their fertilizer use so that the nitrogen runoff also becomes economically optimal from a societal point of view.

Two world views—a world apart

The Simonsen report caused strong reactions in agronomic circles. In a letter to MoA, the chief extension officers in soil science, plant production, and economics claim that the report is based on false premises ("uholdbart beregningsgrunnlag") (Enge et al. 1990).

According to these consultants, Simonsen is using a level for private profitable nitrogen use that deviates strongly from their practical experience with farmers' fertilizer use and the National Agricultural Census 1979-1989. The

Figure 1. Relationship between nitrogen input, yield level, and nitrogen runoff—the economist version

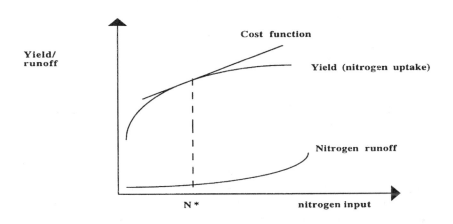

registered fertilizer use is in practice lower than what Simonsen assumes. Furthermore, the variation within a province in fertilizer use levels between farmers is much larger than assumed by Simonsen. They claimed that this is an indicator that many farmers do not maximize profits in their fertilizer use. Second, experience from the extension service is that fertilizer use is strongly adapted to local level variations in soil quality and local growing conditions including climate, slope, and exposition of field, and that it varies with type of crop. All this implies that Simonsen's optimization model and the average figures for yields and runoff do not correspond to what is observed in "real life."

The chief extension officers furthermore refuted the possibility of estimating reliable yields functions presented as mathematical functions. If nitrogen is in less than adequate supply, it will be, according to the consultants, an approximately linear yield increase up to the point of lodging (the kinked point). This point is where other nutrients, moisture, temperature and light set a limit to further growth. At this point, the product function will be horizontal or declining. Following this logic, the runoff of nitrogen will increasingly rise after the lodging limit. Figure 2 shows the relationship between nitrogen input, yield, and nitrogen runoff.

According to this way of thinking, the economic optimal adaptation will be in the kinked point, even with a high tax level. This point gives both a high (maximum) yield and a low level of nitrogen runoff. An ambient tax on nitrogen will thus, following this logic, to a limited degree influence farmers' adap-

Figure 2. Relationship between nitrogen input, yield level, and nitrogen runoff—the agronomist view

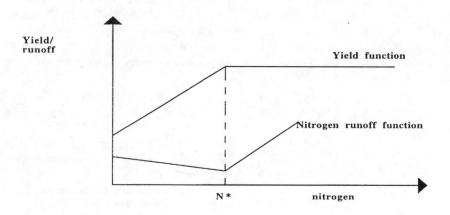

tation and can only be seen as a sheer cost to the agricultural sector—a cost in direct conflict with other agricultural policy aims, according to the chief extension officers (Enge et al. 1990, 1: 9-10).

Aside from these more technical arguments, they also put forward arguments relating to the legitimacy of the instruments—if the instrument is right or wrong, and if it is fair or unfair.

According to the officers, the tax does not differentiate between groups of farmers and types of production. Some groups of farmers do not pollute very much and will be "unreasonably strongly affected" by a tax. The chief extension officers showed, for example, that there is a difference between old and young farmers concerning fertilizer use. They blamed the economists for treating such groups in the same way, regardless of the fact that some groups actually fertilize less than others. They also stressed that the tax be imputed in areas where the pollution is of little or no significance. They therefore concluded:

"It must be wrong to burden agriculture all over Norway with a tax in order to reduce the runoff in a small and confined part of our total coastal waters. It would be much better and much cheaper to intensify information, extension service advice and planning to make farmers fertilize more correctly. This because optimal N-use derived from yield potentials, soil quality, and growth conditions gives the least runoff per produced unit of N, P, and other nutrients. This is documented through a number of trials." (Heie, Tveitnes og Enge 1990)

In response, Simonsen claimed that the tax has much lower abatement cost than other instruments launched in the follow-up of the North Sea Treaty. It is the most cost-efficient instrument of all. Concerning information as an instrument, Simonsen stated (in a subsequent report) that such instruments would only be effective with farmers who are fertilizing more than what is economically optimal and that the farmers lack knowledge about the fact that they fertilize too much (Simonsen et al. 1992:144). This argument was also brought forward much earlier by Holm et al. (1989:54) in a report from the Department of Agricultural Economics, AUN.

Simonsen further stressed that "under any circumstance it is irrelevant to consider what agriculture as a sector may lose in a social welfare cost calculation. The main object is to estimate what adaptation that gives the highest total welfare for society. It is the politicians and not the economist or chief extension officers' job to undertake decisions over distribution of costs and benefits."

Concerning the legitimacy arguments used by the chief extension officers, Simonsen claimed these to be of little relevance for the economic analysis, that primarily discuss the social costs and benefits of environmental taxes (Simonsen 1990:62). Social costs are linked to reduced yields and investments in alternative fertilization systems, whereas the social benefits are linked to reduced runoff (ibid).

The argument that some groups of farmers are not profit maximizers was addressed in a later report (Simonsen et al. 1992) where Simonsen argues that some utility maximizing farmers may fertilize more than profit maximizing nitrogen level, and some less, but that the response to a price increase should still be to reduce fertilizer use. In response to the chief extension officers' assumption that "farmers will reduce the runoff as far as it is practically possible," Simonsen states: "Is it so that the chief advisers in agriculture in Norway claim that farmers' adaptations are independent, not only of what is economically profitable but also of what is economically possible, for the individual farmer?" (ibid).

The kinked point theory is not dismissed altogether by Simonsen, but he points out that over a period of several different years specific kinked points will give a nearly continuous production function. Also, for a field, not all straws fold at the same time. The farmer furthermore bases his/her fertilizer use on an average of expected yield/nitrogen relationships that, roughly speaking, will match the mathematically estimated production functions. According to Simonsen, independent of what kind of theory one might have for the relationship between nitrogen input and yield, the farmer should base his adaptation on a continuous function (where optimization is possible).

Research and bureaucracy

The Simonsen report (1989) met resistance not only in the agrarian research environment at AUN, but also in the "related agrarian environment" in the Ministry of Agriculture. Inside MoA there was scepticism within certain groups of civil servants about the precise causal relationship between increased price on fertilizers, reduced use of fertilizers, and reduced runoff as assumed by the Simonsen report. One month after the submission of the report, it was stated in the Parliament Report No. 1 (St.m. nr. 1 1989, 1990):

> Department of Agricultural Economics at AUN has calculated the effect of a tax on fertilizers. There are many factors imposing uncertainty over these conclusions. Use of taxes as a policy instrument will demand a thorough assessment of different conditions.

It was the MoA that gave this assignment to the Department of Agricultural Economics (Simonsen 1989). Now, the very same ministry assigned the Center for Soil and Environmental Research (previous GEFO) to make a report in order to discuss the factors that may influence the relationship between fertilizer use and yield relative to nutrient runoff. While the Center for Soil and Environmental Research report was being prepared, MoA and MoE discussed policy instruments to follow up the North Sea Treaty. SET rather strongly supported the conclusions by Simonsen, and advocated in a meeting supported for high environmental taxes (150 percent) (SFT 1990). The project leader from MoA stated in this meeting that there is no scientific basis to support taxes of this kind and that tax is not an efficient instrument to reduce runoff.

This is documented in a letter from MoA to MoE (14.02.90), where it is also stated that a new report (Vagstad 1990) on this matter was forthcoming. In a new letter (19.02.90) MoA outlined some major policy instruments to reduce runoff; physical measures are mentioned as the best and that the rate of accomplishment of stated aims can be enhanced by means of information, extension service advice, and planning.

At that time MoA consisted of two sections working with environmental policies. The Agricultural Division operated as a political secretariat and had civil servants with basic agricultural economics backgrounds. The Farming Division basically was in charge of environmental problems and had close contacts with local government institutions in the construction and implementation of measures to reduce pollution. Most of the civil servants there had agronomic backgrounds in soil science, agricultural engineering, veterinary science, and plant production. The formal statements and handling of cases from MoA on environmental policies came from the Farming Division.

After the letter on the "demise of taxes" was sent to MoE, a civil servant from MoA, Landbruks-division, wrote a new letter to MoE stating that taxes were considered a cost-efficient instrument and the only instrument that could lead to a goal fulfillment relative to the North Sea Treaty (MoA, 08.03.90). This civil servant (an agricultural economist) later stated that his letter to MoE was strongly motivated or inspired by a conference where the agricultural economists at AUN put forward their perspective on environmental taxes (Børve).

In December 1990, the Center for Soil and Environmental Research (GEFO) report was submitted (Vagstad 1990). This report concludes that the runoff loss is important only when the nitrogen input exceeds the "agronomic needs of the plant." This report describes the level of runoff to be more closely linked to the ratio between fertilizer input and yield level than to the level of nitrogen use in itself. Weather, climate, and soil quality are also more important than the level of fertilizer use itself.

The variations between areas and between years therefore indicate that the level of fertilizer use must be adapted according to the particular year and plot in question. According to Vagstad, using average figures for yield and runoff, such as Simonsen does, will lead to the most important problems being overlooked relative to reduced runoff from agriculture.

The debate over environmental taxes has continued into the 1990s in journals, research reports, and public reports. In the summer of 1996, Enge, the chief extension officer, wrote in a report about the Simonsen report:

> This is a very dangerous report because it is used by the central government. A lot of work was put in to put the report dead. If we have succeeded or not, I am not too sure about. (Enge 1996:38)

What kind of rationality is characterizing the actors described in this case? Prior to further discussion of this question, we will review some central concepts of rationality.

Two traditions on rationality

Social constructivism

Social constructivism can be linked to Berger and Luckman (1967) but has roots back to the old classic social scientists including, among others, Weber, Durkheim, and Marx. The society and culture are seen as a product of man. A society has no existence other than what is given through man's action and consciousness. But at the same time man is a product of the society and the culture into which he is born. Man cannot exist apart from society. Societies constitute man. This understanding of man and society as mutual or reciprocal products is important.

People are born into a society and a culture. Children at the beginning of life face society in this sense as objective reality. Growing up, they will partly internalize this objective world: language, social values, social norms, etc. As increasingly more is taken for granted, the more objective reality becomes, an objectivization is taking place. Growing up, people are taught and trained in "conventional wisdom." Individuals will, however, also over time influence and partly change the world around them through externalization. These processes of internalization, objectivization, and externalization constitute the social construction.

Such construction takes place at different levels in society. There is to some extent a Norwegian social construction that may be common and distinguish Norwegians from other people. In different parts of the Norwegian society this construction is also different; being an economist or agronomist doing research at AUN or a bureaucrat in MoA or MoE has bearings in this respect.

The existence of different social values, norms, symbols, and institutions implies that different groups of researchers and bureaucrats will have different world views partly because the social context is also a context within which the actors think, value, and behave (not necessarily always in that order). There is of course a tension both in research and in bureaucracies between these internal processes and the external society that may impose dictates and constraints to this internal formation of meaning, and they are as such to some extent different from an "everyday" social construction.

Kuhn's (1969) paradigm approach opened up seeing sciences as social constructions, where processes of internalization, externalization, and objectivization take place, and where "battles are fought over the hegemony" within the science, generating "the group licensed and time-tested way of seeing" the particular problem. The exemplar becomes important: the research process as a social process.

Knorr Cetina discusses the development of knowledge where researchers, bureaucrats, and practitioners develop through action and over time an epistemic community sense of belonging.

In a social constructionist approach, we may talk about an alternative to a universal rationality, namely a social and collectively developed type of rationality. The rationality is socially contingent, socially created and recreated.

According to Chalmers (1989), a relativist approach states that there is no ahistoric, universal, and noncontextual standard of rationality with respect to which one theory may be judged better than another. This is taking a socially constructionist approach rather far. There is a difference between stating that "anything goes" and stating that there is rationality, but it is socially contingent, and found within paradigms as Kuhn states. It is the latter approach that is used in this chapter.

Rationalism

Rationalism, according to Chalmers (1989), may be defined as the belief that there is a single timeless, universal criterion by which the relative merits of rival theories can be assessed. Popper and Lakatos are referred to as proponents of this "scientific ideal."

Extreme rationalism is not supported by many researchers in theory, but in practice, evidence of rationalism in research as well as in politics and governance is abundant. "We must base our policies on facts and figures," is a common expression.

Comparing social constructivism and rationalism

Bernstein (1983) tries to avoid both rationalism and extreme relativism by stating that it is difficult to choose rationally between paradigms as "rationality belongs to the paradigm." Within a paradigm, however, there is common ground for dialogue and comparison of views.

Rational choice versus socially constructed/embedded choice

From rationalism and social constructivism as overarching philosophy of science theories, there are logical and empirical links to different theories for explaining human behavior.

A proponent of rationalism would also use the theory of rational choice as a point of departure for explaining human behavior and pay little attention to the social dimension of behavior. Rational behavior may be defined as the efficient choice of means to reach a stated set of goals. According to Elster (1983a), there is a difference between thin and broad rationality where the broad rationality assumes consistency, not only between aims and means, but that there is also consistency in beliefs and desires, and that beliefs should be "grounded in available evidence" and linked to judgments.[5] Elster does not dismiss the idea that social factors influence people's adaptations, but they cannot be the cause for human behavior in his view, largely because methodological individualism presupposes that man acts from his own will and no social institutions or norms can have explanatory power. Only man can act. (It would lead to functionalist explanations that, according to Elster, should not be used in social sciences.)

A social constructivist approach would see behavior as, first of all, socially embedded. Behavior may be rational in an intentional way, but only when and because it is a dominant feature of that culture in which action takes place. There are also various other types of behavior such as "value rational behavior," "traditional behavior," and "affective behavior" (Gilje and Grimen 1993).

In explaining behavior, the rationality becomes socially contingent, and Common's (1950) concept of "reasonable man" as an alternative to rational man is to the point. Naustdalslid et al. (1994: 88) state that it is not a rejection of the thought that man follows his own aims and tries to act individually rational, but it reflects that man tries to realize goals within a social context and in the light of norms, rules, and social values (Astley and Van de Ven 1982:262).

How to compare world views

Kuhn discusses the problem of comparing paradigms and introduces two concepts to conduct comparisons. Two paradigms are incompatible for a certain topic if they conflict logically. Two views are incommensurable if they do not have a common yardstick by which their relative merits may be assessed.

If differences between sciences are a result of different modes of understanding, one cannot simply add up or put together different perspectives. They can be mutually exclusive or incompatible. Or they can be incommensurable.

The incommensurability thesis is an attack on objectivism, not on objectivity. This thesis calls into question the modern version of objectivism which assumes that there is or must be a common, neutral epistemological framework within which we can rationally evaluate competing theories and paradigms. The incommensurability thesis also questions the existence of a set of rules that will tell us how rational agreement can be reached on what would settle the issue on every point where statements seem to conflict (Bernstein 1983).

The incommensurability thesis implies that two scientific traditions may reach conflicting conclusions analyzing the same case, for example, the choice of efficient politics to reduce nitrogen runoff from agriculture in economics and agronomy, without either of them necessarily being wrong. The two traditions may differ on which aspects of the case they put in focus, comprehending both theories, methods, and the object of the science. Accepting that there are different sciences, different paradigms (within a science), different world views, or different rationalities leaves us with the problem of identifying some key structures by which to conduct a comparison of these rationalities.

How to identify different world views

Kuhn's concept of disciplinary matrix or paradigm could be used in such a comparison (see, for instance, Vedeld 1994, where economic and ecological approaches to natural resource use are compared). We have, however, chosen a less rigid conceptual framework to identify and compare rationalities between economists and agronomists. Discussing the concepts of incommensurability and incompatibility, we will look at the object of the two sciences and how the sciences handle causality and some other areas where there are differences.

The object of the science

Economics is a science analyzing human choices and behavior. Economics is an anthropocentric science. In the neoclassical tradition the profit maximizing man is brought into focus. Nature is regarded as a means to obtain maximum profit or utility. "Our task is not to define what is good for the environment. Concerning protection of environment and environmental concerns, the point of departure is always human perception of beneficial aspects of the environment" (Simonsen et al 1992:41).

Pollution caused by nitrogen runoff may reduce human welfare. This is the main reason agricultural economists at AUN are engaged in the relationship between the size of yields (maximizing the farmer's profit and agricultural production) and the amount of nitrogen runoff (minimizing loss of human welfare in society). According to Simonsen et al. (1992: 41): "It is important to find the right balance between the environmental costs and the social benefits of the nitrogen runoff and the production output. In agriculture, the environmental costs should be assessed against the value of production." Therefore, valuation of the main variables in monetary terms is important in welfare theoretical calculus. Given the North Sea Treaty's aim of 50 percent reduction of nitrogen runoff, the agricultural economists concentrate on developing a policy for reaching the aims of the North Sea Treaty with the least cost solutions for the society.

Natural sciences basically study natural systems and natural processes and do not explicitly deal with human behavior or adaptations. Traditionally, agronomy has also dealt with the same, but mainly from the focus of how human beings could maximize yields based on knowledge of different relationships in nature: between climate, soil, fertilizer, and yields.

"In plant and soil science the researchers are faced with large variations in climate, weather, and soils. One must use this variation consciously, register it, take advantage of it and sort out the different causal relationships" (Njøs 1989:18). Many agricultural economists feel there is an unwillingness among agronomists to generalize and provide simple, strong models for important relationships.

For agronomists, man is a gardener subordinated to natural laws. Demand for plant and soil quality, rather than utility for man, is highlighted. The plant itself is an object for the science; the plant and the field should look proper and be an expression of economically optimal resource use. The task of the gardener is to utilize the natural conditions for agricultural production in an optimal way. Agronomy is developing and sophisticating the knowledge of natural conditions for food production.

While the maximizing man is the object of neoclassical economics, the conditions of proper gardening are the object of agronomy. The objects of science differ in these traditions, but this does not imply incompatibility between the sciences. Rather, knowledge of the conditions of efficient gardening would be of great interest for the profit maximizing farmer. Agronomy and economics may be characterized by different concepts, theories, and methodologies. Studying agriculture, what seems insignificant from one perspective may be crucial when viewed from the other (Vatn et al. 1996). Related to agricultural management, their contributions of knowledge are complementary, though given in distinct professional fields and terms. Regarding the objects of economics and agronomy, this is a case of incommensurability, following Kuhn's (1969) terminology.

Causality in the sciences

Suggesting an ambient nitrogen tax to reduce nitrogen runoff from agriculture, the agricultural economists of AUN intervene into the professional fields of agronomy. Economists assume an explicit causality between nitrogen fertilization and nitrogen runoff. The more nitrogen fertilizer that is added to the fields, the larger the quantity of nitrogen runoff. In addition, the economists connect nitrogen fertilization with the law of diminishing returns of yields. (These causalities are also shown in Figure 1.) Given the explicit causality between nitrogen fertilization and yield returns, a tax on nitrogen fertilizer will force the profit maximizing farmer to reduce the use of this production factor as a result of a calculation of the intersection between marginal income and marginal costs. Given the causality between nitrogen fertilization and nitrogen runoff, the economists can calculate which level of nitrogen tax is required to obtain the aimed reduction of nitrogen runoff.

Assuming an explicit causality between nitrogen fertilization and nitrogen runoff, agricultural economists challenge accepted causalities within agronomy. The agronomists claim that the quantities of nitrogen runoff will vary with different factors, such as soil quality, management of the soils, climate, fertilization practice, plant density, plant species, and use of catch crops in addition to the amount of nitrogen fertilizer. From the agronomic perspective it is untenable to isolate one factor causing nitrogen runoff. Calling attention to a series of experiments, agronomists from AUN emphasize that factors such as weather, climate, and soil quality are more important for the level of nitrogen runoff than the level of nitrogen fertilizer use in itself. In addition, the connection between the amount of nitrogen fertilizer and yields is said to be of greater significance for the level of nitrogen runoff than the amount of nitrogen fertilizer used (Vagstad 1990).

Agronomists emphasize that nitrogen runoff can be reduced by improving professional skills in connection with management of soils and other agricultural practices. Thus, advising through the extension service seems to be a preferable measure to impart knowledge to farmers. Adapting the amount of fertilizer to climate variation, to the plants' demand for nutrients during the growth season, and to soil quality, the plants are given optimal conditions for intake of nitrogen. At the same time, excess use of nitrogen is avoided; thus, the nitrogen runoff will be reduced. Furthermore, nitrogen runoff may be diminished by reduced soil management use of catch crops. Given optimal growth conditions, the nitrogen runoff will be a constant low or even declining, which increases nitrogen fertilization to the point of lodging. From this point, where the nitrogen demand of the plants is covered, the nitrogen runoff will increase (see Figure 2).

299

The assumption of an explicit causality between nitrogen fertilization and nitrogen runoff is incompatible with an assumption regarding the level of nitrogen runoff being more dependent on the relationship between yields and nitrogen fertilization than the supplied amount of nitrogen fertilizer. In the first case, the nitrogen runoff will increase with each extra unit of supplied nitrogen fertilizer (at least in the production area of interest discussing a nitrogen tax). In the second case, the nitrogen runoff will be constant or diminish up to the lodging point, given an optimal relation between yields and use of nitrogen fertilizer. These arguments are in conflict logically.

Other indicators of scientific traditions

Sciences can be distinguished by their use of symbols and metaphors. Searching for a concept for a technical term for the preferred policy instrument to reduce nitrogen runoff, the agronomists alternate between "correct fertilization" and "environmentally optimal fertilization." Agronomists focus on professional skills connected with soil and plant treatment. Economists, adopting the policy measure "reduced intensity," highlight the choice of the economically optimal amount of nitrogen fertilizer use.

While agronomists base their research and advice on series of experiments and firsthand data material, economists often use average figures from secondhand data material as the basis for their calculations. The difference between concrete, practical agronomy and the generally oriented economics is mirrored in the choice of policy instruments. Agronomists are inclined to choose extension and detailed regulations as policy instruments, while economists in this case select environmental tax, which is a typical generally oriented policy instrument. It is in particular the high transaction costs of detailed regulation that makes economists choose the latter. From an economic point of departure, there would also be little point in trying to make the farmer do something he will lose money on.

Economists and agronomists also interact with different groups. In agronomy, by imparting knowledge traditionally, the researchers have interacted not only with the extension service, but also with civil servants and private sector organizations. Through this interaction, a continuous exchange of meaning and social values has taken place resulting in the formation of an epistemological community (Knorr-Cetina 1981). Thus, in the social construction of reality, agronomists relate to patterns for understanding and explanation (as scientific actions) in the community of agronomic science and in the agrarian communities. Environmental tax on nitrogen fertilizer, an expensive policy instrument for Norwegian agriculture seen apart, is not acceptable for agronomists.

The agricultural economists of AUN seem to be influenced by the progress of environmental economics in the 1980s. In the 1970s, when the authors of this article were students at AUN, the then Department of Agricultural Economics still was characterized by the tradition of farm management. The farm management tradition had a close connection with "good agronomy" and practical farming. Throughout the 1980s this tradition was displaced by economists importing mathematical modeling, especially from schools of environmental economics in the U.S. The point of departure is the "rational economic man" (profit maximizer) and the modeling of how this actor will react upon changing economic conditions in production and consumption. This means interaction with mathematical modelers and neoclassical economists. The interests of the agricultural community are not so important when modeling an environmental tax.

The latter discussion relates the debate of rationality to social construction and reconstruction of reality. The rationality within agronomy and economics seems to be dependent on social construction of patterns for explanation and understanding between related interaction partners in the scientific communities.

Summary

The case we present represents on one level a fight over the hegemony between economists and agronomists concerning who is to present the "officially approved story" concerning the problems and solutions relative to diffuse runoff of nutrients from Norwegian agriculture. It is our contention that we also redefine this debate between different groups of bureaucrats in the relevant ministries. Taking this one step further will give insights in the strategic and instrumental use of research (Naustdalslid et al. 1994) by the bureaucracy relative to this case.

What are important implications of the difference?

The debate between researchers

The debate as a struggle between different world views. The disagreements we have described above should clearly indicate that there was a conflict between different world views, where incompatibilities and incommensurable views emerged over what is the relevant area of research, and what are "correct" causal relationships, what are relevant symbols and metaphors, and what are underlying social and political values. Both participants and observers of the debate would most likely agree to this.

A conclusion that follows is that scientists from different disciplines live in different worlds where communication is difficult and constrained and where problem description, explanation, and prescription will systematically differ between sciences for the "same problem."

The struggle as a process. According to Kuhn and Feuerabend, research takes place in a social context. Comprehending the social, interactional construction of reality is important and necessary in order to understand the development of knowledge within a particular field. Knorr-Cetina (1981) emphasizes that generation of knowledge through scientific action is important in additional construction of meaning through social interaction. She finds that the researcher in action does not passively "describe how things are in the real world," but actively construct facts into his (her) own world. Researchers' actions are also governed by local conditions and local factors and not by universal standards or criteria (that are common for the entire scientific community) (Steinsholt 1995).

If we return to the case, Enge and his agronomic colleagues clearly believe in their story, just as Simonsen and his colleagues believe in theirs. As the debate heats up, researchers in both camps take their positions and combine forces against the opponent. Both parties seek and get confirmation and support from colleagues within the discipline. Opponents are mutually accused of being nonscientific, uninformed, and politically biased. It is a fight over hegemony, in a Kuhnian sense, the fight over what should be the official story over the problem of runoff and the choice of intelligible policy instruments.

It is not a fight where the participants seek dialogue and communication—bringing the world forward through "the incommensurability solution"—translating different knowledge into your own field, etc. The process is more like a "Balinese Cockfight" (Geertz 1973), where the main objective is to beat the enemy at any cost. In this media-focused process the participants expose themselves through arguments and become vulnerable also in the sense of academic prestige. "Their hands become tied"—the room for dialogue and for giving opponents credit for arguments is narrowing down. The antagonists consequently bring out what they see as the weakest arguments of their opponents, and there is no reference to areas where one may actually agree. There is no search for common ground.

Was this bound to happen? Are we so well entrenched in our own perceptions and language games that we must fail to see the relevance and importance of other world views and other rationalities?

A scope for interdisciplinarity. Like culture in general, a science is not a straight jacket from which it is impossible to transcend or go outside. As a counterbalance to the debate above, we will refer to an example of successful interdisciplinary cooperation.

At the Department of Agricultural Economics a research program was launched in 1990-1991. Both ecologists and economists were set to work together on basically the same problems as discussed above. However, the Economy/Ecology (Ec/Ec)-program was less policy oriented and aimed toward generating a model framework for handling relationships between soil and plant growth processes, use of fertilizers, and how different policy instruments may be thought to influence runoff rates in different ways.

Considering the experiences from earlier debates between agronomists and economists, the program under its leader, Arild Vatn, realized the potential problems of communication between scientists from different disciplines. By upgrading colloquium groups and working together, the collaborators of the Ec/Ec-program consciously tried to generate a kind of common horizon. According to Vatn, there was a period of trial and error, including frustrations over the cooperation. But over time, through experience and through "knowledge in action" each came to respect and accept one another's positions.

This process was facilitated through the development of a common language in the methodology of mathematical modelling. A common ground was established, a common ground that also might be established between economists and "modern" model-oriented agronomists.[6] On this basis, areas of incompatibility were accepted or left aside. As a result of dialogue and practical research, areas of incommensurabilities became fewer. Thus, compromises could be reached. Explicit debates over problems of interdisciplinarity and cross-scientific communications were carried out, and a common understanding gradually developed. At the end of the Ec/Ec-program the researchers themselves stressed the following experiences from the interdisciplinary team work:

> Certainly natural and social sciences are more complementary than rival. Still, the experience gained both from the above debate (the Simonsen et al.; Enge et al. controversy) and the research to be documented here shows that it is important to provide comprehensive insights into how even complementary disciplines view features, especially along the boundaries between them. What seems indifferent or unimportant from one perspective or discipline may be crucial when viewed from the other. (Vatn et al. 1996: 3).

Still, an unanswered question is: What is right, and what is wrong? Whose knowledge and what kind of knowledge can be trusted? According to prior discussions, scientific knowledge from both disciplines can be based on assumptions of causalities and social construction of patterns for understanding and explanation. However, understanding different perspectives does not help anybody choose the most relevant perspective handling a specific case, for example, in choosing efficient policy instruments to reduce nitrogen runoff.

Research and rationality. We believe that scientific knowledge, explicitly or implicitly, is based on theoretical models that simplify reality. Such models are founded on a number of basic assumptions. Neoclassical economics assumes that human actions can be analyzed as profit or utility maximizing endeavors. According to the neoclassical approach, the role of the extension service is limited to imparting knowledge so that farmers can adjust the use of nitrogen fertilizer toward economic optimum. Does this model comprehend anticipated effects of policy instruments based on information or attitude generation?

According to social constructivism, environmental consciousness as a social norm may transcend the logic of profit maximizing. As a result of altered attitudes in a society, farmers might want to maximize profit without being an "environmental pig." Environmentally oriented farming might even become a characteristic of good agronomy in the agrarian community, an example of good agricultural code, according to Lowe and Ward (1996).

Neoclassical economics and the sociology of science (within which social constructivism is a perspective) attach different importance and potential to information and attitude generation as policy instruments. Clarifying these differences within the scientific communities could facilitate a rational discourse between sciences. As a result, the researchers could make their basic assumptions explicit and accessible in their reports. Subsequently, the decision-makers would get a better foundation for their choice of appropriate policy instruments.

The point of departure for this discussion is rationality within sciences. We have not challenged the internal logic of the scientific models, the internal rationality of the science. Assuming internal rationality, the assumptions of scientific models have become the focus for discourse. Thus, self reflection is seen as one way to clarify perspectives within a science and between sciences.

Developing a common language, as in the Ec/Ec-program, might be another and more active way to bring different sciences together. Instead of using two different approaches to the same problem, one broader, more unified approach can give a wider perspective on a specific problem (Vatn et al. 1996). As communication across scientific borders is maintained over time, communication becomes more precise and less time is wasted on mal-communication. Communication alters character from multidisciplinarity to interdisciplinarity. Interdisciplinarity is thus not first of all a product, but a more or less consciously generated process of interaction and a "wanted" dialogue.

However, we cannot expect that clarification of scientific knowledge will be counterparted by a system of rational comprehensive use of such knowledge in bureaucracy planning. According to social constructivism, scientific knowledge can be used in different ways in the bureaucratic community.

Debate and contact between researchers and bureaucrats: Research use

Throughout this chapter we discuss the difference between a rationalistic approach and a social constructivist approach to research in itself and to the bureaucrats' use of research.

Research in itself. Following a rationalist approach, research results and knowledge are objective. This constitutes a firm and undisputable basis for rational and even reasonable decisions to be made.

In a socially constructed world view this is less obvious. Knowledge is socially constructed. It is generated on the basis of particular axioms and assumed causal relationships. Knowledge applied to different and separately perfect and consistent logics may give different descriptions, explanations, and prescriptions.

Bureaucrats' use of research. In Table 1, we outline a rationalist and a social constructivist approach to the instrumental and strategic use of research. A rationalist approach.

Table 1. Rationalist and social constructivist approaches to instrumental and strategic use of research

	Instrumental use of research	Strategic use of research
Rationalism	Rational comprehensive planning	Public choice
Social constructivism	"Transcendable planning"	Epistemological community choice

- Instrumental use of research. Following this view, one task of researchers is to supply objective knowledge about actual phenomena to neutral bureaucrats. Given stated aims, concrete knowledge concerning the causalities between use of nitrogen fertilizer and nitrogen runoff will give bureaucrats a cognitive foundation to choose reasonable policy instruments to reach political goals. This is in line with Carol Weiss' (1979) "instrumental" function of research. According to Naustdalslid et al. (1994:53), the instrumental use of research "assumes that the research influences the material reality by force of the knowledge it generates." Research is assumed to be neutral and self-acting in the sense that it impacts society independently upon other actors and forces.

With this perspective, instrumental use of research presupposes a bureaucracy without self interests. The role of the bureaucracy is to obtain information and knowledge and to be able to make rational decisions regarding choice of policy instruments. In this connection, the bureaucrats are regarded as obedient servants of the political system.

• Strategic use of research. In the world of politics there are different actors with different interests, also within a public bureaucracy. The public sector is an interactive system, where a number of different games play out (March and Olsen 1989:7). In this sense, research may be used to advocate different groups' own (and private) interests, which is in line with Weiss' concept of the tactical function of research, or what we could call the strategic use of research.

According to public choice theory, bureaucrats will have evident self-interests in the political games. Such interests might be to secure or fortify one's own position in the bureaucracy. According to Sandmo (1990:93), public choice theorists assume that the behavior of bureaucrats can be explained from the motive of maximizing own income, status, power, and influence. In general, the bureaucrats will have incitements to maximize the size of (their share of) the public sector. In the game between different ministries, and departments within ministries, the bureaucrats will seek to enhance the size and power of themselves and their own organization.

As in neoclassical economics, public choice theory assumes intentional rationality. The bureaucrats are conscious of their aims and how these aims are ranked in proportion to each other. Consequently, they will look for means to reach these aims. This intentional rationality characterizes all the bureaucratic players heading for power and influence. This intentional rationality based on self-interest forms a common episteme in the bureaucracy.

Following this theory, representatives for different ministries are apt to look for specific scientific knowledge and research results to support their own case. This can be reflected in two different ways of strategic use of research. In the first place, the bureaucrats will demand research results with arguments and particular logic in agreement with interests in the bureaucracy. Secondly, research results can be used to expand the scope for decision-making in the bureaucracy.

In practical politics, generating smoke screens and creating confusion can be just as valuable as clarity and unanimous scientific conclusions relative to strategic aims. In general, strategic research is easiest to use when experts clearly disagree or when there is a real lack of knowledge in the field (Nilsson 1992).

A social constructivist approach

We link this approach to the theory of new-institutionalism (March and Olsen 1989). According to new-institutionalism, organizations gradually become more independent through the establishment and routinization of knowledge, meanings, and procedures. Inspired by the sociology of knowledge (Berger and Luckmann 1967), new-institutional approaches to bureaucracies emphasize the significance of social norms and rules. Thus, the formation and maintenance of institutional norms and values is understood as a result of social construction of reality.

Assuming the character of institutions, an intrinsic value is attached to organizations, procedures, or routines beyond their instrumental function for the realization of other values. For this reason, institutionalized behavior cannot solely be interpreted as intentional rationality based on an aims/means perspective, but in addition ought to be related to value rational conduct (Gilje and Grimen 1992; March and Olsen 1989). This implies a separation of new-institutionalism from public choice theory.

"It is a commonplace observation in empirical social science that behavior is contained or dictated by cultural dicta and social norms. Action is often based more on identifying the normatively appropriate behavior than on calculating the return expected from alternative choices" (March and Olsen 1989: 22).

Within a bureaucracy the routinized application of solution procedures connected with distinct policy instruments can attain an intrinsic value and achieve status as an aim. Therefore, the new-institutional approach also differs from rational comprehensive planning, because the latter model presupposes a structure where means are subordinated by the stated aims in politics and bureaucracy (Banfield 1973; March and Simon 1960; March and Olsen 1989).

• Instrumental use of research. According to new-institutionalism, the bureaucrats will interpret new knowledge through the culture in their ministry or the department of the ministry (Olsen 1988). Considering a culture of bureaucracy, we will emphasize world views and standard solution routines, i.e., concepts associated with professions and professional standards (Krogh et al. 1996). In addition we regard routinization of "appropriate" problem definitions and standard solution procedures as results of a historical process (Laegreid and Olsen 1988).

The instrumental use of research expresses search for increased knowledge. The existence of a strive for new knowledge and change will supposedly vary between different ministries. Young ministries handling new policy areas, as MoE, will probably be characterized by a high degree of instrumental use of research. As a result of instrumental use of research, present perceptions and knowledge may be transcended in the bureaucracy. Thus, the way might be paved for new knowledge and action, although, according to social constructivism, new knowledge will always be interpreted in a social and cultural context.

• Strategic use of research. Contrary to public choice theory, which presupposes a common bureaucratic episteme, social constructivism suggests the existence of different epistemological communities within the bureaucracy. An epistemological community is much alike what Egeberg calls a segment (Egeberg et al. 1988). A segment may be organized around the politics of fishery, industry, or agriculture. Such a segment may have representatives from the ministry, the parliament, media, certain organizations, and certain research traditions. These representatives will share some basic values and have a common interpretation of certain situations.

This description of a segment is also suitable for an epistemological community. Still, the notion of segments differs from our understanding of epistemological communities. We suppose that shared values and interpretations in the epistemological community partly are tacit or implicit. Again, the pure cultivation of intentional rationality is questioned.

Launching the concept of "transepistemic arena," Knorr-Cetina highlights the social relations between researchers and users of research results, i.e., bureaucrats (Knorr-Cetina 1981). She asserts that these relations are important for the research process and even for the research results. Elzinga (1985) warns against an epistemic drift from the scientific society toward the users and buyers of research. Social constructivism not only focuses on strategic use of research, but even on strategic production of research. This does not necessarily mean conscious control over scientific assumptions, theories, and methodologies, but rather implies development of some common basic values, problem definitions, and solution procedures in an epistemological community consisting of bureaucrats and researchers.

Case application

The process connected with the implementation of the aim of 50 percent reduction of nitrogen runoff stated in the North Sea Treaty shows several examples of bureaucratic use of research results (see Hesjedal and Helland 1995). Different events in our case may be discussed to illustrate both rationalistic and socially constructed use of research in the bureaucracy in instrumental as well as strategic ways. In the following, we will draw attention to discussions concerning the environmental tax on nitrogen fertilizer between the Ministries of Agriculture and Environment by using the rationalistic and the social constructivist approaches.

• Rationalistic approach—instrumental use. To obtain implementation of the stated aims concerning reduction of nitrogen runoff, the two ministries established a cross-ministry organizational working group. At first this group collected information from different research milieus to elucidate and extend the knowledge on this area for decision-making and choice of policy instruments. This intention can be interpreted as a reflection of an ideal of rational comprehensive planning.

- Rationalistic approach—strategic use. After a while, disagreements between SFT and MoA occurred in the working group. Following arguments from the agricultural economists from AUN, SFT suggested environmental taxes on nitrogen fertilizer. The representatives from MoA felt that this policy instrument was in conflict with other political aims (the production aims). In addition, the representatives from MoA criticized NPCA for launching such a proposition. In a formal letter to MoE, MoA emphasized that launching such kinds of propositions is the task of the ministries, not the subordinated directories. MoA claimed earlier that they had not only the duty, but also the right to elaborate appropriate policy instruments, while MoE had the responsibility for developing the aims for the environmental policies. These views can be interpreted as attempts to defend territory and strengthen the position of MoA in the power game. In the context of strategic use of knowledge, this can be interpreted as a way of excluding suggested policies based on unwanted research results.

Recognizing positive attitudes to an environmental tax on nitrogen fertilizer among certain groups, MoA ordered a second report from groups they knew and that had its confidence (Vagstad 1990). There are indications that they may have known the conclusions of this report in advance (see MoA 1990a and 1990b). This is in line with a rational, strategic use of research and also in line with a public choice approach explanation, as the alternative policy instruments to reduce runoff that were suggested in the second report then would be administered by this division in MoA.

- Social constructivist approach—instrumental use. From another point of view, the collection of knowledge in the field may have convinced MoA that the causalities between nitrogen fertilizer and nitrogen runoff underlying the proposal of the environmental nitrogen tax were too simplistic. This was one of MoA's main arguments for ordering the report from the Center for Soil and Environmental Research. According to Vagstad (1996), this was a period when the consciousness of the complexities of the actual causalities was expanding. Even though this research of the Center for Soil and Environmental Research hardly would transcend world views in MoA, the ordering of the report may have been interpreted as an instrumental use of research from within a well-established and common culture.

- Social constructivist approach—strategic use. Using Knorr-Cetina's epistemological community choice point of view, the civil servants in the MoA have a world view sharing much common ground with agronomists and with farmers and with the agrarian community at large. And they also interpret the economists' tax suggestion not as a way to solve the problem, but as a way to cut subsidies in the agricultural sector. Furthermore, it will not make farmers fertilize less, and it has little or no effect on reducing runoff. Along these lines, a report supporting taxes, in what they saw as a biased and narrow analysis, could not be left unchallenged. "Use of taxes as a policy instrument will demand a more thorough assessment of different conditions" as was stated by MoA in the parliamentary report of 1990. This is also strategic use of research, but understood in a more contextualized framework.

Rationality and utilization of research in society

Up to now, we have addressed rationality within and between sciences and different interpretations of the motivations for bureaucrats' use of research in rationalism and social constructivism. In the following, we shall consider some wider implications of the latter discussion.

Increasing use of experts. There is a trend toward increased use of experts by the various public bureaucracies. Underlying this is an opinion that politics should be and can be rational, as demonstrated by Prime Minister Brundtland in a meeting about whaling in Norway: "We must have the facts and figures on the table so that our policies can be made on a rational basis."

The Norwegian Research Council (1996) reports that 50 percent of what is spent in public sector research is controlled directly by the different ministries and that the rest is channeled through the research council. This is also formally institutionalized in Parliamentary Report 35 (75-76), where the different ministries were given formal responsibilities for their own research in terms of financing, formulation of aims, and for establishing priorities.

Political dimension of the increased use of research. In this way research increasingly becomes intertwined in both politics and bureaucracy. According to Giddens (1990, 1991), expert systems are penetrating societies of high modernity through the complexity of social organization and dependence on advanced technology. Even though this presupposes comprehensive trust and faith in expertise, high modernity is also characterized by reflexivity and market economy. As research becomes a public phenomenon and a product for sale, the researcher is no longer only responsible toward himself and his research colleagues, but just as much toward all users of the research who place demands on "useful" research and "relevant" research as seen from their desires or needs (Naustdalslid et al. 1994:12).

Using research results to legitimize policies, the state increasingly takes interest in defining the role of research and its intentions. This commitment for research results is forming a mutual dependency bond between the state and the science. According to the Norwegian Council of Research Policy (1988:21): "Controlling research is an important part of the general steering of the society."

Following rationalism, political use of and control over research is important to secure fulfillment of political goals. The increasing interplay between the public and private sector is a major cause of the public need for governance relevant knowledge. The active state needs to know more about the society it is to govern. As we have seen, the state also has an instrumental interest in using research strategically in policy-making.

In a constructivist perspective, the picture is again more complex. As we have seen, different sets of knowledge can be applied to the same problem. Moreover, research is used by the modern state to exert control. A substantial part of present research is directly funded by and approved by the ministries. Given that research is financed by ministries with well-defined steering interests, research perspectives and results that legitimize and approve standard problem definitions, routines, and steering interests in funding ministries will be given priority.

According to Elzinga (1985), excessive influence of politicians and bureaucrats on the funding of research has caused an epistemic drift from scientific to external extra-scientific criteria. Thus, social interests may heavily influence both formation of research as well as interpretation and utilization of research results (Foucault 1980; Sundquist 1991).

In this way, scientific criteria related to independence and objectivity, criteria that are to the core of rationalism, are threatened. Thus, basic research resting on what may be called transcending cognizing interests can be weakened.

Does it make sense to assume rational politics? According to constructivism, facts and figures are not the reason the MoA failed on the final conclusion regarding the choice of policy instruments in this case. The epistemical community constituted by agrarian politicians and representatives from farmers' organizations, established research institutions, and the MoA seems to find the solution suggested by Simonsen unacceptable. The rejection is caused by different reasons, with regard to both agronomic ideals, to how problems should be solved in "farmer compatible ways" and to the political implications of a cut in agricultural transfers that a tax could have incurred.

The debate within the bureaucratic cycles concerning research products is also important concerning rationality. How important is scientific knowledge? It is stated as a point that the public sector also constitutes a social system with its own world view, with its own social norms and social rules. This social system is partly permeated by a mainstream set of understandings of different scientific contexts. The good argument or research report is not interpreted in a vacuum, but in context in a process.

What are policy implications of a relational and contextual rationality? We cannot talk about one world and one universal rationality. In this sense basing

311

our policy-making on scientific knowledge is not a guarantee for "correct" or "rational" policies. On the other hand, the answer is definitely not to state the opposite, scientific knowledge is not necessary or important for the choice of intelligible policy instruments and policy aims. The bottom line is that research- and knowledge-based policy-making is a necessary but not a sufficient condition for "reasonable policies." And there is always a danger for strategic use of knowledge by the power groups in society.

Conclusion

Among researchers, the scientific point of departure seems to be conditioned by the object of their science. According to our study, the object of the science forms the understanding of nitrogen runoff. In soil and crop sciences, knowledge about soil processes and conditions for plant growth determines the comprehension of causalities and remedies concerning runoff. Neoclassical economists use knowledge from the soil and crop sciences to consider natural conditions for production and runoff, but this knowledge is interpreted and adjusted to be compatible for the rational, profit maximizing actor. Through different points of departure, the same data sets are used to support theoretical assumptions by means of accepted methods in neoclassical economics and soil and crop sciences, respectively.

From our study, the sciences seem to have formal logic in common. But even though they use the same data sets, they also draw different conclusions concerning choice of policies to reduce nitrogen runoff. Seeing the world through the eyes of the profit maximizer or, on the other hand, through the eyes of the gardener occupied with nutrient demand of plants, seems to be one major difference causing incompatibilities and incommensurabilities between the sciences.

According to the crop scientist, it is important to be conscious of the complexity of natural conditions affecting plant growth. The amount of nitrogen fertilizer added to a field is just one of many factors influencing nitrogen runoff. Therefore, to focus on a tax on nitrogen fertilizer will not solve the runoff problem. The economists interpret the data sets in a different way. They claim that nitrogen fertilizer is the main source of nitrogen runoff. Given a tax on nitrogen fertilizer, the profit maximizing farmer will reduce nitrogen fertilization and improve manure use and thus cause a reduction in runoff. The complexity of plant production matched with uncertain data sets for nitrogen runoff makes it difficult to judge who is right and who is wrong.

As opposed to between the sciences, knowledge is created within each science with a relatively high degree of consensus. We cannot choose both at the same time; the stories are partly incompatible and incommensurable. There is no unified scientific base from which rational policies can be deduced directly. We still believe that generating clarity over these different epistemic communities and explicitly generating a self-reflection among both researchers and bureaucrats is important. As also demonstrated, there are scopes for translation and communication in particular through practical research and cooperation.

This opens the choice of research results in the Ministries. One possibility is that research milieus with unwanted perspectives will not be funded, while research collaborators will be chosen among those who produce desired results. Such behavior might be interpreted as a strategic rational choice undertaken by the Ministry. On the other hand, the Ministry might choose from a value rationality point of view reflecting the culture of the institution. In this case, the choice of like-minded researchers who share problem definitions and solutions with the Ministry might constitute the foundation of an "epistemological community" through social construction of the reality. In both cases, the optional character of finding research results and funding research milieus to legitimize policies is striking.

Considering the bureaucrat as a part of a ministerial culture and structure, he or she will have neither the capability nor even the willingness to rationally select the "best available information" and from there make the rational choice. This should not be overlooked as it transcends (the illusion of) a comprehensive rational scientifically based policy-making.

In our society, there is an increased trend to legitimize political decisions by claiming that they are scientifically grounded. In this sense, research has and will increasingly become the arena for hidden political agendas. The ivory tower no longer exists. Research is part of the wider world. We will never return to the days of the ivory tower, if it indeed ever existed.

Notes

1. Intensification of agriculture refers to a "high input/high output" system of production. It implies increased land and labor productivity, in part generated by high capital investments per farm and coupled with decreased labor input.
2. Göran Sundquist (1991) discusses and dismisses a naive "seriousness" argument, along with the "increased knowledge" argument or the "economic capacity argument," as good answers to why certain environmental problems enter the political agenda. He stresses the importance of analyzing this in a context of social embeddedness of science and the role of experts.
3. Institute for Geophysical Investigations and Pollution

4. Quotes are translated from Norwegian. The source is given.
5. There are many things, according to Elster (1983a), that may be rational: beliefs, preferences, choices, decisions, actions, behavioral patterns, persons, and institutions.
6. There is the old school of agronomist researchers working in the field with research trial plots, experimenting with agronomic practices, practically oriented and with a close link to the extension service and a strong commitment toward practical farming. The more modern school of agronomists is often more specialized in its education, working more with abstract and theoretical modeling, less oriented toward practical farming. In many agricultural universities there is at present a fight over hegemony between these two schools of thought. A similar difference can be found between farm management agricultural economists and more microeconomics-oriented schools of thought.

References

Asteley, W.G., and A.H. van Ven (1983): Central Perspectives on Debates in Organizational Theory. Administrative Science Quarterly. June 245-273.

Banfield, E.C. (1973): Ends and Means in Planning. In Faludi: A Reader in Planning Theory. Oxford. New York.

Berger, P. and T. Luckman (1967): The Social Construction of Reality. New York. Anchor Books.

Bernstein, R. (1983): Beyond Objectivism and Relativism. Basil and Blackwell. London. 285 pp.

Chalmers, A.F. (1982): What is this Thing Called Science? Open University Press. Milton Keynes, U.K. p. 179.

Commons, J.K. (1950): The Economics of Collective Action. University of Wisconsin Press. Madison.

Egeberg, M., J.P. Olsen, and H. Saetren (1988): Organisasjonssamfunnet og den segmenterte stat. In J.P. Olsen: Statsstyre og institusjonsutforming. Det Blå Bibliotek. Oslo, Norway. p. 336.

Elzinga, A. (1985): Research, Bureaucracy and the Drift of Epistemic Criteria. In B. Wittrock and A. Elzinga (eds): The University Research System. Stockholm. Almqvist and Wiksell Int.

Enge, R., K. Heie, and S. Tveitnes (1990): Miljøavgifter på kunstgjødsel-N og -P og på plantevernmindler. Landbruksøkonomisk Forum. 2/90. 7 Årgang. pp. 38-43.

Enge, R. (1996): Presentasjon av forsøksresultat. Norwegian Journal of Agricultural Sciences. Supplement 22. 1996. Ås Science Park. Ltd. Norway. pp. 38-43.

Elster, J. (1983a): Explaining Technical Change. Cambridge University Press. Cambridge.

Faludi (1973): A Reader in Planning Theory. Oxford. 1973.

Foucault, M. (1980): Power-Knowledge. Pantheon. New York.

Geertz, C. (1973): The Interpretation of Cultures. Basic Books. New York. p. 457.

Giddens, A. (1990): The Consequences of Modernity. Cambridge Polity Press,

Giddens, A. (1990): Modernity and Self-Identity. Cambridge Polity Press.

Gilge, N. and H. Grimen (1993): Samfunnsvitenskapens forutsetninger. Innføring i samfunnsvitenskapenes vitenskapsfilosofi. Universitetsforlaget. Oslo. p. 266.

Hesjedal, R. and M. Helland (1995): Miljø-tiltak-ansvar. Norges reduksjon av naeringssaltutslipp til Nordsjøen. Hovedoppgave. IØS.NLH.

Holm, Ø., D.P. Sødal, and A. Vatn (1989): Virkemidler for å ivareta miljøhensyn i landbruket. Rapport Nr. 11. Landbrukspolitikk og miljøforvaltning. SEFO.

Knorr-Cetina, K. (1981): The Manufacture of Knowledge. Pergamon Press.

Krogh, E., F. Gundersen, A. Vatn, and P. Vedeld (1996): Spillet om naeringssalter. Teorinotat. Miljø, makt og styring. Discussion Paper Nr. 14.1996 IØS.NLH.

Kuhn, T. (1969): The Structure of Scientific Revolutions. International Encyclopedia of Unified Science. University of Chicago Press. Chicago. p. 210.

Lowe, P. and N. Ward (1996): The Morale Authority of Regulation: The Case of Agricultural Pollution. in Proceedings from the Workshop on Mineral Emmissions from Agriculture Dept. of Economics and Social Sciences. AUN.

Laegreid, P. and J.P. Olsen (1988): Byråkrati, representativitet og innflytelse. En studie av norske departementer. In J.P. Olsen: Statsstyre og institusjonsutforming. Det Blå Bibliotek. Oslo. p. 336.

March, J.G. and H. Simon (1960): Organizations. Wiley. New York.

March, J.G. and J.P. Olsen (1989): Rediscovering Institutions. The Organizational Basis of Politics. Free Press. New York.

MoA (1984): Forurensinger fra jordbruket-omfang og virkemidler, delutredning 1. Rapport nr. 2.1984. LD Oslo.

MoA (1986): Forurensinger fra jordbruket- kostnader ved forurensingsbegrensende tiltak og aktuelle virkemidler. Delutredning 2. Rapport nr. 2. 1986. LD. Oslo.

MoA (1990a): Letter from Aa.Stenrød (MoA) to MoE (Vannmiljøavd.). Kommentar til utkastet til miljøvernministerens notat til regjeringskonferanse. 14.2.1990. p.1.

MoA (1990b): Letter from Aa.Stenrød (MoA) to MoE (Vannmiljøavd.). Norsk rappportering i Haag om oppfølging av Nordsjødeklarsjonen på naeringssaltsiden. 19.2.1990.1p.

MoA (1990c): Letter from K.Børve (MoA) to MoE; Angående Miljøavgifter. 8.3.1990.p.3.

MoE, Miljøverndepartmentet (MD) (1992): St. Melding. 64. 1991-1992. Om Norges oppfølging av Nordsjodeklarasjonene. Norway's Fulfilment of the North Sea Treaty. p. 103.

Naustdalslid, J. and M. Reitan (1994): Kunnskap og styring. Om forskningens rolle i politikk og forvaltning. Tano. NIBR. Oslo. p. 208.

Nilson, K. (1992): Policy, Interest, and Power. Studies in Strategies of Research Utilization. Meddelanden från Socialhögskolan. 1992:1. Lunds Universitet. Socialhogskolan.

Nijøs, A. (1989): Planteproduksjonens utvikling og framtid. In Innstilling fra Miljøavgiftsutvalget. Norges Bondelag. Appendix 3.

Olsen, J.P. (1988): Statsstyre og institusjonsutforming. Det Blå Bibliotek. Oslo. p. 336.

Polanyi, M. (1966): The Tacit Dimension. London. Routledge and Kegan Paul.

Rognerud, B. (1989): Handlingsplan mot landbruksforurensinger. Informasjonskampanje. utproving av tiltak mot arealavrenning. Rapport Nr. 1. GEFO. As.

Rognerud, B. (1990): Tiltak mot forurensing fra landbruket. i K.K. Heje (1990): Håndbok for landbruket. Landbruksforlaget. Oslo. pp. 41-45.

Sandmo (1990): Noen refleksjoner om public choice - skolens syn på økonomi og politikk. Norsk Økonomisk Tidsskrift. 104 (1990) pp. 89-112.

Statens Forurensingstilsyn (SFT) (1989): Langtidsplan 1990-1993. Long Term Plan SFT. Oslo.

Statens Forurensingstilsyn (SFT) (1990): Letter from SFT. Oslo.

St.m.46.(1988-89): Om rollefordeling i miljøvernpolitikken. On the Distribution of Roles in the Environmental Policy Making in Norway. Oslo.

St.m. nr. 1. (1989-90): National Budget 1989-1990. The Norwegian Parliament 1990.

Simonsen, J.W. (1989): Miljøavgifter på kunstgjødsel -N og -P og pa plantevernmidler. Taxes on Chemical Fertilizer - N and -P and on Pesticides. A Report to the Ministry of Agriculture. Dept. of Agricultural Economics. AUN. p. 106.

Simonsen, J.W. (1990): Gjødselavgifter i korn. Landbruksøkonomisk Forum. 2/9O. 7 Argang. pp. 50-65.

Simonsen J.W., S. Rysstad, and K. Christoffersen (1992): Taxes Versus Command and Control. Studies of measures for reducing Nitrogen Pollution from Agriculture. Dept. of Economics and Social Sciences. Agricultural University of Norway. Report No. 10. 1992. p. 199.

Steinsholt, K. (1995): Utsyn mot vitenskap og forskning. En introduksjon. Caspar Forlag. 207 s.

Sundquist, G. (1991): Vetenskapen och miljøproblemen. En expertsociologisk studie. Science and Environmental Problems. An Expert-sociological study. Monograph from Dept. of Sociology. No.46 University of Gothenburg. May 1991.

Sødal, D.P. and J. Aanestad (1990): Tiltak mot Arealavrenning. Miljømessige og Økonomiske virkninger av redusert arealintensitet og endra regional produksjonsfordeling i jordbruket. Measures Against Diffuse Runoff. Environmental and Economic Effects of Reduced Area Intensity and Changed Regional Production Composition in Agriculture. Report 1. 1990. Dept. of Economics and Social Sciences. AUN.

Vagstad, N.H. (1990): Miljøoptimal gjodsling. Nitrogen som miljø -og produksjonsfaktor i jordbruket, spesielt i kornproduksjonen. Environmentally Optimal Fertilization. Nitrogen as an Environmental - and Production Factor in Agriculture, with particular Reference to Grains. Center for Soil and Environmental Research. Dec. 1990. 26pp.

Vatn, A. (1989): Landbrukspolitikk og regional spesialisering. Effekten av kanaliseringspolitikken i norsk landbruk. Agricultural Policies and Regional Specialization. Effects of Regional Policies in Norwegian Agriculture.Melding nr.60 IOS.NLH.

Vatn, A., L. Bakken, M. Bleken, P. Botterweg, H. Lundeby, E. Romstad, P.K. Rorstad, and A.Vold (1996): Policies for Reduced Nutrient Losses and Erosion from Norwegian Agriculture. Integrating Economics and Ecology 1996. Norwegian Journal of Agricultural Sciences. Supplement No. 23.1996. Ås Science Prk. Ltd. Norway. p. 319.

Vedeld, P., E. Gran, and R. Aspmo (1992): Bønder og Gjødsling. Farmers and Fertilizers. A Study of Fertilization Practices among Farmers in Gausdal, Jaeren and Ås. 147 pp. Report No.7.1992. Dept. of Economics and Social Sciences. AUN. Norway.

Vedeld, P. and E. Gran.(1992b): Gjødslingsplanlegging, avling og nitrogenbruk. Fertilization Planning, Yield Levels and Nitrogen Use. A comment to the Norwegian Fertilization Planning Programme. Presentation at Yield Curve Seminar at Ås 4-5.11.92. ECEC-group. Dept. of Economics and Social Sciences. p. 25.

Vedeld, P. (1994): Interdisciplinarity and Environment. Neoclassical Economical and Ecological Approaches to the Use of Natural Resources. Journal of Ecological Economics. pp. 1-13. 10. 1994.

WCED (1987): Our Common Future. The World Commission on Environment and Development. Oxford University Press. Oxford.

21

Soil and Water Conservation Policies in Germany

Peter Weingarten and Klaus Frohberg

In the last two decades, the public in Germany has become more and more aware of the fact that agricultural activities always influence the environment, often in negative ways. In order to reduce negative impacts of agriculture on soil and water, various policy measures have been discussed and implemented in Germany. The next section of this chapter provides information on the extent of soil and water pollution caused by agriculture. An overview of the most important legislation and institutional settings is given in Section 3. Impacts of alternative water conservation policies simulated by using a regionalized agricultural sector model are discussed in Section 4. The chapter ends with some conclusions.

The problems of agriculturally induced pollution in Germany are outlined in 1992 in Frohberg et al. (1994). Hence, this paper focuses on new policy measures implemented during the last few years. Furthermore, the extent of pollution by agriculture is given.

Agriculturally induced soil and water pollution in Germany

Soil Erosion

Although the extent of soil erosion is not very well documented, soil erosion is not considered a major problem in Germany (Frohberg et al. 1994). The average soil loss per hectare of arable land is estimated to be at 8.7 t/ha for the former FRG (Werner et al. 1991). The corresponding figure for the former GDR amounts to 4.6 t/ha (Wodsak and Werner 1994). The minor importance of erosion in the former GDR is largely a result of natural conditions: The topography of the former GDR has low relief. Rainfall is also very low. For an area of nearly 500,000 ha arable land located in northeast Germany, characterized to be rather flat and having annual rainfall between 470 and 570 mm, soil losses due to water erosion and wind erosion were assessed on average to be 1.1 t/ha and 0.5 t/ha, respectively (Deumlich 1995). As illustrated in the soil erosion atlas of *Baden-Württemberg* (Gündra et al. 1995), a hilly state *(Bundesland)* located in southwest Germany, 5.5 t/ha are eroded per year on typical arable land in this state.

Water pollution

As in many countries, modern and intensive agriculture causes many problems in Germany. In particular, the pollution of surface- and groundwater by nutrients nitrogen (N) and phosphorus (P) and by pesticides has been a pressing issue.

Pollution of surface water

In the late 1980s, the blooming of algae which occurred at certain periods in the North Sea and the Mediterranean and which caused negative effects on ecology and tourism placed eutrophication of surface waters on top of the political agenda. Thus, at the 3rd International North Sea Protection Conference in 1990, the countries bordering the North Sea committed themselves to halving the nutrient stream into the North Sea by 1995 relative to 1985. According to a study by the German Federal Environmental Agency (FEA), however, Ger-

Table 1. Estimated load of surface water with nitrogen and phosphorus in Germany

	Nitrogen in 1989/91		Phosphorus in 1987/89*	
	1000 t N	kg N/ha**	1000 t P	kg P/ha**
Nonpoint sources				
Direct inputs	79	2.2	12.0	0.34
Atmosphere, litter	22	0.6	0.9	0.03
Drain	54	1.5	2.9	0.08
Erosion	73	2.0	31.0	0.87
Groundwater	400	11.2	1.2	0.03
Total nonpoint sources	**630**	17.7	**48.0**	1.35
Total point sources	**410**	11.5	**52.0**	1.46
Sum total	**1040**	29.2	**100.0**	2.80

* For the GDR the years 1991/92 were taken.
** Per hectare of the entire land area.

Source: Projektgruppe Nährstoffeinträge.

320

many would comply with this target only with regard to phosphorus, whereas the nitrogen load would be reduced by only 25 percent, i.e., half the required amount (Projektgruppe 1994).

For 1989–1991 the FEA estimated the total nitrogen (N) load of surface water in Germany to be 1.0 million t N (Table 1). Sixty-one percent of the total originated from nonpoint sources, while 39 percent came from point sources. Most of the nonpoint sources are affected by agricultural activities. With an estimated 0.4 million t N entering the surface water via groundwater, the latter is the most important nonpoint source. According to the FEA, 53 percent of all surface waters N pollution originated from agricultural areas. With regard to the estimated load of surface waters with phosphorus (P), the corresponding figure amounts to 47 percent. Whereas erosion is of only minor importance in regard to N levels, nearly one-third of the total estimated P load is affected by this source.

Pollution of groundwater

Since agriculture is the most important nonpoint source of nitrate pollution of groundwater, and data on nitrate concentrations in groundwater representative for Germany are still lacking, N-balances are often used as a proxy for assessing the extent of pollution with this nutrient. In the former FRG, the average N-surplus increased from less than 25 kg N/ha in the 1950s to more than 100 kg N/ha in the early 1980s (Köster et al. 1988). More recent development calculations carried out with the "Regionalized Agricultural and Environmental Information System" (RAUMIS) (cf. section 4.1) indicate a significant reduction of the N-surplus from 1987 to 1991 (85 as compared to 104 kg N/ha in 1987, Table 2).

Table 2. Nitrogen balances of the former FRG (in kg N/ha)

	1979	1983	1987	1991
Chemical fertilizer	113	117	131	117
Manure	101	109	109	105
Other N-input	35	35	35	34
Total input	249	260	275	256
N-removal with crops	119	125	139	138
Ammonia	30	32	33	32
Balance	**100**	**103**	**104**	**85**

Source: Weingarten.

Table 3. N-surpluses and distribution by farming types in the former FRG in 1990–1991 based on FADN data (in kg N/ha)

	25% of farms w/ lowest N-surplus*	50% of farms w/ medium N-surplus*	25% of farms w/ highest N-surplus*	All farms
Cereal farms	73	94	115	96
General cropping farms	88	113	138	115
Dairy farms	101	12	163	129
Drystock farms	96	122	168	126
Granivore farms	172	··	··	224
Mixed farms	95	126	189	132

* In the case of granivore farms every category covers 33 percent of all farms.
·· No data given, due to sampling size falling short of 15 farms.

Source: Brouwer et al.

Regarding pollution by different farm types, Brouwer et al. (1995) provide information on N-balances. Based on the representative survey of farm types by the Farm Accountancy Data Network (FADN) of the European Commission, their findings show that feed grain farms have the highest average N-surpluses in Germany (224 kg N/ha, Table 3). On the other hand, cereal farms were found to have the lowest N-surpluses (96 kg N/ha).

Farm size (measured in ha or in standard farm income) is not correlated with the N-surplus according to a study by Nieberg and von Muenchhausen (1996). Nieberg and von Muenchhausen surveyed 478 farms in the former FRG and 728 in the former GDR. In both parts the differences in the level of N-surplus between regions are considerably higher than between farm size classes within one region.

Wendland et al. (1994) developed a model to trace nitrate flow in the groundwater in Germany at squares of 3 by 3 km. On the basis of hydrological, hydrogeological, and agricultural data, maps with potential nitrate concentrations in the recharged groundwater and in the spring water are drawn. It is demonstrated that variation of potential nitrate concentrations across the squares is more affected by different hydrological and hydro-geological conditions than by variation of N-surplus between agricultural areas. Nevertheless, this finding does not reduce the need for a general reduction of N-surpluses.

Based on the above mentioned N-surpluses for 1991 calculated with the model RAUMIS, Weingarten (1996) assessed the potential nitrate concentration in the soil percolation water (cf. section 4.1). According to this assessment the potential nitrate concentration is less than 10 mg NO_3/l for 10 percent of the total soil percolation water, between 10 and 25 mg NO_3/l for 21 percent, and between 25 and 50 mg NO_3/l for 31 percent. For the remaining 38 percent of the soil percolation water, the potential nitrate concentration exceeds the maximum allowance of nitrate content in drinking water in the EU which is fixed at 50 mg NO_3/l. At the average, the potential nitrate concentration reaches 43 mg NO_3/l. In this assessment only the N-surplus originating from agricultural area is regarded as an input into the soil percolation water. Hence, the figures quoted are likely to be even higher depending on the impact of other sources.

In 1995, a working group of ministries in the *Bundesländer* responsible for water management published a report on the quality of groundwater with regard to nitrate (LAWA 1995). The report was based on some thousands of water analyses carried out in recent years. It is, however, difficult to judge the representativeness of the results obtained. Some of the water samples originated from the surface near groundwater, while other water samples were taken from groundwater in deeper aquifers. Some of the plots from which the water samples were obtained were located within areas that were intensively used by agriculture, and some were located within areas where mainly naturally caused nitrate inputs could be expected.

According to this report, more than one-third of all water samples contained

Figure 1. Distribution of the analyzed nitrate concentrations in Germany in the early 1990s

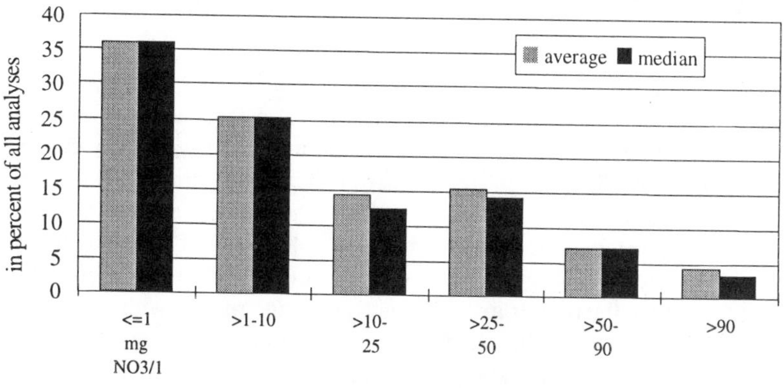

Source: *Länderarbeitsgemeinschaft Wasser (17).*

323

Table 4. Number of pesticide detections in water reported to the Federal Environmental Agency by the *Bundesländer* and water treatment plants

Reporting period	Reported analyses		Without pesticides		—With pesticide detection— concentration <0.1 μg/l		concentration >0.1 μg/l		Total number w/ pesticides	
until Dec. 90	49,736	100%	42,808	86.1%	3,999	8.0%	2.519	5.1%	6,783	13.6%
Dec. 90-91	77,673	100%	69,753	89.8%	6,104	8.8%	2.170	2.8%	8,038	10.3%
Dec. 91-92	68,219	100%	64,043	93.9%	1,592	2.5%	1,640	2.4%	4,203	6.2%
Dec. 92-93	75,472	100%	68,583	90.9%	6,858	10.5%	1,031	1.4%	6,889	9.1%
Dec. 93-94	60,564	100%	57,539	95.0%	2,674	4.6%	691	1.1%	3,363	5.6%
total	331,664	100%	302,726	91.3%	21,227	7.0%	8,051	2.4%	29,276	8.8%

Source: Own calculations on the basis of a study by the FEA.

less than 1 mg nitrate/l and three-quarters less than 25 mg nitrate/l (Figure 1). One-quarter of the analyses showed that there were significantly increased concentrations that were caused especially by agricultural land use. The maximum allowance of 50 mg nitrate/l drinking water was exceeded by one-tenth of all analyses, often correlated with using agricultural land for cultivating special crops such as vegetables or fruits.

Although N-surpluses originating from agriculture have decreased in the last few years on average, increasing nitrate concentrations in groundwater are expected. This increase is attributed to the actual N storage in the soil and reduced capacities for denitrification in the aquifers. In many water treatment plants the nitrate concentration has increased since the 1950s by 0.5 to 1 mg nitrate/l per year. However, in some cases nitrate concentration has stabilized since the late 1980s (LAWA 1995).

Pesticide pollution

According to the EC Drinking Water Directive established in 1980 the concentration of pesticides in drinking water is limited to 0.1μg/l of any chemical substance and to 0.5 μg/l of all substances together. In 8.8 percent of the 331,664 water analyses reported to the FEA between 1989 and the end of 1994 pesticides or their metabolites were found, and in 2.4 percent of them the detected pesticide levels exceeding the limit of 0.1 μg/l (Table 4) (Umweltbundesamt 1995). However, the figures in Table 4 need to be interpreted with caution. It is not possible to draw conclusions from these figures on the overall pollution of both ground- and surface water with pesticides since the detection of, for example, 100 different chemical substances in one water sample is referred to as 100 reported analyses. Interestingly enough although the use of atrazine is prohibited in Germany since 1991, atrazine and its metabolites are responsible for three-quarters of all pesticide detections. Several factors could cause this problem. First, atrazine or its metabolites might still be stored in the soil and leached out into the groundwater. Second, regenerating polluted groundwater may be a long process. Third, deregistered atrazine might still be applied. In some of Germany's neighboring countries farmers still are allowed to use it which makes it rather easy for German farmers to obtain atrazine.

Legislation and institutional settings

This section discusses some relevent legislation and institutional settings implemented in recent years. Due to page limitation, only the most important items will be discussed.

European Union

In 1996, the EC Commission passed its Communication on Community Water Policy (COM (96) 59 final, February 21, 1996). This Communication sets out the principles for Community Water Policy recommending to design and implement a Framework Directive On Water Resources (cf. Olsen 1996). From the commission's point of view, such a Framework Directive would be an important step toward an integrated management of surface water and groundwater taking into account both quality and quantity of river basins as the basis (Olsen 1996). Since this directive is aimed at replacing several existing ones, the EU water policy would become more transparent and coherent.

Also in 1996, the EC Commission proposed an action program aiming at integrating protection and management of groundwater. Four main fields of actions were suggested: a) development of principles for an integrated planning and management of waters; b) implementation of regulations concerning the quantitative protection of water resources; c) establishing instruments to control groundwater pollution by nonpoint sources; and d) implementing instruments to control groundwater pollution by point sources.

See Rude and Frederiksen (1994) for an overview of the nitrate policies of the EU and 7 Member States (including Germany).

Regarding pesticides, the EC Commission proposed in 1995 to amend the Drinking Water Directive, e.g., to cancel the existing maximum allowance of 0.5 μg/l drinking water of all pesticides together (Agra-Europe 1995). Following the arguments of the commission, this limit was deemed unnecessary because, in the past, an excess of this limit always coincided with an excess of the maximum allowance of 0.1 μg/l for a single substance. Furthermore, it is argued that controlling the maximum allowance of all substances together is not practical.

Both the EC Commission and some Member States argue for a growing need to integrate environmental and agricultural policies (Fischler 1996; Bundesministerium 1996). The ongoing discussion on a revision of the Common Agricultural Policy (CAP) of the EU points toward an increasing need to put more emphasis on agri-environmental programs.

Germany

In Germany, environmental as well as the agri-environmental policy is dominated by command-and-control measures. Economic instruments such as tradable quotas or Pigouvian taxes are seldom used although taxation of N fertilizer was often called for by some environmentalist groups, scientists, and political institutions during the last 10 years to internalize negative externalities (cf. Weingarten 1996). However, positive economic incentives are often offered to farmers to promote ecologically sound farming methods.

Federal Soil Act

The federal government aims at enacting a Federal Soil Act containing the most important regulations concerning the soil. At present, such regulations are spread across different acts. Currently, only a draft version of the Federal Soil Act exists, with a goal of protecting the soil from degradation (Agra-Europe 1995). Among others things, it lists the principles for proper land stewardship by farmers. However, these principles are vaguely defined and open to broad interpretation. In general, the measure states that land has to be cultivated in agreement with site specific conditions. The quality of the soil structure has to be preserved or improved. Soil compaction and erosion are to be prevented as much as possible.

Ordinance on sewage sludge

With the Ordinance on Sewage Sludge (*Klärschlammverordnung*), the application of sewage sludge on agricultural and horticultural land is regulated. To be applied sewage sludge has to meet different criteria concerning the content of heavy metals (e.g., less than 900 mg lead per kg dry matter sewage sludge) and organic compounds. Also, the land on which sewage sludge is to be applied has to fulfill some criteria. These relate to the content of heavy metals and the pons Hydrogenous (pH). It is forbidden to apply sewage sludge on forest land, permanent grassland, and within national parks. A minimum distance to surface waters of 10 meters must be kept. The application of sewage sludge is limited to 5 tons dry matter per hectare within 3 years. Interestingly, current problems with applying sewage sludge on farmland are overdoses of nutrients rather than heavy metals or other toxic particles.

Federal Water Act

Due to her federal structure, Germany has several institutions at various state levels which are involved in water conservation policy. While the federal government is authorized to enact so-called framework legislation, the individual *Bundesläender* are required to enact specific laws. The *Bundesläender* further allocate responsibilities to institutions at various regional levels (*Kreise,* *Regierungspräesidien*). The way this is done differs between the *Bundesläender* (Scheierling 1994).

The Federal Water Act (*Wasserhaushaltsgesetz*) constitutes the framework for the individual laws specified by each of the *Bundesläender*. These regulations assign all surface and groundwater resources to public management. Hence, in Germany, there is no private ownership of water resources.

The most important items of the Federal Water Act for agriculture are the declaration of water protection areas (WPA), management restrictions, and compensatory payments within these WPAs. Additional items of importance are the regulations concerning the storing of materials which are potentially dangerous to water bodies.

During the preparation of the amendment of the Federal Water Act in 1986, the discussion of regulation about compensatory payments was very controversial. The final wording that was adopted states that although farmers have no property right to the groundwater, they need to be compensated if management restrictions within WPAs limit the "proper use of agricultural land" (*ordnungsgemäesse landwirtschaftliche Nutzung*). The controversy arose due to the fact that these compensatory payments contradict the "polluter pays" principle. Moreover, no consensus has been reached on the definition of the "proper use of agricultural land."

Since the *Bundesläender* are responsible for implementing federal legislation, the water conservation policies differ between them. In particular, differences occur with regard to the size of designated WPAs, to establishing and implementing management restrictions and compensatory payments as well as to financing these payments. According to a survey by the Federal Ministry of Agriculture (N.N. 1996), the proportion of designated WPAs in total area of a *Bundesland* ranges from 1 percent in *Schleswig-Holstein* to 31 percent in Thuringia. The average for Germany is 10 percent, i.e., 3.7 million ha. An additional 1.5 million ha are planned to be assigned to a similar status. The extent to which this includes agricultural area is not known.

Concerning the financing of the compensatory payments and the management restrictions imposed, two basic approaches can be distinguished:

First, one approach adopted by *Baden-Wüerttemberg* requires that water treatment plants pay a levy on used water to the state government. This levy is used

327

for transferring compensatory payments to farmers which in general amount to 310 DM/ha arable land located in a WPA. The main regulation that is imposed restricts the application of nitrogen fertilizer to 20 percent less than the usual rate. If the mineralized N in the soil exceeds 45 kg/ha in autumn, it is assumed that the farmer was not in compliance with this regulation.

Second, there is another approach implemented in North Rhine-Westfalia. In this *Bundesland* the government supports the cooperation between water treatment plants and farmers. Both parties are required to negotiate management restrictions and compensatory payments to be paid to the farmers by the water treatment plant benefiting from these restrictions.

In 9 of the 16 *Bundesländer*, water use is levied with different amounts, depending on the regulation implemented in the individual *Bundesland*, the type of body the water is taken from (surface or groundwater), and the purpose the water is used for. The levy to be paid ranges from 0.005 DM/m^3 to 1.00 DM/m^3 in these nine *Bundesländer*.

Implementation of the nitrate directive by the fertilization ordinance

After more than 4 years of discussion, the Fertilization Ordinance was finally enacted in 1996. This implemented the EU Nitrate Directive of 1991 in Germany. Since the EU directive was to be implemented in 1993 at the latest, this meant a delay of more than 2 years. It is not uncommon in the EU that EU directives are put into force by Member States considerably later than originally agreed. The main reason for the long delay was a disagreement between the German Federal Ministries of Agriculture and of Environment. The Minis-

Table 5. Limits for the application of manure according to the German Fertilization Ordinance (kg N/ha on farm average)

	Arable land	Grassland
(1) Upper limit for manure application	170.0	210.0
(2) plus max. 20 percent for application losses	212.5	262.5
(3) plus 10 percent storing losses for slurry and dung water (solid dung: 25 percent) = maximum allowed N from animal excrement	236.1 (283.3)	291.6 (350.0)
(4) Max. allowed "losses" for slurry and dung water (solid dung) [(3)-(1)] (= ammonia emissions)	66.1 (113.3)	81.6 (140.0)

Source: Weingarten.

328

try of Agriculture was more concerned with farmers' interest, the Ministry of Environment with that of ecologists. An additional reason was the need to modify the German Fertilizer Act implemented in 1994 before enacting the Fertilization Ordinance in order to provide the necessary legal basis.

This ordinance defines codes of good agricultural practice with regard to using fertilizer. It comprises regulations concerning the application of fertilizer, the peculiarities of manure application, the calculation of fertilizer requirements, and the obligation to record nutrient balances.

In agreement with these regulations, fertilizer has to be applied in such a way that the nutrients can be used readily by plants in order to minimize nutrient losses. Furthermore, machines employed for spreading fertilizer have to function properly with regard to spreading and evenly distributing the amount intended. Application of manure is quantitatively restricted and allowed only in certain periods. With the exception of solid dung, using manure on arable land after harvest of the main crop is only permitted under specific conditions. If manure is applied to uncultivated land it must be worked into the soil immediately. Between November 15 and January 15, manure application is in principle forbidden.

For manure in general, the application per farm is limited to an average of 170 kg N/ha on arable land (without set-aside) and 210 kg N/ha on grassland. A former draft of the Fertilization Ordinance also contained limitations regarding the application of phosphate and potash (K) which are, however, no longer included in the version currently enacted.

Maximum application limits are set to 236.1 kg N/ha on arable land (291.6 kg N/ha on grassland) in the case of slurry and dung water and to 283.3 kg N/ha on arable land (350 kg N/ha on grassland) in the case of solid dung (Table 5). These differences arise from storage and application losses which need to be taken into account. For example, in the case of slurry and dung water 10 percent of the N contained in the excrement can be considered as storage losses. For solid dung this proportion amounts to 25 percent. In addition, a maximum of 20 percent of nitrogen contained in manure before application can be assumed to be unavoidable application losses. The ammonia emissions allowed for by the Fertilization Ordinance, however, exceed the critical load tolerated by many ecosystems (Isermann 1994).

The upper limits for manure application in the Fertilization Ordinance are less restrictive than those of the manure ordinances (*Gülleverordnungen*), which existed in some of the *Bundesländer* and which have been substituted by the Fertilization Ordinance. Fertilizer requirements (N, P, K) have to be calculated for each plot taking into account such factors as type of crop to be planted and the nutrient availability in the soil. Since nutrient losses are to be mini-

329

mized, it is prohibited to apply more (inorganic or organic) fertilizer than necessary. This implies that manure application is not only restricted by these upper limits in the ordinance mentioned above but also by the case-specific fertilizer requirements.

As an exception, however, soils with a high P or K content have no P or K fertilizer requirement. In order to allow farmers to spread manure, its application to these soils must be equivalent to the nutrient extraction by plants. According to an evaluation of soil quality by Isermann (1993), this exception would be valid for 27 percent of arable land (12 percent of grasslands) in the former FRG and 39 percent (25 percent) in the former GDR with a high phosphate content. For potash, these shares are 26 percent (40 percent) for the former FRG and 53 percent (57 percent) for the former GDR. These figures indicate how important these exceptions are.

Fearful of possible disadvantages for their competitiveness as compared to the farmers in other EU Member States, the German farmers' association has criticized the Fertilization Ordinance for also regulating the application of P and K (Leser 1995). It is argued that the influx of P into surface water is caused by erosion rather than by fertilizing and that potash does not pose an environmental problem. In addition, it is pointed out that in accordance with EU legislation covering P and K is not necessary. Nevertheless, the attempt to define codes of good agricultural practice related to fertilizer application in the Fertilization Ordinance means positive impacts on water conservation can be expected if the regulations are adhered to. This latter precondition, however, raises some doubt about the efficacy of the Ordinance. In particular, doubts are raised with respect to the experience made with the manure ordinances (*Gülleverordnungen*) and the fact that many of the regulations of the Fertilization Ordinance are defined rather ambiguously.

Pesticide legislation

Since no new regulations have been enacted during the last few years, see Frohberg et al. (1994) for a brief overview of the relevant pesticide legislation.

Agri-environmental programs

The accompanying measures of the Common Agricultural Policy reform of 1992 offer Member States the opportunity to promote ecologically sound farming methods (EC Directive 2078/92). In Germany, for almost 5 million ha, i.e., 29 percent of the total utilized agricultural area (Agra-Europe 1996), farmers voluntarily adopt agri-environmental protection programs. These programs mainly include rules regarding the application of smaller amounts of chemicals or none at all, constraining the animal density to a certain level and the protection of the landscape as well as the maintenance of the countryside (cf. Frohberg 1995). In some *Bundesländer,* minimum tillage and underseeding row crops in order to reduce erosion and nitrate leaching are also elements which can be supported by these programs. In 1994, payments for ecologically sound farming methods made to farmers who signed up for these programs amounted to 417 million Deutsch Mark (DM), in 1995 to 705 million DM, and are likely to reach 826 million DM in 1996 (Agra-Europe 1996). The money for the payments comes from three sources: the EC Commission, the federal government, and the government of the corresponding state.

Assessment of the effects of alternative water conservation strategies

In this section, the effects of two water conservation strategies will be assessed. They are compared to a reference run which excludes any water conservation measure. The analyses are carried out with the model RAUMIS which is briefly described in the following section.

Overview of the Model RAUMIS (Regionalized Agricultural and Environmental Information System for Germany)

Commissioned by the Federal Ministry of Agriculture, the model RAUMIS (Weingarten 1996; Henrichsmeyer et al. 1992, 1996) was developed at the Institute of Agricultural Policy of the University of Bonn. RAUMIS is designed for quantitative analyses of agricultural and environmental policies. It is currently used by several entities, including the Institute of Agricultural Policy, the Federal Ministry of Agriculture, and the Federal Agricultural Research Center (Henrichsmeyer et al. 1996).

RAUMIS depicts the main interdependencies between agriculture and the environment. It consists of several modules, the most important of which is a system of Linear Programming (LP) modules at the county level. Since RAUMIS is designed for the former FRG, it consists of 240 of such LP modules. Together, they depict the agricultural sector in consistence with the Economic Accounts for Agriculture. Restrictions of data availability explain the chosen time differentiation. Thus, ex-post RAUMIS is based on the model years 1979, 1983, 1987, and 1991 abbreviated denotations for the three-year periods were investigated. The target year of the comparative-static simulation analysis of water conservation strategies is 2005.

In RAUMIS agricultural production is divided into 29 crop and 12 animal activities.

For the ex-ante analysis, flexibility constraints are used to control the scale of adjustment. The intensity of chemical application is determined by profit-maximizing behavior in these LP modules. For this purpose, empirical response functions were included into these LP modules.

The interdependencies between agriculture and the environment are modeled using environmental indicators, of which the N-balance is the most important one. On the input side, chemical fertilizer, manure, symbiotic and asymbiotic nitrogen fixation, and input of the atmosphere are taken into account. Removal rates of N from fields vary according to the crop planted. Ammonia emissions are calculated endogenously. The N-balance represents that amount of N which will be denitrified, leached out into the groundwater, or accumulated in the soil. Assuming that N-storage in the soil is in a long term equilibrium, accumulation of N is neglected in the RAUMIS analysis.

Table 6. Analyzed water conservation strategies

Strategy 1 (Regionally differentiated groundwater conservation policies)

1. **outside water protection areas:**
   no farming regulations and no compensatory payments

2. **inside water protection areas:**
   farming regulations and compensatory payments
   1. Maximum utilization of N-fertilizer (in kg N/ha):

winter wheat	110	sugarbeets	150
rye, winter barley	85	winter rape	135
winter and summer maslin	85	summer rape	90
oats	95	grassland	140
summer wheat, summer barley	75	silage maize	140
grain maize	120	mangel-wurzel	150
potatoes	120	legumes	0

   2. Maximum livestock density: 1.0 manure units (MU)/ha
      (i.e., the equivalent of 120 kg N/ha)
   3. Prohibition of plowing up grassland
   4. Regionally differentiated compensatory payments

Strategy 2 (Blanket coverage of groundwater conservation policies)

1. Nitrogen levy on chemical fertilizer: 0.66 DM/kg N at current prices
2. Nitrogen levy on slurry surpluses (slurry IN) exceeding
   1.5 MU/ha: 0.66 DM/kg slurry-N
3. Prohibition of ploughing up grassland
4. Uniform compensatory payments per hectare utilized agricultural area
   according to N-levy revenues

Source: Weingarten.

On the basis of the regional N-surplus and the regional soil percolation water, potential nitrate concentrations of the soil percolation water are assessed on the assumption that 50 percent of the N-surplus leaches into groundwater. A second assessment includes not only N-leaching from agricultural land but also from forests (30 kg N/ha).

Depending on the potential nitrate concentration and information about groundwater raising, potential costs associated with treating the nitrate-polluted groundwater for drinking purposes are estimated. Figures available in the literature on the costs of treating nitrate pollution per additional cubic meter groundwater differ widely. Hence, two variants of cost figures were used in the analysis. However, in this paper only the scenario with the more likely one is reported. Groundwater raised 0,01 DM/m^3 are calculated for groundwater with a potential nitrate concentration in the soil percolation water between 10 and 25 mg NO_3/l. If the potential nitrate concentration amounts to 25 to 50 mg NO_3/l, 0,20 DM/m^3 are assumed. Given a potential nitrate concentration of more than 50 mg NO_3/l, this figure raises to 0,70 DM/m^3 (Buetow and Homann). In order to facilitate the assessment of the monetary value of the nitrate pollution in the soil percolation water, hypothetical costs are estimated on the assumption that the entire soil percolation water would be conditioned as groundwater used for drinking purposes.

Water conservation strategies

Depending on the objective pursued, water conservation strategies are to be designed differently. Using the model RAUMIS, two water conservation strategies were investigated (Weingarten 1996). Strategy 1 aims at protecting ground-

Table 7. Nitrogen balances of the former FRG for 1991 and 2005 (in kg N/ha)

	1991	Reference run	Strategy 1 total land	Strategy 1 WPA land	Strategy 2
			2005		
Chemical fertilizer	117	112	106	70	80
Manure	105	82	79	65	79
Other N-input	34	33	33	33	33
Total input	256	227	219	168	192
N-removal with crops	138	131	128	113	121
Ammonia	32	24	23	19	21
Balance	85	73	67	36	50

Source: Weingarten.

Figure 2. Potential nitrate concentration of the soil percolation water in 1991 and in 2005 under different scenarios

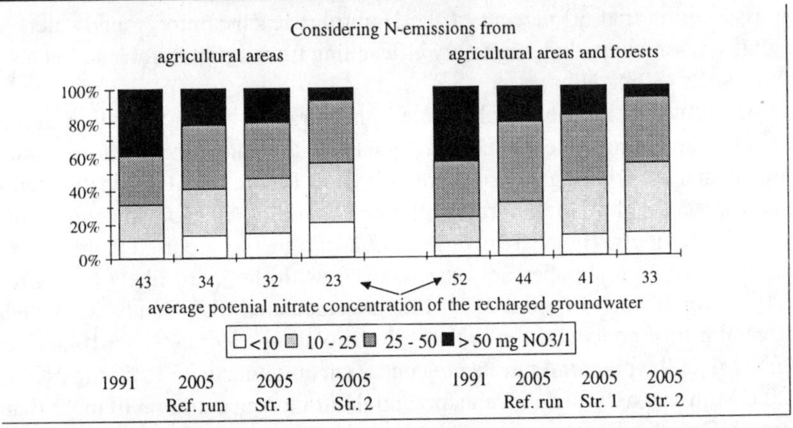

Source: Weingarten (28).

water as a resource for drinking water. Thus water protection areas are defined where groundwater is used for drinking purposes. Details of restriction on farming imposed and compensation payments provided are listed in Table 6. Within WPAs, farmers have to limit fertilizer application and are restricted on plowing up of grassland. They are compensated by regionally differentiated payments so that they do not suffer any income losses. Outside the WPAs no regulations for applying special practices exist.

Strategy 2 is designated as to protect groundwater everywhere because of its functions within ecosystems and in the water cycle. Hence N levies on chemical fertilizer and slurry surpluses are introduced and the plowing up of grasslands is forbidden (Table 6). Farmers get the revenues of the N-levies paid by them refunded on a uniform basis per hectare.

Assessed impacts on ecological and economic characteristics

Based on the assumption of a continuation of current policies up to 2005 and on other specifications made in the model, the results of the reference run indicate a further decline in average N-surplus (Table 7). The reduction amounts to 73 kg N/ha by 2005. This is mainly due to less manure application, higher efficiency in plant uptake of manure-N, and a considerable reduction of N application on grassland. Restrictions on farming practices as specified in strategy 1 lead to a substantial lowering of N-surplus in WPAs which is 36 kg N/ha.

Table 8. Potential costs of water treatment plants caused by nitrate (ref. extracted groundwater) **and the hypothetical costs** (ref. soil percolation water) **for 1991 and 2005** (in billion DM per year at current prices)

	Extracted groundwater		Soil percolation water	
	Land from which N-emissions are considered			
	agriculture	agr. a. forest	agriculture	agr. a. forest
1991	1.24	1.61	17.24	21.08
		——2005——		
Reference run	0.90	1.35	11.51	18.01
Strategy 1	0.19	0.56	10.08	16.32
Strategy 2	0.47	0.85	6.25	10.45

Source: Weingarten.

The average for all of the cultivated land amounts to 67 kg N-surplus/ha which is only 8 percent lower than that in the reference run, but 34 percent higher than if the policies of strategy 2 were implemented. The latter reaches a surplus of 50 kg N/ha.

The potential nitrate concentration of the soil percolation water averages 34 mg NO_3/l in 2005 (reference run) as compared to 43 mg NO_3/l in 1991 (see the left side of Figure 2). Since the measures of strategy 1 restrict farming only within WPAs, the potential nitrate concentration of the entire soil percolation water is only slightly lower than in the reference run. However, the potential concentration of the groundwater extracted is drastically reduced to 19 mg NO_3/l on average. Strategy 2 reduces the average potential nitrate concentration in the soil percolation water to 23 mg NO_3/l. If an assumed 30 kg N-surplus per hectare forest is taken into account in the calculation, the potential nitrate concentrations are always higher than the figures mentioned above (see the right side of Figure 2).

Due to lower potential nitrate concentrations (especially in hilly regions), potential savings for water treatment plants at current prices will be lower in 2005. Whereas the potential costs based on the N-surplus of 1991 are estimated to be 1.2 billion DM per year, the corresponding figure for the reference run in 2005 amounts to 0.9 billion (Table 8, considering N-emissions from agricultural land). The potential costs are lowest if strategy 1 is realized (0.2 billion

Table 9. Impacts of the water conservation strategies on economic indicators compared to the reference run without water conservation measures (real, in billion DM/year at current prices)

	2005 Reference run	2005 Strategy 1	2005 Strategy 2
(1) Compensatory payments induced by strategies	—	+0.40	±0˙
(2) Other subsidies	—	-0.02	-0.05
(3) Agricultural income˙˙	—	±0	-1.54
(4) Interim sum [= -(1) - (2) + (3)]	—	-0.38	-1.50
(5) Potential costs of water treatment plants (extracted groundwater)˙˙˙	—	-0.71	-0.43
(6) Hypothetical costs soil percolation water˙˙˙	—	-1.43	-5.23
(7) Total effect considering the potential costs (extracted groundwater) [= (4) - (5)]	—	+0.32	-1.06
(8) Total effect considering the hypothetical costs (soil percolation water) [= (4) - (6)]	—	+1.05	+3.78

˙ The compensatory payments (612 million DM) are financed by the farmers paying the N levy, which for this reason are excluded here.

˙˙ The change of agricultural income (net value added in factor costs in the agricultural sector) includes the change of the compensatory payments and of the other subsidies.

˙˙˙ Cost variant A, only considering N-leaching from agricultural areas.

Source: Weingarten.

DM). This figure, however, does not cover compensatory payments for farmers. Strategy 2 reduces the potential costs to half of those in the reference run.

To get an estimate of the financial costs caused by the nitrate pollution of the entire groundwater, hypothetical costs were calculated, assuming that the entire soil percolation water would be treated similar to that water used for drinking purposes. Based on the N-surpluses of 1991 hypothetical costs of 17 billion DM per year were estimated (Table 8). The reference run 2005 results in hypothetical costs of 12 billion DM. Since the WPAs cover not more than 15 percent of total agricultural area, the farming restrictions induce only a slight decrease in the hypothetical costs. In contrast, the hypothetical costs are cut into half in strategy 2.

In the reference run agricultural income at current prices measured as net value added at factor costs decreases in 2005 as compared to 1991 by nearly 50 percent to 10.1 billion DM. The decrease is mainly due to lower output prices and a reduction in meat production. If farmers were to be compensated for their income losses resulting from the farming restrictions imposed within WPAs, an additional 0.4 billion DM would have to be paid. Although the levies on nitrogen (0.6 billion DM) are refunded to farmers in strategy 2 the net value added at factor costs still decreases to 8.5 billion DM. However, one has to keep in mind that, according to the model results, labor input declines by 40 percent in 2005. Therefore, income per full-time equivalent of farm labor decreases not as much as total income does.

If one compares the decrease in the potential costs for water treatment plants with the payments to compensate farmers for losses incurred and the loss in agricultural income respectively and taking also into account the (minor) change of other subsidies, strategy 1 results in a plus of about 0.3 billion DM (Table 9). Strategy 2 induces in this partial analysis a welfare loss of about 1.1 billion DM. Considering the hypothetical costs as an indicator for valuing the agriculturally induced nitrate pollution of the entire soil percolation water, strategy 1 causes a welfare increase of 1.0 billion DM and strategy 2 one of 3.8 billion DM per annum.

However, these figures should be interpreted with caution. This comparison is not a complete cost-benefit analysis. Additional aspects would have to be taken into account such as using shadow and not market prices as done in this analysis. Furthermore, transaction costs of various amounts caused by the two strategies and impacts of the strategies on environmental resources other than groundwater also have to be taken into account.

Conclusions

The analysis of nitrogen balances at county level provided in this chapter is indicative of reductions of nitrogen surpluses beginning with the second half of the 1980s. This is a result of changes in the CAP of the EU and growing awareness of farmers with regard to protecting the environment. Any further adjustment of the CAP is likely to strengthen this development.

During recent years, additional restrictions on farming were also imposed. These restrictions increased the intensity of control measures applied to agriculture. The impact of the Fertilization Ordinance on water pollution critically depends on how precisely farmers calculate fertilization requirements for crops and grassland. The impact also depends on how strictly the farmers follow these constraints.

The two water conservation strategies presented on imposing restrictions on nitrogen use show how important it is to clarify whether the entire groundwater or only that used as drinking water shall be protected. The welfare change depends critically on this question.

References

Agra-Europe. 1/2/1995. Änderung der Trinkwasserrichtlinie vorgeschlagen. Europa-Nachr. p.1.

Agra-Europe. 41/1995. Vorentwurf des Bundesumweltministeriums zum Bodenschutzgesetz. Dokumentation pp. 1-62.

Agra-Europe. 34/1996. Agrarumweltprogramme: Deutsche Landwirte mit an der Spitze. Länderberichte pp. 1-3.

Buetow, E., and H. Homann. 1992. Endbericht zum Forschungsvorhaben "Quantitative Analyse von Vorsorgestrategien zum Schutz des Grundwassers im Verursacherbereich Landwirtschaft" im Rahmen des TA-Projektes "Grundwasserschutz und Wasserversorgung." Institut für wassergefährdende Stoffe, Berlin.

Bundesministerium für Ernährung, Landwirtschaft und Forsten. 1996. Perspektiven der Agrarpolitik im kommenden Jahrzehnt - Konzeptionelle Überlegungen. Agra-Europe 26/96. Dokumentation pp.1-22.

Brouwer, F.M., F.E. Godeschalk, P.J.G.J. Hellegers, and H.J. Kelholt. 1995. Mineral balances at farm level in the European Union, Den Haag (draft report).

Deumlich, D. 1995. Landschaftsindikator Bodenerosion. In: Bork, H.-R., C. Dalchow, H. Kächele, H.P. Piorr, and K.-O. Wenkel. Agrarlandschaftswandel in Nordost-Deutschland unter veränderten Rahmenbedingungen: ökologische und ökonomische Konsequenzen. pp. 241-263. Ernst, Berlin.

Fischler, F. 1996. Erste Überlegungen zur gemeinsamen Agrarpolitik im 21. Jahrhundert. Der Förderungsdienst 44:1-8.

Frohberg, K. 1995. Programs for Stimulating Extensive Agriculture in Germany. In: Hofreither, M.F., and S. Vogel (Ed.). The Role of Agricultural

Externalities in High Income Countries. pp. 161-179. Wissenschaftsverlag Vauk, Kiel.

Frohberg, K., J. Dehio, B. Strotmann, and P. Weingarten. 1994. Agriculturally Induced Pollution Problems in Germany. In: Napier, T. L., S.M. Camboni, and S.A. El-Swaify. Adopting Conservation on the Farm. An International Perspective on the Socioeconomics of Soil and Water Conservation. Soil and Water Conservation Society, Ankeny, Iowa, p. 269-288.

Gündra, H., S. Jäger, M. Schroeder, and R. Dikau. 1995. Bodenerosionsatlas Baden-Württemberg. Agrarforschung in Baden-Württemberg 24.

Henrichsmeyer, W., C. Cypris, W. Löhe, and M. Meudt. 1996. Entwicklung des gesamtdeutschen Agrarsektormodells RAUMIS96 am Lehrstuhl für Volkswirtschaftslehre, Agrarpolitik und Landwirtschaftliches Informationswesen der Universität Bonn. Agrarwirtschaft 45: 213-215.

Henrichsmeyer, W., J. Dehio, R. von Kampen, P. Kreins, and B. Strotmann. 1992. Endbericht zum Forschungsvorhaben "Aufbau eines computergestützten regionalisierten Agrar- und Umweltinformationssystems für die Bundesrepublik Deutschland" - Modellbeschreibung - (BMELF 88 HS 025). Institut für Agrarpolitik, Bonn.

Isermann, K. 1993. Nährstoffbilanzen und aktuelle Nährstoffversorgung der Böden. Berichte über Landwirtschaft, SH 207: Bodennutzung und Bodenfruchtbarkeit, Bd. 5: Nährstoffhaushalt 15-54.

Isermann, K. 1994. Studie E: Ammoniak-Emissionen der Landwirtschaft, ihre Auswirkungen auf die Umwelt und ursachenorientierte Lösungsansätze sowie Lösungsaussichten zur hinreichenden Minderung. In: Enquete-Kommission "Schutz der Erdatmosphäre" (Ed.). Studienprogramm, Band 1: Landwirtschaft, Teilband 1.

Köster, W., K. Severin, D. Mühring, and H.-D. Ziebell. 1988. Stickstoff-, Phos-phor- und Kaliumbilanzen landwirtschaftlich genutzter Böden der Bundes-republik Deutschland von 1950 - 1986. Hannover.

Laenderarbeitsgemeinschaft Wasser (LAWA). 1995. Bericht zur Grundwasserbeschaffenheit Nitrat. Stuttgart.

Leser, H. 1995. Düngeverordnung: Stand der Diskussion. Deutsche Bauern-Korrespondenz 48: 93-94.

N.N. 1996. Wettbewerbsbeeinflussende rechtliche Rahmenbedingungen (I). Agra-Europe 41/95: Dokumentation pp. 1-75.

Nieberg, H., and H. von Muenchhausen. 1996. Zusammenhang zwischen Betriebsgröße und Umweltverträglichkeit der Agrarproduktion - Empirische Ergebnisse aus den alten und den neuen Bundesländern, Schriften der

Gesellschaft für Wirtschafts- und Sozialwissenschaften des Landbaues 32: 129-140.

Olsen, A.M. 1996. Water Policy in the European Union - Consequences for Agriculture. Paper contributed to the Workshop on "Mineral emissions from Agriculture," held in Oslo, Norway, January 25 - 28, 1996.

Projektgruppe "Nährstoffeinträge in die Nordsee" des Umweltbundesamtes. 1994. Stoffliche Belastung der Gewässer durch die Landwirtschaft und Maßnahmen zu ihrer Verringerung. Berichte Umweltbundesamt 2/94, Berlin.

Rude, S., and B.S. Frederiksen. 1994. National and EC Nitrate Policies - agricultural aspects for 7 EC countries. Statens Jordbrugsökonomiske Institut. Rapport nr. 77. Kopenhagen.

Scheierling, S.M. 1994. Overcoming agricultural pollution of water: the challenge of integrating agricultural and environmental policies in the European Union. World Bank technical paper no. 269.

Stahr, K., and D. Stasch. 1996. Einfluß der Landbewirtschaftung auf die Resource Boden. In: Linck, G., H. Sprich, H. Flaig, and H. Mohr (Ed.). Nachhaltige Land- und Forstwirtschaft: Expertisen. pp. 77-119. Springer, Berlin.

Umweltbundesamt. 1995. Jahresbericht 1994. Umweltbundesamt, Berlin.

Weingarten, P. 1995. Das "Regionalisierte Agrar- und Umweltinformationssystem für die Bundesrepublik Deutschland" (RAUMIS). Berichte über Landwirtschaft 73: 272-303.

Weingarten, P. 1996. Grundwasserschutz und Landwirtschaft : Eine quantitative Analyse von Vorsorgestrategien zum Schutz des Grundwassers vor Nitrateinträgen. Vauk, Kiel.

Weingarten, P. 1996. Quantitative Analyse von Maßnahmen zur Verringerung von Nitrateinträgen ins Grundwasser - eine Anwendung des Modellsystems RAUMIS - Schriften der Gesellschaft für Wirtschafts- und Sozialwissenshaften des Landbaues 33:555-567.

Wendland, F., H. Albert, M. Bach, and R. Schmidt. 1994. Potential nitrate pollution of groundwater in Germany: A supraregional differentiated model. Environmental Geology 24:1-6.

Werner, W., A. Hamm, K. Auerswald, D. Gleisberg, W. Hegemann, K. Isermann, K.H. Krauth, G. Metzner, H.W. Olfs, F. Sarfert, P. Schleypen, and G. Wagner. 1991. Geschässerschutzmaßnahmen hinsichtlich N- und P-Verbindungen. In: Hamm, A. (Ed.) Studie über Wirkungen und Qualitätsziele von Nährstoffen in Fließgewässern. pp. 653-830. Academia Verlag, Sankt Augustin.

Wodsak, H.-P., and W. Werner. 1994. Eintragspfade Bodenerosion und Oberflächenabfluß im Festgesteins- und Übergansbereich. In: Werner, W., and H.-P. Wodsak. Stickstoff- und Phosphoreintrag in die Fließgewässer Deutschlands unter besonderer Berücksichtigung des Eintragsgeschehens im Lockergesteinsbereich der ehemaligen DDR. pp. 85-87. DLG-Verlag, Frankfurt a.M.

22

Soil and Water Conservation in the Former East Germany

Monika Frielinghaus and Hans-Rudolf Bork

The loss of soil fertility and the inability to achieve ecological soil functions is an important problem in many areas that have been used for agricultural production in the last 25 years. Enhanced concentration, intensification, and specialization of crop and livestock production without sufficient consideration of the natural site specific soil and climate conditions caused pronounced degradation and partly irreversible damage to the soil. However, under these conditions many soils are less and less able to fulfill their ecological functions within the system pedosphere/hydrosphere/biosphere/atmosphere.

Therefore, soil conservation management includes the prevention of water, air, and nature pollution. Although conserved land use complexes offer a high potential for adequate management, they are rarely adapted. New strategies in politics, extension of experiences, and subsidy are needed urgently.

Landscape and land use characterization of the former East Germany

Northeast Germany is characterized by mainly young morainic areas. Sandy soils with a low groundwater level, loamy sandy soils, and clay soils with bound water change with peaty soils. In this region, the climate is marked by low precipitation (about 450-600 mm.a^{-1}). In the old moraine region, poor sandy soils are predominant. The importance of the loess region is based on the high quality of the loamy soils. Precipitation is also low. In the mountain and hill region there are different weathering soils with attributes of moisture. The different soil regions of eastern Germany are detailed in Figure 1.

Table 1 indicates the area in the former East Germany that is revealing of anthropogenic soil degradation due to various management processes over a long period of land use.

It is obvious that soil compaction is by far the most significant damage process (Dürr et al. 1995), but water and wind erosion also are causing considerable damage to soil fertility and ecological functioning, while chemical degra-

Figure 1. Soil regions of the former East Germany

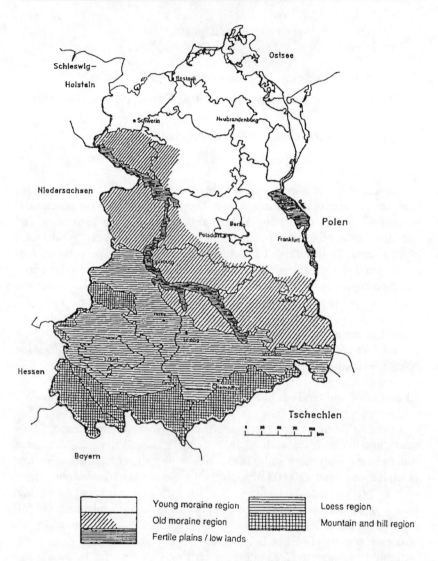

Young moraine region

Old moraine region

Fertile plains / low lands

Loess region

Mountain and hill region

dation is of minor importance. However, local acidification may be of great significance (Hüttl et al. 1994).

In Table 2 the prevailing degradation processes of the various regions are presented together with the impacted main areas indicating clearly that the glaciated region suffers from the greatest degradation problem.

Together with unfavorable climatic conditions this represents a heavy historic burden when new concepts for agricultural management practices or other forms of land use are discussed.

Table 1. Areas with anthropogenic soil degradation in relation to the major damage process

Regions	Young and old moraine regions	Fertile river plains/ Peat areas	Loess region	Mountain and hill region
Agricultural land (ha)	2,226,000	2,774,000	1,343,000	839,000
Damage processes*				
Water erosion	16	0	21	16
Wind erosion	8	6	4	0
Soil compaction	47	21	27	6
Water logging	5	5	1	1
Chemical degradation	2	4	0	2
Humus reduction	0	7	0	0
Set-aside program	9	n.d.	10	18

n.d.: not determined*proportion of land degraded (%)

Source: Schmidt, 1991.

Table 2. Regional centers of soil degradation

Region	Process of degradation	Area (ha)
Young and old moraine region	Erosion, compaction, water logging, nutrient loss, acidification	940,000
Fertile river plains/ Peat areas	Humus and peat reduction, water logging, compaction	107,000
Loess region	Erosion, compaction, nutrient loss	179,000
Mountain and hill regions	Compaction, acidification, nutrient loss, erosion	98,000

Source: Schmidt, 1991.

Figure 2. A view on the heterogeneous areas of northeast Germany

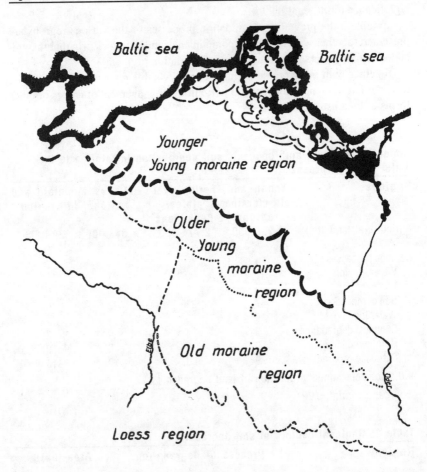

To illustrate this statement, results from a study in the young morainic region are briefly discussed.

The landscapes of northeast Germany are characterized by the greatest share water protection areas in all of Germany and a high relief energy with a dense net of "thalwege" (Figure 2).

Sandy soils with a low aggregate stability, a negative water balance in summer season, and a strong variation in the annual precipitation predominate.

Land use is characterized by large areas of arable land without barriers for surface runoff and wind. Maize and oil seeds are grown on endangered areas; deep drainage on heterogeneous areas results in an increased wind erodibility. A high compression by heavy crop harvest and transport vehicles increases water erosion.

A strong horizontal and vertical inhomogeneity of the soils dominates and largely defines the status of soil fertility. However, due to intensive management since the beginning of arable farming, reference to soils being man-made is relevant.

A result of the glaciation is more than 40,000 lakes and thaw lakes, rivers, streams, and other water protection areas. The pollution of these areas was increased by intensive agricultural land use. About half of the nonpoint phosphorus and 9 percent of the nonpoint nitrogen in the North and the Baltic Sea is caused by runoff and water erosion (Deumlich et al. 1994a; Weingarten et al. 1996).

Potential of water and wind erosion

The estimation method for water erosion risk is based on the soil substrate type, the inclination type, and the inclination type of the mesoscale agricultural mapping 1:25.000 (MMK 1983). The data are available on the basis of communities or catchments for the former East Germany (Deumlich et al. 1994b). The result of the calculation of the water erosion risks is a map with six classes (Figure 3). The result of the calculation of the wind erosion risks is also a map with six classes (Figure 4).

In the next step, the potential "thalwege" emerging from the morphology in endangered catchments for the estimation of preferred runoff and erosion paths is determined (Frielinghaus et al. 1996).

The calculation of the wind erosion risk requires the determination of the main wind direction for a long-term climate data basis and the estimation of the endangered landscape wind openness. The change of landscape structure that occurred from 1826 to 1985 is represented in Table 3.

Figure 3. Potential water erosion risk in the country—Brandenburg

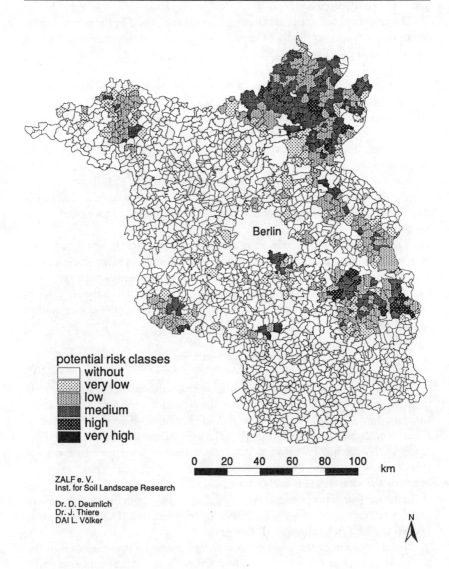

potential risk classes
- without
- very low
- low
- medium
- high
- very high

Berlin

0   20   40   60   80   100   km

ZALF e. V.
Inst. for Soil Landscape Research

Dr. D. Deumlich
Dr. J. Thiere
DAI L. Völker

N

Figure 4. Potential wind erosion risk in the country—Brandenburg

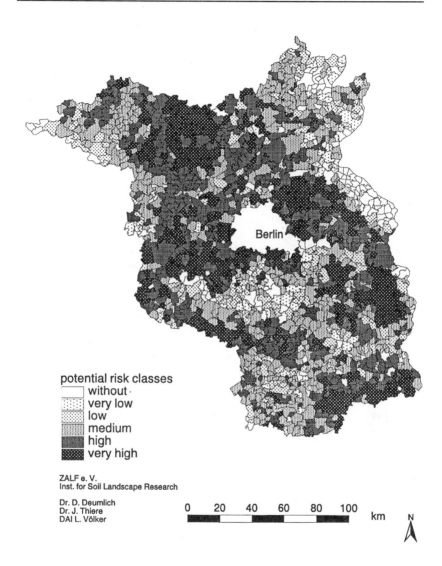

Berlin

potential risk classes
☐ without ·
▨ very low
▨ low
▨ medium
■ high
■ very high

ZALF e. V.
Inst. for Soil Landscape Research

Dr. D. Deumlich
Dr. J. Thiere
DAI L. Völker

0   20   40   60   80   100

km   N

Man-made erosion and compaction

The primary soil degradation problems are soil erosion and soil compaction on intensively used arable land. Soil loss by erosion increases with the intensity of soil compaction and technical stress.

The most important paths for runoff and sediment losses are the wheel lanes and ruts arising from intensive production on arable land (Frielinghaus et al. 1994).

The structure stability and the vertical pore system of the topsoil and of the subsoil are often damaged as a result of too frequent traffic with heavy machines and transport vehicles, which exert high specific ground pressures because their wheels have to support high axle loads (Dürr et al. 1995). The situation worsens with operations during autumn or spring periods due to a high soil instability resulting from specific moisture conditions. To predict offsite damages, the man-made paths of erosion in the agricultural land use systems have to be estimated. The application of the USLE (Wischmeier et al. 1978; Schwertmann et al. 1990) with the factor C for management and land use was not successful because the intensity and dynamics of linear erosion are not accounted for in the USLE. The rain erosivity factor R does not include winter precipitation, which is important due to soil water saturation and decreased storage capacity. Soil loss in traffic lanes is demonstrated for the winter period in Table 4 and for the summer period in Figure 5.

Table 3. Change from land use and wind breaking elements from 1826 to 1985

Kind of use		1826	1926	1985
built up area	%	2	2	2
forest	%	7	9	8
grass area	%	16	6	4
agriculture area	%	73	81	82
water area	%	2	2	3
water-courses	m.ha^{-1}	9	11	10
ways	m.ha^{-1}	23	27	16

Source: Frielinghaus et al., 1994.

Table 4. Soil loss in traffic lanes on areas with winter wheat after different crops

Crop before winter wheat	Total soil loss t.ha⁻¹	Soil loss in lanes t.ha⁻¹	Soil loss in lanes %
clover	50	27	54
clover	111	24	22
sugar beets	94	14	15
sugar beets	112	28	25
grass	15	5	33
potatoes	34	15	44
sugar beets	18	6	31
maize	8	8	100
maize	50	3	6
maize	25	21	85

Source: Saupe 1990.

The stress by normal crop production systems is very great (Table 5).

About 50 percent of the tracked area is heavy pressured. The total tracked area for sugar beet management is about 981 percent, i.e., all parts of this area were stressed more than nine times (Frielinghaus et al. 1994).

Soil conservation possibilities in the former East Germany

Differences in soil loss depend on the degree of organic soil surface cover. The temporal and spatial distribution of the soil cover by plants or plant residues is important. About 2t.ha⁻¹ dry matter or above 50 percent proportionate surface cover is effective to protect the soil (Table 6) (Frielinghaus 1988).

In periods without soil cover, the erosion risk is high, whereby gradations must be made after the roughness of the soil surface (Helming 1992).

A simple method is a direct appraisal of soil cover with crops and crop rotations (Figure 6).

A simple code, which correlates with the current cover as well as the plant development stages in the vegetation year, is the organic material distributed on the soil surface. A monitoring of soil cover based on remote sensing can supply weekly up-to-date information for risk prediction.

Important information about the relevance of soil cover for soil conserva-

Figure 5. Soil and nutrient losses between and in traffic lanes of potatoes and corn

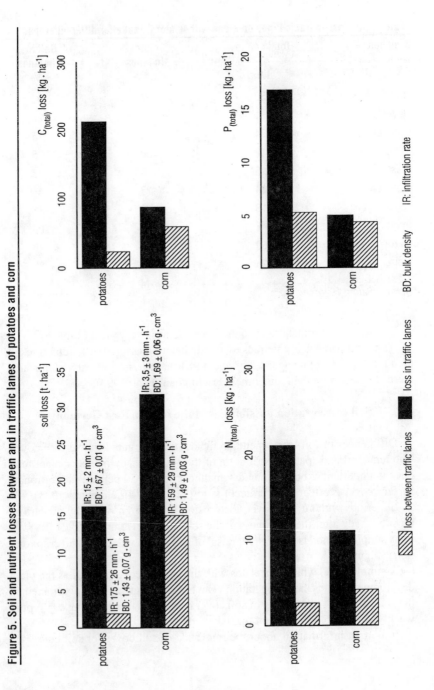

Table 5. Traffic characteristics represented for common production methods
(proportions of wheel tracks after harvest of winter wheat, potatoes, sugar beets, and maize, in relation to the total area, without primary tillage)

crops	reference width of mechanical system*	tracked area	of that moderate pressured	of that medium pressured	of that heavy pressured	tracked area total**
	m	%	%	%	%	%
potatoes	4.5	98	15	29	56	552
sugar beets	2.7	89	0	53	47	981
maize	5.0	66	12	22	66	344
winter wheat	5.0	61	19	34	47	481

* machine width of the main systems in the former East Germany
** considering repeated wheeling of traffic lanes

Table 6. Influence of plant residue cover on runof and soil loss

soil cover %	plant residues t.ha⁻¹	runoff %	soil loss %
0	0	45	100
<20	0.5	40	25
20-<30	1	25	8
ca.50	2	0.5	2.5
ca.70	4	0.1	2.5
>90	8	0	<1

tion was published for German conditions by Brunotte (1991); Sommer et al. (1996); Funk (1995); and Frielinghaus et al. (1996).

Recent soil erosion is human-induced (Bork 1988), which provides the opportunity to react with efficient erosion protective land use systems.

Further protection strategies can be summed up in four complexes:
- Middle- and long-term strategies for optimal wind and water breaking infrastructure and landscape planning;
- Short- and medium-term strategies for soil protection by plant cover, by crop rotation, and by plant residues;
- Short- and medium-term strategies for conservation tillage and management;
- Short- and medium-term strategies for the decrease of compression and track burden on intensively used arable land.

Elements of the four strategy complexes and their evaluation with regard to their efficiency and acceptance are presented (Tables 7 to 10).

353

Figure 6. Appraisal matrix for soil cover

	Time span to establish soil cover	Distribution of soil cover	Time period of soil cover in summer	Time period of soil cover in winter	Time period without soil cover due to tech. system	Evaluation of soil protection
			Soil cover > 50%			
Crops						
grass	1	1	1	1		1
winter barley	1	1	1.5	1		1
winter wheat						
seeding before Oct. 1	2	1	1.5	1.5		1.5
seeding after Oct. 1	3	1	3	2.5		2.5
summer barley	1.5	1	2	3		2
potatoes	2.5	3	3	3		3
sugar beet	3	2.5	2.5	3		3
corn / maize	3	3	2.5	3		3
sunflower	3	3	2.5	3		3
Crop Rotations						
grass rotation	1	1	1	1		1
maize, winter wheat, winter barley	2.5	2	2	2.5	2.5	2.5
sugar beet, winter wheat, winter barley	2.5	2	2	2.5	2.5	2.5
Special Crop Rotations						
winter barley, cash crops, mulch (frozen cash crops), maize, winter wheat, winter rye with underseed	1.5	1	1.5	1.5	1.5	1.5
winter barley, cash crops, mulch (frozen cash crops), maize, winter wheat, winter barley	1.5	1	1.5	1.5	1.5	1.5
winter barley, cash crops, mulch (frozen cash crops), sugar beet, winter wheat, winter barley	2	1	1.5	2	1.5	2

Appraisal:

1)	quickly	even	protected	protected	without	sufficiently protected
2)	moderate	moderate	poorly protected	poorly protected	moderate	poorly protected
3)	slowly	uneven	not protected	not protected	long	insufficiently protected

A key subject in decreasing soil erosion and soil compaction is soil conservation management. Soil compaction, soil erosion, and surface sealing should be minimized by a generation of soil aggregates with a high stability during long periods without any agricultural traffic on the fields and with a plant or plant residue cover (Sommer et al. 1996).

Possibilities to increase the acceptance of this tillage strategy should be checked individually for all regions of Germany.

Table 7/Complex 1. Infrastructure retarding erosion by wind and water

Measure	Description and Effects	Acceptance and Evaluation
Arrangement/ rearrangement of the land/field	Optimal fitting of fieldsize and fieldshape for soil protection Minimization of harmful factors (paths of flow, transport, blowing, wheeling ways) from transportation traffic on fields Orientation of certain fields rectangular to slope and main wind direction Segregation of strongly endangered areas (thalwegs, hollows, steep slope sections) for specific use	*Regionally high, differentiated* Change in ownership, type and system of farming make acceptance difficult Administrative standardized procedure does not always preserve effective structural elements for Germany Offices and regulations in preparation, tedious start
Alteration of land use type	Drastic reduction of surface runoff and soil erosion by adaption of landuse to erodibility and compactibility (surface protecting crops in spite of erosion enforcing spring crops, insertion of perennial forage growing, permanent grasslands as well as setting aside or afforestation)	*Low* because of market conditions and EU funding rules Intervention by extra supported nature and water conservation is possible if input into biotopes or water areas is detected
Construction of farm roads and country roads	Leading water off from traffic lines to decrease accumulation effects Cultivation of sections rectangular to water flow and main winds Reduction of wheeling ways for transportation traffic on the fields Shelter-belt and windbreak planting at all traffic lines	*Low* because highly expensive, change in ownership type and system of farming make acceptance difficult
Filtering edges, shelter-belts and windbreaks amid endangered fields and at waters	Interruption of surface runoff and sediment transport by water or wind Sedimentation at margins and strips Filter sections of sufficient extension depend on erosion and emission risk along waters and biotops Influence on microclimate	

Table 8/ Complex 2. Soil covering methods in crop rotations and endangered crops

Measure	Description and Effects	Acceptance and Evaluation
Selection of covering crops	Soil surface protection by increasing the time of cover about 50% after evaluation (appraisal matrix with 5 criteria) Reducing broadstanding row crops (corn, sugar beets, potatoes, winter rape, sun flower) Increasing winter cereals and perennial forage crops	*Low* due to contradictory regulations at the market and EU
Planting of cash crops	Cropping forage and green manure frozen dead to increase surface cover especially during the erosive winter season Plowless cultivation wherever possible for avoiding of uncovered fallow Soil amelioration for structure stabilization by organic matter	*Low* due to more cultivation procedures and costs *Recommended for support in future
Underseeds	Increasing surface cover during winter season and autumn by cropping of grass or underseeds in winter cereals and clover or clover/grass mixture in corn Increasing surface cover by start of rotating fallow land or set aside land	*Low* lack of experience *Recommended for support in future
Strip cropping	Decreasing sheet surface runoff velocity (often overestimated) by strips of grass or cereals parallel to isohypses Lifting up the windfield and reducing the wind power	*Low* regional implementation, effective just on gentle slopes
Strip cropping	Parallel strips of erosion reducing and erosion fostering crops rectangular to main slope or wind direction for diminishing erosive slope length and wind blowing paths Reducing the flowing and blowing velocity by increasing surface roughness	*Low* in Northern/Eastern Germany, although extensive experiences in other countries with large fields *Unsuitable against water erosion at highly inclined areas in Northeast Germany *Recommended for support in future at wind erosion endangered areas

Table 9/ Complex 3. Conservation tillage and special tillage

Measure	Description and Effects	Acceptance and evaluation
Design of crop rotation	Medium term planning and realization of erosion reducing plant production for: *reduction of tillage operations during crop rotation for decrease if mechanical soil loading and wheeling *reducing agrochemical operations by positive natural effects *increase of soil covering during crop rotation *precondition for soil conservative setting aside of fields	*High* in stable farms, part of BMP *widely practiced in former GDR *nowadays almost impossible because of the conditions of leasing, market and EU subsidies *One of the most efficient measures of soil conservation*
Soil conservation tillage Mulch-seed: *with seedbed preparation *without seedbed preparation	Seeding into soil surface covered with dead crops/crop residues (dead frozen cash crops, chemical/ mechanic treated cash crops, harvest residues left at the surface or mixed into the upper soil layer) for: *extended soil cover in time and space for reduction of raindrop splash effect and wind impact, increase of rain infiltration and soil structure stability/soil load capacity*	*Low, regionally high, very differentiated* (2-3%, increasing) *Key procedure for soil conservation* *Recommended for support in erosion endangered areas without seedbed preparation
Direct seeding	Seed by special machines without soil tillage into stubble of corn, stubble of cereals or residues of cash crops (effective by soil cover of >2t/ha dry matter for: *avoiding erosive critical conditions of the seedbed preservation of macropores reducing the sealing effect of raindrop splash or wind impact)*	*Low* *in Germany rarely practiced (few pressure by weeds), heavy harming of the soil surface by wheel tracks after harvest *deficit of precipitation
Contour farming	Cultivation adapted to the relief, rectangular to the inclination Tracks of cultivation provide barriers against surface runoff for: *avoiding preferential erosional pathways along main inclination*	*Low till medium, regionally differentiated* *Practicable just to the limit of 10% inclination *Opposite effect at extreme precipitation amounts*
Round field surface	Creating a heterogenous soil surface by row seedbed preparation, satisfying field emergence by fine and recompacted seed rows for: *increased infiltration and aggregate stability against water and wind by coarse and loose interrow material*	*Low* *regionally differentiated, depending on cost reduction

Table 10/ Complex 4. Conservation management for reduction of soil compaction and wheel tracks

Measure	Description and Effects	Acceptance and Evaluation
Changing plant production management	Combining cultivation operations Shortening uniformly cultivated slopes and wind blowing ways Network of farm roads to reduce transport on the fields	*Medium* component of BMP *Realization is limited by high costs
Changing the vehicular parameters	Enlarging the wheel contact area, reducing the inflation pressure, reducing of the axle load for: *diminishing of erosion along wheel tracks and preservation of a high infiltration and percolation capacity level*	*Low* currently not practicable *Heavy machinery is preferred because of higher efficiency, special wheels are very expensive
Increasing the soil load-carrying capacity	Plant cover (winter cereals, perennial feedings, catch crops, residues) Conservation tillage for: stabilization of soil structure and macropore system	*Low, but increasing,* when same yield level is guaranteed and costs are reduced
Site-dependent soil tillage and cultivation	No-tilling and cultivation under wet field conditions (>70% soil water content) Avoiding wheel tracks before winter and in early spring Post-emergence treatment with pesticides and mineral fertilization for: *decreasing of soil sealing and crusting, reducing of potential erosion and off-site damages by soil compaction and trafficked wheels*	*Medium* at satisfying yield potential *Element of BMP, but contradictory acting is quite common *Lack of advisory services
Extensivation by set aside and rotation	Set aside at field parts, along endangered waters, biotopes Extensive use of river plain sites and peatland for: *erosion and compaction control*	*High* at the recent support situation, realization for erosion control remains an exception

Table 11. Possibilities and instruments for realization of the complexes 1 to 4

Complex No.	General adminis-tration rules	Planning			Education, Consulting	Instruments for realization
		long-term	medium-term	short-term		
Complex 1 (infrastructure)	X	X			X	Landscape planning supporting programs Cooperation with water and nature con-servation services
Complex 2 (Soil covering)			X	X	XX	Advisory services BMP (part of Soil Protection Law)
Complex 3 (Conservation tillage)	X	(X)	X	X	XXX	Advisory services Supporting pro-grams
Complex 4 (Conservation management)	X	(X)	X	X	XX	Advisory services BMP (part of Soil Protection Law) Lack of cooperation with farm machinery industry

*Water Protection Law, Nature Protection Law, Soil Protection Law

Although all complexes offer a high potential for adequate management, they are only rarely adapted due to insufficient political strategies, extension of knowledge, and insufficient subsidies for soil protection measures.

The possibilities to realize the complex 1 to 4 are given in general adminis-tration rules, short-term, medium-term, and long-term planning methods, and education and consulting forms (Table 11).

A protection of soils from water and wind erosion and soil compaction also results in reduction of water pollution. Corresponding guidelines should be es-tablished under the common term "good agricultural management" (compa-rable with "Best Management Practice"—BMP) in the soil protection rules as well as in rules for environment checkup.

The recent short-term financial regulations of EC confuse the farmers' selection of conservation management systems. The necessity for further activities must be emphasized. The pressure of reducing costs, which the farmers face, can lead locally to an increase of erosion effects. On the other hand, the acceptance of systems of reduced soil management with high protection effect may grow because of their ability to reduce costs. In the long run a sufficient payment to the farmer for soil and water protection management is recommended. The funds required can be provided by a modification of the current support policy.

Conclusions

- Regional differentiated methods for the estimation of the potential of erosion and compaction onsite and offsite and methods for the identification of runoff paths in the landscape provide the opportunity to react with efficient protective land use systems.
- Protecting soil also results in prevention of offsite water pollution by water erosion and air pollution by wind erosion.
- A key subject in decreasing soil erosion and compaction is soil conservation management increasing soil surface cover. Possibilities to improve the acceptance of this tillage strategy should be checked individually for all regions of Germany, evaluated economically and ecologically, and supported financially.
- Although four complexes offer a high potential for adequate management, they are rarely adopted. The reasons for this behavior are insufficient strategies in politics, lack of extension of knowledge and experience, and lack of subsidies offered for soil protection, but not a lack of research results.
- Corresponding guidelines should be established under the common term "best agricultural management" (BMP) in the soil protection rules as well as in rules for environmental checkup.
- The recent short-term financial regulations of EC must be changed soon. In the long run a sufficient payment to the farmer for soil and water protection management is recommended.

References

Bork, H.R. 1988: Bodenerosion und Umwelt — Verlauf, Ursachen und Folgen der mittelalterlichen und neuzeitlichen Bodenerosion. Landschaftsgenese und Landschaftsökologie, 13, Braunschweig.

Brunotte, J. 1991: Maßnahmen zum Bodenschutz im Zuckerrübenanbau. KTBL-Arbeitspapier 159, Darmstadt.

Deumlich, D. and M. Frielinghaus. 1994. In Werner, W.; H.-P Wodsak, 1994: Stickstoff- und Phosphateintrag in die Fließgewässer Deutschlands unter besonderer Berücksichtigung des Eintragsgeschehens im Lockergesteinsbereich der ehemaligen DDR. Schriftenreihe Agrarspectrum, Band 22, Frankfurt/Main.

Deumlich, D. and L. Völker. 1994. Nutzung der Alklgemeinen Bodenabtragsgleichung zur Bodenabtra gsabschätzung im Zuge des Agrarlandschaftswandels in Nordostdeutschland.Mitteilgn. Dtsch. Bodenkundl. Gesellsch., 74, 175-178.

Dürr, H.J., H. Petelkau, and C. Sommer. 1995. Literaturstudie "Bodenverdichtung," Umweltbundesamt FB 95-036.

Frielinghaus, Ma. and L. Müller. 1994. Ableitung von Zielen für die Landschaftsgestaltung eines Flußniederungsstandortes aus alten Kartierunterlagen. Arch. Acker-Pflanzenbau, Bodenkunde 38, pp.115-126.

Frielinghaus, Mo. 1988. Wissenschaftliche Grundlagen für die Bewertung der Wassererosion auf Jungmoränenstandorten und Vorschläge für die Einordnung des Bodenschutzes. Habilitationsschrift, Berlin, Akademie der Landwirtschaftswissenschaften. 146 pp.

Frielinghaus, M., H. Petelkau, and C. Roth, 1994. Evaluation of increased erosion risk on slopes with wheel tracks. In: International Soil Tillage Research Organization ISTRO, Proceedings of 13th International Conference, Aalborg, Denmark, pp. 347-352.

Frielinghaus, M. 1995. Abtrag von Böden/ Erosionsformen. In: Autorenkollektiv: Handbuch für Bodenkunde, ecomed- Fachverlag, Landsberg/ Lech, Chapter 6.3.1.2.

First European Conference & Trade Exposition on Erosion Control, Sitges - Barcelona. Conference Book.

Funk, R. 1995. Quantifizierung der Winderosion auf einem Sandstandort Brandenburgs unter besonderer Berücksichtigung der Vegetationswirkung. Dissertation TU Berlin, Zalf- Bericht Nr. 16, Müncheberg, 1995.

Helming, K. 1992. Die Bedeutung des Mikroreliefs für die Regentropfenerosion. Dissertation TU Berlin D 83.

Hüttl, R.F. and M. Frielinghaus. 1994. Soil fertility problems- an agriculture and forestry perspective. The Science of the Total Environment 143, pp. 63-74.

Schmidt, R. 1991. Anthropogene Veränderung und Degradation landwirtschaftlich genutzter Böden in den neuen Bundesländern Deutschlands. Z.f. Kulturtechnik und Landentwicklung 32, pp. 282-290.

Schwertmann, U., W. Vogl, and M. Kainz. 1990. Bodenerosion durch Wasser: Vorhersage des Abtrags und Bewertung von Gegenmaßnahmen. Stuttgart, Ulmer.

Sommer, C. and J. Brunotte, 1996: Fachliche Einführung zur Konservierenden Bodenbearbeitung. Bornimer Agrartechnische Beriche, H. 19, Potsdam Bornim Institut für Agrartechnik., pp. 13-25.

Weingarten, P. and K. Frohberg. 1996. Soil and Water Conservation in Germany. Paper contributed to the symposium on Soil and Water Conservation Policies, Prague.

Wischmeier, W.H. and D.D. Smith. 1978. Predicting rainfall erosion—a guide to conservation planning. USDA, Agric. Handbook Nr. 537 Science and Education Administration, Beltsville, Maryland.

23

Landscape Protection and Economic Interests

Rita Kindler

The worldwide decrease of farmland, meadows, and forests is typical of the Industrial Age and leads to the irreversible conversion of natural resources. The ecological efficiency of the landscape of areas used for settlement, traffic, and industry is restricted or ceases altogether. Environmental damages can be compensated by enhancing the ecological effects of the remaining countryside. Such measures, however, are very expensive and must be financed by governments.

It is an important task to preserve and restore the ecological stability of the countryside. The better the ecological systems function, the less it costs society to rectify unintended damages and negative external effects. It is therefore essential to minimize the loss of farmland and forests, because these lands have the highest ecological efficiency.

Global ecological problems culminate in the conversion of land. The general transition to the principles of a farsighted "sustainable economic management" would reduce the conversion. This demands more than merely a correct ecological and economic evaluation of land and its planning. It demands a reformation of the framework of society, a reform toward a "sustainable economic system."

In societies not governed by use of sustainable methods, the annual conversion of landscape is very high. In the former West Germany, degradation amounted to 0.27 percent between 1960 and 1989; the corresponding figure was 0.44 percent per year in the former East Germany (because of higher mining of brown coal). Between 1989 and 1993, the conversion of agricultural and forest land increased in the former West Germany to 1.3 percent (0.32 yearly). In the former East Germany it was to 1.88 percent (0.47 yearly) because of a high accumulated demand as to land for infrastructure and industry. Today, it is estimated that in Germany there is a daily loss of 90-100 acres of agricultural and forest lands.

The German constitution guarantees that general policies of nature and water conservation can be enacted. Correspondingly, the federal building legislation states that the conversion of agricultural and forest lands must be reduced to the absolute lowest level. Supposedly this is to preserve natural and cultural lands. Regional Planning Law asserts that people must protect natural resources—especially water and soil. The Federal Nature Conservation Law states that an intervention into nature and land has to be compensated by financing substitutes or providing other ecological areas. These regulations are enforced by the nature preservation ministries of the states that must approve the loss of agricultural and forest areas. Since the enactment of this progressive legislation, it has not been possible to slow down the land conversion.

This shows that the economic interests of the landowners and the investors are maintained in spite of the legislative duty toward ecological policies of compensation. Also the Bundestag and the Bundesrat, the parliaments of the states and the communities, often make decisions which contradict the legal requirements of utilizing nature and culture lands sparingly. The Bundestag is in the lead with its decisions, e.g., the construction of the fast train Transrapid from Berlin to Hamburg.

Although preservation of land is guaranteed by law, economic interests hinder preservation. The federal tax and sponsor policy is based on economic interests, which motivate companies, private persons, and communities to convert farmland for nonagricultural purposes.

The landowner's economic interests

Since land for construction is worth much more than agricultural and forest areas, the owners of forests have an enormous interest in the land conversion. By doing so and without having to perform any work, they obtain an increased income tending to be multiplied by 40 or 50. Here the development of speculative deals finds its origin. Sometimes land is purchased in order to make a good investment. More and more people who are not farmers buy agricultural land, although they do not intend to use it for farming. Due to the resulting higher demand of land in the market, its price is heightened unreasonably—especially when the land sits close to a city. Even farmers are driven to sell their land in case they are in need of money (e.g., for investment or to ensure cash reserves).

Tax laws encourage investment in land. Non-farmers, who acquired land by speculation, pay only a very low agricultural land tax. Not only does the invested capital grow by a sudden increase in land value, but it is also almost spared from tax laws.

The economic interests of investors

The public sector, companies, and also private persons, who want to build houses, are investors. Their interests vary. Cities and communities want to see that new jobs are created and that the tax revenues rise. Higher profits either by the price rise of the acquired land, by tax advantages, or state aids are in the foreground for private persons and companies. The relocation of companies and the building of privately owned homes are profitable. Companies tend to move to sites "on green meadows," because the building land is cheaper there. They sell their ground located in commercial districts of a town and use the surplus for the expansion and modernization of their company. By comparison, it is much more expensive to build on former industrial ground because buildings would have to be pulled down and the polluted soil would have to be cleaned.

As to state aid and to depreciation, there is no difference between building on "a green meadow" or on former industrial ground. It is more profitable for the investor to convert agricultural land than reclaim industrial land.

Economic analysis of land conversion using this approach will suggest policies required to reduce the loss of farmland and forests. Developed land is very expensive. The result is that landowners make profit simply by turning agricultural and forest land into building sites. The starting point to reform the landowner's interest is the reduction of such an income without performance. In the 1950s and 1960s, there were attempts in Germany to tax away those profits without performance. However, the experiments failed; perhaps it is now time to reexamine this approach. Seen from an environmental perspective, environmental taxes may have a better chance to be effective now than in the past.

The taxing away of profits without performance offers a means which could be used to protect the environment. This approach would limit landowner interest in land conversion and would contribute to environmental protection.

There are a number of factors that could affect the economics of land conversion:

• Tax advantages and state aid for building land should be limited. This would put an end to the so-called "black subventions" for measures which initiate land conversion, harm the environment, and for which the society has to pay dearly. These measures could introduce a "housecleaning" of tax and state aid policies.

• The price of agricultural and forest land for building sites should increase in value. This could be done by applying an environmental tax. Investors themselves would pay for the consequences of land consumption. The money collected should be accumulated in an "environment fund" so society would have enough means for compensation.

If those two steps are taken, people who convert agricultural land would be required to assume costs of the necessary ecological restoration. This is how it is meant in 8a of the Federal Nature Conservation Law. The investor's expenses must be extended by the so-called "external secondary costs." Pigou (1995), Binswanger (1981), and Von Weizsäcker (1989) have defined it as necessary to calculate and to measure with "true prices." True prices of land conversion must contain the secondary costs of ecological damages caused elsewhere in society.

An environmental tax

Examples of secondary damages include the following:

• Extra costs of water supply. This supply is caused by the progressing decimation of meadows, woods, and fields which naturally accumulate water.

• Extra costs to filter carbon dioxide. Money is needed for an adequate ecological cultivation and for other operations necessary to balance the climatic function of agricultural and forest areas.

• Extra costs to save plants and animals from extinction. If habitat is transformed into areas used for housing, traffic, and industry, society must acquire nature preserves.

• Restoration of landscape degraded by land conversion.

Only the non-capitalistic systems of Eastern Europe have experimented with a tax for the protection of agricultural and forest land. For example the former GDR introduced a so-called "Fee on Land Use" in 1968. The fee amounted to a fixed sum per hectare. It was assessed after nonagricultural use of land had been permitted by administrative planners. A distinction was made between farmland, forest, and pasture; the quality of the soil was also used as a criteria. All this was done in order to protect especially the productive soil from transformation. The ecological value of the land was not considered. The fee ranged from 15 marks per square meter for sandy farmland up to 60 marks per square meter for black-earth soil. The average fee was 25 marks. If the ground was used for special purposes, such as public housing or agricultural buildings, the fee was lowered.

Officials soon realized that the desired effect of the fee was not achieved. After 10 years, people demanded the abolition of the fee. The failure was obviously the result of the manner in which the fee was implemented. The environmental tax was passed on as costs included in the prices. Whenever decisions were made, the economy played only a minor role. Also, a certain redistributive effect could not be denied. The revenues of this tax accumulated in the national budget were not used appropriately, but they provided funds designed to increase the productivity of agricultural land.

While some may suggest that the tax should be abolished, Kindler (1985) argued that the tax not be abolished. On the contrary, it was concluded that the fee should be increased. The rationale was that the fee would act as a substitute for missing land prices. The funds from the taxes could be used to replace the productive potential of converted areas.

- alternative extra costs in agriculture in order to compensate the missing potential income of the transformed areas; and

- alternative extra costs that our society has to pay for rebalancing the missing ecological potential of the transformed areas (renewal of groundwater, conversion of carbon dioxide into oxygen, infiltration and decomposition of toxic agents, recreation areas, etc.)

This concept corresponded to theoretical discussion advanced by Chatschaturow (1969, 1972) and Strumilin (1968) in the Soviet Union, and Graf (1980) and Roos and Streibel (1979) in the former GDR. The topic being discussed was the economic evaluation of natural resources. The discussion was focused on the costs of differential use of natural resources. The goal of the discussion was to direct future decisions toward conservation of natural resources. During the 1920s Pigou (1995) argued that "every piece of bread has its price," and that we could not use nature without paying for it—the idea fell on fertile ground.

If you compare recent discussions about an ecological tax with those of the 1960s in Eastern Europe concerning the economic valuation of natural resources, the similarities are evident. The goal is to affect investor actions. In the present system, ecological taxes are used to complement the market forces and correct failures of this market system to adequately value natural resources. If there are any supervised decisions in the non-capitalistic systems that were made on the basis of inadequate prices and did not take into account lasting ecological interests, then these decisions are to be criticized by economic evaluations.

To determine the magnitude of the ecological tax, it is necessary to calculate the costs of restoration of ecological damage caused by the conversion of land. These expenses must be increased to the amount required to preserve the ecological potential of landscape.

Principally, there are two possible ways to determine the environmental tax:

- **Market economy approach.** The fee is added as a surcharge on the purchase price of the converted land. For example: if a surcharge of 50 percent is determined as a reasonable fee, the investor must pay more for the future building site. It is a decentralized approach, which considers local conditions. The demand on land and its location dictate the price level of the fee. Unfortunately, the amount of the fee frequently is not related to the environmental value of specific land holdings.

- **Central authority approach.** This is similar to how it was done in the former GDR. The fees are determined by established criteria (productivity and fertility of the soil). The ecological potential of land can be considered. This, however, requires previous scientific data. Such an approach can consider fertility parameters of the soil; therefore, very productive agricultural land can be protected with high fees. This system can be enforced easily, yet tends to be inflexible.

As you can see, both approaches have advantages and disadvantages. Consideration should be given to use of a combination of both approaches. The environmental tax could be partly dependent on local prices, and partly on a centrally determined rate.

The environmental tax does not solve the problem of "incomes attained without performance." The problem can be solved only by a reform of the tax system. Methods must be found to maintain the principle that every citizen should be treated equally. The taxing away of "incomes without performance" will meet with less resistance if the investors—who have made a surplus in the past—are affected as well. This would mean a reform of the land tax and other taxes.

Although it is just a small step, the introduction of the environmental tax will not be realized easily. It is probable, however, that the proposed fee would contribute to the protection of land resources. The creation of an Environmental Fund provides the resources to implement needed environmental programs.

References

Binswanger, H-C. 1981. Wirtschaft und Umwelt - Möglichkeiten einer ökologisch verträglichen Wirtschaftspolitik; Verlag Kohlhammer, Stuttgart.

Chatschaturow, T.S. 1972. Über die ökonomische Bewertung natürlicher Ressourcen in Intensivierung und ökonomische Reserven; Akademie Verlag, Berlin.

Chatschaturow, T.S. 1969.Über die ökonomische Bewertung der natürlichen Ressourcen in Sowjetwissenschaft/Gesellschaftswissenschaftliche Beiträge Nr. 7; Berlin.

Graf, D. 1980. Ökonomische Bewertungen von Naturressourcen im entwickelten Sozialismus; Akademie Verlag, Berlin.

Kindler, R: 1985. Zur ökonomischen Bodenbewertung in der Landwirtschaft der DDR in Wirtschaftswissenschaft Nr. 9; Berlin.

Kindler, R. 1988. Zur Weiterentwicklung der ökonomischen Bewertung des landwirtschaftlich genutzten Bodens unter den Bedingungen der umfassenden Intensivierung in der DDR., Dissertation; Humboldt-University, Berlin.

Paucke, H. and G. Streibel. 1981. Zur ökonomischen Bewertung von Naturressourcen in Wirtschaftswissenschaft Nr. 9; Berlin.

Pigou, A.C. 1995. In: Voss, Gerhard: Ökosteuern. Über die schwierige Umsetzung einer guten Idee; Deutscher Instituts-Verlag, Köln.

Roos, H. and G. Streibel. 1979. Umweltgestaltung und Ökonomie der Naturressourcen; Verlag Die Wirtschaft, Berlin.

Schramm, E. 1969. Die Bodennutzungsgebühr—ein Stimulus zum Schutze des land—und forstwirtschaftlichen Bodenfonds. In Wirtschaftswissenschaft Nr. 5; Berlin.

Strumilin, S. 1968. Über den Preis der 'unentgeltlichen' Naturgüter. In Sowjetwissenschaft/Gesellschaftswissenschaftliche Beiträge Nr. 3; Berlin.

24

Soil and Water Conservation Policies In Austria

A. Klik and O.W. Baumer

Soil and water quality problems caused by industrial and agricultural production practices are receiving increased national attention in Austria. Agricultural production has been identified as a major source of nonpoint source pollution in groundwater and surface water bodies. Therefore, water conservation is perceived by society as a national goal. Although soil erosion is increasing in Austria, it is not yet considered a major problem. The extent of soil erosion is not very well documented.

Soil and water quality are inherently linked; conserving or enhancing soil quality is a fundamental step toward improving water quality. Soil and water conservation policies made by local and federal governments are very important factors. Integrating the activities on various levels of government with various federal agencies has become an increasingly important element of environmental policy for agriculture in Austria.

Present state of water quality in Austria

In Austria, 98 percent of the water used for human consumption comes from groundwater sources. Quality of groundwater therefore has a direct impact on human health. The Water Rights Act of 1959 mandated that all water sources, including groundwater, are to be kept clean, and ground- and springwater can be used for drinking.

Since the latter part of 1991, under the supervision of the Austrian Ministry of Agriculture and Forestry, water quality is being systematically checked. The objective of this program is the inventorying of the state of ground and running waters in Austria, so that in case of recognized polluting trends, appropriate measures can be taken on the basis of hard data. In 3-month intervals, water is tested for 60 physical, chemical, and bacteriological properties.

Before 1991, only drinking water was routinely analyzed. Surface waters have been investigated since the 1960s. Originally only biological properties were sought, but in the 1970s chemical analyses for fertilizer and oxygen content were added. Soon after, the investigation for heavy metals was included in the program. Cost for the establishment of 1,316 measurement sites was funded by the federal government, two-thirds of the sample collection and analyses expenses are paid by the federal government, and one-third by the states.

Groundwater

Nitrate content in the groundwater, next to atrazine and atrazine derivative desethylatrazine, because of its high concentration over large areas, presents the biggest problem. Main areas of nitrate pollution are located in agricultural lands in the east and southeast, where 93 percent of the tilled farmland is located.

In the last 25 years, farm management and crop production have changed. Grain and row crops (mainly beets and potatoes) decreased, and legumes and oilseeds (mainly rape) increased. Also, the proportion of grassland changed significantly (Table 1). During the same time span, the use of (commercial) mineral fertilizers decreased significantly due to decreased application of P and K (Table 2). The amount of mineralized nitrogen increased, but total applied nitrogen decreased because organic fertilizers are only available in small amounts.

Table 1. Proportions (percentage) of various crops in Austria between 1970 and 1994 (Präsidentenkonferenz der Landwirtschaftskammern Österreichs,1995)

	Percentage of cropland in		
Crop	1970	1980	1994
small grains	62.2	71.8	58.6
legumes	0.3	0.1	3.6
oilseed	0.5	0.7	11.9
row crops	15.7	14.8	12.7
vegetables	0.8	1.0	0.9
grassland	19.9	10.6	7.5
not used	0.6	1.0	4.8

Table 2. Comparison of fertilizers used (in kg.ha-1) in 1970, 1980, and 1994 (Präsidentenkonferenz der Landwirtschaftskammern Österreichs, 1995)

Year	Nitrogen	Phosphorus	Potassium	Total
1970	46.1	45.0	58.5	149.6
1980	60.9	34.9	52.7	148.5
1994	50.2	24.7	30.3	105.2

In 1993-1994, mean nitrate content of groundwater for all measurement sites in Austria was about 26.8 mg/liter, with differences between highest and lowest measurements (variances) of 11.8 mg/l (Table 3; Bundesministerium für Land- und Forstwirtschaft 1995). In four Austrian counties with intensive agriculture, nitrate contents were between 24.4 and 54.4 mg/l.

Nitrate content differences of groundwater for various regions are large. Nitrate content of groundwater in western Austria is low because of high precipitation, proportion of grasslands, and rates of groundwater renewal. As expected, higher nitrate concentrations are being measured in these areas, where agricultural land use is concentrated and precipitation is lower and less favorable for groundwater renewal (Bundesministerium für Land- und Forstwirtschaft 1993).

The combined application of pesticides, including herbicides, has been reduced by about 11 percent to 3,980 tons (Bundesministerium für Land- und Forstwirtschaft 1994a). Currently, about 270 different chemicals for crop protection are being used, and their potential for pollution varies widely. Use of atrazine has been prohibited since January 1, 1994, when in about one-quarter of all measurement sites atrazine and desethylatrazine were found (Table 4). In about 2 percent of the measurement sites, desisopropylatrazine was found. With occurrence of heavy metals in the groundwater, no problems are expected.

Table 3. Mean nitrate contents and variances for observations made in 1993-1994

State	number of measurement sites	mean value mg.l-1	variance mg.l-1
Burgenland	121	54.44	42.34
Niederösterreich	238	35.42	14.07
Oberösterreich	259	24.40	7.65
Steiermark	234	26.05	12.55
Austria (Total)	1316	26.82	11.80

Source: Umweltbundesamt, 1995.

Table 4. Occurrence (percentage of all sites) of atrazine, desethylatrazine, and desisopropylatrazine

Pesticide	concentration class		
	> 0.1 - 0.5 µg.l-1	> 0.5 - 2 µg.l-1	> 2 µg.l-1
atrazine	24.0	4.7	1.0
desethylatrazine	28.2	8.1	0.3
desisopropylatrazine	2.0	0.0	0.0

Source: Umweltbundesamt, 1995.

Surface waters

Inflow of fertilizers or pesticides from farms, particularly in areas used for crop production, is one reason for diminished water quality, especially in eastern Austria. An aggravating factor is low water levels in these regions. In western Austria, relatively high nutrient concentrations in surface water bodies can be related to high population density in these areas.

According to calculations of the Austrian Ministry of Agriculture and Forestry (Bundesministerium für Land- und Forstwirtschaft 1994b) agricultural runoff accounts for 45 percent of nitrogen and 30 percent of phosphorus of the total input to surface water. From these amounts, substantial quantities are transported to surface waters by sediments. Total area of erodible lands in Austria is about 380,000 ha. Potential soil loss from this acreage is nearly 8 million tons per year. Eroded sediment carries 16,000 tons of nitrogen and 8,000 tons of phosphorus (Stalzer 1995). Considering the total area used for agriculture, horticulture, and viticulture in Austria occupies about 15,000 square kilometers, this area contributes about 100 tons of undissolved solids per square kilometer per year to surface water. Total nitrogen and phosphorus amounts are 0.2 and 0.1 tons per km^2, respectively. Estimates are based on an export coefficient by Klaghofer et al. (1994).

Present condition of Austrian soils

Assessment of Austrian soils started a few years ago, and unfortunately did not cover the entire country. The inventory includes 6 of the 9 states. Objectives of these inventories are description of soil resources and their ranking with respect to availability and content of plant nutrients, content of pollutants, erodibility, and compaction (Blum et al. 1989). Analyses have shown above average heavy metal contents at several locations, which were not agriculture related, but caused by nearby industries. In areas where grapes are grown, increased copper contents can be found due to the spraying of copper based chemicals for many decades. Although this practice no longer exists, soil copper content persists. No other detriments of soil health could be found.

Many changes in the Austrian agricultural landscape during the last 30 years have increased the susceptibility to erosion. These changes include enlargement of fields, removal of field boundaries, changes in rotation of crops, land use, drainage patterns, and tillage practices. The Federal Environmental Agency (Umweltbundesamt 1988) estimated that under the pedologic and climatic conditions in the eastern part of Austria where most of the croplands are situated the amount of soil erosion can reach more than 80 tons per hectare and year.

Existing water conservation regulations

Protection of Austrian surface and groundwater resources has been mandated through the 1959 water rights law. This federal law contains certain exceptions for some states. The amendment of 1990 is a fundamental reform of the water rights law (Oberleitner 1990). This amendment contains regulations which are focused on the protection, preservation, and renewal of water quality of all waters. Water is seen as an ecological system and is protected as such. The standards of quality of surface and ground waters are regulated through rules and regulations issued by the Austrian Ministry of Agriculture and Forestry. Groundwater must be protected as a source of drinking water.

The present drinking water standards (Fischinger 1995) are:

- For nitrate: until June 30, 1999: 50 mg Nitrate per liter;
  from July 1, 1999: 30 mg Nitrate per liter
- For pesticides: 0.1 mg per liter (Atrazine)
- EC-(European Community) goal: 0.1 mg per liter for each component; 0.5 mg per liter for the sum of all.

As a legislative tool that allows the government to order water conservation measures when concentrations of pollutants in groundwater are increasing, precautionary tolerance levels for organic and inorganic compounds were issued by the government in 1990. In areas where groundwater tolerance levels are reached or exceeded, extensive and thorough measures for water quality rehabilitation have to be implemented (Rossmann 1993). Usually these tolerance levels are 60 percent of the drinking water standards.

Groundwater tolerance levels are:

- for nitrate: until June 30, 1997: 45 mg Nitrate per liter;
  after: 60 percent of the drinking water standard;
- for pesticides: 0.1 mg per liter

The amendment recognizes that conventional use of soils by agriculture and forestry together with other factors has in various ways led to problems concerning water quality. As a consequence limits for the application of nitrogen have been imposed: 175 kg N per ha on tilled land and 210 kg N per ha for fields that have cover crops through the winter period. For higher applications special permission must be sought.

Therefore the phrase "agriculture according to the law" is coined. It means that not only regulations concerning the use of fertilizers, pesticides, and other agrochemicals have to be observed, but also local conditions such as soil, climate, proximity to water, etc. need to be taken into account when using land for agriculture. Regulations may be adjusted for regional differences.

375

Existing soil conservation regulations

Protection of ground and surface water has been regulated by law for quite some time; soil protection has been emphasized only in the last few years. By federal legislation that covers the protection of the environment as a whole, soils are the responsibility of the states and not federal legislature. Therefore, criteria and regulations differ among states.

Laws for soil protection should serve as a guarantee for the maintenance and improvement of lasting soil fertility of agricultural lands. These laws should enhance the protection of soil health, and prevent detrimental deterioration caused by erosion, compaction, and import of harmful chemicals. These laws are specific for the application of polluting organic substances, such as sewage sludge and compost on agricultural lands. Precursors of these laws were the "sewage sludge laws," which contained few points that address soil erosion and compaction.

Soil conservation legislation enacted by the states, stipulates that if soil erosion had occurred during several years of monitoring by the authorities, installation of protective measures could be ordered. Conservation plantings, tillage methods, and protective structures (terraces) are cited. To inform farmers and farm workers about soil conservation the state makes funds available for continuing education.

It was further decided to establish permanent areas for observation of soil properties and efficiency of various management systems and conservation practices. Criteria for the selection of these sites should be characteristic of Austrian landscape units, with different land uses (tilled land, forest, grassland, orchards, etc.), different soils, and various degrees of the severity of agrochemical use. With optimal selection of these areas and uniform and standardized investigation methods, it is expected that results from these investigations can help formulate environmental policies for certain regions and for all of Austria (Umweltbundesamt 1995). At this time no soil loss tolerance level is mandated in any soil protection law.

Continuing policies

It is estimated that the present drinking water standard of 50 mg per liter can be reached for most part only if more time is allowed for rehabilitation work (Cepuder et al. 1995).

Existing laws as well as some special incentives for promotion of new management practices are a good foundation to enable protection and, where needed, improvement of water quality. Furthermore, the media (TV, radio, newspapers, and magazines) have now for several years made the public aware of the importance of water quality. In particular, younger farmers are trying to change to an environmentally safe and soil conserving land management. About 5 percent of all Austrian farms have switched to organic-biological production systems (Umweltbundesamt 1995).

Soil conservation in Austria is not yet seen as an important issue. The nitrate problem is recognized as being caused by human activity. Erosion conversely is viewed by most as a "natural" event. Austria has presently not established soil loss tolerances. Before developing criteria to establish T-values, careful consideration must be given to all adverse effects of erosion that need to be prevented or at least diminished. Agronomic, economic, ecologic, and social considerations of soil loss on-site as well as off-site must be taken into account. T-values must be linked not only to soil quality but also to consequences of their impact on aquatic ecosystems, and the need for Austrian farmers to sustain production at affordable prices must not be forgotten. At this time of development towards sustainable and ecologically protective farm management systems, a rigid T-value concept could do more harm than good. This is especially true in mountainous areas, where the erosion hazard is high. There are different proposals on how to quantify tolerable soil loss. In the political arena, soil protection regulations based on a T-value concept can only be successful if they can be locally administrated.

To formulate a strategy for soil protection from water erosion it is necessary to find out first where, when, why, and how much erosion occurs. After the establishment of a strategy suitable measures to protect land from soil erosion can be initiated.

Information and education of farmers must be further improved. For many years, Austria has had an agricultural schooling system, where students between the ages of 14 and 17 are being taught methods of farming (plant production, raising and feeding of animals, and horticulture), food production and preparation. In 1994, 112 of these schools existed in Austria (Präsidentenkonferenz der Landwirtschaftskammern Österreichs 1995).

In the last few years, some of these schools started long-term trials to demonstrate the effectiveness of soil and water conservation practices. In these schools seminars are held where specialists in soil and water conservation give lectures to the local farming communities about the benefits of soil and water conservation.

Data for environmental assessments are now available in Austria. In every political district (community), land use (types of crops etc.) for each parcel of land is being recorded every year. Soil survey of agricultural lands in Austria is nearly complete. Information on soil and water quality for most parts of Austria are available in the Soil and Water Condition Inventories and will be put into GIS when positions for GIS specialists are created and filled (Blum and Wenzel 1989).

Presently protective measures for large areas are nonexistent. Successful soil protection is practiced in vineyards, mainly consisting of grass sodding between rows and construction of bench terraces. Most of the time costs for these structures range between $10,000. and $15,000 per hectare, which is not affordable for farmers. Also in cropland areas, farmers are concerned about soil loss from their fields. Various types of conservation practices and tillages are being tried in different cropland areas. Mainly conservation tillage and mulching are used. Because the average size of farms in Austria is only about 20 ha, many conservation methods that are successfully used in the U.S. cannot be used without costly adaptations.

Lately, more buffer zones along small ditches are being developed in rural areas. It is estimated that they cause a noticeable reduction of delivery of eroded solids and chemicals adsorbed on the solids and in solution to open water bodies. Nevertheless they should only be viewed as an additional measure. The main emphasis of soil conservation should be to curb erosion where it most often starts, in fields that are used for agricultural production.

Conclusions

In 1995, a National Environmental Plan was published. In it, problem areas of agriculture and environment are discussed, and suggestions for solutions for sustainable and environmentally friendly agriculture are offered. Priority given for action to respond to problems of water and soil resources ranges from high to intermediate. For improvement of water quality, legislation is in place, and funding for several schemes (only some of them will prove to be successful) has been provided. For soil protection, legislation is not yet complete, and work needs to be done to make it comprehensive.

Our conservation efforts should be guided by the thought that we are the stewards of the land that we borrowed from our children. Fertile, healthy soils and clean water are fundamental to the quality of life in Austria.

References

Blum, W.E.H., H. Spiegel, and W.W. Wenzel. 1989. Bodenzustandsinventur. Konzeption, Durchführung und Bewertung. Bundesministerium für Land- und Forstwirtschaft, Vienna. 95 p.

Blum, W.E.H. and W.W. Wenzel. 1989. Soil Conservation Concepts for Austria. Bundesministerium für Land- und Forstwirtschaft. Vienna. 153 p.

Bundesministerium für Land- und Forstwirtschaft. 1993. Gewässerschutzbericht '93. Vienna.

Bundesministerium für Land- und Forstwirtschaft. 1994a. Bericht über die Lage der österreichischen Landwirtschaft 1993. 35. Grüner Bericht. Vienna.

Bundesministerium für Land- und Forstwirtschaft. 1994b. Solidarpakt. Vienna.

Bundesministerium für Land- und Forstwirtschaft. 1995. Wassergüte in Österreich - Jahresbericht 1994. Bundesministerium für Land- und Forstwirtschaft, Vienna.

Cepuder, P., A. Klik, and E. Klaghofer. 1995. Water and Solute Transport in the Unsaturated Soil Zone. Österreichische Wasserwirtschaft, 47(7/8): 145-150.

Fischinger, G. 1995. Kodex des österreichischen Rechts: Lebensmittelrecht. Verlag Orac, Vienna.

Klaghofer, E., K. Hintersteiner, and W. Summer. 1994. Aspekte zum Sedimenteintrag in die Österreichische Donau und ihre Zubringer. Sammelband der XVII. Konferenz der Donaulaender in Budapest. Hungarian National Committee for IHP/UNESCO and OHP/WMO, Budapest, Vol. II: 585-590.

Öberleitner, F. 1990. Die Wasserechtsgesetz-Novelle. Österreichische Wasserwirtschaft, 42(7/8), Supplementum 1.

Präsidentenkonferenz der Landwirtschaftskammern Österreichs. 1995. Zahlen '94 aus Österreichs Land- und Forstwirtschaft. Vienna.

Rossmann, H. 1993. Das österreichische Wasserrecht. Handausgabe österreichischer Gesetze und Verordnungen. Gruppe III, Band 6. Verlag der Österreichischen Staatsdruckerei. Vienna.

Stalzer, W. 1995. General Requirements for a Water Conserving Land Use. Schriftenreihe des Bundesamtes für Wasserwirtschaft, Band 1: 1-24.

Umweltbundesamt. 1988. Bodenschutz - Probleme und Ziele. Bundesministerium für Umwelt, Vienna. 279 p.

Umweltbundesamt. 1995. Umweltsituation in Österreich. Vierter Umweltkontrollbericht des Bundesministers für Umwelt an den Nationalrat. Bundesministerium für Umwelt, Vienna.

25

Agricultural Pollution Abatement in Poland

Andrzej Sapek and Barbara Sapek

Agricultural production under the communist system in east and central European countries was directed toward obtaining the highest yield without accounting for the financial and environmental costs. The prices of commercial fertilizers were low, and the value of animal wastes as manure was ignored. This resulted in an overuse of fertilizers and excessive livestock densities on many state-owned farms. The large state-owned or cooperative farms were financially supported by the government, and, in many cases, they were well provided with buildings and machinery. However, in Poland, only about 30 percent of agricultural land was state-owned, and the remainder was left as private family farms with a mean area of less than 7 ha. The state-owned farms in Poland were assisted by the government and were generally in good condition. The condition of private farms was quite the opposite. They were not supported by government. Private farmers had no right to acquire agricultural machinery and construction materials, which were allocated under a special distribution system. The farms were poor and most farmers, particularly those from small farms, had an additional job in industry or services. Thus, improvement in farm construction was especially connected with animal husbandry and storage of animal wastes. The technical knowledge of farmers, especially those from small farms, was not great and the efficiency of production was and remains low. The national mean yield of four cereals is only 3.0 t/ha and mean yield of milk is below 3,000 kg/cow/year.

State-owned farms have now disappeared, and many private farmers are facing significant environmental problems resulting mostly from the inefficient organization on farmsteads and poor management of farm wastes. Such farms pose a risk to water quality.

The aim of this chapter is to discuss a project sponsored by the U.S. Environmental Protection Agency (EPA).

Project activity

The Poland Agriculture and Water Quality Protection (PAWQP) project was a cooperative effort between the EPA and the Polish Ministry of Agriculture and Food Economy performed from 1992 to 1996. The overall goal of the project was to create a social, economic, and political climate that would encourage both the recognition of agriculture-related water quality issues and also the development of solutions to these problems. The project goal was to protect and improve rural water quality while assisting farmers to improve their production efficiency. Primary features of the project were the establishment of demonstrations of sustainable farming and waste management practices, education of rural communities, and assistance in policy development.

Demonstration farms were established in selected watersheds in northeastern and northwestern Poland. On each farm, facilities for farmyard manure and urine storage were constructed and an array of best management practices was demonstrated. The results of these practices were assessed by monitoring of water and soil quality.

The education activity is primarily directed at advisers of the Polish extension service and to high school agricultural teachers. Advisers and teachers so trained then transfer their knowledge to farmers and rural communities. The training materials were developed in the form of tutorial bulletins and presented in a series of workshops. The main education activity was a series of field days organized each year on demonstration sites.

The experiences gained in demonstrations, monitoring, and training are being used to develop policy proposals designed to abate the water pollution from agricultural sources.

Water quality background

About 180 million tons of crops are harvested in Poland each year (1995). From that amount, only about 8 million tons are sold on the market or to the food processing industry. Of the amount sold, nearly 3 million tons return to the farms as concentrates in the form of bran, oil cake, or marc. So, just 4 percent of agricultural plant products provided the population with food or fiber. The remainder, viz. 96 percent of crops, remains on, or returns to, farms and is used mostly as fodder or bedding for livestock. Virtually all the animal products, some 14 million tons, are delivered to the food processing industry; however, a significant part of it returns to agriculture in the form of bones, blood, whey, etc. Livestock produce a huge mass of wastes, amounting to 125 million tons annually. The mass of product losses due to spoilage, consumption by small

Table 1. Production of animal wastes in Poland (1994) and their nutrient content

Kind of wastes	Total Tonnage (million tons)	Nutrient content (thousand tons)		
		N	P	K
feces	87.6	356	108	149
urine	41.7	263	15	471
poultry feces	0.3	4	2	2
total	129.6	623	271	747

animals, etc., is not known. The biological active organic matter of animal wastes and their nutrients are the main agricultural sources of water and atmospheric pollution (Table 1).

Calculating a balance of nutrients in agricultural production is a good method to estimate their losses. That balance can be studied on either a farm, regional, and/or national basis, if it is assumed that the country is a closed system.

The inputs of nutrients can include: commercial fertilizers, fodder imported (purchased from other systems), seeds and other materials containing nutrients, wet and dry precipitation, and in the case of nitrogen also symbiotic and asymbiotic fixation. The outputs of nutrients from the system include: sold goods from plant products (grain, root crops, fiber crops, etc.) and sold goods from animal products (meat, eggs, wool, etc.). The difference between such estimated inputs and outputs is the surplus of nutrients not utilized in production, which accumulate on farm or can be dispersed into the environment. The nutrient balance is computed on a yearly basis and expressed in kg of nutrient per hectare that give the opportunity to compare different regions or farms. National nutrient balance has been calculated in different European countries (Isermann 1991).

National Statistical Office data were used to calculate the nitrogen balance in Poland (Sapek and Sapek 1993; Sapek et al. 1996). The same data were used to estimate the phosphorus balance.

The nitrogen surplus has decreased, evidently after the changes in the economic system, and reached a minimum in 1991–1992. Since that time, nitrogen surplus has been increasing by several percentage points each year. The top nitrogen input derives from commercial fertilizer followed by precipitation. The chief outputs are moved out with sold cereals (Table 2). The national mean nitrogen surplus amounts to more than 1.3 million tons, which is about 75 kg N/ha of agricultural land, and is not too high when compared with such surplus in some west European countries. But the input of nitrogen used is rather low efficiency, about 13 percent. That is the main economic and environmental problem, as yearly nitrogen losses exceed 1.3 million tons of N, with a market value

Table 2. Nitrogen balance in Polish agriculture (kg N/ha)

Agricultural year	1984/85	1990/91	1991/92	1993/94
Inputs	117.40	84.80	78.40	84.30
Commercial fertilizers	65.40	39.40	33.00	40.30
Imported fodder	9.30	3.60	7.60	6.70
Legume	15.70	14.80	10.70	10.40
Atmospheric deposition	17.00	17.00	17.00	17.00
Biological fixation	10.00	10.00	10.00	10.00
Outputs	15.70	11.40	10.20	10.00
Plant products	9.49	7.05	6.17	6.57
Animal products	6.22	4.31	4.05	3.46
Surplus (kg/ha)	101.70	73.50	68.10	74.30
Surplus (thousand tons)	1,923.00	1,378.00	1,276.00	1,387.00
Nitrogen efficiency (%)	13.40	13.40	13.00	11.90

Table 3. Phosphorus balance in Polish agriculture (kg P/ha)

Agricultural year	1984/85	1990/91	1991/92	1993/94
Inputs	21.80	10.30	6.30	6.60
Commercial fertilizers	20.52	9.73	5.19	5.59
Imported fodder	1.31	0.52	1.09	1.00
Outputs	2.80	2.10	2.00	1.90
Plant Products	1.51	1.15	1.18	1.26
Animals products	1.28	0.89	0.79	0.67
Surplus (kg P/ha)	19.10	8.20	4.30	4.70
Surplus (tausend tons)	356.00	153.00	80.00	88.00
Phosphorus efficiency (%)	12.80	20.70	31.50	29.20

of higher than $600 million.

The Polish food market received about 180,000 tons of N, and, therefore, that is the highest rate of nitrogen which can migrate into the municipal waste water systems or damping ground.

The phosphorus balance was calculated in the same way (Table 3). The change of economic system in agriculture has resulted in the decrease in phosphorus application and its surplus. Nevertheless, existing phosphorus surplus in agriculture amounted to 80,000 tons of P and is two times higher than the phosphorus content in food and washing materials used by the population, which is about 1 kg P per person.

The fate of nitrogen surplus differs significantly from that of phosphorus. Nitrogen not used in agricultural production is dispersed into the environment, while phosphorus is accumulated in soil.

Nitrogen lost from agriculture disperses to the environment by volatilization into the atmosphere due to denitrification, ammonia emission, and washing out into surface water or leaching into groundwater. At present, there are no observations performed in Poland to measure the nitrate losses due to denitrification or to estimate the nitrous oxide emission. Only an inventory was made and simulation performed (Sapek and Sapek 1995b, Sapek 1996c). Denitrification losses of nitrogen from agriculture are expected to be between 200,000 to 600,000 tons of N per year. The public in Poland is not well informed about the agricultural impact on greenhouse effect and ozone layer depletion due to nitrous oxide emission. All the public awareness and some activity is concentrated on abating the carbon dioxide emission.

The emission of ammonia from agricultural sources has not been defined to

Table 4. Ammonia emission from agricultural sources in Poland

Source	Emission thousand tons NH_3/year			Percent of emission
	1985	1990	1992	1992
Dairy cows	154	137	118	28.4
Other cattle	69	64	50	12.0
Hogs	90	99	113	27.2
Sheep	9	8	4	1.0
Horses	18	12	11	2.6
Poultry	16	12	10	2.4
Total animals	356	332	306	73.6
Commercial fertilizer	147	151	73	17.5
Crops	38	38	37	8.9
Total agriculture	**541**	**521**	**416**	**100.0**

Table 5. Share of ammonia in acid rain and eutrophication of ecosystems as compared with other emitted gases

	Emission of gases in Poland — thousand tons/year					
	Sulfur oxides		Nitrogen oxides		Ammonia	
	1985	1992	1985	1992	1985	1992
Calculated on SO_2	4,300	2,817				
Calculated on NO_2			1,500	1,130		
Calculated on NH_4					541	416
Calculated on N			456.0	344.0	444	342
Calculated on proton	134.2	87.9	32.6	24.6	63.6	48.9
Share in acid precipitation %	56.9	53.1	13.8	14.9	27.6	30.3

be a problem in Poland. In spite of that, ammonia in precipitation has a crucial part in acidification of environment and eutrophication of natural ecosystems. For example, about 50 percent of nitrogen loading into the Baltic Sea originates from ammonia emission. The Baltic Sea is extremely sensitive to nitrogen concentration. The estimated ammonia emission from agricultural sources in Poland amounts to about 400,000 tons (Sapek 1995b). The main sources are animal production corresponding to about 75 percent of emission (Table 4). The share of ammonia from the agricultural sources in environmental acidification in the form of acid rain amounted to about 30 percent in 1992 (Table 5). The yearly mean ammonia concentration in rainwater collected in Poland is between 1.0 to 2.7 mg $N-NH_4/dm^3$, which is more than 50 percent of total nitrogen in precipitation (Sapek 1996c). The deposition of ammonia with rain in Poland is between 4 to 10 kg/ha per year.

The prevailing assumption is that more than 50 percent of nitrogen and more than 30 percent of phosphorus in surface water originate from agricultural sources (Roman 1995). The impact of agriculture on deep groundwater has not been determined, but more than about 15 percent of the deep groundwater in use has concentrations exceeding permissible levels for drinking water - 10 mg $N-NO_3/dm^3$ (WHO 1995).

The impact of agricultural sources on shallow groundwater wells is much better known. About 70 percent of the rural population obtains water from wells located in the farmyard area. Most of these wells are hand dug with a water table of 1-8 meters. In many cases, the wells are unprotected and exposed to contamination. Analyses of more than 1,300 samples of drinking water taken from farm wells were used as an educational tool in the PAWQP, where farmers were involved in the sampling activity. The analysis results, along with a short explanation, were sent by mail directly to the farmers. Each sample was analyzed for pH, nitrate, ammonia, phosphorus, chloride, sulfate, sodium, potassium, magnesium, and calcium. Microbiological contamination was not analyzed. From the more than 1,300 samples analyzed, the greatest number of wells were contaminated with nitrate, and in more than 50 percent of the wells, the nitrate concentrations were higher than the Polish standard for drinking water (10 mg $N-NO_3/dm^3$). In many cases, concentrations higher than 40 mg $N-NO_3/dm^3$ were found. The mean concentration of nitrate in all samples was 22.3 mg NO_3-N/dm^3. The highest nitrate concentration was found in water from hand-dug wells, but about 20 percent of drinking water samples from wells drilled down to the depth of 10 to 40 meters was also polluted (Table 6).

Potassium concentrations in unpolluted groundwater in Poland do not exceed 5 mg K/dm^3. Thus, the concentrations found in the analyzed drinking water samples indicate that most are contaminated with this nutrient (Table 7).

Table 6. The nitrate nitrogen concentration in samples of drinking water from farm wells

Different Region in Poland	No. of samples	Percent of samples in N-NO$_3$ concentration range (%)				
		<5.0 mg N/dm^3	5.0 - 10.0 mg N/dm^3	10.0 - 20.0 mg N/dm^3	20.0 - 40.0 mg N/dm^3	>40.0 mg N/dm^3
Total	1324	36.5	13.2	13.9	20.4	15.9
Hand dug wells	244	16.1	11.1	14.0	39.2	19.6
Drills deep wells	151	66.2	13.9	10.5	8.0	1.4

Table 7. The potassium concentration in samples of drinking water from farm wells

Region	Number of samples	Percent of samples in K concentration range (%)				
		<12.5 mg K/dm^3	12.5 - 25 mg K/dm^3	25 - 50 mg K/dm^3	50 - 100 mg K/dm^3	>100 mg K/dm^3
Total	1324	39.1	11.8	11.8	153	22.0

The mean concentration of potassium was 78.8 mg K/ dm^3, in most cases, higher than that of sodium (mean 32.1 mg Na/dm^3). This proportion K:Na = 2.45 indicates that the primary source of pollution in the investigated wells was animal wastes and a secondary source was domestic wastewater.

The issue that receives the most research attention in Poland is the role of plant production and agricultural lands as the main sources of water pollution from agriculture. In the case of groundwater pollution with nitrate, however, generally animal production and farmstead waste management are the crucial issues in water pollution. The farmstead is where chemicals and farm products are stored and where most products from fields and meadows are fed by livestock. Products stored on the farmstead undergo different changes. Part of them are spoiled, dispersed, leaked, eaten by pests, etc. There is little information on the quantity of nutrients lost or dispersed during these events. According to nutrient balances empirically determined on farms, these losses should be considerable.

Much better recognized are the nutrient losses from storage animal wastes. The load of nutrients in animal wastes is several time higher than domestic wastes. Animal production is low efficiency. Only about 15 percent of plant protein is converted into animal protein. Animals excrete the remaining nutrients and organic matter in the form of feces, urine, and gases. A dairy cow produces more than 20 times more nitrogen in waste materials than a human being.

The storage of animal wastes is currently a very significant issue of agriculture impact on water quality in Poland. Under the communist economic system, the development of agriculture was primarily based on large state-owned farms, where animal wastes were stored in the form of slurry. The remainder of the agricultural land, covering about 80 percent of the total land area, was left as small private farms with an average area of 7 hectares. Private farms did not adopt the slurry technology and continued to use the solid manure technology. Private farms have spent little if any money to improve the systems of manure storage and utilization. The survey made during the project activities showed

Table 8. The content of nitrogen compounds in ground profile near manure pile

Depth cm	NO_3-N mg/dm³	NH_4-N mg/dm³
0-20	13.5	123.6
20-40	25.1	87.3
40-60	28.3	67.2
80-100	45.9	37.9
100-120	87.0	17.3
120-160	178.0	4.5
1600-200	234.0	2.1

Table 9. Nutrient concentration in groundwater near new build manure pad

Sampling date	pH	$N-NO_3$ mg N/dm³	$N-NH_4$ Mg N/dm³	P mg P/dm³	K mg K/dm³	CL mg CL/dm³
06-29-93	6.7	23.9	0.12	0.27	20.7	153
08-30-93	6.7	13.0	1.16	1.90	42.7	80
11-09-93	7.0	40.2	0.16	0.76	4.2	427
01-08-94	7.0	71.4	1.34	0.35	68.0	736
03-08-94	6.9	164.0	0.43	0.12	131.1	531
05-09-94	6.9	61.7	0.13	2.28	151.6	328
07-05-97	7.0	13.5	1.13	0.23	61.7	301
10-08-94	6.9	25.1	0.04	0.65	107.3	337
12-05-94	6.6	167.0	0.06	0.01	88.1	562
02-06-95	6.5	65.3	1.28	0.84	106.8	134
04-06-95	6.3	70.3	0.07	0.49	283.7	163
06-08-95	6.6	84.0	0.08	0.17	220.0	268
08-07-95	6.5	88.0	0.17	0.04	364.1	328
10-08-95	6.7	9.6	0.09	1.72	182.4	65
12-10-95	6.8	9.5	0.09	0.34	260.2	300
04-14-96	7.7	4.1	7.83	0.83	645.6	366
05-13-96	7.1	3.5	70.0	10.00	1196.0	556

that only 2 percent of about 5,000 questioned farmers on demonstration areas have concrete manure pads and 25 percent have urine tanks. It is quite common for farmers to keep the manure directly on the ground. Urine and manure water infiltrate into the groundwater and/or run off to the nearest stream or drainage channel. The capacity of the functioning urine tanks is generally too small and its content should be removed rather often. Urine is, in many cases, being applied on a bare or frozen soil or often directly on snow cover. The lack of proper manure storage has a significant effect on the quality of groundwater.

The amounts of inorganic nitrogen compounds accumulated in the vicinity of manure piles can sometimes exceed 300 mg N/dm^3. Ammonium accumulates in the surface layer and nitrate in deeper layers (Table 8). Removing of manure pile and building a concrete manure pad has only a minor effect on the groundwater quality. The results of groundwater analysis of samples taken from a point located only 1 m from manure pad built on the site where a manure pile was formerly situated show only a small improvement in water quality, if any, during a 3-year period (Table 9).

Regional differences

The development of agricultural activity varies widely throughout Poland, primarily as a result of differing historical associations. Until 1918, Poland was divided among Austria (South Poland), Prussia (West Poland), and Russia (East Poland). Small, poor farms (mean area 3-4 ha) were left in territory formerly occupied by Austria. The best agriculture in Europe at the end of the nineteenth century was in the former Prussian territory. An undeveloped farming system remained in the territory occupied by Russia. However, this was not the end of changes in Polish agriculture. After World War II, the victorious allies had moved Poland from the east to west for more than 300 km. The new territory regained in the western and northern parts was adopted mostly as state-owned farms with areas ranging from 500 to 5,000 ha each. The majority of state-owned farms were situated in this part of Poland. These historical facts have an impact on the present structure of Polish agriculture and influence regional threats to water quality.

The data published by the Polish Statistical Office for each province were used to tabulate the threat of water pollution from agricultural sources. The method of cluster analysis was adopted.

Nitrogen surplus in agricultural production was assumed to indicate the potential threat of water pollution with nitrogen compounds. The nitrogen fertilizer input, livestock density, number of cattle, four cereal yields, percent of permanent grassland share of agricultural land, and agronomic categories of land were used as additional variables in cluster analysis. The four risk classes, in descending order, are high, potential high, average, and low.

In the western part of Poland, where state-owned farms dominated during communist time (now the agriculture is disordered and is in a state of deep transformations), the risk of water pollution with nitrogen is the lowest (Figure 1). That is a result of the use of small quantities of commercial fertilizers (only about 37 kg N/ha per year) and low livestock density (less than 0.29 AU/ha). Potential high risk is observed in the mountain area, where small farms prevail which are involved in tourism. The main risk there is high livestock density, mostly dairy cows that provide tourists with milk. Average risk pre-

Figure 1. The regional nitrogen surplus in agricultural production, assumed as a potential risk of water pollution with nitrogen compounds

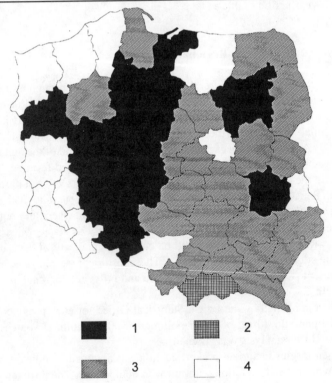

Risk classes: 1 - high; 2 - potential high; 3 - average; 4 - low

dominates in the eastern part of Poland, where agriculture is in a lower state of development.

In the part of Poland where traditions of good agriculture prevail (the central belt from the Baltic to Silesia), the highest risk of water pollution with nitrogen exists. The high-risk classes were additionally divided into four according to the kind of risk:

• Highest use of commercial fertilizer and mean livestock density and small percent of permanent grassland,

• High use of commercial fertilizer and high livestock density and greater percent of permanent grassland,

• High use of commercial fertilizer and low livestock density and the highest yield of four cereals,

• Low use of commercial fertilizer and the highest livestock density and the greatest percent of permanent grassland.

Numerous farms in these provinces are highly productive and their organization and applied technologies are similar to west European countries. The highest risk is seen in the provinces of Toruń and Lublin, where the mean nitrogen use is higher than 85 kg N/ha per year, and livestock density of 0.56 AU/ha (Figure 2). The close proximity of this area to a nitrogen factory resulted in a cheaper price of fertilizer. The farmers in provinces with second degree of risk use only about 65 kg N/ha per year, and have the same livestock density. However, the area of permanent grassland is much higher. The provinces with third degree of risk are characterized by nitrogen use of 55 kg N/ha per year and rather low livestock density amounting to about 0.35 AU/ha. Two provinces of highest proportion of permanent grassland and light sandy soils with high dairy production were considered to be in the fourth degree of risk.

The phosphorus surplus in agricultural production, assumed as potential phosphorus accumulation in soil, was used to indicate the potential threat of water pollution with phosphate. The phosphorus fertilizer input, available phosphorus content in soil, livestock density, number of cattle, four cereal yield, percent of permanent grassland proportion of agricultural land, and agronomic categories of land were used as additional variables in cluster analysis. Three risk classes (high, mean, low), were further divided into subgroups: a group with a large proportion of agricultural soils rich in available phosphorus, and a group with a large proportion of agricultural soils poor in available phosphorus.

The highest surplus of phosphorus and high content of its available form in soil is observed in provinces with traditionally good agriculture (Figure 3). This surplus is the smallest in provinces where former state-owned farms prevailed, but the content of available phosphorus in soil remains high.

Figure 2. The high risk class of nitrogen losses divided in four degrees according to the kind of risk

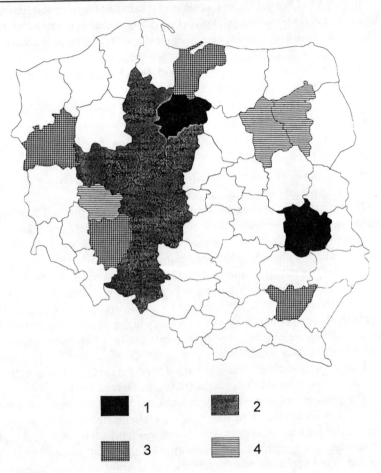

Degree of risk:
1 - the highest use of commercial fertilizer and mean livestock density and small percent of permanent grassland,
2 - high use of commercial fertilizer and high livestock density and greater percent of permanent grassland,
3 - high use of commercial fertilizer and the highest livestock density and the highest yield of four cereals,
4 - low use of commercial fertilizer and the highest livestock density and the greatest percent of permanent grassland.

National policy

The pollution of surface and groundwater in Poland is mostly related to industrial and municipal sources. The slow development in the building of the wastewater treatment plants was evident during communist rule. Most of the large Polish towns and cities, including the capital, are still without such treat-

Figure 3. The regional phosphorus surplus in agricultural production, assumed as a risk of potential phosphorus accumulation in soil

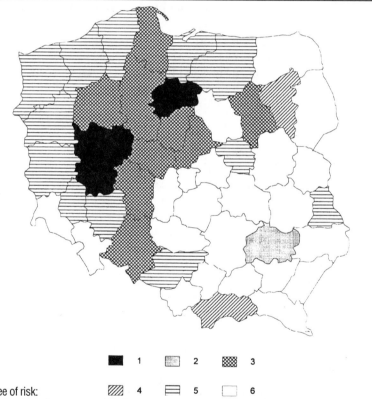

Degree of risk: 1   2   3   4   5   6

1 - high, with soils rich in available phosphorus
2 - high, with soils poor in available phosphorus
3 - middle, with soils rich in available phosphorus
4 - middle, with soils poor in available phosphorus
5 - low, with soils rich in available phosphorus
6 - low, with soils poor in available phosphorus

ment plants. This situation focuses public opinion on solving the industrial and municipal threat. This focus is also evident in rural areas, where the building of the domestic wastewater treatment plants is advertised, and some progress has been achieved. However, the consequence of this focus is that the problems of domestic wastewater obscure other threats to water quality, particularly in rural areas.

In Poland, there is no coherent policy on the issue of agriculture and environmental risk. Agriculture, in the commonly held view of the Polish public and environmental managers, is perceived as a special kind of environment which has to be protected. Agricultural production is commonly believed to be environmentally sound and requiring no special considerations. So far, little has been suggested except a few announcements to replace commercial fertilizers with farmyard manure and to forbid the use of pesticides. Such a naive view does not mobilize policy-makers. They have so far ignored agricultural pollution issues, and, as a result, there is no clear policy connected with abating the impact of agriculture production on water quality. This situation also results in difficulties with education activities aimed at the rural and agricultural communities, as there is no distinct program of action which can be presented and developed.

In recent years, a much higher concern with agriculture and environmental issues has developed in Poland, and a perspective to develop an action plan is close to being achieved. Some proposals resulting from PAWQP project activity will be presented in a broad discussion.

The following topics describe the state of agro-environmental issues:

Most farmers have only a meager understanding of the impact of their activity on the water quality (even on the quality of their drinking water).

Nearly all farms lack proper facilities for storage and distribution of the animal wastes:

- Improper management systems on such farms increase the risk of dispersing the wastes all around the farmstead and its vicinity.

- Most farmers do not have enough money to build manure pads or urine tanks. They first use their money to directly increase the production capacity of their farms.

- Some farmers are too poor to make any investment in their farms.

- The government has insufficient funds in the budget to support farmers in building such facilities.

There is only one regulation dealing with use of fertilizer and manure in agriculture. Slurry is addressed in the same manner as domestic wastewater, and that creates many problems for farmers.

The production system on many farms is out of date and the water quality, as well as the farmer's economic status, can be much improved if farmers are able to adopt recent conventional technologies:

- The rates of fertilizer applied on many farms are below what is required to produce the highest yield. Nevertheless, farmers often use fertilizers ineffectively.

- The extension service is not equipped with proper educational materials to advise farmers on how to abate the risk of water pollution and concurrently increase the farmers' income.

- The issue of agriculture and environment, and particularly the issue of water quality, is not methodically included in the education curriculum in about 300 agricultural high schools.

The discussion presented above does not show the full picture of questions and actions needed, but after 50 years of agriculture based on family farms, some extended period of time is necessary to expand the productivity of farms, increase the farmers' standard of living, and give the rural population a willingness to protect its own and the surrounding environment. The greatest expectations of success rely on the process of education, first the trainers, and then the whole rural population. The activities conducted in the PAWQP project are a modest beginning, but also a real experience, which will be helpful in future activity.

Conclusion

- Nutrients applied to agricultural production are only partly used. The nutrient surplus can disperse into the environment.

- Nutrients can be best utilized with fewer losses if farmers produce on the basis of proper knowledge, particularly with respect to recommendations of fertilizer and animal nutrition.

- Disorder in farmstead, carelessness, and bad management on farms are significant sources of water pollution.

- The lack of proper facilities for storage and handling of animal wastes is also a significant risk to water quality in Poland.

- Scientific authorities in Poland have useful recommendations for farmers, which could help them to protect water quality and to increase their economic status, but the development of a good system of extension and education is needed.

- There is no national program for abating the water and atmospheric pollution from agricultural sources in Poland. Such a program is in the developmental stages.

Summary

Polish agriculture is in a state of deep transformation to a free economic system, which has a significant impact on the environment, particularly on water quality. There are great differences in farming systems and production intensity. In the area where good agriculture existed before the communist system was introduced, the transformations are much quicker, and many farms are close to western European standards. The risk of water pollution depends mostly on farmer knowledge and attitudes. The main cause of water pollution on Polish farms is the lack of a proper system of animal waste storage and handling. The cheapest and most effective way to improve the water quality in rural areas is a good system of education for farmers and entire rural population.

References

Isermann, K. 1991. Nitrogen and phosphorus balances in agriculture - A comparison of several Western European Countries. Proceedings from International Conference on Nitrogen, Phosphorus and Organic Matter, May 13-15, 1991. Helsingør [Denmark]. 1-20.

Roman, M. 1995. Water protection from nonpoint agricultural pollution. Biuro Programu UNEP/WHO, Warszawa. 8-22.

Sapek, A. and B. Sapek. 1993. Assumed nonpoint water pollution based on the nitrogen budget in Polish Agriculture. Water Science and Technology, V. 28. No. 3-5. 483-488.

Sapek, A. 1995a. Ammonia emission from agricultural sources (in Polish). Postêpy Nauk Rolniczych. Nr 2/95. 3-23.

Sapek, A. and B. Sapek. 1995b. Application of Nitrogen Balance and CREAMS Model to Describe and Foresee the Nitrogen Losses in Polish Agriculture. Proceedings of 9th International Symposium on Computer Science for Environmental Protection CSEP '95 Space and Time in Information Systems, Part II. 214-222.

Sapek, A. 1995c. Activities in the Poland Agriculture and Water Quality Project. Proceedings of the Second International IAWQ Specialized Conference and Symposia on Diffuse Pollution, Brno & Prague, Czech Republic, August 13-18, 1995. 150-155.

Sapek, A., B. Sapek, and W. Foster. 1996a. Water quality protection in Polish agriculture. Baltic Basin Agriculture and Environmental Series. Iowa State University, Ames. Report 96-BB 5. April 1-16.

Sapek, A. 1996b. Ammonia emission and ammonia in precipitation in Poland. In: Transactions of the 9th Nitrogen Workshop, Braunschweig, September

1996. 551-554.

Sapek, A. 1996c. Nitrogen balance in permanent grassland. In: Abstracts of papers. International Conference Nitrogen Emissions from Grasslands, May 20-22, 1996, North Wyke.

WHO 1995. Control of non-point pollution from agricultural sources in the Vistula basin. Final report. Warszawa, May 1995.

26

The Lithuanian Karst Area Management Plan:
An Innovative Risk-Based Approach
to Rural Water Quality Issues

Walter E. Foster and Vilija Būdvytienė

In 1991, a proposal that the Lithuanian Ministry of Agriculture concentrate its resources on a small, environmentally sensitive territory won the approval of the nation's parliament. The karst zone of Northern Lithuania was chosen as the pilot area for preparation of a rural environmental protection program.

Once the area for the pilot was selected, it was necessary to delineate the boundaries within which management restrictions would be introduced and also to decide how to deal with the economic effects of such restrictions. A committee composed of technical staff from various government agencies studied the idea for two years before drafting and presenting a resolution to the parliament for discussion.

After the boundaries of the karst region and the set of restrictions to be applied to it had been confirmed by the government, an interdisciplinary work group, comprised of scientists from different institutions, was formed to develop an implementation plan for the program. In 1992 this group designed and presented to the government the "Targeted program on groundwater protection and sustainable agriculture development in the intensive karst zone" (Gutkauskas 1992). This document describes a complex environmental protection program, detailing the implementation of measures for stopping both point and nonpoint source pollution, not only in the intensive karst zone, but also in a surrounding protection zone. The approach is to support the implementation of sustainable and organic agriculture in the region as a means to reduce contamination of groundwater.

While these complex environmental protection measures are designed to be initially implemented in a small territory—the karst region—it is anticipated that similar measures will be extended to all of Lithuania as economic recovery and transition progresses.

In 1993, the targeted program was finally confirmed by Parliament and received financing through the budget. Start-up activities included issuing interest-free credits to some farmers for implementing the transition to organic farming and to others to provide agroservices for organic farming. During the first year, only a few farmers were ready to develop organic farming and thus be eligible to receive financial assistance. The design and construction of four wastewater treatment plants were also begun. Some other measures, integral to the program, were also implemented: several reports were published, seminars and workshops for farmers were organized, an organic farming control and certification system was developed, and an environmental monitoring program was implemented. In 1994, work on the program was continued on a larger scale.

This chapter describes the context, development, and implementation of the Karst Region Management Program and issues associated with its implementation.

The karst zone—agriculture and associated water quality

The active karst zone in Lithuania extends up to 2,000 square kilometers in the Birzai and Pasvalys Districts. A thick layer of gypsum together with dolomite, marl, and clay layers, amounting to several tens of metros, are characteristic of the surface strata of the region. Numerous sinkholes were formed by the process of gypsum dissolution. The karst phenomena continue to progress. This geology results in the contamination of the groundwater through the broken karstic terrain.

An area with thin surface deposits is almost twice as large as the active karst zone, so the groundwater is easily polluted in an area up to 4,000 square kilometers. In the remainder of Lithuania the groundwater is comparatively well protected by the nature of the geology. The surface sediments are rather thick, and reach up to 300 meters. The karst zone crosses the Lithuanian border and extends into Latvia, creating a problem of international character.

The obvious feature of the active karst zone is the presence of sinkholes. Density varies widely depending on the intensity of karstic activity. The range in number of sinkholes per hectare in three demonstration watersheds is shown in Figure 1.

The intensive karst zone was defined as those areas where the density of sinkholes is greater than 80 per square kilometer (.8/ha). This also corresponds to the area of greatest groundwater vulnerability to contamination (Gutkauskas 1992).

Groundwater depths in the area are very shallow, mostly less than 5 meters

Figure 1. Sinkhole densities on individual farms in the karst zone

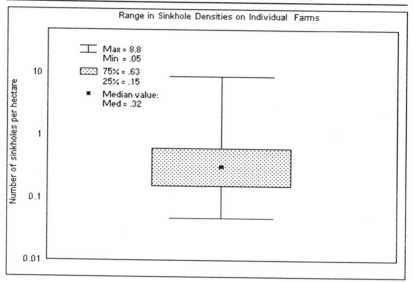

as measured in wells in use. Coupled with the extremely rapid groundwater recharge characteristic of karst activity, this makes drinking water resources in the area extremely vulnerable to contamination. Soils in the region are predominantly acidic, contributing to the karst activity.

Based on a survey conducted by the authors in 1995, two main types of agricultural operations in the karst zone are partnerships and individual farms. Individual farms average about 14 hectares, while partnership size averages over 500 hectares. Size varies widely, however, ranging from 2 to 91 hectares in the individual farms surveyed and from 51 to 1457 hectares in surveyed partnerships.

The partnerships are composed of members and employees (who may also be members). The membership varies from 14 to 645 while employee numbers vary from 12 to 200 in sampled watersheds.

Based on the survey, it appears that crop production in the karst zone is oriented toward cereal grains and feed crops on partnership operations and, on a smaller scale, similarly on individual farms. However, on individual farms, there are also grown cash crops such as flax and subsistence crops such as vegetables (Foster, et al. 1995).

Fertilizer use is indicated by the annual purchases as shown in Figure 2. After a period of decline during the early part of economic transition, the use of

Figure 2. Fertilizer use on individual farms

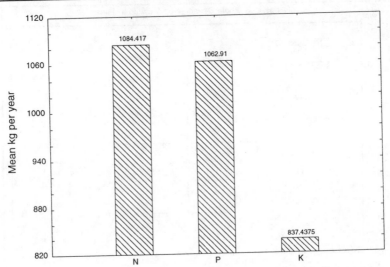

mineral fertilizers is once again rising—a situation that has serious implications for groundwater resources in the karst zone.

Livestock production is focused on pork and dairy operations among both groups of farmers, albeit at considerably different scales.

Much of the equipment in the hands of the individual farmers was obtained from state farms that were dissolved. Although many farmers appear to be well-equipped, the truth is that most of the equipment is ill-suited to small-farm operations. Results of this situation that directly affect the environment include improper doses of fertilizers and pesticides due to inaccurate mixing and application equipment and soil compaction due to oversize and needlessly heavy equipment (Sileika 1995).

Agriculture in the karst zone is generally reflective of that in the rest of Lithuania. The difference is in the vulnerability of the setting—the unique geological characteristics that makes contamination of surface and groundwater resources a serious risk to human and ecological health in the region. Many current practices can be altered under the karst zone management plan to both increase production and, at the same time, decrease the risk of water contamination. In addition, however, significant investment in point source control— such as wastewater treatment for communities, rural households, and large agricultural operations—is a necessity.

Figure 3. Potential contamination sources for drinking water supplies

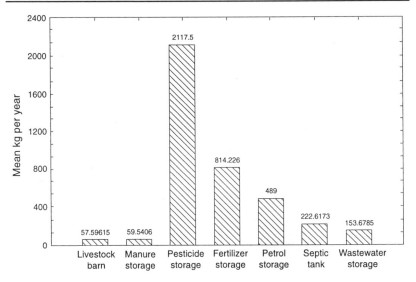

The breakup of large, centralized state farms into large numbers of small operations has seen a concurrent increase in the number and dispersal of point sources for wastewater. Point source pollution is the major contamination source for the karst zone. It is necessary to build new, or reconstruct old, waste treatment plants in the town of Birzai, and in 23 settlements or villages, where the total population amounts to 100,000. For the rest of the rural population and production centers there are no waste treatment facilities at all.

The primary point source impacts come from production units. There are 170 cattle farms with a capacity ranging from some tens to several thousand animals. There are also 30 fertilizer storage facilities, two dumping sites for home wastes, storage facilities for chemicals, machinery yards, etc., in the intensive karst zone.

Both individual farms and large-scale operations need to pay closer attention to locating potential sources of contamination relative to drinking water supplies (Figure 3). The relative proximity, particularly on individual farmsteads, of household wells to livestock barns, and manure storage is an invitation to contaminated drinking water supplies.

From Figure 4, it can be seen that more than 50 percent of farmers dispose of household sewage on the farm and that about 90 percent of all wastewater is transported to surface and/or groundwater supplies.

Figure 4. Surface and groundwater vulnerability to contamination

Receiving Water Body for Wastewater

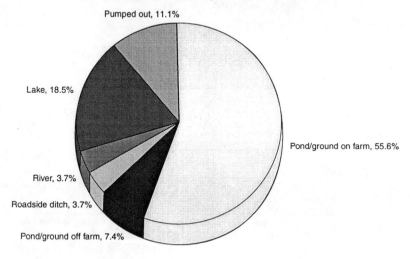

Household Sewage Disposal

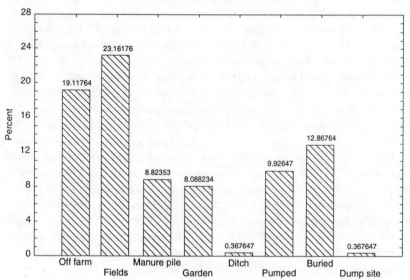

The karst zone management program

The government's karst program development and implementation activities have been conducted in three stages. As a first step, it was necessary to define the size of karst zone territory and to make decisions about appropriate farming activities in it. Although interests and attitudes of geologists, agricultural scientists, and land managers differed, decisions were made through coordination. According to the 1991 resolution setting up the karst zone program, the intensive karst zone was set at 29,000 hectares, and the protective zone at 164,000 hectares.

The intensive karst zone is divided into four groups defined according to the number of sinkholes per square kilometer. Sinkholes were chosen as an indicator, as their density provides a measure of karst activity and groundwater vulnerability.

In the fourth zone, where the density of sinkholes is more than 80 per square kilometer, farming activities are severely restricted. In zone 4, the use of all kinds of fertilizers, including manure, as well as all kinds of chemicals for plant protection is prohibited.

In the first zone, the density of sinkholes is not higher than 20 sinkholes per square kilometer. The restrictions for that land group are minimal because of the landowners' poor economic situation; they cannot afford chemicals, and their use is therefore not an issue. Activities around each sinkhole are severely restricted in all zones.

There are crop specifications for traditional farming set in each zone to control fertilizer and chemical utilization. For example, in the fourth zone it is permitted to have natural meadows, woodland, and various curative herbs and plants. In the third zone, in addition to the above listed crops, it is permitted to have managed pastures and meadows. When the plan and control system for certification of ecologically clean agricultural production is introduced, restrictions concerning crops will no longer be needed. If farmers agree to undertake sustainable (in zone 1) or biological farming (in all zones), they will be permitted to grow all kinds of plants, as well as to raise all kinds of animals and birds.

Farmers are informed of the restrictions on agricultural activities in the karst zone when they apply to obtain land for farming. They receive the land ownership documentation only if they agree to observe the restrictions.

The goals of the program are as follows:

* To develop measures to stop point source pollution (from cities, settlements, production units, and farmsteads) and nonpoint source pollution (from agricultural fields), i.e., to create a program for sustainable and biological agriculture implementation designed to solve environmental problems, while at

the same time producing ecologically clean production to meet the market's demands.

- To create economic incentives for the program's implementation during the transition period.
- To propose ways to establish an infrastructure for ecological farming and to estimate the required investments.
- To establish an ecological education and training system for specialists.
- To establish an environmental monitoring program.

Implementation of the karst zone management program

At present 30 farmers (16 of whom have received credits from the Karst Fund "Tatula") are transferring their farms to biological and sustainable farming. The Karst Fund "Tatula" spends 300 Lt. per year per farmer to finance their participation in the certification program. The number of farmers participating in that program is increasing, as they begin to understand that they don't have a favorable future with traditional agriculture in the karst zone. They understand that by complying with restrictions and participating in the activities of the Karst Fund they will make better profits as they will have a guarantee that their production will be purchased without any middleperson and at a better price.

Most of the educational activities are concentrated in the Joniskelis agricultural school, which is central to the karst region. A new specialty of bio-organic agriculture has been started for students. Seminars and training courses are being organized for the karst zone farmers. Several informational publications have also been prepared. Preliminary market research was conducted by the Institute of Agricultural Economics, and there are plans to begin a program in agroeconomics at the Joniskelis school, so that graduates can later help the local farmers who may apply to them for assistance.

Iowa State University, together with the U.S. EPA Region 7 and the Lithuanian Rural Sociology Association, in February 1995, conducted a survey of farmers and partnerships, which provided a database on their farming practices, potential pollution sources, and attitudes towards environmental protection in the region, among other things. The farms and partnerships surveyed were mapped by the Lithuanian Institute of Land Reclamation and the data digitized by the Institute of Geography. Additional natural data have been collected by the Institute of Land Reclamation at Kedainiai. After all of this data were integrated, the best management models for different karst zone groups were created and presented to the Karst Fund and the Lithuanian Government by the end of the year 1995. The above mentioned institutions also have plans to join in the support of educational activities at Joniskelis agricultural school.

In 1995, very successful fairs on bio-organic production were held in Vilnius, Kaunas, and Panevezys. Bio-organic and sustainable farming and production were advertised in an effort to get more and more farmers interested in biological and sustainable farming. The goal was to get more farmers involved not only in the karst zone but all over Lithuania.

All the implementation work that is carried out by the Karst Fund is being organized by means of competitions. Any organization or individual is free to participate in those competitions.

The extension of bio-organic farming today is not only an expression of a concern for environmental protection, but also an expression of the current economic state, related to the loss of markets for traditional agricultural production and the price increase in material and energy resources. In bio-organic agriculture, the utilization of those resources is reduced to a minimum or changed with organic fertilizers, with biological means of plant protection or with field rotation. Bioproduction is highly rated in the world and its supply is constantly growing and is not able to meet the actual demand. In many countries, the prices for bioproduction are 20–100 percent higher than for traditional production. High requirements for production, quality, and sales are applied to bio-production.

With reduction of the amount of fertilizers used, productivity decreases. Nevertheless, the ecologically pure production is of a higher value and enables compensation of losses due to reduction of productivity.

For the farmers of the region to transfer to bio-organic or sustainable farming more easily, it is necessary to choose proper farming models that work under different restrictions, and it is further necessary to develop the economic structure by which such farms can compete with other producers.

The Karst Fund currently has 95 members, about half of which are individual farmers, who are preparing to produce ecologically pure production. Together, they control 1,117 ha of land. The majority of those who are ready for the transformation process to biological and sustainable farming have less than 10 ha or more than 30 ha.

There are favorable conditions for the karst zone farms to raise plants and obtain high harvests without applying mineral fertilizers. According to calculations the most profitable cultures are vegetables and caraway seeds. If apple tree breeds are properly selected, orchards might bring high income as well. But the productivity indicated in the tables will be reached only by the fifth year of farming. Due to high costs and low prices, cattle breeding results in low income (nevertheless the prices for ecologically pure production are increased by 20 percent in comparison to current producer prices). Therefore, in order to increase farm profitability, it is necessary to utilize a part of the farm for growing vegetables and fruits and berries along with livestock and grain. A survey conducted by the Institute of Agrarian Economics indicated that ecologically pure vegetables, fruit, and dairy products have the greatest demand.

Vegetable production is highly labor intensive, therefore it should be supplemented with poultry production, cattle breeding, or pork production. Such diversified operations have an additional advantage in that the operator can supply himself with organic fertilizers to produce good compost, which are necessary for growing vegetables. In the first zone of the karst region it is possible to raise more animals in order to expand dairy production, as the planting practices recommended for this zone are favorable for fodder production.

In order to reduce production costs, farmers should specialize, keeping higher numbers of animals in one place. In that case barns, manure storage, and utilization equipment would be used more intensively and would yield higher profits. On the third group of the karst zone, the possibilities for fodder production are lower; therefore, it is recommended that sheep production, bird production, fruit orchards, and grain production should occupy no more than 30 percent of the arable land.

Farm profit analyses indicate that there are large variations in profitability. For example, on horticulture—gardening farms of 10 ha size, by growing caraway seeds, cucumbers, cabbages and early potatoes, it is possible to make profits 3–4 times higher than on nonspecialized and dairy-cattle breeding farms.

Farmers in the karst region will have to monitor the demand for ecologically pure production and select agricultural specializations and mixtures to both make high profits and benefit the environment. The Karst Fund "Tatula" guarantees those farmers who have signed the agreements with it that their ecologically pure production will be purchased during the next several years.

According to research of the Institute of Agrarian Economics, in the karst region, where the majority of farms are small, it is very important to develop agroservices and cooperative services for production processing, and to expand marketing services. The karst zone farmers' survey, conducted by the above mentioned institute, indicates that only 13 percent of all respondents are satisfied with the equipment they have. Most of the farmers would prefer to acquire all the needed machinery in spite of the farm size, and to use it for their own needs, as it is not profitable to provide services for others. Harvesting services have the greatest demand. Forty-seven percent of all respondents need such services.

Farmers' equipment holdings vary widely. Some of them have several tractors and all needed machinery, while others have almost nothing. Therefore, there is a big demand for soil tillage, harvest processing, and other farming services.

Development of more cooperation among farmers would help in solving that problem. The machinery and the proper base for its maintenance could be more effectively maintained at cooperatives. The farmers would spend less money for the purchase of needed technology and it would be utilized more efficiently.

Cooperation would reduce the need for investments, which are hard to obtain. Therefore, the Karst Fund supports cooperation based on a business plan, and the money assigned to these enterprises is utilized more efficiently than it would be if it were assigned to single farmers.

Karst region farmers will have to use the services of cooperatives for production processing and marketing. That is confirmed by the survey data, indicating that farmers would join cooperatives for production processing and sales. Every farmer of the Karst Fund would like to participate in one or another cooperative, several farmers would like to join 4–5 cooperatives. One-third of farmers would like to join meat and diary cooperatives, 17 percent grain cooperatives, 23 percent fruit and vegetable processing cooperatives, and half of the farmers questioned would like to join production sales cooperatives. Agricultural cooperatives, engaged in production processing, sales, supply, credit, etc. are very popular all over the world. Their expansion is supported by the profit that members of the cooperative are able to generate, especially those having small farms. The members of the Fund would like to participate in the activities of such cooperatives.

Based on the survey data of the Institute of Agrarian Economics, the following cooperatives could be established in the karst region: meat, milk, grain, vegetable, and fruit processing and marketing; production supply (with seeds, fodder, fuel, veterinary, and technical services) and production cooperatives, joining together several farms but not violating the private ownership rights of cooperative members. The karst region farmers understand the benefits of cooperation and are ready to use all services provided by mechanization, marketing, and production processing cooperatives.

Karst zone residents attitudes and participation in associations

Based on a comprehensive socioeconomic survey of karst zone agricultural operations, both individual and partnership, it is now possible to identify areas where attitudes may hinder implementation of the karst zone management program, and also areas where intervention might prove fruitful.

As was previously mentioned, farmers in the karst zone must comply with many restrictions. The survey data showed that about 30 percent of the surveyed farmers were aware of these restrictions (Figure 5). They were asked if

Figure 5. Individual farmers' awareness of karst zone restrictions and problems with compliance

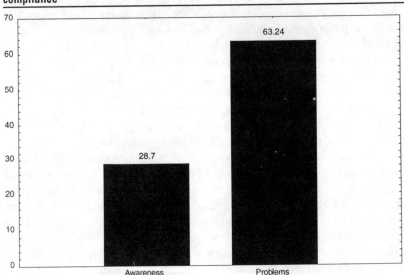

these restrictions cause them some problems; 64 percent of the surveyed farmers indicated that restrictions applied for the karst zone caused them some problems.

Nevertheless, the main hindrances for successful farming (Figure 6) can be attested to by farmers not only in the karst zone, but also by all farmers in Lithuania: 1) the price offered by processing enterprises does not cover production expenses (20 percent of respondents); 2) the shortage of financial possibilities to acquire the needed technology, fertilizers, seeds, etc. (18 percent); 3) no information about prices for farmers production (16 percent) (Foster, et al. 1995).

For the improvement of water quality, respondents believed the following proposals would be the most effective (Figure 7): 1) provision of the safe fertilizer/pesticide storage (85 percent of respondents) 2) reduction or elimination of the amounts of pesticides used (correspondingly 84 percent and 77 percent); 3) reduction or elimination of the amounts of fertilizers used (84 percent and 64 percent).

In assessing the proposals helping to reduce contamination from the farm, the most helpful for farmers would be: 1) financial assistance for sustainable and biological farming development; 2) provision of equipment to avoid use of

Figure 6. Main hindrances to successful farming in the karst zone

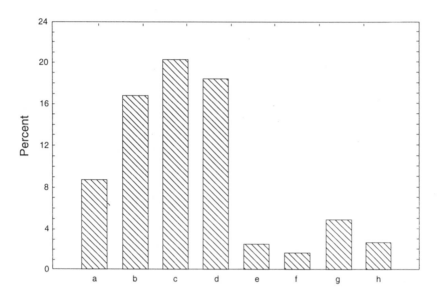

What are the main hindrances to your successful farming:
a) no information where to sell the production
b) no information about prices for your production
c) the price offered by processing enterprises doesn't cover production expenses
d) shortage of financial resources to acquire needed technology, fertilizers, seeds, etc.
e) low level of agroservice development
f) marketing infrastructure is not developed enough for production purchasing
g) the services offered by agroservices and other organization are too expensive
h) other

Figure 7. Farmers' opinions on ways to improve water quality

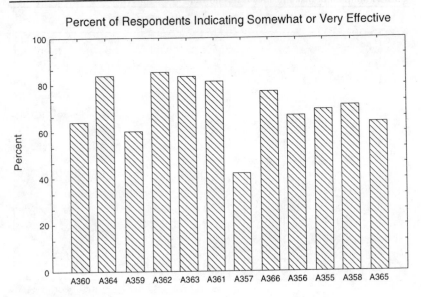

How effective are the following proposals to improve water quality
1) Non effective   2) Rather effective   3) Very effective

355) Introduce compact systems to handle household wastes
356) Introduce central systems to handle wastes in settlements
357) Reconstruct waste systems that are old and function poorly
358) Separate food processing factory wastes for separate use and
      treatment
359) Isolate waste stream of some industries from other municipal waste
360) Change production and waste handling of large concentrations of
      livestock
361) Improve manure storage and distribution
362) Provide safe storage for fertilizers and pesticides
363) Reduce the amounts of fertilizers used
364) Reduce the amounts of pesticides used
365) Eliminate the utilization of fertilizers
366) Eliminate the utilization of pesticides

412

pesticides; 3) and financial assistance to build adequate manure storage facilities.

Karst zone farmers were asked about government activities in making it easier to comply with restrictions (Figure 8). The most valuable in their opinion would be: 1) reduction or release of production taxes (87 percent); 2) reduction or release of land taxes (85 percent); 3) guarantee for processing and purchase of biological production (83 percent); 4) subsidy for organic products (82 percent); 5) compensation for money lost in complying with restrictions (81 percent).

In the majority of farmers' opinions, traditional agriculture in the karst zone solves the following problems: 1) Offers desirable family life and rural living; 2) produces reliable supplies of food for consumers; 3) provides habitat for wildlife; and provides a safe environment for the producer and family. Thus, with the exception of financial returns, karst area farmers have a rather high opinion of the benefits of traditional agriculture. Nevertheless, as indicated in subsequent questions, they are willing to consider new practices to protect their environment.

While assessing the main problems that farmers face in the karst region, the majority of respondents indicated material problems with limited resources as the most important. Most frequently cited shortages were for money and credit to buy needed machinery and buildings, needed seeds, fertilizers, and pesticides. Also cited was limited gene pool for better livestock. Nevertheless, farmers consider improper disposal/storage of livestock manure, surface water pollution and groundwater pollution as serious problems. Results from this question generally reflect the prevailing economic conditions but also indicate an underlying concern with risks to environmental health.

In making decisions (Figure 9), farmers use the following sources: 1) listen to radio or TV (57 percent); 2) discuss decision with his spouse (48 percent); 3) talk with or seek advice from a relative or family member (32 percent). Answers to this question show the most likely avenues for intervention by karst program managers, that is, how education and information programs can be designed to effectively reach the most residents.

Farmers believed that over the next 20 years new ideas for better farming will mostly arise from on farm experimentation by farmers or universities or schools of agriculture. However, a majority also think that the Karst Fund will provide new ideas. This question shows what institutions farmers look to for their farming ideas and practices and thus indicates where the karst program should focus its education and technology transfer components.

The majority of farmers surveyed (Figure 10) agreed with the following statements: 1) I worry about the purity of my family's drinking water (88 per-

Figure 8. Farmers' ideas on assistance needed for compliance with karst zone restrictions

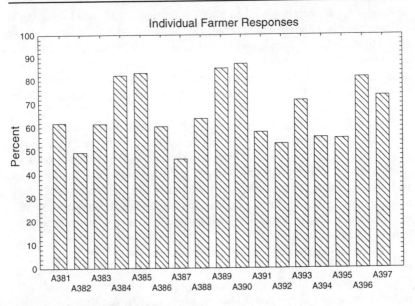

Which of the following government activities would make it easier to comply with restrictions in the karst zone?

381) Instruction about sustainable and biological farming
382) Demonstrations of sustainable and biological farming
383) Market information about organic products
384) Subsidy for organic products
385) Guarantee for processing and purchase of biological production
386) Certification and control of organic production
387) Regular tests of your well water
388) Regular tests of your soil for nutrients
389) Reduction or release of production taxes
390) Reduction or release of land taxes
391) Designs for manure storage
392) Information about best manure use
393) Provide best seeds and plants for biological and sustainable farming
394) Setting special agroservices
395) Information and assistance on nontraditional farm-based rural activities
396) Compensation of money, lost in order to comply with restrictions
397) Provide low interest loans to comply with environmental protection restrictions

Figure 9. Sources of information in farmers' decision-making

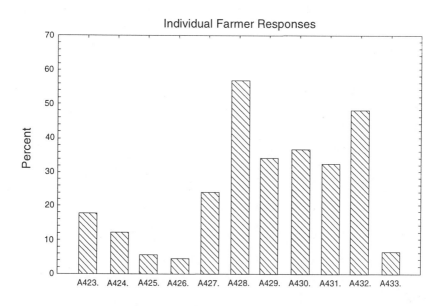

Which of the following sources do you use in making your decision?

423) Seek or obtain materials from Department of Agriculture in Regional Municipality
424) Talk with someone from DARM
425) Talk to someone at the Department of Environmental Protection
426) Talk to someone at the agricultural school or university
427) Listen to specialist discussions and lectures
428) Listen to TV or radio
429) Talk with or seek advice from nearby neighbors
430) Talk with or seek advice from a respected farmer who is not a personal friend
431) Talk with or seek advice from a relative or family member
432) Discuss decision with my spouse

Figure 10. Farmers' attitudes on water and agrochemicals

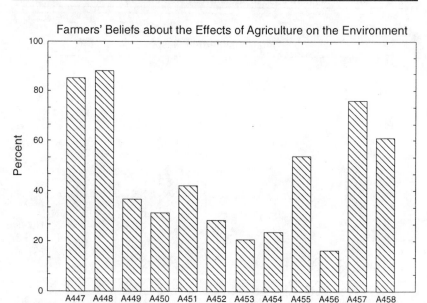

Farmers' Beliefs about the Effects of Agriculture on the Environment

Indicate if you 1) "agree" with the following statements; 2) "disagree"; or are 3) "undecided" about how you feel.

447) I worry about the purity of drinking water for the regional inhabitants.
448) I worry about the purity of my family's drinking water.
449) I am confident that agricultural pesticides, if used as directed, are not a threat to the environment.
450) Agricultural chemicals are the best alternative we have to control weeds, insects, and plant diseases.
451) Modern farming relies too heavily upon insecticides and herbicides.
452) There is too much attention about the harmful effects of pesticides.
453) There is too little attention about the benefits of pesticides.
454) The soil blocks most pesticide movement to my drinking water.
455) In this area, animal manure is not a significant factor affecting water quality.
456) When mixing and applying pesticides, a slightly richer mix or application than the manufacturer recommends is often beneficial.
457) With proper management of livestock manure there is little need for commercial fertilizer on my farm.
458) I am optimistic about the future of farming in Lithuania.

Figure 11. Farmers' relationships with Karst Fund "Tatula"

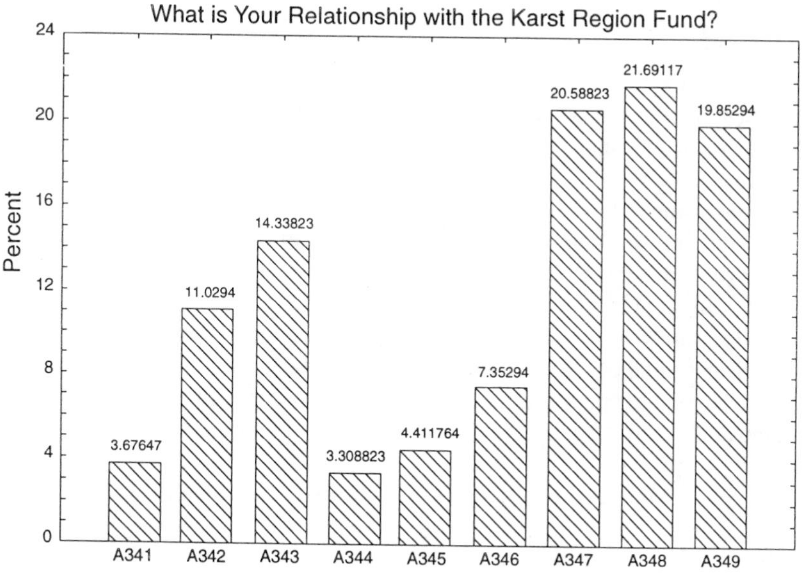

What is Your Relationship with the Karst Region Fund?

What is your relationship with the karst region fund?

341) I am one of its founders.
342) I am a member.
343) I get consultations from it.
344) I have gotten credit from it.
345) I participate at the competitions for credit assignments.
346) I am going to participate in the competition for credit.
347) I'm going to use its service to process and sell my production.
348) I do not need its assistance.
349) I have not heard about it.

cent); 2) I worry about the purity of drinking water for the regional inhabitants (85 percent); 3) With proper management of livestock manure there is little need for commercial fertilizer on my farm (75 percent). The results from this question clearly indicate karst area residents' concern with water quality and their willingness to consider alternatives to traditional high-input practices.

Farmers were asked to indicate their plans for the future. More than half of the questioned farmers intended to farm conventionally as in the past. However, more than a quarter (26 percent) of the farmers were willing to implement sustainable farming, and 28 percent were willing to implement biological farming in the future. Thus, a significant number of the farmers are currently willing to attempt to implement agricultural practices in line with karst program requirements.

An individual farmer's relationship with the Karst Fund is shown in Figure 11. Interestingly, more than 40 percent of the respondents have either not heard of the Karst Fund or state that they need no assistance from it. This would tend to indicate that the plan is not being advertised and/or promoted sufficiently, a conclusion that was supported in interviews with program managers.

Management officials and policy-maker attitudes

For a complete assessment of the potential for success of the karst zone management plan, it is necessary not only to understand farmer's attitudes and knowledge—as described in the previous section—but also to have some understanding of the knowledge, attitude, and commitment of the management officials at the national and local levels who are responsible for the plan's implementation.

A survey of a representative sample of these officials was conducted to determine the answers to these questions. The following is a summary of the responses (Foster et al. 1995):

All respondents agreed that the plan is very useful for addressing the critical situation in the karst zone, but the implementation process is too slow. The respondents believe that, while the general organization of the plan appears to be economically sound given the prevailing conditions in the karst zone, more funding should be assigned to address such issues as production practices and the purchase of organic production, ecological education, and monitoring.

Underfunding is the universally stated problem with plan implementation. While the Ministry of Agriculture has allotted considerable funding, the Ministries of Environmental Protection and Health should be more involved, particularly to address the critical problems of point source pollution, water monitoring, and continued scientific research.

One major shortcoming of the program lies in its lack of coordination with the design of land reform legislation. The respondents believe that land in the most intensive karst zone should not have been privatized but should have been placed in a nature protection reserve. Because of this, the karst program on protection of sinkholes and protection of the most intensive karst zone lands can now be implemented only with some difficulty. For that reason, these lands should now be purchased from farmers and transferred to the regional park.

The respondents generally believe that plan implementation will be successful if by 2005 a significant portion of the karst zone population (in the intensive karst zone) is involved in the activities. There are about 20 certified biological farms and more people are being attracted to the program.

The managers state that the organizational, rather than governmental, nature of the implementing institution is very positive in getting cooperation from area residents. The distribution of funds through competitions is also a positive procedure. To avoid compromise in fund distribution, the fund has set up a procedure of competitions and statutes governing their publicity.

The respondents believe that the karst program already has had some positive environmental effects, though not nearly enough. After production went down, pollution decreased and water quality improved in many places. Nitrogen amounts are reduced in separate testing places, but the total average in shaft wells is still rather high (N-60 mg in Pasvalys and 80 mg in Birzai water wells).

Summary

It is evident that nonpoint source pollution from agricultural practices and point source pollution from rural households and villages present a serious threat to surface and groundwater resources in the karst zone. It is equally evident that the Karst Zone Management Plan represents a comprehensive attempt to address these problems in a regional, holistic manner.

Agriculture in the karst zone is reflective of Lithuanian agriculture as a whole with large state-run enterprises being broken up into numbers of small individual operations. As a result, waste management has become a dispersed problem, much more difficult to address than in former, centralized times. Agricultural input and management practices are also more difficult to address as large numbers of individual operators are more difficult to reach than smaller numbers of state farm management teams. Nevertheless, the attitude of the individual farmers appears to reflect much more concern with environmental issues than that demonstrated by the former enterprises. The karst area management plan reflects a similar concern at the national level.

419

The plan appears to have been well designed and takes into account both the prevailing economic conditions and the geophysical characteristics of the karst zone. Difficulties in implementation of the plan derive from those same economic conditions as there is a chronic lack of funding for critical major projects such as wastewater treatment plants for towns and cities in the area. Other needs, such as a widespread, ongoing monitoring program are also expensive and have yet to be implemented. Farmers in the zone, while generally aware of the plan and the unique conditions in the area that require special practices, appear to have an unfulfilled need for education and advisory services that will enable them to implement the required practices. While the farmers appear to be concerned with their environment and willing to implement practices to protect it, the agricultural economy of the area, as in the rest of Lithuania, is such that there also is a need for the government to develop an incentive and/or indemnity program to induce farmers to comply with restrictions that may affect their somewhat fragile economic security.

The authors believe that the karst zone plan can succeed, if, as improving economic conditions allow, the government provides the resources to construct the necessary infrastructure and support the economic integrity of area residents upon whom, ultimately, the success of the plan depends. If successful, the plan can serve as a model for regional approaches to agroenvironmental problems throughout the Baltic Region.

References

Budvytiene, V., S. Budvytis, and N. Kazlauskiene. 1995. Environmental Policy: Institutional and Legal Framework of Environmental Protection in Agriculture—The Lithuanian Case. in K.H. Pederson (ed.) The Institutional and Legal Framework for Environmental Policy in the Agricultural Sector. Report No. 1 Incentives and Obstacles to the Implementation of More Sustainable Methods in Agriculture in Estonia, Latvia and Lithuania. EU Socio-Economic Environmental Research No. CIPDCT930030.

Carlson, G. 1993. Agricultural Run-off Management Study in Lithuania, Final Report.

Foster, W., V. Budvytiene, A.S. Sileika, and A. Gutkauskas. 1995. The Karst Zone, Agriculture and the Karst Area Management Plan. Center for Agricultural and Rural Development, Iowa State University, Ames, IA.

Gutkauskas, A. 1992. Targeted Program on Groundwater Protection against Pollution and Sustainable Agriculture Development in Intensive Karst Zone in Lithuania. Ministry of Agriculture. Vilnius.

Johnson, S.R., A. Bouzaher, W.E. Foster, and S.C. Gordon. 1993. Joint United States Agricultural Run-Off Program: Poland Agriculture and Water Quality Protection Project. CARD Report 93-50 62, Iowa State University, Ames, IA.

Sileika, A.S. 1995. Agricultural and Environmental Issues in Lithuania. Unpublished manuscript.

Tamosaitiene, A., S. Seselgiene, and A. Gutkauskas. 1995. On Ecologically Pure Production Market Research. Zemes Ukis, June, 1995.

Vaikutis, V. 1993. Lithuanian Agriculture. Lithuanian Ministry of Agriculture. Vilnius.

Vytautas, S. and A. Ladyga. 1995. Agriculture in Lithuania. Valstybinis Leidybos Centras. Vilnius.

27

Soil and Water Conservation Policies in Yugoslavia

Stanimir C. Kostadinov, Miodrag D. Zlatic, and Nenad S. Rankovic

Erosion control works projects (ECP) within the territory of the Federal Republic of Yugoslavia started by the end of the nineteenth century; however, ECP started in Serbia in 1907 and in Montenegro in 1880. Since then, programs have been performed with different intensities depending on the degree of professional support, socio-economic conditions, soil and water conservation policies, and available finances. This chapter examines the scope of ECP in Yugoslavia, with special reference to the effect of soil and water conservation policies on the type and scope of programs in the Republic of Serbia which occupies about 90 percent of the Federal Republic of Yugoslavia territory.

Method of research

The discussion includes the following stages:
1. Scope and type of ECP;
2. Analysis of laws, as the expression of the state policy, which regulated ECP;
3. Quantity of investments made in ECP;
4. Analysis of the research.

The data on the scope and type of ECP in the period before World War II were taken from various reports of the Departments of Banovina-Government Administrations and other sources of archive documents. The data for the period 1941–1960 were obtained in a similar way. In 1961, Federal Bureau of Statistics started publishing annual bulletins that also include the data on the performed ECP. Some of the information provided is as follows:

1. Technical works in the channels:

   a)   Transversal structures (number of structures and quantity of works, in m^3),

   b)   Longitudinal structures (number, length in km and quantity of projects, in m^3);

2. Bio-engineering works (small walls, terraces, contour ditches, bench terraces in *m)*,

3. Biological works (forestation, forest reclamation, pasture and meadow reclamation, establishment of orchards in hectares).

These data are available for the period 1961–1993 for the Republic of Serbia. The data on the total ECP in Serbia (1907–1993) and Montenegro (1880–1981) have been presented in separate tables.

The scope of projects in Montenegro is much smaller (only about 10 percent of the total projects in Yugoslavia). These projects were performed mainly before 1918 in the districts which were then under Austro-Hungarian rule. They were predominantly technical in stream projects with very little forestation.

In Serbia, ECP were mostly performed in the period after World War II, when the legislative activity was the highest. More detailed analyses of ECP refer to the works in Serbia, where the scope of ECP is more than ten times greater than in Montenegro.

Laws have been in existence for Yugoslavia (as the Federal State) and for the Republic of Serbia for many years. Since 1952, the republics have had the responsibility for such projects.

Net National Income (NNI) in Serbia has affected the quantity of investments made in erosion and stream control. This analysis covers the period 1961–1988, because the Federal Bureau of Statistics has been publishing these data since 1961. Namely, the period 1961–1988 was characterized by relatively stable economic and socio-economic conditions. The period 1989–1993 (the last year with published data) was not included because this was a period of high inflation in Yugoslavia.

Legislation in the field of erosion control

Yugoslav laws

Until 1930, there were no regulations or laws in the field of erosion and flood control in the Kingdom of Yugoslavia. Although the ECP were performed during this time period, projects were implemented using existing regulations from the period of Austria-Hungary, the Kingdom of Serbia, and the Kingdom of Montenegro.

The Law of Torrent Control was passed on February 20, 1930. This was the first law that was valid throughout the Kingdom of Yugoslavia. This law established the definitions of stream and watershed and provided financing. The responsibilities of the ministries were defined and the primary service agency was named "Forest-technical Department for Torrent Control." Many of the provisions and resolutions could be applied today. It is valuable because it was the foundation for developing the Service for Erosion and Torrent Control.

After World War II and the liberation of all the Yugoslav nations, Yugoslavia was constituted as the Federal Peoples' Republic of Yugoslavia (the name was changed to the Socialist Federal Republic in 1963) consisting of six republics (Socialist Republics from 1963) including Serbia and Montenegro.

Such a governmental and social system defined the two types of legislation (Federal and Republican). Many activities (including erosion and torrent control) were regulated by the Republican legislation. Thus, erosion and stream control (soil and water conservation) in the period after World War II were regulated in different ways, depending on various specifications of individual republics. Federal legislation passed two laws that were valid for the whole of Yugoslavia.

The Fundamental Law of Waters of Yugoslavia was passed in 1965 and supplemented in 1968. The importance of this law is that erosion and stream control works were included in flood protection works, which also affected the method of financing. By this law, communities have the responsibility for the maintenance, reconstruction, and construction of protection plants and structures. The Republican Funds for Waters were established and financed erosion and torrent control works in addition to water management.

The Fundamental Law on Agricultural Land Use, passed in 1965, was significant because of the section found on Soil Conservation and Erosion Control. By this law, communal assemblies were authorized to prescribe erosion-control measures and works for the protection of endangered soil. This law defined the erosive region, erosion and stream hazard, as well as the measures that can be prescribed. The costs of on-site stream and erosion control were paid by the landowners, or land users. However, the law was not significant in practice, because it was not practically obeyed.

Laws of the Republic of Serbia

The first regulation after World War II was passed in 1952 in the form of the Law of Soil Protection from Leaching and Rockslide in the Region of Grdelièka Klisura (Gorge) and Vranjska Kotlina. This law resulted from the need to implement the systematic action for the control of very intensive erosion processes in that region that contributed to frequent torrential floods. Local inhabitants were to be moved from the region (to Vojvodina) to prevent damage to roads and railways caused by tributaries of the Juzna Morava River.

The Law of Erosion and Torrent Control was passed in 1954 by the Assembly of the Peoples' Republic of Serbia. This law was designed to carry out ECP on the land threatened by erosion regardless of ownership. In 1960, the Assembly of the Peoples' Republic of Serbia passed the Law of Erosion and Torrent

Control and by putting it into force, the previous two laws were annulled. In 1965, during the coordination of the laws with the new constitution, the Law of Erosion and Torrent Control was repealed and since then, the issues of erosion control have been regulated by other laws. Erosion as a phenomenon was only partially governed by these regulations. During this time period, there were several laws that governed the field of erosion, including the following:

- Water Resources Law (1967) with amendments passed in 1967 and 1968 and amendments and supplements in 1970 and 1971;
- Law on Agricultural Land Use in 1974 had a special Chapter VI (Conservation of Agricultural Soil) which regulated the types of erosion-control measures that can be prescribed (prohibit the plowing of meadows, crop rotation, growing of perennial crops, method of soil cultivation, establishment and maintenance of shelterbelts, etc.); however, expenses were charged to landowners and land users;
- Forest Law in 1974;
- Law of Capital Construction in 1973;
- Railway Law in 1975;
- Mining Law in 1978;
- Law of Security in Railway Traffic in 1977;
- Law of Water Resources (1975) and amendments in 1976, 1977, 1982, and 1986 was the most complete legal regulation of erosion control. Chapter VIII of that law regulated this issue. The interesting provisions of the law were those found on large projects. Dams, roads, etc. could not be built without funds being provided for erosion control. The structures for erosion control and stream control were classified as protective structures and were defined as equal to other water management structures (protection of water, water protection, water utilization, drainage, irrigation, etc.). Thanks to such a treatment, erosion and torrent control works were classified into the group of priority works and the funds for their building were provided by "self-managing communities of interest" by the principles of solidarity and mutual assistance. Regulations of the law dealing with erosion control were enforced and implemented by local communities. This was problematic because most community groups had no organizational or financial capacities to execute the law;
- Water Resources Law in 1989 established seven Water Regions and the Public Water Management Enterprise "Dunav." The financing of erosion and stream control projects was by the Funds for Water Resources of Serbia, supplied by priorities, in which erosion control was the second priority after flood control and before water supply;
- Erosion and stream control problems were included in the latest Law of

Water Resources in 1991. Compared to the previous Water Resources Law, the latest law was a step backward. Namely, the problems of erosion and stream are mentioned in only three articles (37, 38, and 39). There are no clear provisions from the former law that prescribe that any greater works (dams, roads, etc.) cannot be built without previous projects and without funds provided for ECP. According to this law, three Public Water Management Enterprises were formed: "Danube," "Sava," and "Morava," and the finances from the Fund for Water Resources can be used for erosion control only as the fourth, final priority.

Of all the above laws, the practical significance for ECP is attributed only to the Water Resources Law from 1967 and 1975. The provisions of the remaining laws that had some relationship to erosion control were not enforced.

Erosion control works in Yugoslavia

In Yugoslavia, the first projects were stream control and channel training to protect railroads. The projects were in streams of the Grdelièka Klisura, a gorge in South Serbia where the international railway line (Belgrade-Skopje-Athens) passes, and in the upland watershed at the Montenegro Littoral. These projects consisted of one or several check dams constructed upstream from railroads. Protective linings (masonry revetment or concrete) were constructed along rail lines for rapid removal of water and sediment. There was nothing for the control of streams and erosion to prevent floods. Such programs were present throughout the period until 1945.

Erosion and stream control projects in Yugoslavia republics are presented in Tables 1 and 2.

All the works have been classified in two groups:
- masonry (including all the construction engineering works made of concrete, stone masonry and dry laid masonry, in stream channel, transverse and longitudinal works),
- biological works (including all the areas where biological works were carried out in the watershed area, such as forestation, reclamation of forests, grassing, reclamation of meadows and pastures, establishment of orchards, etc.).

Tables 1 and 2 show that the scope of technical ECP in Serbia was ten times greater than in Montenegro, and the extent of biological ECP in Montenegro was negligible. Owing to the economic crisis, during the last 5 years in the Federal Republic of Yugoslavia investments have been significantly reduced. This resulted in the reduction of projects carried out.

Erosion control works in Serbia

There are seven stages of development and realization of ECP in the Republic of Serbia.

The first stage, 1907–1940, is characterized by the predominance of stream control projects. The extent of technical works was 56,194.00 m³. On-site soil conservation was performed only by small-size forestation (only about 575 ha). Direct protection of arable soil was not applied because the land was in private ownership and also because the knowledge in this field was insufficient.

The Torrent Control Service was organized together with the Hydrotechnical Service, as the Department of the General Direction of Waters at the Ministry of Agriculture and Waters. In 1927, the Department of Torrent Control separated from the General Direction of Waters and joined the Ministry of Forests and Mines. The reason for joining the Department of Forestry was due to the specific work in watersheds that required forestation. The French system of torrent control was adopted. From 1927 to 1930, the development was more intense, preparatory field sections were organized for carrying out the works,

Table 1. Erosion and Torrent Control Works in Serbia in the Period 1907-1993

Stage	Period	Masonry		Biological works	
		Total m³	Annual Average m³/y	Total ha	Annual Average ha/yr
I	1907-1940	56,194.0	1,652.80	575.50	16.90
II	1941-1944	1,301.0	325.20	5.00	1.25
III	1945-1954	56,774.0	5,677.40	457.00	45.70
IV	1955-1966	386,334.0	32,194.50	16,008.00	1,334.00
V	1967-1977	476,505.0	43,318.64	16,194.00	1,472.18
VI	1978-1988	421,234.0	38,294.00	55,011.00	5,001.00
VII	1989-1993	86,105.0	17,221.00	16,443.00	3,228.60
	Total	1,484,447.0	17,062.61	104,693.50	1,203.37

Table 2. Erosion and Torrent Control Works in Montenegro in the Period 1880-1981

Stage	Period	Masonry		Biological works	
		Total m³	Annual Average m³/y	Total ha	Annual Average ha/yr
I	1880-1941	110,090.0	1,894.83	—	—
II	1945-1963	17,571.0	976.17	266.30	14.79
III	1964-1981	17,744.0	1,043.76	—	—
	Total	145,405.0	1,530.58	266.30	2.80

and the projects were prepared in the Department of the Ministry. The projects were co-financed by the Ministry and Railway Administration.

In the second stage (1941–1944) in the occupied territory of Serbia, very little was done regarding stream control. Nine streams were treated with 1,301 m³ of masonry and 5 ha of forestation.

The third stage, 1945–1954, was characterized by only the most necessary stream control projects (construction works in the torrent channel) in order to protect the railway line and roads. The extent of works was very small (56,774.0 m³ masonry and 457 ha forestation).

The period between 1945 and 1951 can be considered as the age of progress of the Erosion and Torrent Control Service in the sense of organization, because the first signs of independence could be seen. During 1945–1946, the service was an office in the Department of Bareland Forestation and Torrent Control. In 1950, the Administration for Torrent Control and Civil Engineering was formed in the Ministry of Forestry of the Republic of Serbia. ECP were financed by the Ministry of Forestry and, to a lesser extent, by the budget of Railway Administration in Belgrade. However, due to the lack of finances, there were comparatively few projects. During this time period, the funds were directed to the building and reconstruction of the country devastated by the war.

The fourth stage, 1955–1966, is characterized by significant legislative action, development of the service organization, participation of a higher number of professionals, expanded use of technical equipment, increasing investments and the application of the new technologies: American (contour ditches, bench terraces, strip farming, etc.), French (Algerian banquettes with the establishment of orchards, etc.), Italian (terraces, gabions, etc.). There was a sudden rise of investments that affected the extent of erosion and torrent control works (masonry 386,334 m³ and biological works 16,008 ha, Tables 3 and 4). During this stage of development, on-site erosion control on farmland operated by forestry and agriculture experts occurred for the first time and developed rapidly. This action was terminated in 1962. The method of financing was reorganized, as well as the service and the companies which carried out the works.

Based on the Law of Torrent and Erosion Control of the Republic of Serbia, in the period 1954–1957, 10 Regional Sections for Erosion and Torrent Control were formed with the aim of carrying out erosion and stream control works. In the financing of these projects, along with the budget of the Republic of Serbia, districts and communities (national budget), the Directorate for Roads, mines, electric power industry, Railway Administration (also previously), and other interested business firms started to participate.

After the Law of Torrent and Erosion Control in Serbia was adopted in 1954, for the better coordination of erosion control works, the Directorate for Erosion and Torrents was formed. The agency had the task to manage, supervise, and control the work of the Erosion and Torrent Control Service.

During this period, the Department of Erosion and Torrents was formed at the Faculty of Forestry in Belgrade by the decision of the Executive Council of the Republic of Serbia in 1959. The first generation of students was enrolled in 1960–1961. This department resulted from numerous problems of water erosion and streams, as well as from the awareness of the political authorities that it is necessary to educate the professionals who will deal with soil conservation and torrent control. The decisive role was played by the 1954 Law of Torrent and Erosion Control which provided the legal base and the material basis for the employment of the new professionals.

The fifth stage, 1967–1977, was characterized by the repeal of the Law of Erosion and Torrent Control when the direct conservation of arable soil stopped, except for a very modest cooperation with individual owners of farmland, by giving them free seeds for grass or forest tree seedlings, with the obligation that they must not plow the meadows on steeper slopes for 8 years.

In the fifth stage, there were no engineering projects in the watershed (contour ditches, various types of terraces, etc.) except small terraces for forestation. Bare (untillable) areas of the watersheds were forested with or without previous technical works (small terraces, etc.). Transversal structures were built in the stream channels and the lower courses were regulated. The projects amounted to 476,505 m^3 of masonry and 16,194 ha of biological works, which was an increase from previous stages.

Considering the 1967 Law of Waters in Serbia, the Service for Erosion and Torrent Control was reorganized. The former independent Regional Sections for Erosion and Torrent control were integrated with Water Communities into Water Management Enterprises which coincide with the largest rivers: Morava, Sava, Dunav, Drina, Timok, Mlava, and Pek. The Directorate for Erosion and Torrent Control was also repealed. In this way, the previous independent service was now governed by the Water Management Department

In the following period, ECP were performed in the interest of water management. Thus, the previous policy was abandoned. In the previous stage (1955–1966), there were at least some elements of soil and water conservation (ECP on arable land). In the fifth stage, it could deal with erosion control on the slopes of the watershed only from the aspect of the adverse effect on water regime of the water courses. The works were financed by the Republican Fund for Waters.

The sixth stage, 1978–1988, began at the time when the regional and republican "self-managing communities of interest" were established. They succeeded the former Republican Fund for Waters and continued to finance all the water-engineering works, including erosion and torrent control works.

430

Table 3. Bio-engineering works in the watersheds of Serbia

Year	Walls, terraces, etc.	Afforestation	Forest reclamation	Pastures and meadows	Orchards
	km	ha	ha	ha	ha
1961	481.447	501	64	383	59
1962	358.507	743	88	325	58
1963	254.790	489	120	177	21
1964	290.258	428	76	138	1
1965	174.390	253	108	51	—
1966	615.500	896	215	117	—
Total 1961-1966	2,174.890	3,310.00	671.00	1,191.00	139.00
Average	362.48	551.67	111.83	198.50	23.17
1967	1,052.397	1,045	231	265	91
1968	219.467	1,570	42	567	—
1969	220.967	532	100	241	2
1970	233.833	843	—	594	—
1971	309.921	724	—	901	—
1972	827.440	733	55	246	—
1973	224.834	528	96	253	—
1974	619.553	578	—	68	—
1975	697.746	634	363	186	—
1976	446.955	1,057	485	235	—
1977	1,128.787	1,706	854	369	—
Total 1967-1977	5,981.900	9,950.00	2,226.00	3,925.00	93.00
Average	543.810	904.55	202.36	356.82	8.45
1978	812.620	1,127	835	434	—
1979	182.120	1,699	1,000	575	—
1980	611.148	1,035	—	1,586	—
1981	544.200	421	35	998	—
1982	325.000	1,038	—	645	—
1983	346.200	799	—	1,090	—
1984	709.800	1,072	2,992	—	—
1985	448.763	1,459	10	2,460	—
1986	1,339.500	7,146	13	3,190	262
1987	64.132	1,758	10	8,642	—
1988	69.353	1,906	10	10,764	—
Total 1978-1988	5,452.840	19,460.00	4,905.00	30,384.00	262.00
Average	495.710	1,769.09	445.91	2,762.18	23.82
1989	62.453	1,906	10	2,704	—
1990	3.900	913	—	1,893	—
1991	—	801	905	1,678	—
1992	4.761	846	—	1,877	—
1993	71.961	2,909	—	1	—
Total 1989-1993	143.075	7,375.00	915.00	8,153.00	—
Average	28.615	1,475.00	183.00	1,630.60	—
Total	**13,752.705**	**40,095.00**	**8,717.00**	**43,653.00**	**494.00**
Average	**416.749**	**1,215.00**	**264.15**	**1,322.82**	**14.97**

Table 4. Technical works in the stream channels

Year	Transversal structure		Longitudinal structures		
	Total pcs	Quantity m³	Total pcs	Length km	Quantity m³
1961	243	31,595	29	127	64,941
1962	267	15,194	22	48	19,609
1963	311	31,228	33	9	11,718
1964	480	20,640	27	7	15,523
1965	142	14,997	23	6	18,858
1966	184	20,864	30	5	11,796
Total 1961-1966	1,626.00	20,864	30	5	11,796
Average	271.00	22,419.67	27.33	33.67	23,740.83
1967	225	18,511	36	123	27,015
1968	138	16,228	21	6	31,642
1969	87	11,258	23	9	80,568
1970	137	26,084	16	10	38,983
1971	47	4,721	9	2	19,677
1972	66	11,381	10	3	2,108
1973	74	2,683	14	5	10,819
1974	87	17,191	15	3	4,828
1975	77	10,761	17	5	14,976
1976	56	13,258	28	9	56,142
1977	74	20,858	28	9	56,142
Total 1967-1977	1,068.00	152,994.00	225.00	187.00	323,511.00
Average	97.09	12,908.55	20.45	17.00	29,410.09
1978	42	9,583	32	10	137,170
1979	46	6,651	43	10	44,102
1980	87	9,034	32	7	54,350
1981	16	3,979	37	154	27,354
1982	63	6,610	43	13	23,277
1983	19	3,715	18	5	8,489
1984	13	2,598	58	19	23,302
1985	46	6,423	43	14	12,165
1986	83	10,361	30	18	10,519
1987	50	5,513	50	20	5,900
1988	34	3,313	43	15	6,826
Total 1978-1988	499.00	67,780.00	429.00	285.00	353,454.00
Average	97.09	13,908.55	20.45	17.00	29,410.09
1989	162	3,850	37	19.8	1,900
1990	1	230	14	26.3	9,830
1991	5	281	11	37	68,466
1992	6	510	—	—	—
1993	5	500	1	0.32	538
Total 1989-1993	179.00	5,371.00	63.00	83.42	80,734.00
Average	35.80	1,074.20	12.60	16.68	16,146.80
Total	3,372.00	360,663.00	881.00	757.42	900,144.00
Average	102.20	10,929.18	26.70	22.95	27,277.09

Technical works were no longer performed on the slopes of the watersheds (except for small terraces for forestation). Forestation was carried out on bare (untillable) parts of the watershed, with or without making small terraces. Some transverse structures (check dams) were made in torrent channels, but much less extensively than in the previous stage. During this stage (even more than in the previous stage), longitudinal channel-control works were massively performed to protect towns. In this stage, there were somewhat less technical works (masonry 421,234 m³), but there was a rise in biological works—55,011.0 ha.

The seventh stage is 1989 to the present. In this stage, the extent of technical works (masonry only 86,105 m³) and biological works (only 16,443 ha) decreased compared to the previous stages. This is the result of the economic crisis in the country due to well known reasons in the period 1991–1996 (disintegration of the former SFR Yugoslavia war in the neighboring countries of FR Yugoslavia). In addition, the lower extent of ECP is by all means also affected by the laws adopted in this period (Republican Laws of Waters in 1989 and 1991), because the issues of erosion and erosion control are regarded in an inadequate manner.

Investments in erosion control works in Serbia

The value of erosion and stream control projects for the period 1961–1988 is presented in Table 5. Because of the change of money value and inflation, the prices have been reduced to the real value in 1987 as the basis of the deflation index (index of gross prices). Average annual rate of growth of the investments in this period amounts to 8 percent. It should be noted that the investments in 1961 were very low and grew significantly until 1967. Then investments decreased until 1974. In 1977, investments increased, followed by a marked drop in the period 1979–1988. Table 5 also presents the annual range of net national income and the percentage of investments in ECP in the value of NNI.

The percentage of investments in ECP in the value of NNI was very low in 1961 (about 0.05 percent), followed by a mild lowering until 1965, and the maximum in 1967 (0.62 percent). After that, there was another drop of less than 0.2 percent in the period 1971–1974, followed by 0.5 percent. Investments decreased to about 0.2 percent in 1988.

In this respect, the statistical analysis was performed so as to determine the effect of individual factors on the amount of investments in ECP in Serbia. The analysis covers the period 1961–1988. All the prices were reduced to the prices in 1987 by the deflation index.

The effect of NNI (X_{NNI}) on the amount of investments in ECW (Y_{IECW}) was analyzed via a simple correlation. The following regression model has been obtained (t-statistics in brackets):

$$\ln Y_{IECW} = -13.267 + 1.989 \cdot \ln X_{NNI},$$
$$(-5.885) \quad (6.221)$$

$R^2 = 0.60 \quad R^2\text{adj}. = 0.58 \quad R = 0.78 \quad SE = 0.684 \quad F(1,28) = 38.705.$

The obtained regression model is characterized by a high percentage of explained variance (60 percent). The parameter with NNI shows that an increase of investments in ECP can be expected to be about 2 percent, with the increase of 1 percent in NNI.

We have also analyzed the mutual effect of NNI, technical works in the stream channels (X_{TW}), and biological works in the watershed (X_{BW}) on the amount of investments in ECP. The following regression model has been obtained (t-statistics in brackets):

$$\ln Y_{IECW} = -17.595 + 1.975 \cdot \ln X_{NNI} + 0.322 \cdot \ln X_{TW} + 0.143 \cdot \ln X_{BW}$$
$$(-5.077) \quad (5.335) \quad (1.624) \quad (0.787)$$

$R^2 = 0.64 \quad R^2\text{adj} = 0.60 \quad R = 0.80 \quad SE = 0.672 \quad F(3,28) = 14.301$

characterized by a high percentage of explained variance (64 percent). It can be concluded that the rise of 1 percent in NNI leads to the increase of investments in ECP of about 2 percent. The increase of 1 percent in technical and biological works causes the increase of investments in ECP of about 0.3 percent, i.e., 0.14 percent, with the possibility of greater deviations. The degree of the homogeneity of the obtained model $(r = 1.975 + 0.322 + 0.143 \approx 2.4)$ shows that, with the increase of 1 percent in all the observed factors, we can expect the rise of investments in ECP for *cca* 2.4 percent.

The analysis of research results

The research results show that ECP were not uniformly distributed throughout SR Yugoslavia. The total in Montenegro in the period 1980–1981 was 145,405 m^3 of masonry and only 266.3 ha of biological works. The extent of technical works is about 10 percent of technical works in Serbia (1,484,447 m^3) which corresponds approximately to the relationship of surface areas occupied by these two republics. Of the total extent of technical works in Montenegro, about 76 percent were implemented prior to 1941 (I stage). After World War II, the annual average in Montenegro was only about 1,000 m^3 of masonry. The extent of biological projects in Montenegro is negligible—only 266.3 ha in the II stage (1945–1963).

Table 5. Economical and technical parameters of ECP

Year	National income mld. din.	Net-national income mld. din.	Investment in ECP mld. din.	Part IN NI %	Part IN NNI %	Technical works m^3	Biological works ha
1961	408.033	378.777	0.201	0.05	0.05	96086	1007
1962	468.910	432.797	0.236	0.05	0.05	34803	1214
1963	573.270	502.920	0.283	0.05	0.06	42946	807
1964	778.333	717.800	0.295	0.04	0.04	36163	643
1965	1027.633	957.067	0.240	0.02	0.03	33855	412
1966	985.050	919.525	4.984	0.51	0.54	32660	1228
1967	1027.400	943.675	5.879	0.57	0.62	45526	1632
1968	1078.325	987.225	3.188	0.30	0.32	47930	2179
1969	1024.060	936.400	3.962	0.39	0.42	91826	875
1970	1171.320	1072.080	3.798	0.32	0.35	65067	1437
1971	1282.400	1179.667	2.220	0.17	0.19	24398	1625
1972	1306.400	1189.686	2.319	0.18	0.19	13489	1034
1973	1446.738	1320.100	2.950	0.20	0.22	13502	877
1974	1383.536	1256.109	2.146	0.16	0.17	22019	646
1975	1316.593	1207.436	3.335	0.25	0.28	25737	1183
1976	1480.173	1335.667	2.839	0.19	0.21	50011	1777
1977	1714.963	1561.163	7.774	0.45	0.50	77000	2929
1978	1845.789	1680.767	5.888	0.32	0.35	146753	2396
1979	2071.767	1894.514	4.070	0.20	0.21	50753	3274
1980	1869.768	1711.074	2.566	0.14	0.15	63384	2611
1981	1712.841	1572.588	3.175	0.19	0.20	31333	1454
1982	1734.106	1563.261	3.851	0.22	0.25	29887	1683
1983	1747.966	1573.149	2.507	0.14	0.16	12204	1889
1984	1643.448	1487.267	2.387	0.15	0.16	25900	4064
1985	1517.889	1374.444	2.290	0.15	0.17	18588	3929
1986	1611.765	1462.475	3.263	0.20	0.22	20880	10611
1987	1714.600	1518.600	3.159	0.18	0.21	11413	10410
1988	1671.846	1485.415	2.552	0.15	0.17	10139	12680

In Serbia before World War II, only 4 percent of technical projects and an insignificant percentage of biological projects were implemented. The greatest extent of technical works in Serbia was in the period 1967–1977, 32 percent of the total, with the annual average effect of 43,318 m^3. The greatest scope of biological works was in the period 1978–1988, 52.5 percent of the total biological works, with the annual average effect of 5,001 ha.

In the observed period, forestation in Serbia was continuous, with a steady increase until 1968 (1,570 ha), followed by a decrease until 1973 (528 ha), and another increase, with a maximum of 1,706 ha in 1977. Forestation fell to 421 ha in 1981. From that date until the end of the observed period there was an increase in forestation with the highest being in 1986—7,146 ha. Forest reclamation had a continuous upward trend until 1979 (1,000 ha) with the maximum in 1984 (2,992 ha). Since then, the trend has been downward (except in 1991— 905 ha). Pasture and meadow melioration show a similar trend to forest reclamation. The largest were in 1980 (1,586 ha) and 1983 (1,090 ha) followed by a sudden increase in the following years and another maximum in 1988—10,764 ha. It should be noted that, in the period 1978–1988, there were no biological works, and water management organizations used to distribute grass seeds free of charge to landowners, who sowed the grass and reclaimed the land.

As for check dams and sills in the period 1961–1993 in Serbia, there was a constant decline in average annual values from 22,420 m^3 in the period 1961– 1966 to 6,162 m^3 in the period 1978–1988, to the drastic reduction to 1,074 m^3 in the period 1989–1993. The opposite is the case for the building of longitudinal structures, which increased from the average annual value of 23,741 m^3 in the period 1961–1966, to 32,132 m^3 in the period 1978–1988, with a decline to 16.147 m^3/yr^1 in the period 1989–1993. This can be explained by the fact that transverse structures prevailed in Serbia from 1907 to 1961 because they were built for the protection of roads. The building of technical structures prevailed after 1961, when Yugoslavia and Serbia were economically stronger.

Erosion and torrent control works in the region of Serbia, especially during the last 30 years, have had very significant on-site and off-site effects. First of all, they improved the state of ecosystems in erosive regions. Furthermore, through the reduction of erosive sediment yield, ECP significantly reduced the transport of sediment throughout the greater part of the drainage pattern in Serbia. From the water management aspect, this effect is especially important because the hazard of silting of reservoirs is reduced. The effects of ECP on sediment transport reduction in the drainage pattern cannot be determined more exactly because of the lack of systematic measurements of sediment in the major water courses in Serbia. Elementary effects of ECP have not been systematically recorded, i.e., sediment volume captured in the siltations of all the torrential check dams in Serbia. In the absence of reliable and detailed data, only a global assessment of the effects of ECP can be given.

A rough quantitative assessment of ECP effects on the reduction of sediment transport in the drainage pattern in Serbia can be given based on the observed fact that the annual transport in the rivers Juna, Zapadna, and Velika Morava in the period 1974–1994 was two times lower than in the previous period. This reflects the collective effect of ECP and storage reservoirs built in the watershed. Bearing in mind that the storage of the hydroelectric power station "Derdap I" is the recipient of the sediment from the watershed of the V. Morava River, the reduction of average annual sediment transport of about 2.5 x 10⁶ tons means that during the period 1974–1994, the volume of sediment deposited in the storage was 50 million tons less than would have been the case without ECP. Approximately 50 x 10⁶ tons of storage area was preserved, which is significant.

In addition to the storage of the power plant "Derdap," the ECP in Serbia also affected the other existing storages. However, they cannot be quantified because of the lack of measurements. Consequently, it would be desirable to undertake the systematic monitoring of siltation in the mayor storages in Serbia.

The effect of ECP on town and traffic protection is especially important. Many streams were controlled in urban zones. The most efficient examples of traffic protection are the gorges Grdelièka Klisura along Juzna Morava and Ibarska Klisura. In the watersheds and tributaries in the sectors of these gorges, extensive ECP were carried out, resulting in quite a satisfactory degree of traffic safety was achieved on important roads.

The highest annual investments in Serbia were in 1977 (about 0.5 percent of NNI), then in 1978 (0.35 percent of NNI), in 1966 (about 0.54 percent of NNI), and in 1967 (0.62 percent of NNI). This is the order of absolute values of investments in ECP, but it can be seen that it does not coincide with the percentage of these investments in NNI of Serbia. In the latter case, the order would be 1967, followed by 1966, 1977, and finally 1978. In the remaining years, the percentage of ECP values in NNI of Serbia varied mostly by about 0.2 percent. The development of erosion control and ECP in Serbia, as well as the method of their financing, was by all means affected by legislation, because the laws were the basis for the formation of the necessary funds. The analysis of legal regulations and the values of investments shows that from 1965, the funds increased suddenly, resulting from the 1965 Basic Law of Waters in Yugoslavia.

According to the new Law of Waters adopted in 1975 with amendments in 1977, the financing of ECP was through regional and Republican "self-managing communities of interest" for water management. This method of financing ensured much higher funding than previously. In the analyzed period, the investments reach the maximum in 1977, with 7.8 million dinars. Later on, the investments were restricted and reached the lowest level in 1983, with 2.5 million dinars. This affected the extent of ECP, especially the more expensive projects (building of stream structures), and their scope was reduced. From then until the end of the analyzed period, there was a mild increase, amounting to 2.6 million dinars in 1988. The positive solutions of this law did not produce the adequate results, because its provisions were not obeyed in practice.

The marked rise of investments in ECP in Serbia in 1977 and 1978, was influenced by the National Loan for Water Management, which ensured much higher funds for ECP in 1977 and somewhat less in 1978. The loan was launched after the catastrophic floods in 1975 and 1976 in Serbia.

The period 1955–1966 was the "golden period" of the Service for Erosion and Torrent Control. That was the period of active work on erosion and stream control. Regional sections were formed and supplied with machines, equipment, and professional staff. Sections were independent, however they operated under the supervision of the Direction for Erosion and Torrent Control in Serbia. After study tours of numerous erosion control experts to the United States, many useful experiences of the U.S. Soil Conservation Service (SCS) were transferred. Laboratories were established in the sections, and modern erosion control works were applied (contour ditches, bench terraces, strip farming, Algerian banquettes, gabions, etc.). Conservation of farmland was implemented for the first time by forestry and agricultural experts. In order to protect the roads, transverse structures, as well as stream channelization, were put in place (Vucicevic, 1987).

It can be concluded that the erosion control activities in Serbia can be characterized as lacking systematic programming and effort. Attitudes changed from period to period. Only in the interval 1955–1966 did issues of erosion and erosion control have partially adequate treatment.

After 1967, the Service for Erosion Control is governed by water management, so the type and extent of ECP are influenced by water management requirements. The 1965 Basic Law of Waters resulted in an increase in investments in ECP in 1967. The 1977 National Loan produced an increase in ECP. After 1967, investments in ECP declined seriously. Since 1989, little attention has been focused on erosion control.

In Montenegro, in the period after World War II, little ECP was carried out. Only one Section for Erosion and Torrent Control existed and it was canceled in late 1970s. This proves that the Republican legislation and government in Montenegro did not pay attention to this problem.

Present state of soil erosion in Serbia and future prospects

Listed below are levels of erosion derived from the erosion map included in the Water Management Plan, adopted in Serbia in 1996:

Excessive erosion (I category)	2,880.0 km^2	3.27 percent
Severe erosion (II category)	9,138.0 km^2	10.34 percent
Medium erosion (III category)	19,386.0 km^2	21.94 percent
Weak erosion (IV category)	43,915.0 km^2	49.70 percent

| Very weak erosion (V category) | 13,035.0 km² | 14.75 percent |
| Total | 88.361.0 km² | 100.00 percent |

The erosion map was produced using the S. Gavrilovic (1972) method. The mean coefficient of erosion is $Z_{se} = 0.44$ (the value of Z ranges between 0.01 and 1.50), which means that the processes of medium erosion prevail in Serbia.

Considering the distribution of erosion processes, the quantity of sediment can be assessed. The total annual sediment yield in Serbia is about 37×10^6 m³ (in average years), while the specific yield amounts to about 490 m³ $\cdot$ km² $\cdot$ yr¹. Annual transport of erosive sediment in Serbia is 9.4×10^6 m³ (in average years), while the specific transport amounts to about 122 m³ $\cdot$ km² $\cdot$ yr¹.

If it is assumed that 90 percent of the transported sediment is suspended sediment eroded from farmland. Assuming that the thickness of top soil is 0.25 m, it means that each year about 3,400 ha of farmland is irrecoverably lost by water erosion. This is just the first visible part of the damage, and off-site damage caused by sediment must not be neglected.

According to the Water Management Plan, to protect the future and the existing storage reservoirs, altogether 1,700,000 m³ (340,000 m³ for the existing, and 1,350,000 m³ for future storages) of technical projects should be carried out. The total amount of biological works is about 204,000 ha (43,000 ha for the existing, and 161,000 ha for future storages). If the average annual scope of building is about 40,000 m³ $\cdot$ yr¹, it will take about 40 years to accomplish. Biological projects to protect storage facilities will take about 70 years.

Conclusions

The results of the research show that in Yugoslavia relatively good results have been achieved for erosion and stream control. The development of this activity has not been uniform in both republics. The positive results of ECP are primarily reflected in the protection of transportation and towns from floods. Good results were achieved in the reduction of sediment transport; however, some water storage facilities have been destroyed by deposited sediment.

The scope and the type of ECP in Serbia and Yugoslavia is largely dependent on state policy regarding the solution of erosion issues.

The entire period since the beginning of ECP has been characterized by the absence of the necessary understanding by the state and its control, which is reflected in its investments in ECP compared to the annual value of NNI.

Soil and water conservation policy of the state has not been consistent, and ECP was carried out only to solve the acute problems. Sometimes good legal solutions were adopted but were not enforced. Erosion was neglected for a long time in the period before World War II, which was followed by war destruction. The constant lack of finances and their irregular inflow made the situation even worse.

For almost 90 years of ECP in Serbia, erosion control works were not performed on farmland on the slopes, except in the IV stage (1955–1966) when these works were on a small scale. Considering the importance of farmland in food production and almost equally important as a factor affecting the regime of water courses and water quality in them, the absence of soil and water conservation works for sustainable agriculture should not be permitted.

Successful erosion control in Serbia and Yugoslavia is based on specialized enterprises with a sufficient number of professionals, institutions for education, and scientific research, as well as the necessary organizational structure which is distributed so as to cover uniformly the whole territory of Central Serbia. Such a network should be formed in Montenegro. In addition, successful ECP requires adequate state policy expressed by laws, regulations, and financial support.

References

Gavrilovic, S. 1972. Inzenjering o bujicnim tokovima i eroziji (Engineering of Torrents and Erosion), Izgradnja, Beograd.

Izvorni materijali za vodoprivredu. 1989-1993. (Source publications for water), Republicki zavod za statistiku Republike Srbije (Bureau of statistics of the Republic of Serbia), Beograd.

Nacrt vodoprivredne osnove Srbije. 1996. (Draft of Water Management Plan of Serbia 1996, Institute "Jaroclav Cerni" (Institute "Jaroslav Èerni"), Beograd.

Pavic, Lj., 1987. Zakonski propisi u oblasti uredjenja bujica (Legal Regulations of Torrent Control), Zbornik radova sa Prvog Jugoslovenskog savetovanja o eroziji i uredjenja bujica (Proceedings of The First Yugoslav Conference on Erosion and Torrent Control), Lepenski Vir (161-163).

Vucicevic, D. 1987. Organizacija i finansiranje antierozionih radova u SR Srbiji od 1907-1987 godine (Organisation and Financing of Erosion Control Works in SR Serbia 1907-1987), Zbornik radova sa Prvog Jugoslovenskog savetovanja o eroziji i uredjenja bujica (Proceedings of The First Yugoslav Conference on Erosion and Torrent Control), Lepenski Vir (89-98).

Pertovic, D., Petrovic, J.,1959. Uredjenja bujica i borba protiv erozije (Torrent Control and Battle Against Erosion), Monograffija "Posleratni razvitak Sumarstva Srbije" (Monograph "Past War Development of Forestry in Serbia"), SIT Sumarstva i prerade drveta Srbije (The Society of Forestry and Wood Industry Engineers and Technicians of Serbia), Beograd.

Statisticki bilten "Sumarstva" (Statistical Bulletin "Forestry"). 1961-1983. Savezni Zavod za Statistiku (Yugoslav Federal Bureau of Statistics), Beograd.

Statistièki bilten "Vodoprivreda" (Statistical Bulletin "Water Management"). 1984-1988. Savezni Zavod za Statistiku (Yugoslav Federal Statistical Bureau), Beograd.

Razliciti materijali i dokumenti (Miscellaneous publications and documents), Arhiv Republike Srbije (Archive of the Republic of Serbia), Beograd.

28

The Need for Soil and Water Conservation Policies in Central Europe

Jon Hron

In recent years, strict water laws have been adopted in Central Europe to protect water from pollution by wastewater. The reason why soil protection lags behind water protection is that soil is an ecological medium fixed to the surface of the Earth that remains permanently damaged after the cause of contamination or degradation has been removed. It is easier to assess the state of water contamination and provide a remedy than to evaluate the level of soil degradation and contamination and suggest a remedy.

While soil protection presents a number of problems at the scientific level, it is even more difficult to convince politicians that soil must be viewed as an ecological medium. For many years, people in most European countries viewed soil in terms of its agricultural function, while the environmental perspective was confined to the U.S. Concern was expressed in both the U.S. and Western Europe for pollution resulting from the use of sewage sludge for land reclamation. Later, concern was expressed for water contamination by pesticides, nitrogen, and sediments. During the last 15 years, the U.S., the Netherlands, Germany, and Canada have significantly reduced agricultural and urban pollution.

In central European, postcommunist countries, the development of soil and water conservation programs and policies, apart from common features, were often quite different. The postwar trend to increase agricultural land, which was a result of a relative market isolation, continued until the time of the collapse of the system in 1989.

Intensive use of agriculture land was justified by the need for the highest production possible in the moderate climate zone. In the former Czechoslovakia, the Act on the Protection of Agricultural Land (No. 93,1996) emphasized protection of arable land not only against transfer for the purposes of non-agricultural development, but also for different biological uses. Changes in cultures

443

and forestation required administrative action. The transfer to nonbiological use was and still is accompanied by restrictive fees determined by soil quality. Originally, recultivation was required as well on land with low soil quality.

A gradual increase in application rates of industrial fertilizers as well as the use of pesticides were used to achieve the highest yields possible. Fertilization leading to optimal yields in economic terms was slowly adopted. The use and storage of agrochemicals was limited in protective water resources zones but these zones were defined only in general terms. These measures, together with a strict Water Act (No. 138, 1973), failed.

Poor land management in large areas of arable land resulted in increased land erosion and reduced biodiversity. It also contributed to the loading of streams with sediments, nitrates and phosphates, and resulted in excessive draining of land to increase technological accessibility of land tracts. The use of heavy mechanization caused compaction of many soils. The concentration of live-stock production was accompanied by problems of manure management and an uneven flow of carbon into the soil. The extreme use of organic fertilizers (manure, straw) resulted in the reduction of fertility and in the increase of negative effects on the environment.

Plant and animal production mainly in Czechoslovakia, Poland, and the former East Germany were exposed to the influence of polluted air near metallurgical and chemical plants and highways. In some areas, such as the North Bohemian region, emissions from power plants and other industries affected larger areas. Acid emissions caused by burning brown coal rich in sulfur negatively affected forests mainly in mountain regions with conifers.

Before political changes in 1989 it was thought that soils in the black triangle (Czech Republic, East Germany, Poland) were strongly contaminated by emissions. High consumption of agrochemicals, and long-term use of pesticides and phosphates enriched with Cd contributed to the concern about highly contaiminated soils.

While many negative effects of agriculture on the environment can be identified, some positive features can be observed. For example, systematic research on agricultural and forest soils took place in the former Czechoslovakia and was supplemented by the land evaluation and the forest typology evaluation. After political changes, work continued by soil survey and monitoring of soil quality. This research revealed many features about the degradation and contamination of soils. Results proved that the problem of contamination and soil loss is not as critical as had been thought.

After the political changes in 1989, the ownership structure of land was adjusted for agricultural resources, forests, and streams. In the first stage of euphoria, it was expected that the elimination of large land holdings would result in reduced degradation of the environment while maintaining high levels of farm output. This did not take place in Poland where there was a significant transformation to ecological management in small farms. Later on, when fertilization and liming were markedly reduced and yields were significantly reduced, opinions changed. These findings strongly suggest that societies must accept the fact that production will have to be decreased. Apart from the expected remedy of some qualities of degraded soils, some negative consequences can be expected, for example, the acidification and mobilization of hazardous elements. The solution to this problem in large areas is the financing of grazing land and forestation.

At present, scientists are able to solve the issues associated with soil and water conservation with the help of the legal system and specialized institutions. Many countries (including the CR) must address the issue of competencies of individual ministries. It is difficult to enforce conservation of soil, forests, and hydrosphere without a permanent solution to the clash between owner interests, who prefer market production within a short-term horizon, and the public interest to protect the environment. In spite of all the problems, the environmental policy adopted by the Ministry of Environment in 1995 mentions the necessity to complete water conservation laws and to include soil conservation among environmental issues.

After the concept of soil conservation was formuulated in 1985 and after the first conservation act was adopted for Baden-Würtenberg lands (1991; 1992; 1993), a more comprehensive soil law is currently under preparation in Germany. The Soil Conservation Law for the CR, which is now being prepared, will include all soils and will be coordinated with other laws which are indirectly related to soil. The law will protect productive and ecological functions of soil, primarily against actions which lead to degradation and contamination. By means of soil conservation, water resources and the biosphere will be protected. Types of soils, and sources of contamination, are in a stage of elaboration.

Proponents of the law suggest that an effective protection of land resources will be possible if land is divided into several categories reflecting the priorities and specific features of soil conservation.

The main categories of land resources proposed are as follows:
- soils of national parks and protected areas;
- soils of zones for water protection—used for agriculture;
- soils of zones for water protection—forest;
- soils for forest conservation;
- productive agricultural land resources;
- soils for commercial forests;
- soil affected by erosion;

- agricultural land resources with low production;
- contaminated soils;
- forest soils (and successive stages after felling, affected by acid pollutants);
- urban soils, soils of industrial areas, and infrastructures together with those included in land-use plan.

The categories reveal preferences assigned to soil functions. Each type has specific limitations on the use and management of soils. Consequently, different legislative norms, economic measures and administrative measures will be required to protect these resources.

The first four categories are focused on the protection of nature and on water resources. Here, lands have production, are unburdened by land taxes, and are subject to strict management. Lands used for agricultural purposes in zones of water protection will be more accurately specified in a smaller area, which will ensure that conservation will be employed. Productive land will more likely result in profitable production, if land is taxed according to the quality of the site. The agricultural land resource is also protected to ensure food security and preservation of best quality lands for future generations. Unlike previous categories, the two following categories represent a low degree of protection. Lands in the last two categories are damaged by external factors. On the lowest level they cannot be used for production agriculture, on the higher level they require a change to produce trees, and in extreme cases they must be decontaminated or isolated. Attention must be paid to lands that have been contaminated in the past to prevent future degradation of usable resources. The law is designed to limit shift between categories by means of restrictive fees in case of the first six categories. The magnitude of the fees depends on the quality of the subcategory and stage of protection.

Further development of soil conservation policy and programs face many problems such as the following:

- The absence of a commitment to soil and water protection which would be a springboard for the new legislative structure;
- Achieving agreement both on the scientific and political level to advance soil and water conservation. This is prevented by unclear competencies and the fact that the role of land for society, agriculture, forestry, environment, and development of urbanization is undervalued;
- Lack of agreement within the scientific community regarding which activities would be more effective than legal measures to resolve environmental problems.

446

29

Soil and Water Conservation Policies in the Czech Republic

F. Dolezal, M. Janecek, J. Lhotsky, and J. Nemecek

There is not yet a clearly elaborated policy of soil and water conservation in the Czech Republic. It is only recently that the country's political system has changed. The Czech Republic has undergone a transitional period during which the objectives, methods, and tools of conservation policies have been reformulated. Lengthy debates have taken place on these themes at various levels. Virtually everything that the society has inherited from its socialist past has to be rethought and reevaluated.

The Soil Resource (also referred to as "the Soil Fund") of the Czech Republic, expressed in hectares per capita, is given in Table 1. The total area of the country is 7,886,651 ha (Soil 1995). The country is characterized by a large area of arable land (73.75 percent of the total agricultural land, cf. Soil 1995), a considerable area of forests (33.3 percent of the total country's area, cf. Report on Forestry 1995), and a relatively small area of permanent grassland (20.71 percent of the total agricultural land, cf. Soil 1995). The area of agricultural land has been declining for several decades (cf. Table 2). The hilly terrain and the climate, which varies between the oceanic and the continental types, result in quite a high potential risk of soil erosion by water. The potential risk of erosion (not considering the vegetation cover and the erosion control measures) was evaluated according to the modified Universal Soil Loss Equation (Janecek 1995; Tippl 1996). The overall high potential erosion risk in combination with the high proportion of arable land results in high levels of water erosion and sediment runoff rates (judged by Central European criteria); e.g., the actual runoff of sediments from the Czech part of the Labe catchment was evaluated as 902,084 tons in 1987 (Dolezal and Vasatko 1992). Wind erosion occurs only locally in drier parts of the country. The soils of the country cover a whole spectrum of texture types from extremely light sands to extremely heavy clays, but the medium-textured soils prevail, representing 84 percent of the total agricultural land (Soil 1995). A consequence of the intermediate character of the climate is that the majority of soils are mineral while the proportion of peat bogs and other organic soils is quite small. There are virtually no saline and alkaline soils in the country, either naturally occurring or man-made. Morphologically, most of the Czech soils are Cambisols and Luvisols (according to FAO/UNESCO 1988). Virtually all agricultural land is now used by private farmers or other private subjects (shareholders' companies, limited liability companies or cooperatives).

The hydrological balance of the country, determined by the country's orography, is such that the only significant input is atmospheric precipitation. All relevant rivers and streams originate either inside the country or not far beyond its border. Therefore, it is our country which is responsible for quality and quantity of its water resources. The average annual rainfall is 693 mm, which gives about 55×10^9 m^3 of water per year, of which about 15.2×10^9 m^3 is the surface runoff in rivers and 1.4×10^9 m^3 is the groundwater runoff (Moldan et al. 1990). The consumption of water by all sectors is mainly provided by the surface wa-

Table 1. Per capita area of the soil resources of the Czech Republic in 1994

Resources	Area (ha per capita)
Total country's area	0.76
Arable land	0.31
Total agricultural land	0.41
Forest land	0.25
Non-biologically used land	0.10

Source: Statistical Yearbook 1994.

Table 2. The development in time of the per capita area of the agricultural soil resources on the territory of the present Czech Republic

| Year | Area (ha per capita) of | |
	Total agricultural land	Arable land
1845	0.79	0.57
1936	0.47	0.36
1950	0.56	0.43
1960	0.48	0.35
1970	0.45	0.33
1980	0.43	0.32
1990	0.41	0.31

Source: Lhotsky 1994.

ter resources (streams and reservoirs). Only about a quarter of the total consumption (mainly the drinking water supply) is abstracted from groundwater sources (cf. Indicative 1995). Used water and wastewater, treated or untreated, is mainly disposed of by being released into surface streams.

The purpose of this chapter is to provide a reader with a brief overview of the soil and water conservation policies currently applied in the Czech Republic, with a reference to the policies recently abandoned as well as those planned for the future. We found it difficult to draw a sharp division between the soil conservation on one side and the water conservation on the other side. We focused on soil problems and water problems which are connected with the soil.

The pre-revolution state

The "socialist" soil conservation policies consisted of not allowing any other use of agricultural lands except forest (according to the Land Resource Protection Act 75/1976). Whenever this was not possible, a comparable land was provided to the land owner. An unforeseen consequence of this policy was that industrial enterprises, dwelling houses, and roads were being built on good quality land while land parcels returned to agriculture were very often infertile plots.

Large-scale destruction of the land resources due to the surface mining of lignite and other raw materials was in process for several decades. However, devastated areas were systematically reclaimed with enormous expenditures.

Some attention was given to erosion control and protection of soils against chemical pollution. However, such efforts were not successful because of inappropriate use of incentives.

Protection of water sources, both on the surface and underground, was based on the Water Act (no. 138/1973) and accompanying regulations. The legislation was very strict in setting requirements which a user of a water source had to fulfill before he/she was allowed to use it. In practice, however, many water users were exempted from the law requirements. Without such exemptions, the country's industrial production and the development of housing estates would have been heavily retarded. As a result, the quality of water sources deteriorated or stagnated at a very bad level (Moldan 1991).

Irrigation and drainage systems were usually designed and constructed in order to maximize the first order effect; i.e., to irrigate (drain) as much land as possible and supply (remove) as much water as possible. The true needs of agriculture and the environmental aspects were often neglected. No regard was paid to the boundaries between properties of different owners, because the lands had been physically consolidated. Boundaries between different owners' parcels, although formally valid, were no more visible in the field.

Present approaches to soil conservation

The soil, being a valuable means of production as well as an essential part of the ecosystem and a cultural asset of the nation, has to be protected against degradation caused by natural processes and against the activity of man (admitting that some unfavorable natural processes are also generated or exacerbated by the activity of man). The innovative Land Resource Protection Act (no. 334/1992) follows principally the same line as the previous land resource protection acts, i.e., protects the agricultural soils quantitatively against being exempted from agricultural use. The principle of obligatory compensational land reclamation has been abandoned; the process of exempting land parcels from the agricultural land resource is now controlled by a mixture of administrative and financial tools. The quantitative protection of the forest land resource is treated in a similar way by the Act on Forests (no. 289/1995) (cf. The Act on Forests 1996). A proposal has been elaborated for a new Soil Protection Act (Nemecek 1995) of which the principal ideas are discussed below. This proposal gives more emphasis to the qualitative protection of the soil itself and pertains to all kinds of land use.

Today's concept of soil conservation relies very much upon the general trend of European agriculture which the Czech Republic is assumed to imitate, i.e., the one of suppressing the surplus agricultural production. Many believe that an adequate soil conservation will occur automatically if only the agricultural sector is forced, in one or another way, to produce less. In more precise terms, this means that only the most fertile sites shall go on being cultivated intensively as arable lands while the submountainous and mountainous regions, as well as the zones of protection of water sources, shall be turned into grasslands and forests of which the use will only be extensive. Recommendations have been elaborated for subsidizing the transition from arable lands to grasslands based upon the potential risk of water and food chain pollution (e.g., Sklenicka 1994; Kvitek 1996a, 1996b; Nemecek 1995). This idea is positive also in the sense that soil conservation comes here hand-in-hand with water conservation.

The manner in which the land husbandry is to be made less intrusive is a matter of dispute. The easiest way to do it is to abandon land completely. This presently happens on quite a large scale but is not defined as being a recommended procedure. A much better approach would be to convert arable land into grassland or forest which could be managed (e.g., mowed, grazed, or felled) while agricultural inputs (fertilizers etc.) are minimized. The mere abandonment of the land would lead to depopulation and desolation of many regions of the country. The conversion of arable land into grassland was subsidized by the state in the early 1990s but was not motivated by an explicit intent to protect soil and water (Orientation 1992). It led to some increase of the proportion of grasslands in the marginal submountainous areas, while an opposite trend was observed in the regions of intensive agriculture. It is only the maintenance of grasslands and forests which is presently subsidized (Rules 1996). New rules for subsidizing transition from a less conservative biological land use to a more conservative one are being prepared (Kvítek 1996).

Application rates of mineral fertilizers and lime applied on agricultural lands have been reduced substantially during the post-revolution period (cf. Table 3). The decline seems to have stabilized at present levels. Even though the rates are environmentally favorable, they are unsustainably low. Somewhat higher average rates can be expected in the near future, but application rates will probably remain low (except for lime) on marginal and environmentally vulnerable land.

The contamination of soils has been monitored regularly by several institutions (cf. Results of check 1995). Study findings show that only small areas around some industrial point sources (mainly in northwest Bohemia) and alluvial soils along large and medium rivers in the areas downstream of industrial enterprises and large cities are seriously contaminated (cf. Podlesakova et al. 1995a, 1995b).

The protection of soil against compaction caused by agricultural machinery as well as reclamation of lands so afflicted (representing about 40 percent of the total arable land) are not considered significant issues. It is hoped that farmers will become concerned about such issues and that they start to regard themselves as true owners of the land. Less intensive agricultural production could also produce such an outcome.

What motivates farmers to adopt conservation production systems are the consequences that soil erosion and the subsequent accumulation of eroded sediments produce in ditches, on roads, in inhabited areas, and in streams and reser-

Table 3. Average application rates of mineral fertilizers on agricultural soils of the Czech Republic

Year	Avg. annual doses of mineral fertilizers (kg of elemental nutrient per ha of agricultural land per year)			Avg. annual doses of liming materials (ton/ha)
	N	P	K	
1989	99.24	65.56	57.98	0.62
1990	86.82	52.34	47.41	0.61
1991	46.65	10.77	10.76	0.16
1992	46.69	10.76	8.61	0.06
1993	40.00	13.00	10.50	0.05

Source: Green Report 1995.

voirs. In the last few years, a considerable number of catastrophic floods have occurred on small streams all over the country. The reasons for these floods are now being investigated and the measures to be taken in order to avoid similar disasters in future are being proposed. Opinions regarding the cause of this flooding are divided. Some suggest that the floods have been caused by extraordinary rainfall events of which the return period is 1,000 years or more, remarking that the disastrous consequences of such events could not be avoided even if the runoff and erosion control measures were optimal. Others argue that the floods and their devastating effects were primarily caused by inappropriate land use in the catchments involved. Still others argue that bad design of structures (houses, roads, streambed fortifications) did not allow for safe passage of the flood waters.

As the land starts to be recognized as private property of the land owner, it also starts to be increasingly difficult to put soil and water conservation measures in practice. The basic question is: How can farmers be forced to practice conservation measures? The two extreme answers to this question can be formulated as follows:

1. Do nothing, the farmers will help themselves as soon as they realize that the land is really theirs and worth taking care of.
2. It is the state that must control the process by keeping sufficient number of supervisors who distribute the state money in order to support conservation measures and are entitled to withdraw the support or even collect a fine when recommendations are not followed.

A realistic answer has yet to be found and will lie somewhere between the extremes. Most of the present agriculturists became landowners or land tenants quite recently. Therefore, their attachment to the land is not as strong as that of the traditional family farmers. A competent extension service for land and water conservation must be built. It must be able to demonstrate that a viable solution exists to each practical problem. Some of the existing institutions (e.g., the State Bureau of Reclamation) may become the core of such an extension service. It is also the legislation side of the problem which has to be treated anew. Clear rules must be set forth specifying the conditions under which farmers be compensated for losses which they will incur when they are not allowed to practice intensive agriculture or continue to farm on the lands on which the intensive agriculture cannot be permitted. A strict application of the "polluters pay" principle is difficult to implement under these circumstances.

To support decision-making at all levels, soil information systems must be built and refined. Developed countries can contribute to the solution of our problems by forcing information providers to provide relevant information in an understandable way.

Main features of the proposal for a new Soil Protection Act

Three basic types of soil protection were outlined in the proposal for a new Soil Protection Act (Nemecek 1995). They were: 1) protection against degradation and destruction; 2) protection against contamination; and 3) protection against a nonbiological use.

Degradation is understood in its broadest sense, as the process resulting in reduction or loss of some of the soil functions. Destruction means a complete loss of all soil functions. However, the soil shall only be protected by law against those types of degradation that could exert negative influence on adjacent lands or water sources, such as the erosion by water and wind, landslides, and degradation of humus in forest soils which leads to extensive leaching of pollutants into groundwater sources. Protection against other forms of soil degradation (e.g., compaction of subsoil, breakdown of soil structure, loss of humus in agricultural soils, reduction of biological activity of soil), as long as they do not affect the surrounding pedo-, bio- or hydrosphere, shall be left to the land owner's judgment.

The chemical degradation of the soil (cf. Hurni et al. 1996) is treated in the Soil Protection Act proposal under a separate chapter, being referred to as "soil contamination" or (somewhat less explicitly) as a "soil load." The following main categories of soil load are distinguished:

- surplus nitrogen load (that leads to pollution of water sources by nitrates),
- surplus accumulation of pesticides in the soil,
- the load of emission acidifiers, causing mobilization of aluminum (particularly in forest soils),
- the load of persistent pollutants (mainly trace elements and xenobiotic organic compounds), differentiated qualitatively into three stages: a) significant excess of background concentrations of pollutants in the soil; b) significant input of pollutants into food chain and water sources; and c) significant health risk for humans and animals.

The use of the land for structures, pavements, transport lines and junctions, mines, military training sites, waste disposal sites, etc., causes harm by reducing the area on which the soils are capable of fulfilling their productive and ecological functions, reducing the infiltration capacity of the soil on considerable parts of some small catchments, and producing additional point or diffuse sources of pollution afflicting the adjacent land and water sources as well as the atmosphere. Therefore, land used to maintain biological diversity must be protected by law.

In order to make soil protection viable, the Soil Protection Act proposal suggests that the lands be grouped into seven categories:

453

- nature reserves,
- zones of protection of water resources,
- productive soils,
- soils subject to erosion risk,
- soils of low productivity,
- contaminated soils,
- lands used for intensive land use.

The nature reserve areas must be delimited explicitly by law in order to protect their natural ecosystems and other valuable components of the landscape. Land management in these areas should focus on ecosystems.

Water protection zones comprise recharge zones of important groundwater aquifers and protective strips around surface water reservoirs and adjacent streams. Agriculture and forestry in these zones has to be restricted so that dissolved pollutants (like nitrate or pesticides) and eroded soil particles and the pollutants attached to them (like phosphates) cannot reach water sources. In the past, protective strips around surface water sources (reservoirs and streams) were determined mainly on the basis of distance. The resulting zones of restricted agricultural production were unnecessarily large. When agriculture becomes a sphere of private entrepreneurship and is by far less subsidized than in the past, there will be a tendency to minimize the size of the protection zones. This, however, must not be done blindly. Terrain morphology has to be taken into account when delineating the boundaries of the protection zones. The slopes adjacent to the streams and reservoirs must certainly be protected, as well as the alluvia around them. Slopes not adjacent to streams and the flat upland plains can be excluded from protection. The best protection can be achieved by turning the protective zones into extensively used meadows or forests (cf. Kvitek 1996).

The productive soils are the main source of nation's wealth and security in food, animal feed, wood and other renewable raw materials. For this reason, they must be protected against uncontrolled exploitation. It is especially around large cities and in the recreation areas that highly productive agricultural soils are endangered by urbanization. The Soil Protection Act proposal suggests requiring an approval of the soil conservation authority before changing land use.

Soils which are subject to a considerable erosion risk shall be used as grasslands or forests. A matter of dispute is whether this land may be used as construction sites without limitations. Considerable erosion may take place during construction and afterwards.

The category of soils of low productivity has been introduced in order to comprise "the other" soils, i.e., the soils which cannot be classified under any other category. No special protection of these lands is envisioned. However, one should not forget that these soils present most environmental problems on the world scale (Hurni et al. 1996).

The lands contaminated by persistent pollutants must be registered and treated in a special way, according to their degree of contamination. Namely, the three subcategories must be treated as follows:

- inputs of chemicals (fertilizers, conditioners, pesticides, etc.) shall be restricted, and the certified clean food production shall not be allowed on the soils in which the background concentration of one or several pollutants has been exceeded,
- as soon as a significant input of pollutants into the food chain or water sources has been found, all necessary steps shall be taken in order to interrupt the contamination chain (for example, no edible crops shall be grown on the land or the water source involved shall not be used for drinking water supply),
- soil which might cause significant health risk for humans and animals shall be inspected in detail and, if necessary, isolated from their environment, removed, or remediated.

The lands which are already used for intensive use purposes or have been approved for such use shall be inspected from the following viewpoints:

- "old loads" (zones already contaminated shall be identified and made harmless),
- gardens and greenhouses in or near the existing urbanized or industrial areas shall be tested to determine if they produce contaminated food,
- impervious pavements and other impervious land surfaces shall be minimized in order to allow precipitation water to infiltrate and thereby reduce the risk of floods,
- landfills and other waste disposal sites shall be safely designed and controlled.

An approximate extent of the first five categories of lands as estimated by Novák (1996) is given in Table 4.

Present approaches to water conservation

New water legislation is being formulated because the old Water Act (no. 138/1973) is no longer relevant to changed circumstances. The situation is complicated by the fact that water-related affairs are presently administered by three different ministries (agriculture, environment, and economy). The distribution of responsibilities among the ministries is also being reconsidered.

Much of the present discussion is focused on water ownership. While there is a broad agreement that water running in the streams is a public property, streambeds, banks, and the accompanying plant areas are regarded by many as objects of private ownership. This may lead to conflicts between private and public interests. Conflicts also arise at the boundaries between water reservoirs (publicly or privately owned) and the adjacent private land parcels where the owners have to put up with the reservoir users' stepping onto their land in order to maintain the reservoir borders. Adequate protection of water against pollution produced by the adjacent land, as well as that of the land against being polluted or damaged by the adjacent water, may be impossible if ownership rights are not adequately resolved.

Streams of all types are generally recognized as being essential elements of the landscape. Several broadly based programs of landscape stabilization have been launched in recent years. These are as follows:

1. The program of the territorial systems of ecological stability, leading to the design of additional forest, grassland, and other green areas in the landscape, sufficient to support the required diversity of flora and fauna. Windbreaks and the greens on stream banks are indispensable parts of these systems.

2. Complex land consolidation and land use planning projects are not only to make the agricultural production more efficient but also to clear up the ownership rights and to improve the existing land use patterns. These schemes include water and wind erosion control measures as well as protection of surface water and groundwater sources. Handbooks have been published which specify in detail how the complex schemes should be made (e.g., Dumbrovsky 1995). Some projects of this type have been designed and implemented. In most cases, however, the land and water ownership situation is not mature enough for implementing complex schemes. Instead, simple and temporary land consolidation projects are currently being accomplished all over the country in order to compromise the land ownership with the land usership.

3. Revitalization of river systems (Program of revitalization 1995), in which the streams themselves and the living organisms in them and around them is a major issue. The shape of the stream route, the streambed cross-section, depth and speed of water in the stream, fortification material and accompanying vegetation have to be redesigned in order to make the stream and the surrounding banks more natural and biologically diversified. This program is being gradually implemented and has a chance to be successful. The most serious obstacle to its progress is inadequate funding, as the financial resources are presently insufficient even for the current maintenance of small streams and drainage canals (Runstuk and Stransky 1995).

Water sources of all kinds, in particular water supply streams and reservoirs, have enjoyed special protection according to the Water Act (no. 138/1973) and the accompanying regulations. New regulations for the protection of water

456

sources are still under preparation (e.g., Kvitek 1996).

Enterprises owned by the state but formally private are responsible for managing large rivers (Labe, Vltava, Ohre, Morava and Odra) and the quality of water in them. Because of poor water quality in the largest Czech river, the Labe (called "Elbe" in Germany), an extensive international effort was focused on determining the actual state, identifying the sources of water pollution, and designing and implementing measures to improve the situation (The Labe Project 1991). Several large and many small wastewater treatment plants were built after the revolution of 1989. These facilities, together with the decline of some branches of industry, have already resulted in improvement of water quality in the Labe and some other rivers. The irrigation of lands in drier parts of the country is still proceeding but its extent, presently about 50 percent of the pre-revolution level (Runstuk and Stransky 1995), may decline considerably as soon as the state ceases to subsidize it.

Irrigation systems capable of surviving are being privatized (Slavik 1996). Water for irrigation is primarily drawn from large rivers. As the water quality in these rivers (particularly in the Labe) decreases, concern arises about whether or not the irrigation systems may go on using this water supply. Investigation is being performed (e.g., Zavadil 1994) into the effect polluted water can have upon the content of pollutants in plants and animals (both domestic and wild). The drainage systems are regarded by many people as being harmful. Many people feel that irrigation should be discouraged. Such systems are an essential part of the Czech landscape. Even in the cases when they speed up surface runoff of water, much skill and care will be needed to "switch them off" harmlessly (in order to produce artificial wetlands or to recreate the natural ones) or to convert them into double-purpose structures performing both drainage and water retention (Soukup 1996).

Forests cover 33.5 percent of the Czech Republic (Basic principles 1994) and represent an essential component of the Czech landscape and perform im-

Table 4. Approximate extent of categories of soils according to the Soil Protection Act proposal

Category	Area (ha) of the total agricultural land
1. Nature reserves	522,000
2. Zones of protection of water resources	750,000
3. Productive soils	1,950,000
4. Soils subject to erosion risk	450,000
5. Soils of low productivity	1,650,000

Source: Novak 1996.

portant conservation roles. Since 1960, Czech forests have been categorized and selectively administered according to their prevailing functions. One can distinguish, among other categories:

- Forests important for protection of water sources (around the surface water reservoirs and the recharge zones over important aquifers and springs).
- Forests in areas declared as "the protected regions of natural water accumulation" mainly in mountainous areas, characterized by high precipitation and low evaporation. Most of our streams begin in these regions. If these areas were not forested, the risk of floods and water erosion would be very high.
- Forests surrounding spas, urbanized, and industrial agglomerations in which recreational functions prevail.
- Forest in the officially declared national parks and protected natural reserves.
- Wood production forests.

Forest areas by different categories are given in Table 5. Since categories overlap, the total percentage of the forests representing other production functions is 57 percent. A more detailed categorization was introduced by the Act on Forests no. 289/1995.

All the above-mentioned functions of forests were impaired by atmospheric emissions from mining, power production, industry, and traffic. The proportion of forests damaged by emissions was 5 percent in 1970 but as much as 57 percent in 1989 (Basic principles 1994). While the production and recreational functions of forests were damaged heavily by emissions, soil and water conservation functions continue to be fulfilled more or less successfully by successive stands of low quality trees, bushes, and herbs. Reforestation of the damaged areas is difficult because most of the unfavorable environmental factors persist (even though the emission load itself is declining).

Systematic monitoring of water quality in small streams running through agricultural areas (Slesinger et al. 1995; Svoboda 1996) seems to suggest that

Table 5. Areal percentage of different categories of forests in the Czech Republic

Prevailing function	% of the total afforested area
Protection of water sources	12
Conservation in the regions of natural water accumulation	16
Recreation	23
National parks and reserves	25
Wood production	43

Source: Basic principles 1994.

the municipal wastewater from villages is a more serious source of water pollution than the agricultural production itself. Fortunately, increasing interest of local communities in building wastewater treatment plants is emerging (Kockova et al. 1994). As most of the small streams either begin in the forested areas or pass through them, a coordination is needed between forestry and agriculture in most small catchments (Pasak 1996).

Conclusions

The situation in the domain of soil and water conservation in the Czech Republic as analyzed above suggests that a combination of free-market mechanisms, competent state policies, and skilled conservation programs may make sustainable land and water management in the Czech Republic. Problems should continue to be discussed openly and matter of factly.

References

References to the acts and regulations published in the Collection of Laws of the Czech (or Czechoslovak) Republic are only made in the text using the serial number and year. They are not repeated in this list of references except when they are parts of other publications.

The Act on Forests and accompanying regulations. In Czech: In: Practical Handbook, No. 12/1996, Ministry of Agriculture - Agrospoj, Prague, 130 pp.

Basic principles of state policy in forestry. In Czech. Ministry of Agriculture - Agrospoj, Prague. 1994, 41 pp. plus appendices.

Dolezal, L., and J. Vasatko. 1992: The regime of sediments in selected rivers of Bohemia. In Czech. Prace a studie CHMU 21: 52-64.

Dumbrovsky, M. et al. 1995: Specificity of the complex land use planning projects in the zones of hygienic protection of surface water sources. In Czech. Guide (Metodika) no. 17/1995. Research Institute for Soil and Water Conservation, Prague, 64 pp.

FAO/UNESCO. 1988: Soil map of the world. Revised legend. World Soil Resources Report 60. FAO, Roma, 119 pp.

Green Report. 1995: A report on the state of Czech agriculture - 1995. In Czech. Ministry of Agriculture - Agrospoj, Prague, 264 pp.

Hurni, H. et al. 1996: Precious Earth: From soil and water conservation to sustainable land management. International Soil Conservation Organization (ISCO), and Centre for Development and Environment (CDE), Berne, Switzerland, p. 89.

Indicative Water Management Plan of the Czech Republic. 1995: In Czech. Water Resources Bulletin 1994. Publication SVP no. 42. Ministry of Environment - T. G. M. Research Institute of Water Resources, Prague, 176 pp.

Janecek, M. 1995: The potential risk of water and wind erosion on the soils of the Czech Republic. Scientia Agriculturae Bohemica 26(2): 105-118.

Kockova, E., P. Kriz, V. Legat, J. Salek, and Z. Zakova. 1994: The vegetation root-zone wastewater treatment plants. In Czech. Ministry of Agriculture, Prague and Brno, 70 pp.

Kvitek, T. 1996: A draft solution of surface and undergroundwater protection in the Czech Republic by delimitation of vulnerability zones. In Czech. Rostlinna vyroba 42(8): 349-356.

The Labe Project. 1991: Intentions, targets and first results. In Czech. T.G.M. Research Institute of Water Resources, Prague, 100 pp.

Moldan, B. et al. 1990: Environment of the Czech Republic. In Czech. Academia, Prague, 284 p.

Nemecek, J. 1995: Principles of the Soil Protection Act. In Czech. Manuscript, Prague, September 1996, 18 pp.

Orientation, principles and rules of the policy of subsidies in the agro-food complex, forestry and water management of the Czech Republic for the year 1992. In Czech. Agrospoj, Vol. 3, Special issue, February 1992.

Pasak, V. 1996, private communication.

Podlesakova, E., and J. Nemecek. 1995: Retrospective monitoring and inventory of soil contamination in relation to systematic monitoring. Environmental Monitoring and Assessment 34: 121-125.

Podlesakova, E., J. Nemecek, and G. Halova. 1995a: Nonpoint contamination and pollution of soils by persistent hazardous substances in reference to water pollution. In: Janal, R. (ed.): Topical questions of contemporary agriculture and silviculture (from the ecological viewpoint). Proceedings of International Conference, UZPI, Prague, pp. 63-68.

Podlesakova, E., J. Nemecek, and R. Vacha. 1995b: Contamination and pollution of soils in the Czech Republic. In: Proceedings, Third International Conference on the Biogeochemistry of Trace Elements, Paris, May 1995.

Program of revitalization of river systems. 1995: In Czech. Ministry of Environment - Enigma, Prague.

Report on Forestry of the Czech Republic. 1995: In Czech. Compiled by the Forest Management Institute, Brandys nad Labem. Ministry of Agriculture, Prague, 174 pp.

Results of check and monitoring of hazardous substances. Situation in 1994. In Czech. A yearbook of Ministry of Agriculture. Vol. II (1995), Prague, 83 pp.

The rules for granting the investment and non-investment subsidies by the Ministry of Agriculture for the year 1996. In Czech. Agrospoj, Supplement, February 1996.

Runstuk, K., and V. Stransky. 1995: An interview of the month. In Czech. Vodni hospodarstvi 45(1): 2-4.

Soil. A report about situation and perspectives. In Czech. Ministry of Agriculture - Research Institute of Agricultural Economy, Prague, July 1995, 54 pp.

Sklenicka, P. 1994: Zoning of the territory of the Czech Republic for the purposes of dumping the agriculture. Scientia Agriculturae Bohemica 25(2): 85-94.

Slavik, L. 1995, personal communication.

Slesinger, J., J. Pokorny, J. Novakova, and J. Kotrnec. 1995: Monitoring surface water [quality] and investigating the transport of pollutants in small catchments. In Czech. Ministry of Agriculture, Prague, 60 pp.

Soukup, M. 1996, personal communication.

Statistical Yearbook of the Czech Republic for the year 1994. In Czech. Czech Statistical Office, Prague.

Svoboda, Z. 1996, personal communication.

Tippl, M. 1996, personal communication.

Zavadil, J. 1994: Contamination of soil, crops, and groundwater by hazardous substances from irrigation water. In Czech. Partial Research Report (Program "Healthy Nutrition"). Research Institute for Soil and Water Conservation, Prague, 32 pp. plus appendices.

30

Some Possibilities of Impact Modeling
of Soil and Water Conservation Policies
Upon Economic Development
in the Czech Republic

Jiri Tvrdon

Agriculture is one of the oldest human activities, which historically has gone beyond its most immediate mission—survival. It has molded human thinking, been a bearer of culture, and helped to create the environment in which people live and cultivate the land. In the Czech Republic, however, this natural development was interrupted by the planned economy for a period of 40 years during which valid relationships between human work and its fruits were torn apart.

Absence of a correlation between work and output became a crucial factor of low efficiency of production processes and a lack of concern for environmental problems arising in this time (Tvrdon and Havlicek 1995). It can be shown by low labor productivity that during the planned economy the number of people occupied in agriculture was approximately twofold relative to the same output. Regarding care of the environment, it can be documented, for example, by a loss of agricultural land during the planned economy. In the 1980s, 133,000 ha of agricultural land was diverted for other uses, of which 41,000 ha was arable land. This figure was even larger during the 1950s and 1960s. During the 1980s, 4,133 ha were diverted for industrial development, 11,933 ha for civic and apartment construction, 5,457 ha for agricultural construction, 5,692 ha for dam and water work construction, 8,387 ha for remaining construction, 10,306 ha for coal mining, 3,402 ha for remaining output, 11,753 ha were forested, and approximately half of the total amount of hectares (54.2 percent) was not statistically specified. Approximately 26.7 percent of the diverted land was for construction, 10.3 percent for production, and 8.8 percent for forests (Lux 1994). The biggest portion represents land losses due to direct violation of land and water conservation laws and rules valid even in the planned economy. The other damages were caused by leaking pipelines transporting drinking water. It is estimated that 30 percent of the total drinking water supply was lost through pipeline leakage.

Characteristics of present land and water management in the Czech Republic

- Land use (as well as water) is primarily dominated by agricultural and forest activities.
- Forest and agricultural land cover about 90 percent of the country, of which agriculture land represents approximately 54 percent, that is, 0.41 ha of agricultural land per capita of this 0.31 ha of arable land.
- The country is composed of 14 natural zones, which can be divided into four main groups with the following characteristics:
  1. The lowland warm natural zone, which includes about 5 percent of the agricultural land available, is characterized by average annual temperatures 9–10°C and average annual rainfall of 500–550 mm. The height above sea level is 150–250 m. This is a warm region, medium to very dry.
  2. The lowland natural zone, which includes approximately 35 percent of the agricultural land available. This area of the Czech Republic covers a region that is mildly dry, relatively flat and low-lying, 200–300 m above the sea level and occasionally higher.
  3. The lowland hilly natural zone. This is the most extensive zone, which includes 44 percent of the farming land available. The region lies mostly at 400–600 m above the sea level in a mildly warm to mildly cool climatic region, mostly with a good supply of moisture.
  4. The highland natural zone includes about 16 percent of the agricultural farming land available and is from the climatic point of view a less favorable territory for farming activity. It is a territory with very diverse horizontal and vertical relief, mostly at heights more than 650 m above the sea level, in a region with a humid climate and many slopes, even at heights below 550 m.
- From the point of view of the natural fertility of the soil in the Czech Republic, only 26 percent of the agricultural land available has very good productive properties, 42 percent has good properties, 32 percent has preconditions for low production, and 3 percent of the soil is not very suitable for farming purposes.

The inappropriate merging of land into large landholdings at the time of agricultural collectivization led to a tenfold increase in intensity of erosion processes. Potential erosion threatens 56 percent of arable land. A significant limiting factor, especially in crop production, is the burden of emission, at present affecting 793,000 hectares of farm land. This is not only the effect of local sources of pollution but also the significant impact of long distance transfer of pollutants mainly from Germany and Poland. Pollution flows in the opposite direction. Roughly 35,000 hectares of land in the Czech Republic are affected by mining activity, and its gradual recultivation and incorporation into the landscape is one of the major environmental issues, especially in the basins of the North Bohemian and West Bohemian regions. All of these environmental issues affect the development of and intensity of farming activity. In this region it will be necessary to adopt specific production systems to protect water resources. A number of regions have other functions which restrict farming activity and these often overlap. Legislation expresses the interest of the state, especially in regions significant for environmental protection. There are 1,170 reserves of small areas. Their overall acreage is rather small (0.7 percent of the area of the Czech Republic) but significant for the preservation of a number of rare and endangered species. The preservation of the variety of species, however, depends on the ability of human-exploited land to provide suitable conditions for the life of wild animals and plants. It seems that only 35 percent of plant ecosystems are not endangered. Increased protection of species is practiced in protected reserves and national parks. There are 22 such areas—three of which are national parks: Giant Mountains National Park, Bohemian Forest National Park, and Dyje Basin National Park. Large reserves encompass approximately 14 percent of the area of the Czech Republic. The intention behind the establishment of national parks is not only the strict protection of the natural environment but also the preservation of the typical features of the landscape created over many years. For this reason, it is not possible to wholly exclude farming and forestry on the respective territory but to adapt it to the need to preserve these valuable regions.

Forest activities follow the same pattern in the Czech Republic. The Czech Republic is one of the most densely forested countires in industrial Europe. The forest covers the needs of the national economy in timber and at the same time fulfills other useful purposes, such as protection of soil from erosion, and recreation.

The positive consequence of the long purposeful care of forests in the past is that 700,000 hectares of forest presently exist, which is more than at the beginning of the last century. Wooded areas have increased by 10 percent. The forests in the Czech Republic cover one-third of the country, which is more than 2.6 million hectares of land. Approximately half of the forests remain the property of the state.

The total stock of timber is approximately 564 million m³; of this 486 million m³ are coniferous and 78 million m³ are deciduous. The stock in nature stands is approximately 190 million m³. The annual volume increment in our forests is 17 million m³ of timber annually.

The data on timber exploitation provide an overview of the production capacity of the forests. Annually, 12–13 million m³ of timber are felled in the forests of the Czech Republic. Of this amount 9.5 million m³ of timber (8.55 million m³ coniferous, 1 million m³ deciduous) are from final cutting; the rest is thinning. Annually, almost 28,000 hectares of forest are reforested. According to data on timber stock (216 m³/ha) and annual felling (approx. 4.8 m³/ha of forest and more than 1 m³ per capita), the Czech Republic ranks high among countries with advanced forestry. The share of forestry in the creation of the national income represents only one percent; nevertheless, its significance in the life of the nation is much higher. Among others, forests in the Czech Republic create a suitable base for wildlife.

Annually in the Czech Republic, 166,000 head of large game are hunted; of this number 80,000 are roe, 23,000 deer, and 48,000 wild boar. Small game are hunted, too. The most popular are pheasant (600,000), wild duck (250,000), and hare (130,000).

Human influence on forests has not always brought the most beneficial results. Earlier efforts to achieve the highest timber yield have influenced the generic structure of forest growths in favor of coniferous woods. The unfavorable generic structure of forests, climatic influences, and pollution are the main causes of the lesser resistance of our forests to vermin, fungus diseases, and other damaging factors.

A consequence of the poor forest management was random lumbering (i.e., unsystematic lumbering of wind-fallen trees, withered, dead, and dying trees), which in the period after 1980, amounted to more than 50 percent of total lumbering.

The other important factor influencing the quality of the environment in the Czech Republic is water management. Water management is essential in the Czech Republic due to relatively poor water sources in Central Europe. The Czech Republic is crossed by the upper courses of rivers with small volumes of water and unstable regimes which are polluted by wastewater. Sources of groundwater, too, are limited and unevenly distributed.

The development of society in the present period places great demands on water supplies, and the situation arises in which certain regions are affected by the scarcity of water. This unfavorable trend of development in the national economy has not yet been fully addressed.

The main source of water in the Czech Republic is surface water. The total length of the approximately 3,600 economically significant streams is 18,000 km, and other small streams represent more than 55,000 km.

There are more than 120 dams in existence, with the total volume of over 3.1 billion m³ of water in reservoirs. Reservoir water balances the unsteady flow of the rivers and allows their use for the needs of the population. In addition to these sources of water, there are about 24,000 small reservoirs and ponds in operation, with a total area of over 510 km², which have a significant impact on meeting the needs of farming and fishing. More than 2.6 billion m³ of water are taken from the surface sources annually. Of this amount 750 million m³ are used for public water use and more than 70 million m³ are used in agriculture and forestry.

Good quality groundwater is approximately 45 m³/s. The total annual use is 750 million m³, 550 million m³ is used for public consumption and more than 65 million m³ is used by agriculture and forestry.

In the Czech Republic, 8.5 million people, or more than 81 percent of the population, are now supplied with safe water from the public water systems. The remaining population uses groundwater from public and private wells.

There are more than 3,500 public water systems in operation of a total length of 46,000 kilometers. In recent times, regional water systems are becoming more popular due to a more rational use of interconnected water sources. At present there are 500 such systems in operation with 30,000 km of water mains pipelines. The important ones are the Prague and Central Bohemian systems, with the main sources in Kadan, Podolí, and the Zelivka reservoirs, North Bohemian (Krimov, Fláje, Jirkov and Prísecnice reservoirs), North Moravian (Morávka, Zermanice and Sance reservoirs), South Bohemian (Rímov reservoir), and others.

As far as sewage collection and water treatment are concerned, most of this type of wastewater is collected from towns, villages, and industrial and agricultural plants through public sewage networks, and treated in wastewater treatment plants.

There are 215,000 kilometers of sewage pipelines in operation. Seventy-two percent of the population of the Czech Republic are connected to this network. There are 2,620 sewage treatment plants in operation which treat 92 percent of the discharged sewage, amounting to 900 million m³ of communal and 600 million m³ of industrial and agricultural wastewater annually. Nevertheless, the remaining wastewater discharged untreated into the rivers causes considerable pollution and endangers the health of Czech people. A special problem is the pollution of water, due to incorrectly applied intensive farming methods. This is the reason that one of the main objectives of water management in the Czech Republic is the acceleration of the construction of treatment plants and the promotion of environmentally friendly production processes in industry and agriculture. It is hoped that these goals can be achieved in the framework of the national environmental program to restrict further pollution of surface and groundwater and achieve a fundamental improvement in water quality.

The Czech government, being aware of the environmental problems inherited from the previous economic regime, accepted in 1992 the "Basic principles of the agricultural policy of the government of the CR up to 1995 and for a further period" (Lux 1994). According to this document, the Ministry of Agriculture and the Ministry of Environment, in view of the role of agriculture and forestry in molding the countryside and their close relationship to the ecology of the countryside, initiated ecological programs supporting the rational development of the devastated regions of the country. Special attention was focused on the border areas. The government created and announced the principles of its agricultural policy, including its relationship to regional problems, and the infrastructure and problems of country settlements. The government also modified its regulatory instruments. Government fulfills this task mainly by supporting ecological investment in the economy, as noted in Table 1.

The data in Table 1 show that ecological investment at the present time is approximately ten times higher than in 1989. According to some authors (Vith 1995), these investments are even higher than what would be expected given the state of the Czech economy. Others provide evidence that the magnitude of investment is appropriate (Sejak 1995).

Lower industry emissions have a direct impact upon agricultural vegetation and water quality. During the 1990s the amount of heavy metals in the food chain decreased significantly. This process began with less contaminated feed.

Since 1990, the situation has improved. Similarly, the quality of water is improving. The system for monitoring the quality of drinking water includes both the inspection carried out by the Hygiene Service at the outlet from the waterworks and in the water distribution network, and the inspection provided by the supplier of drinking water. These investigations in 1994 showed that

Table 1. Ecological investment

Year	GDP in current prices	Ecological investment (EI)	Share of EI from GDP
1989	524.5	3.602	0.69
1990	567.3	6.048	1.07
1991	716.6	9.376	1.37
1992	803.3	16.954	2.11
1993	911.0	19.890	2.18
1994	1037.0	28.272	2.73

Source: Czech Statistical Office.

approximately 5–6 percent of the drinking water supplied does not meet all the requirements of the Drinking Water Regulation. Another problem is water from household wells. Quality of water from household water wells frequently does not meet the required standards (particularly microbiological standards).

The above mentioned results were achieved partially by increased use of environmentally friendly farming systems, which is greatly supported by the Ministry of Agriculture. Table 2 gives information on the most important part of the subsidy program oriented on environmental improvement.

Evaluation of soil and water conservation policy in the Czech Republic

There is always discussion whether an economic policy is efficient, what are its advantages and weaknesses. To answer this question satisfactorily is difficult because of the multiple outcomes produced by a specific policy.

For economic policy evaluation, there are several methods, of which the most promising is the application of econometric modeling. If this approach is suitable for evaluation of economic policy as a whole, it can be applied for evaluation of soil and water conservation policy. This assumption is realistic, based on experience with the application of econometric modeling to the solution of problems such as health care policy, transport policy, energy policy, etc. (Intriligator 1978). The starting point is to formulate the structural form

Table 2. Subsidies directly provided by the Ministry of Agriculture of the Czech Republic

Direct nonreturnable subsidies for agricultural production	
Purpose	Amount
Seeding grass	6000 CZK/ha
Maintenance of grass	3000 CZK/ha
Afforestation	4.86–7.26 CZK/bare-root plants
	12.84–16.64 CZK/container-grown plants
Renewal of vineyard	Up to 250,000 CZK/ha
Renewal of hop-garden	Up to 100,000 CZK/ha
Renewal of orchard	Up to 100,000 CZK/ha
Non-dairy cow covered by bull of beef cattle breed	4,000 CZK/ha
Calf of beef cattle breed	3,000 CZK/ha
Bee colony	2,000 CZK/ha
Purchase of genetic material	50 percent of purchase costs
Embryo transfer	Up to 1,000 CZK/pregnant cow

of the econometric model in which predeterminated variables are divided into three groups:

$$By_t + \Gamma_1 y_{t-1} + \Gamma_2 z_t + Ar_{t-1} = u_t \tag{1}$$

y_t = vector of g endogenous variables of dimension (g x 1)
y_{t-1} = vector of g lagged endogenous variables (g x 1)
z_t = vector of m exogenous variables (m x 1)
r_{t-1} = vector of l exogenous variables that are subject to control of policy-maker
　　　　(l x 1). Policy variables in vector r_{t-1} are also called instruments.
A ....matrix of structural parameters of instrument variables (g x l)
B ....matrix of structural parameters of endogenous variables (g x g)
G_1....matrix of structural parameters of lagged endogenous variables (g x g)
G_2.....matrix of structural parameters of exogenous variables (g x m)
u_t....vector of stochastic variables (g x l)

The current endogenous variables in y_t are also called targets, for which the policy-makers have certain goals.

Structural form determines their values if B is not singular from the reduced form of model:

$$y_t = -B^{-1}\Gamma_1 y_{t-1} - B^{-1}\Gamma_2 z_t - B^{-1}Ar_{t-1} + B^{-1}u_t \tag{2}$$

Endogenous variables in reduced form are functions of lagged variables, exogenous variables, instrument or policy variables, and stochastic variables.

If estimated matrixes in equation 2 are B, Γ_1, Γ_2, and A and policy variables have forward effect on endogenous variables, Equation 2 can be rewritten:

$$y_{T+1} = -B^{-1}\Gamma_1 y_T - B^{-1}\Gamma_2 z_{T+1} - B^{-1}Ar_T + B^{-1}u_{T+1} \tag{3}$$

Equation 3 is a starting point for various methods used for policy evaluation. With limited space, only the instrument, the target, and the simulation methods will be described.

The instrument–target method

The method originally developed by Tinbergen (1956), is based on two assumptions described in (Tvrdon and Petrova 1996). The first is that there exists a certain desired level of each of the endogenous variables of Equation 3 given as

$$y^0{}_{T+1} \qquad (4)$$

Values in relation 4 express the fixed targets of policy. In the case of the soil water policy evaluation, it can be desirable characteristics of environment—limits of foreign substances in soil and water, share of arable land from agricultural land, forest or park areas, maximum limit of agricultural land decrease p.a., etc.

The second condition is a sufficient number of policy variables–instruments—exceeding or equating to the number of endogenous variables.

$$l \geq g \qquad (5)$$

The difference l–g is called the policy degrees of freedom. If we consider the first version with zero degree of freedom then $l = g$ and A is a square matrix. Multiplying equation 3 by A^{-1} enables us to find a solution for optimal values of instruments variables:

$$r^*{}_T = -A^{-1}By^0{}_{T+1} - A^{-1}\Gamma_1 y_T - A^{-1}\Gamma_2 z_{T+1} + A^{-1}u_{T+1} \qquad (6)$$

In the case of the model for soil and water conservation policy as instruments, variables can be chosen that are used as tools of environment regulation—for example, subsidies for ecological farming, penalties for environment violation, permission for restricted pollution, credits and interest for ecological investment, fines for conversion of agricultural land into non-agricultural use, etc.

Equation 6 determines the basic relationship between policies–instruments and policy goals.

When $l > g$, values of number of instruments variables equal to the difference l–g can be set up a priority at a specified level and optimal values of remaining instruments can be solved according to Equation 6.

Simulation method

Generally, simulation using an econometric model means determining system behavior by means of computing values of variables from a model of this system (11). The system's behavior expressed quantitatively is then specified—simulated under different assumptions for analyzing its response to a variety of inputs. Each simulation is an experiment where the values of endogenous variables differ according to the values of instruments or policy variables, other exogenous variables, and stochastic variables.

Simulation can be realized in several forms. The historical simulation calculates values of endogenous variables observed during data collection using real values of exogenous variables including policy–instruments variables. Historical simulation can be called forecast ex-post. The simulated values can be compared with their real counterparts in order to test forecasting ability and properties of the model. There is a set of testing procedures based upon calculation of so-called normative disturbances taking into account also the different variability of endogenous variables.

A second type of simulation is projection calculations for deriving values from collected data when extrapolated values of exogenous variables are used. This is forecasting ex-ante.

A third type of simulation is policy simulation specifying values of endogenous variables for different sets of values of policy variables corresponding to the considered alternative policies.

For purpose of policy simulation it is most suitable to use the reduced form of econometric model:

$$y_{T+1} = -B^{-1}Ar_T - B^{-1}\Gamma_1 y_T - B^{-1}\Gamma_2 z_{T+1} + B^{-1}u_{T+1} \qquad (7)$$

Matrixes and vector are at dimensions defined for Equation 1.

Policy-makers have access to different variants, the most convenient of which can be selected when taking into consideration the other circumstances.

Findings of simulations are expressed in Table 3.

The simulation method enables a test to determine the impact of different soil and water policies that cannot be tried in reality. Simulation substitutes for experimentation in economic reality. It is a lab for policy-makers. In the case of the Czech Republic, there is only one serious problem—data scarcity due to the short history of the market economy. Nevertheless, simple models of market

Table 3. Simulated impacts of different policies

Option	Alternative values of policy variables	Corresponding values of endogenous variables
1	r^1_T	y^1_{T+1}
2	r^2_T	y^2_{T+1}
.	.	.
.	.	.
n	r^n_T	y^n_{T+1}

behavior specified for the Czech economy provide very promising results, and within a short time it will be possible to construct environmental models using a philosophy originally developed only for economic phenomena. This tool can be valuable in conditions where there is a longer history of environment of conservation.

References

Intriligator, M.D. 1978. Econometric Models, Techniques and Applications, Prentice-Hall, Englewood Cliffs, NJ, pp. 538-561.

Lux, J. 1994. Basic principles of the agricultural policy of the government of the Czech Republic up to 1995 and for a further period. Ministry of Agriculture, Agrospoj, p. 5.

Sejak, J. 1995. Our environmental expenses, Economy Journal, October 3, 1995, Prague, p. 15.

Sine. 1994. Czech Agriculture 1992-1994, Ministry of Agriculture, Agrospoj 1992-94, pp. 6-7, 10, 38-39.

Sine. 1994. Results of foreign substances control and their monitoring, Statistical Yearbook of Ministry of Agriculture, Prague, p. 15.

Tinbergen, J. 1956. Economy Policy: Principles and Design, North-Holland Publishing, Amsterdam.

Tvrdon, J., and J. Havlicek. 1995. Adapting conservation on the farm: An international perspective on the socioeconomics of soil and water conservation. Ted L. Napier, Silvana M. Camboni, and Samir El-Swaify (eds.), Ankeny, IA, SWCS Press, pp. 323-329.

Tvrdon, J., and J. Peterova. 1996. Some possibilities of market modeling, Agricultural Economics, Czech Academy of Agricultural Science, Volume 42, September 1996, pp. 415-421.

Vith, J. 1995. Environmental expenses are above level, Economy Journal, August 19, 1995, Prague, p. 14.

31

Private Farmers and Contemporary
Conservation Subsidy Programs in the Czech Republic

Miloslav Lapka, J. Sanford Rikoon, and Eva Cudlinova

Soil and water conservation policies and programs begin with human definitions and ecological visions of the landscape. Rural and farmer constructs of
the environment, however, are too often developed in isolation from governmental policy processes and by groups whose political influence is limited. As
a result, the economic, social, and ecological values that characterize farmer
orientations are often not reflected in national policies or programs. To successfully implement conservation programs, we must develop strategies that meet
the requirements of diverse value systems, especially those of the groups whose
behaviors have the most direct impacts on rural environments. We must also
consider environmental value systems as heterogeneous. In many cases, the
ecological concerns and values of minority groups are often overlooked or ignored (Rikoon 1995). But a successful national conservation policy must be
flexible enough to incorporate and combine diverse conceptions. In this paper,
our focus is on Czech private farmers, a minority albeit a critical agricultural
group, and their opinions of constraints in the present system of conservation
subsidies now available to farmers in the Czech Republic.

The main physical characteristics of the Czech landscape are the heterogeneity of its ecosystems and the great number of ecotones within the country's
relatively small space of 78,894 square kilometers. The ecological and geographic diversity of the Czech environment has had a profound influence on
settlement and population growth, agricultural systems, and societal institutions. Since the very beginning of settlement, the ecological heterogeneity of
rural areas has been connected with cultural diversity. Within the European
environment, Czech patterns of cultural systems on the landscape illustrate the
mutual links between human and natural resources (Lapka 1994).

The loss of Czech national independence in the years 1620–1918 in part
motivated citizens to turn their interests away from political questions and in-

stead to focus on the country's cultural and ecological resources. National character and Czech identity became linked more closely to relations with the land, to the countryside, to wildlife, and to nature. By the close of the nineteenth century, the landscape had assumed special importance to the Czech people as something "owned," a national heritage and symbol that stood in place of national political independence (Lapka et al. 1995). The ability of rural landscapes to carry such strong symbolic meaning was due in part to the large percentage of the population that lived in rural areas and to the domination of agriculture by private family farms. At the close of World War II, there were more than 1.5 million private farmers in Czechoslovakia (Bartos et al. 1992).

The Socialistic period from 1948 to 1989 led to deep social destruction of the rural community and ecological destruction of the landscape. As a result of collectivization and the expansion of cooperative farms, average farm size rose from about 7 hectares in 1948 to close to 3,000 hectares in 1980. In terms of rural environments, the story of the degradation of soil and water resources is generally well known (Bartos 1987). Yet, however large are the environmental problems resulting from large-scale farming, perhaps even more significant are the consequences of the losses in terms of holistic social connections to the environment. The removal of many people from farming and the loss of personal connections and knowledge of the land has meant an increasing spatial separation of the aesthetic and symbolic functions of the landscape from its productive and economic functions. Urban dwellers often retain strong symbolic attachments to the *idea* of the Czech countryside. On the farm, many of those involved with changes over the past 50 years have come to believe that aesthetic and nonutilitarian values can be found only in Czech landscapes located "somewhere else." In the main, agricultural production and economic values have come to dominate (Cudlinova and Lapka 1996). In terms of private farmers, their number decreased from over 1,400,000 in 1950 to approximately 2,000 in 1988. Over the last 7 years, however, rural transformations have witnessed an increase to approximately 75,000 private farmers.[1]

In this chapter, we view the Czech landscape as a social, cultural, and ecological phenomenon. Within this landscape, our concern is with the roles of private farmers in terms of their participation in conservation programs, adoption of long-term management strategies for soil and water conservation, and maintenance and restoration of ecologically valuable and fragile landscapes. Our interest in private farmers is based on these four trends: private farmers have been the dominant influence on the rural landscape for most of Czech history; private farmers persevered in some areas under socialism and in these areas maintained an ecologically viable agriculture; areas of greatest soil erosion and water degradation can be correlated with the decrease in private farms as a structural agricultural unit; and our belief that a system of private farmers represents the best hope for maintaining environmental resources and viable rural communities. Private farmers now operate approximately 25 percent of agricultural land and their number comprises close to 33 percent of all people employed in agriculture (Czechoslovak Institute of Statistics 1950 and 1988; Czech Institute of Statistics 1994).

Methodology

If we understand the landscape as having social, cultural, and ecological dimensions, then one's choice of research methods must be adequate to understanding all of these dimensions, particularly as they influence peoples' behaviors on the land. A variety of researchers have pointed to the need for using innovative qualitative methodologies in order to give interviewees opportunity to fully explicate their rationales for conservation behavior (Kraft 1989; Rikoon et al. 1996). Such approaches are intended to reveal the world hidden behind a number or a simple answer on a questionnaire and to counter some of the limitations of quantitative approaches in terms of understanding the social and material environments in which decision-making occurs. Explication of the social environment in which decision-making occurs is not self-evident in lists of practices or behaviors but must be addressed as an independent phenomenon affecting human interactions with the environment.

Our empirical research with private farmers actually couples qualitative methodologies and quantitative instruments. We started field investigations in 1991 with the use of questionnaires and more recently have turned to qualitative techniques in order to probe wider dimensions of data collected through initial survey instruments (cf. Fitchen 1990). Our qualitative work emphasizes the use of the "dialogue," a discursive methodology with antecedents dating back as far as Socrates. Most contemporary uses are influenced by the cultural studies work of Mikhail Bakhtin, who suggests (1979, 1981) employment of an open-ended dialogue that gives greater voice to a diversity of viewpoints and to the values that underlie behavior. Our research thus relies on intensive interviewing and long-term fieldwork with individual farmers that provides opportunities to learn about farmers' values and attitudes and then to situate their conservation practices within both cultural and social contexts (Richards 1993).

The first stage (1991–1994) of our research included person-to-person interviews with 52 private farmers as well as structured surveys of 194 private farmers. We initially used a random selection procedure to identify 500 potential research participants from the pool of all private farmers identified by these four sources: lists compiled through previous research by the Institute of Landscape Ecology, members of the principal political party of private farmers, members of dairy cooperatives formed by private farmers, and lists provided by regional offices of the Ministry of Agriculture. This number is almost 10 percent the total number of private farmers estimated by the Agricultural Census in 1992. The 246 farmers included in this research represent a 49 percent participation rate from the identified pool of respondents. Given the difficulties of assessing the number and membership in the group of all private farmers in the Czech Republic (Librova 1994; Sokol 1994; Note 1), we believe that our sample is most representative of private farmers residing in southern portions of the Czech Republic, particularly in the Sumava Mountain Region, South Bohemia, and South Moravia (Cudlinova and Lapka 1994; Lapka et al. 1995).

The second source of data is comprised of intensive case studies of farmers selected from the first round of research. Case study methodologies are increasingly used by social science researchers to provide additional depth to data and to understand the rich nature of social action, to generate theoretical and evaluative criteria for program evaluation, and to develop interpretation and concepts within new or special areas of investigation. In examining farmer responses to new conservation programs in the Czech Republic, case studies provide the best opportunity to generate valid assessments of individual situations through the use of multiple interviews and a dialogic method of collecting information. Initial interviews focused on general discussions about farming and conservation; subsequent sessions concentrated on specific environmental, landscape, and rural issues of concern to the interviewees.

Case study results are derived from 20 completed studies and a similar number nearing the end of the investigative process. We used a "snowball" sampling method (Babie 1995; Bailey 1992; Ostrander 1984) to select farmer participants from survey respondents in two "pilot" regions: agriculturally marginal areas around the Sumava Mountains and the more fertile region of incorporating Bosilec Marshland and Taborsko. "Snowball" sampling (also called network, chain referral, or reputational sampling) was a relevant procedure to use in this special circumstance because it allowed us to identify and include farmers identified both by our research and by other farmers as economically viable operators and community leaders. The rationale to target our first phase of qualitative work at these operators was that they were a group whose participation in these programs would be a key factor in their success. As a group, the case study participants thus represent successful private farmer networks in two critical southern regions of Bohemia, and while not a comprehensive sample of all private farmers in the Czech Republic, they provide the opinions and values of successful operators in the initial post-Socialist period.

The case study sample is rather evenly divided between older farmers who had been operating private farms before 1989 and younger farmers who began their operations after the Velvet Revolution. The former group averages about 60 years of age; the mean for the latter group is close to 20 years younger. All of our case study farms include multiple generations and a strong belief that there will be a succeeding generation of operators from within their families. Farm sizes in this sample range from 38 to 150 hectares. All case study participants are active leaders in their local communities and in both formal and informal groups. Some have official leadership roles in the Agrarian Chamber and others serve as elected officials of local government. Their farmsteads are open hosts to foreign visitors and to government officials, including two families who have hosted the Minister of Agriculture.

Findings

This chapter focuses on private farmer opinions about shortcomings in current subsidy programs targeted directly at soil and water conservation. Two major types of subsidies are presently offered to farmers. First, the Agricultural and Forestry Support and Guarantee Fund (AFGSF), enacted in 1994, has the basic mission of supporting entrepreneurial activities in agriculture and forest management through the provision of loan guarantees and subsidies on interest rates. The fund may be used for operational support (e.g., providing loans during periods of low financial resources) and farmer initiatives (e.g., for support of production activities in agriculture and forestry aimed at increasing efficiency and improving productivity). A portion of AFGSF moneys are also allocated according to the sensitivity of regional ecosystems. For example, special compensations are given to operators in regions with fragile ecosystem characteristics (e.g., protected areas, water resources, and national parts). The second program, and one providing more direct subsidies for landscape protection, began in 1994 as a joint effort of the Ministries of Agriculture and Environment. In comparison to the AFGSF, this newer program is less production-oriented; rather, it seeks to enhance conservation by sharing some of the costs involved in specific protection practices, including the conversion of fragile cropland to pasture, agroforestry, and improvements of the genetic potential of livestock and plants (Ministry of Agriculture 1995).

Before turning to the responses of farmers to these programs, we want to first note the opinions of the Ministry of Agriculture and its Research Institute of Agroeconomy regarding obstacles faced by current programs.[2] There are two important reasons for understanding the attitudes of these sources. First, they have responsibility for conceiving, introducing, and evaluating programs from an agency perspective. Second, we can compare administrators' attitudes with those of private farmers faced with choices of adoption and participation. The following listing summarizes agency representatives' views of the primary obstacles of current subsidy programs to enhance and affect conservation:

- The subsidies programs change each year and annual developments often provide little continuity with the previous year's program; as a consequence, agency personnel have a difficult time administering programs and farmers do not know what programs are available;
- The programs specify a limited set of generic conservation practices and implementation; they therefore provide little flexibility for support of practices appropriate to regions or locality or to the social and cultural dimensions of implementation;
- Specific practices supported by the subsidy programs often result in economic losses to operators; as a consequence, there is little incentive for operators to participate;
- A complex program administration and set of guidelines often prevent farmers from making decisions and participating in these efforts;
- The complex system of land tenure (e.g., many absentee owners, some land ownership titles still unclear) makes it difficult to know who should take the lead in participation (e.g., owner or renter) or who should be targeted for assistance.

Farmers' critiques of ecological subsidies in agriculture

All of our case study farmers have participated in the general subsidy programs and one-half have direct experience with the conservation payments programs. Even nonparticipants in the latter have knowledge of these programs, either through membership in regional Agrarian Chambers or because of personal informal contacts with other farmer participants.

All of the private farmers we interviewed emphasized their belief that the most basic and general problem with the government's overall conservation efforts is their lack of integration with other policies and programs aimed at maintaining viable farms and strong rural communities. The absence of an integrated policy that supports rural infrastructure, decent market practices, and the general social and economic well-being of farm families contributes to a general lack of incentive and motivation to protect rural environments.

Farmers recognize a variety of problems in rural areas. They see their villages decaying and young people leaving. Towns do not "buzz" with activity; rather, they are losing trade and services. Public transportation between villages has declined since 1989 and the level of cultural activities is eroding in many rural locales. Conservation practices are part of wider farming and living systems in which such factors as long-term horizons on the land and attachment to place are important contributing factors. While subsidies for environmental protection and restoration are important and useful, private farmers consider the renewal of the village and support of a viable farm economy as significant related issues. In essence, environmental protection requires a holistic approach to the conservation of ecological, social, and cultural systems.

Farmers also call attention to the fact that market policies and commodity subsidies are not aligned to the practices favored by conservation programs. For example, subsidies for pasture and meadow maintenance often result in higher yields of hay and an ability to increase livestock numbers. But what will farmers do with more hay and livestock in an economy where prices for these commodities are decreasing and do not cover costs? Economic and environmental programs require greater coordination or farmers will have little incentive to improve their pasture land.

Private farmers believe that the conservation programs can be improved through greater monitoring and emphasis on the quality of practice implementation. According to our respondents, present programs do not support a change in the conservation ethic of farmers, their relationships to the landscape, or their orientation toward farming. In fact, as one farmer put it, "subsidies discourage people farming." Many payments are given simply for carrying out a specific task without any quality requirements. For example, farmers can receive payments for mowing their pastures but there are no standards concerning the quality of the grass or the application of the practice. As a result, some participants agree to mow simply for the payment and without concern about the timing of the mowing or what happens to the grass. In many cases, mowing practices (because of poor timing and procedures) have actually encouraged weed growth and the spread of weeds from meadow outskirts and ridges.

Farmers also view some supported practices as undesirable, while environmentally sound alternatives remain unsupported. For example, many private farmers view the subsidized option to convert arable land to forests as irrelevant because it symbolizes a degradation of their farming system and conflicts with their sense of what constitutes relevant farm work. They associate forestry with a lowering of their own social status (which, we should add, they feel is already low within the Czech Republic). According to most farmer world views, forests are a reserve or hedge against economic downturns but not a central or cherished occupational enterprise. On the other hand, there are no government incentives to support production of organic products or for many practices associated with more sustainable agricultural production systems (e.g., intercropping). Farmers view this situation as a paradox. Ecological subsidies for environmental protection do not include farming systems that are most in harmony with the land and that promise to offer the least amount of ecosystem disruption. Interestingly, the number of organic farms and production of organic foods (called, respectively, "bio-farms" and "bio-foods" in the Czech Republic) is increasing, albeit independent of the government's conservation title programs.

Farmers would like programs to be more responsive to differences in regional and local environments. A regional approach would specify practices and incentives relevant to local ecological conditions and sociocultural norms and not, as is presently the case, solely on the basis of soil types and agronomic zones. Farmers would also like to see more local control of program administration, although they recognize this scenario carries its own set of problems. The benefits of local control include an increased ability to monitor the results and quality of practice implementation, a more flexible and accurate evaluation of farm plans, objectives, and impacts, and a strengthening of local governance and priorities. Private farmers see potential problems in this last area, however, because it offers opportunities for existing structures to deny participation of deserving, yet excluded (for political, class, ethnic or other reasons) peoples. Private farmers believe this exclusionary scenario would be detrimental to their own futures in villages where social structures remain largely unchanged from the situation prior to the Velvet Revolution. In areas, for example, where former collectivist structures persist (even under new economic structures and names), local governments may be unwilling to support the efforts of private farmers.

Finally, administrative and program rules are a source of discontent. For example, farmers cite diverse administrative obstacles that constrain wider participation. The whole process proceeds very slowly, they claim, and each step is loaded with a number of official documents that must be completed by a specified deadline. Missing a single deadline requires a one-year delay in participation in that particular program. Most importantly, under the present system the government views subsidy payments to farmers as taxable business income. Farmers believe this situation inhibits program participation and landscape protection, because the system is oriented toward the support of the economy, with environmental protection serving only as a vehicle to increase the flow of money through banks and financial institutions. Under AFGSF, for example, bank profits are assured through the receipt and administration of government loan guarantees.

Discussion

Rural landscapes are complex systems with their own internal logic. Conceptualizations of effective conservation programs and policies in part emanate from how different groups understand the environment, perceive conservation problems, and base their actions on these understandings and perceptions. To program officials, environmental restoration and maintenance subsidies address critical ecological concerns, but to many farmers such subsidies focus on the superficial manifestations of deeper sociocultural problems. To the latter group, rural landscapes have ecological, social, and cultural dimensions, including the farmers, farms, and villages. Further, this landscape has depth, having been created as the result of activities over several hundreds of years.

While this chapter cannot begin to detail the bases of farmers' opinions and attitudes towards the landscape, a few salient responses from the larger farmer survey (see Appendix 1) provide relevant background context and support for farmers' claims about conservation programs. For example, almost all private farmers strongly agree that maintaining the landscape is a high priority. Further, they tend to believe that individual farmers must take responsibility for this objective if it is going to be achieved. Less than one-quarter agree that nongovernmental organizations can effectively achieve this aim and only 8 percent believe that care of the landscape can be "ensured by state administration." Farmers' phenomenological views about who will best protect the landscape are associated with their ideas on how individuals acquire a responsible environmental ethic. As the last question in this table demonstrates, 43 percent of respondents stress the social links between generations as most crucial. It is thus not surprising that farmers view local social and cultural contexts as significant factors affecting landscape protection. Social and economic changes that disrupt intergenerational continuity on the farms and social stability in rural communities threaten the most critical and viable form of learning. In contrast, only one-fifth of respondents believe a "love for the soil" can be acquired through education, and thus by implication technical assistance may provide short-term panaceas but not instill the type of long-term human-nature relationship most conducive to preserving soil and water resources.

As a result of their more holistic approach to rural landscapes, farmers favor strategies that include strong components of economic survival and viability. In spite of some predisposition to view farmers as having some sort of inborn stewardship ethos or "ecological sentiment," the overwhelming majority are neither professional ecologists nor do they possess an inherent predilection to act as protectors of nature. They are farmers whose livelihood depends on production. Some operators, particularly private farmers, have a long-term horizon on the land. Based on particular systems of knowledge and consciousness, these operators develop long-term occupational strategies. In this sense, a farmer understands economic profit multidimensionally and not only lineally or short-term (Rikoon, 1988). Conservation programs and practice alternatives are thus evaluated as to how well they also support characteristics and strategies of family farming. Among these variables are such items as cycles of work and labor, long-term intentions, quality and type of work, and the status and profitability associated with various sorts of agricultural production.

In other words, farmers balance three major areas of concern: their agricultural operation, the biophysical environment, and their quality of life, including their home rural community. If we conceive of these as interdependent points on a triangle, we would suggest that, for private farmers, farming occupies one peak (alpha) of the triangle, with the other two peaks being village (beta) and landscape (gamma). This tripartite scheme is why Czech private farmers often state that environmental protection begins (but does not end) with renewal of the village and rural social structure. These are the contexts in which conservation takes place.

The administrating orientation contrasts with that of farmers. Concerned most immediately with environmental management, the former group logically views the purpose of conservation programs as directly impacting the biophysical landscape. They thus favor conservation payments and subsidies in support of the creation and maintenance of relevant landscape structures. The triangle of these relations is thus constructed as follows: landscape (alpha), farming (beta), and village (gamma). Their vision also is predicated on a conviction that economics is the dominant motive underlying farmers adoption of appropriate conservation behavior, and therefore the success or failure of programs rests on such techniques as the guarantee of financial credits and the provision of payments for conservation-oriented practices. Long-term social and cultural factors are at the same time not addressed.

We should note here one interesting consequence of present conservation programs and agricultural trends. Financial subsidies and payments to farmers for conservation have resulted in the creation of a new rural social group comprised of a small number of "new experts" (similar to the structural role of "custom farmers" in the U.S.). These individuals are typically owners of production resources who commonly lease large areas of land (typically accumulating over 500 hectares under their control). The "new experts" focus on the performance of those operations for which they can receive conservation subsidies. For example, the availability of mowing subsidies in their first year means that they will cut pasture grasses. During the second year they may well convert this land to forest if such activities are subsidized. And in the third year they will be ready to plow down forests to plant crops if that provides the most viable opportunity for government subsidy. In essence, these entrepreneurs are "experts" on government policies and programs. How they "farm" depends on what practices and systems bring the highest levels of subsidy.

Many longer-term rural residents have an instinctive and great resistance to such patterns and behavior. There is no Czech precedent for a group of farm operators dependent on leased land. But "new experts" are simply adapting to the system of rewards set by government programs, and their behavior corresponds with the visions and short-term objectives of administrative groups. Landscape change occurs quickly, and program promoters can point to high levels of hectares affected. We cannot say how long this trend might last, for much depends on the nature of future agricultural and conservation policies and the introduction of monitoring and requirements for longer-term commitments to particular practices. As in the U.S. (Constance et al. 1997; Rikoon et al. 1996), however, an important contributor to this trend is the high number of hectares owned by absentee landlords who cash rent their acres and have little knowledge of the management of their land. As this class of owners remains high in the Czech Republic, opportunities for "new experts" may expand.

In regard to the politics of conservation subsidies, we can identify two important paradoxes. First, the government offers programs without consideration of the social structures characteristic of Czech agriculture and rural villages. Private farmers, for example, comprise a minority group that, during the period of 1948-1989, was significantly marginalized in terms of political and social power. As a consequence, small and medium-sized farm operators typically are on the fringes of both central and local governmental systems. Part of the tragedy of this situation is, of course, that the landscape features that conservation agencies so desire to support are those that were originally created, developed, and maintained by private farmers.

Second, and on a general theoretical level, the clash of administrators and private farmers in their opinions on conservation subsidy programs represents a difference of two models based, respectively, on postindustrial and postmodern paradigms. The former is characterized by such features as concentration, centralization, and specialization. All of these, of course, are powerful political and administrative tools. If conservation subsidy programs are to be successful, however, we believe they must effectively support characteristics of the private farmers' models, including diversification, decentralization, and individualization. (Cudlinova and Lapka 1994, de Haan 1994). Unfortunately, the dominant features of the models used by those who construct conservation programs run counter to both the systems favored by private farmers and the traits characteristic of ecosystems themselves. These sorts of contrasts are not confined to conservation efforts in the Czech Republic, of course, but are characteristic of both general critiques of development and conservation programs in countries such as the U.S. In the latter case, we are witnessing, in part as a response to these criticism, the rise of local, community-based watershed projects and a movement away from centralized planning and policy.

Conclusion

An analysis of private farmers' opinions on contemporary programs for conservation in Czech farming environments reveals three significant functional shortcomings. The most important of these is the fact that banks and other financial institutions derive the greatest economic profit from the present organization of conservation programs. This not only discourages farmers and decreases their motivation to participate, but it does not contribute to the effectiveness, viability, or sustainability of conservation. Banks are rarely connected with the rural landscape and have no horizon beyond the length of the financial management of subsidy funds. They are also, in most cases, not knowledgeable about farmers or farming, or connected with rural life.

A second shortcoming is the construction of conservation subsidies as a monofunctional economic intervention that focuses exclusively on the "face" of the landscape. Including only generic lists of environmental criteria and practices prevents consideration of the system of rural settlements and the nature of rural and local knowledge. This narrow focus also excludes agency consideration of the relationships between, on the one hand, social and cultural systems, and, on the other hand, farmers' desires and abilities to affect conservation. As we noted, many rural residents place conservation as a function of two initial processes—strengthening of the village and implementation of viable farming operations. The present administration construction of conservation subsidies that begins and ends with the environmental landscape thus omits (or assumes a kind of "trickle down" effect upon) issues of farming and rural life.

Finally, the current emphasis on technical and administration aspects of conservation programs acts more as a stress factor than as a motivation for participation. Effective conservation subsidies for protection of the landscape cannot function if at the same time there exists no supporting social structure in the village and country. The way to reach conservation goals often leads through the social and cultural factors of rural communities and family farms.

In conclusion, we would like to suggest a series of recommendations to enhance the conservation of agricultural/rural environments. The following recommendations address the interconnection of rural social systems (connected narrowly with renewal of the village), farming (connected with family farmers), and ecological systems (connected with renewal of the environmental landscape). Conservation of rural resources will occur only if all dimensions of the system are addressed in simultaneous fashion.

Social system recommendations

- Create and develop coordination between agricultural conservation subsidy programs and other policies, subsidies, and programs affecting farming and rural life. Cooperation is needed among agencies responsible for agriculture, environment, economy, and social services.
- Enact programs in support of rural and village development in order to maintain viable and compatible social structures for environmental protection. Programs enhancing the creation and survival of diversified small and medium-sized businesses are necessary to maintain services and to stave off village deterioration. Support of transport, service providers in the village, small industrial enterprises, and tradesmen are appropriate first targets.

Farming

- Target programs especially to family farms because of their longer-term connections to the land and heightened knowledge of the ecological conditions of the landscape.
- Focus subsidy programs on long-term practices, quality work, and the reforestation of marginal and nonarable lands.
- Turn some subsidy resources away from inputs and more toward policies and programs to support the outputs and products of sustainable agricultural systems.
- Remove rules that identify conservation subsidies as taxable income. Ending taxes on subsidies will usually mean additional economic impacts on rural areas.
- Focus support on small and medium-sized farm units. Such efforts will have both bioecological and socioecological impacts. The latter includes an enhanced likelihood of more families in the village and more time and energy available for conservation activities.
- Change administrative policies from the present system that fosters stress to one that acts as a motivation for local development and participation. Local authority should be strengthened and empowered where such capabilities already exist. In other contexts, the appropriate governmental oversight may be on the regional level.

Landscape

- Focus on conservation practices that reflect and support existing sociocultural patterns of work, labor, and social relationships. Such practices are most sustainable as they reinforce existing local patterns.
- Introduce special programs in areas in which structural and occupational trends suggest lack of strong sociocultural concern for landscape system

(e.g., areas where the previously discussed group of "new experts" dominate). In these areas, increased amounts of oversight, requirements of long-term commitments, and perhaps greater regulation are appropriate.

- Change subsidy to allow participants up to 3 years to wholly implement a new practice or landscape change. Longer implementation periods will provide not only the time to implement and maintain practices, but will also allow additional time for operators to incorporate new ideas and practices into their farming systems and to develop knowledge of the ecological relationships between practices.

Notes

1. There are a host of problems in accurately assessing the number of private farmers in the Czech Republic, not the least of which is that the government has not conducted any reliable national agricultural census on ownership and operator status since 1992. Additionally, estimated figures vary according to the definition one employs of a "private farmer" and the overall dynamic situation in Czech agriculture in which occupational and legal status is in constant flux (Hudeckova 1995; Sokol 1994). The estimated 75,000 private farmers includes all private individuals who raise and sell agricultural products from their own land. This figure thus includes close to 40,000 individuals who "farm" less than one hectare of land. In this chapter and our research, we include as private farmers those individuals who (1) own at least some of the property they farm; (2) derive a majority of their income from farming; (3) operate primarily through the labor of their families (although they may have hired workers as well); and (4) are themselves involved in daily management of the farm.

2. Opinions of program officials were gathered from diverse sources. First, eight of the ten administrators of the two relevant programs were interviewed by researchers of the Institute for Landscape Ecology. Second, we have analyzed reports of the Ministry of Agriculture, Ministry of Environment, reports of the Research Institute of Agricultural Economics (Ministry of Agriculture), the annual "Green" reports of the Ministry of Agriculture, the "Farmer" periodical (a publication for farmers produced by the Ministry of Agriculture), and various other written and documentary sources.

References

Babie, E. 1995. The Practice of Social Sampling. Belmont, CA: Wadsworth.

Bailey, K.D. 1992. Methods of Social Research. New York: Free Press.

Bakhtin, M.M. 1979. An Aesthetics of the Literary Art. Moscow: Isskustvo. (In Russian)

Bakhtin, M.M. 1981. Discourse in the novel. In The Dialogical Imagination, ed. Michael Holquist, pp. 259-442. Austin: University of Texas.

Bartos, M. 1987. Influence of large-scale farming methods on soil exploitation in Czechoslovakia. In Land Transformation in Agriculture, eds. M.G. Wolman and F.G.A. Foumier, pp. 319355. London: Routledge.

Bartos, M., V. Mejstrik, and J. Tesitel. 1992. Changes in agricultural landscape in Czechoslovakia during forty years of collective farming. In Agriculture & Environment in Eastern Europe and the Netherlands, ed. J.L. Meulenbroek, pp. 11-18. The Hague: Mouton.

Constance, D.H., J.S. Rikoon, and J. Ma. 1997. Landlord involvement in environmental decision-making on rented Missouri cropland: Pesticide Use and Water Quality Issues. Rural Sociology 61(4): 577-605.

Cudlinova, E. and Lapka, M. 1994. The potential role of small-scale private farmers in the ecological restoration of the Bohemian landscape. Ecological Economics 11: 179-186.

Cudlinova, E. and Lapka, M. 1996. Cultural capital, rural ethics and approaching sustainability. In Ecology, Society, Economy: In Pursuit of Sustainable Development. Tome I, Session IV.

Czech Institute of Statistics. 1994. The Statistical Yearbook of Czech Republic, 1994. Prague. (In Czech)

Czechoslovak Institute of Statistics. 1950. The Statistical Yearbook of CSSR, 1950. Prague. (In Czech)

Czechoslovak Institute of Statistics. 1988. The Statistical Yearbook of CSSR, 1988. Prague. (In Czech)

de Haan, H. 1994. In the Shadow of the Tree: Kinship, Property and Inheritance among Farm Families. Amsterdam: Het Spinhuis.

Fitchen, J. 1990. How do you know what to ask if you haven't listened first: Using anthropological methods to prepare for survey research. The Rural Sociologist 29 (Spring): 15-22.

Hudeckova, H. 1995. Privatization in agriculture and rural regeneration. Sociologie Venkova Zemedelstvi 41: 307-315. (In Czech)

Kraft, Steven E. 1989. Is there a place for ethnomethodology in the analysis of agricultural public policy—the case for the environment. Paper presented at the Second Biennial Conference of the Agriculture, Food, and Human Values Society, Little Rock, AR.

Lapka, M. 1994. Czech landscape as an object of ecological research. In Czech Landscape Today, ed. H. Svobodova, pp. 39-47. Prague: Karolinum. (In Czech)

Lapka, M. et. al. 1995. Land, Culture & Crisis. Final Field Report - Czech Republic. Report of European Union Project, coordinator Simon Miller, University of Manchester. Institute of Landscape Ecology. Ceske Budejovice. 127pp.

Librova, H. 1994. Colorful and Green. Brno: Blok. (In Czech)

Ministry of Agriculture. 1995. "Green Report" 1995. Praha: Agrospoj. (In Czech)

Ostrander, S. 1984. Women in the Upper Class. Philadelphia: Temple University Press.

Richards, P. 1993. Cultivation: knowledge or performance. In An Anthropological Critique of Development: The Growth of Ignorance, ed. M. Hobart, pp. 61-78. London: Routledge.

Rikoon, J.S., D.H. Constance, and S. Galetta. 1996. Factors affecting farmers' use and rejection of banded pesticide applications. Journal of Soil and Water Conservation 51(4): 322-334.

Rikoon, J.S. 1995. Conflicts of knowledge and the cultural construction of the environment: implications for the remediation of ecology and democracy. Sociologie Venkova Zemedelstvi 41: 161-179. (In Czech)

Rikoon, J.S. 1988. Threshing in the Midwest, 1820-1940. A Study of Traditional Culture and Technological Change. Bloomington: Indiana University Press.

Sokol, Z. 1994. Determining who is a private farmer. Ekonomika Polnohospodarstva 17 (1): 18. (In Czech)

Appendix 1. Selected options of private farmers (N=246)

Survey statement	Responses	% of responses
The mission of Czech farmers is to produce healthy products in a market economy and to maintain the good condition of the landscape.	Agree strongly	87
	Agree somewhat	11
	Do not know	2
What does "healthy soil" mean to you?	Health of people	31
	Health of landscape	36
	Change of farming	25
	Other	8
Who will best ensure care for the landscape?	Farmers	62
	No one	14
	Do not know	14
	No response	2
Care of the landscape will be ensured by non-governmental organizations (NGOs).	Agree	24
	Disagree	43
	Do not know	25
	No response	8
Care of the landscape will be ensured by state administration.	Agree	8
	Disagree	71
	Do not know	18
	No response	3
How do farmers acquire a love for the soil?	Inborn	33
	Passed through generations	43
	Education	20
	Do not know	5

Source: Based on Lapka and Gottlieb 1993.

32

Soil Conservation Policies in Australia: Successes, Failures, and Requirements for Ecologically Sustainable Policy

Ian D. Hannam

Land degradation has expanded in each Australian State and Territory in the 60 years since soil conservation legislation has existed in Australia. This is a general indication that soil conservation has failed. Traditionally, government policy concerning land use has not been based on integrated natural resources decision-making, but has concentrated principally on reasons and strategies for further land development with an unclear role for the legislation, leading to disastrous ecological consequences (Australia 1996b; Boer and Hannam 1992). The dominant characteristics of Australian soil conservation policy are similar to the dominant characteristics of national soil conservation policy in other parts of the world (Musgrave and Pearse 1984), where the disciplines of agricultural science and soil science have had the major influence. These disciplines together cannot effectively manage land and soil degradation because each is too fragmentary and incomplete in relation to basic rules of ecology. Moreover, recent major land management initiatives promote the adoption of ecologically sustainable land use practices and nonland degrading practices but with an emphasis on community service practices rather than regulation. The dominant paradigm for soil management relies mainly on the practical field management approach, implemented and influenced through land user advisory schemes (Aveyard and Charman 1991).

Global soil policy has existed since the early 1980s (FAO 1982; UNEP 1982). Even though this soil policy has limitations, at that time it was far more advanced than Australian developed soil policy and could have provided a sound basis to improve the Australian policy (Boer and Hannam 1992; Hannam 1993). There is now a substantial body of global environmental policy and law which should be drawn on to improve Australian soil conservation policy and law (Gardner 1994). Further, national and state policy material concerning ecologically sustainable land management, biodiversity conservation, forest management, and threatened species conservation, for example, is based on ecological concepts and principles that are adaptable to soil policy and legislative reform in Australia (Boer 1995). Soil conservation policy in Australia now requires substantial reform with greater emphasis on ecologically sustainable land management strategies and practical standards (Australia 1992a). However, in the first instance the legislation needs a fresh, ecologically oriented approach for an objective of effective sustainable policy to be achieved. Legislation must also be able to define sustainable limits of land use and prescribe the circumstances for intervention.

Unsustainable soil

The European settlement of Australia has been synonymous with the degradation of its natural ecosystems. Until 200 years ago, the continent was inhabited at a sustainable rate by its Aboriginal population, which was low in numbers in keeping with the carrying capacity of the land. The degradation of the soil has been a feature of European land use processes (McTanish and Boughton 1993), and the greatest impact on the soil was associated with the rapid expansion of technology. The first recognized soil conservation legislation was introduced in 1938. Although there is an inconsistency in format, all states now have some form of soil conservation legislation (Bradsen 1988; Hannam 1993).

In the 60 years since legislation was introduced, all forms of land degradation have expanded and many natural ecosystems have become dysfunctional (Australia 1996b; McTanish and Boughton 1993). Wind and water erosion are the most obvious forms of degradation, but dryland salinity, soil structural decline, woody weed invasion and mass movement have been slower to appear and have equally significant effects on the soil (Charman and Murphy 1991). Farming systems in Australia have evolved over nearly two centuries, beginning with the European methods, followed by American ideas (Prately and Rowell 1987). The view that agricultural land use was "development" was firmly entrenched by the mid-1930s but the degradation of the soil was not recognized at the time as a limitation to agricultural progress (Wadham and Wood 1939). The rate of soil and land degradation accelerated in the 1960s when the land used for cropping expanded greatly, intensity increased, and larger machinery allowed more frequent cultivation (Prately and Rowell 1987). Cropping activities continue to extend into the drier inland and steeply sloping, moister marginal areas, and it is questionable whether lessons of the past have been learned (Australia 1996b; Shelly 1990).

Considerable work has been done on the development and adaptation of farming systems that, ecologically, maintain or enhance the natural resource base (Charman and Murphy 1991), but many landholders are reluctant to adopt these sustainable land use practices (Bradsen 1994). The first national survey of land degradation was undertaken in the mid-1970s (DEHCD 1978) and in 1987-88 the first systematic land degradation survey was undertaken of New South Wales (Soil Conservation Service of NSW 1987a), and it indicates that almost every part of the state is affected by one or more forms of land degradation (Graham 1992). Recent research has established that 2 percent of Australia has soils regarded as excellent quality and 70 percent of Australia is comprised of soil not usable for agriculture (CSIRO 1996).

Soil conservation definition

It is argued that one of the reasons for the ineffectiveness of soil conservation policy and legislation is that the definition and understanding of soil conservation generally encourages the processes which cause the degradation of soil. The current Australian definition of soil conservation (Houghton and Charman 1986) emphasizes the prevention, mitigation, or control of soil erosion and degradation through the application to land of cultural, vegetative, structural, and land management measures, either singly or in combination, to enable stability and productivity to be maintained for future generations. This definition is consistent with global soil conservation literature which asserts (Charman and Murphy 1991; Morgan 1986) that the aim of soil conservation is to obtain the maximum sustained level of production from a given area of land while maintaining soil loss below a threshold level which, theoretically, permits the natural rate of soil formation to keep pace with the rate of soil erosion (Edwards 1991). This thinking is enshrined in Australian soil conservation legislation and policy through their adherence to the conservation with agriculture theme. The western Australian Soil and Land Conservation Act, 1982, is the only piece of Australian soil conservation legislation to include a legal definition of soil conservation, but it adheres to the same approach. History shows that land management practices based on these definitions have, for most agricultural areas, generally not been ecologically sustainable (Australia 1996b).

Soil conservation legislation

Despite there being sufficient soil conservation legislation in Australia throughout the period when land degradation has expanded, there has been little policy forthcoming from the legislation seriously challenging the reasons why degradation was expanding, and to argue forcibly for major changes in land use practice (Boer and Hannam 1992). Although legislation is one main area in which policy can be formulated and enacted (Fowler 1984), the fact that little policy has eventuated is not entirely a deficiency of the legislation, but also a matter of perception and attitude by administrators and the community (Bradsen 1994; Hannam 1993). The lack of specific objectives and targets in the acts, in conjunction with many inadequate provisions (Bradsen 1988), has been a major limitation to policy development. The acts have several basic characteristics in common, including administrative features, powers, duties and functions, interagency relationships, cooperation and coordination, objectives, land use planning, and financial assistance provisions. The south Australian Soil Conservation and Land Care Act, 1989, the most recently reformed act, is an exception in that it contains the provisions for a land capability-based approach to land use planning and land degradation control (Bradsen 1991).

When viewed as a specific group or class of environmental law, soil conservation legislation, by comparison to other types of Australian environmental law (e.g., forests, endangered and threatened species, environmental planning), does not have an adequate ecological strategy. The legislation is characterized by practical agricultural provisions rather than the provisions that can adequately determine the ecological constraints of land and to take sustainable land use action. Major attitudinal change is required to create an ecological philosophy for soil and to introduce modern principles and provisions that are able to assess the ecological limitations of the soil environment and to confidently share the responsibility for its conservation between public and private interests (Gardner 1994).

Commonwealth of Australia and legislation

The Commonwealth government of Australia has specific power through the Australian constitution (Australia 1986b) to legislate with respect to soil conservation. It has never done so, but it has been argued that it could have a very significant and direct role in soil conservation (Bradsen 1991). Commonwealth input has been mainly restricted to broad policy statements and some specialized environmental agreements (Australia 1992b). Wherever the role of the Commonwealth in the environment arises, the debate usually quickly turns to which level of government can better perform the tasks of devising and implementing the various aspects of environmental responsibility (Bradsen 1991). There has been a number of situations where constitutional powers have been applied to "intervene" in a state environmental dispute (Fowler 1993), indicating that the Commonwealth has significant power to act in the interests of a national environmental problem as a whole. Some commissions have considered a constitutional amendment, to include a specific environment protection and conservation power (Constitutional Commission 1987), but without these changes, it is unlikely that constitutional powers will be used in a national sustainable land management and policy program to control land degradation in Australia.

Soil conservation policy and agricultural policy

The effects of agricultural activities on land degradation have not always been clear-cut. Over the decades, management practices specifically to deal with land degradation in many cases have not had proper policy support or have generally been ignored (Australia 1989). Government policy as it affects land degradation has been separated into two distinct categories: general policy associated with commodity and input programs, and those providing for specific policies directed at alleviating land degradation such as concessions for soil conservation expenditure (Blyth and Kirby 1985). The tendency has been for soil conservation policy to concentrate on development of technology to enable the land to continue to support agricultural production objectives and economic-based targets (Charman and Murphy 1991). However, there has been some consideration for conservation resources to be primarily directed to prevention and restoration measures that are economically justifiable. In the past 10 years, the recommendation has been made frequently for soil conservation agencies to be part of state departments of agriculture (Boer and Hannam 1992), thus confusing the issue of whether soil conservation is truly a "conservation" or an "agricultural" activity. In some states where soil conservation and agriculture have been institutionally amalgamated, it was acknowledged as being the primary reason for failure of soil conservation to prevent land degradation (Bradsen 1987). In 1992 when the New South Wales Soil Conservation Service was integrated with the State Lands Administration, the era in Australia of specialized and independent soil conservation agencies ended.

Land development policy

Agricultural policy has been dominated by government promotion of crop expansion, new land development and settlement programs, stabilization of commodity prices, and protection from cheap imports for decades (Balderstone 1982). The appropriateness of this agricultural economic attitude has now been challenged by the Commonwealth Ecologically Sustainable Development Policy (Australia 1992a). However, without formal commitments between the Commonwealth and the states to policies for land degradation control, governments will continue to use politically convenient occasions to introduce policies aimed

497

at demonstrable improvement in national economic performance and social welfare, and to maximize farm sector contribution to overall national economic performance. This occurs despite the existence of environmental policy, including the soil policies, and the strategic integrated resource policies (Australia 1986a; 1996a). Land tenure and land settlement policies have played a significant role in development and expansion of land for agriculture (Bates 1992), and the policy associated with the Australian system of land tenure has often created or led to the incentive for land users to adopt practices that initiate or accelerate soil degradation (Boer and Hannam 1992; McTanish and Boughton 1993; Parliament of NSW 1984). With approximately 80 percent of Australia under Crown Land (leases and other types of Crown tenure), governments have had substantial power to restrict and enforce land management conditions through the leasehold provisions of the Crown Lands Acts and other associated acts (Bates 1992; Fisher 1986). In fact, Crown Land legislation in some Australian states includes comprehensive environmental assessment provisions (e.g., New South Wales Crown Lands Act 1989) which can be used to achieve sustainable land management of vast pastoral and arable areas.

Vegetation degradation

Continuous removal and modification of Australian native vegetation has been a major contributor to soil erosion and dryland salinity (Australia 1995a; Woods 1983). Some specific pieces of legislation and government policy encouraged tree removal and substantially affected land degradation (Australia 1984; 1987), for example, the effect of taxation concessions on land clearing of the Commonwealth Income Tax Assessment Act 1936, to deduct the full cost of land clearing in the year in which it was incurred, was devastating for the soil. It was aimed at reducing the cost of clearing, thereby making it profitable to clear greater areas of land. A lack of comprehensive surveys makes it difficult to obtain accurate figures about the extent of vegetation loss (Australia 1995a) but most Australian states have now taken positive action to protect and manage native vegetation, especially in the drier inland areas (Dendy and Murray 1996). Special "protected land" powers were inserted in the NSW Soil Conservation Act 1938 in 1972 to control the removal of vegetation on steep erodible topography (Hannam 1995) and in South Australia, native vegetation protection legislation, first introduced in 1986 and substantially modified in 1991 as a Native Vegetation Act, has eliminated broadscale clearing of agricultural land. In all cases, native vegetation protection legislation has support policy which forges the critical ecological link between soil and vegetation conservation, and this is evidenced in particular in the NSW Native Vegetation Conservation

498

Act of 1997. South Australia and New South Wales have been through comprehensive reform programs, including extensive public involvement, to ensure that legislation and policy combine effectively to manage nature conservation values, biodiversity, and the soil.

Drought management policy has acted against land resource sustainability generally aiming at retention of livestock resources rather than protecting the soil, which is the least renewable component (Blyth and Kirby 1985; McDonald and Hundloe 1993). The national report on the effects of drought assistance measures and policies on land degradation (1988) agreed that general purpose loans and specific purpose loans as well as subsidies resulted in detrimental effects to the soil and generally did not encourage sustainable land management in the rangeland and cropping land. The attitude to drought management strategy has begun to change and greater consideration is now being given to the ecological sustainability of soil and vegetation. Important policy reform includes a phasing down of in-drought assistance measures, increase in incentives for restructuring and improved pasture and water management, adoption of conservation farming techniques (McTanish and Boughton 1993), and measures to improve awareness of climatic uncertainties.

Commonwealth soil policy

Numerous national soil conservation policy and legislative inquiries in the past two decades have been unable to influence the adoption of land management practices to prevent serious soil degradation. The Senate Standing Committee on Science, Technology and the Environment (Australia 1984) recognized the general failure of land use policy in coming to grips with land degradation. It pointed out that proposals for the development of a national approach to land use policy to provide an overall framework for land use and resource and environmental management, and the various support mechanisms required to plan and implement such a policy, have been made by numerous committees of inquiry over the years, but little or no action has been taken. The recommendations of this report (Australia 1984) place emphasis on the funding-based feature of past policy approaches, and failed to consider the need to change attitudes of policy-makers and land-users.

In 1989 a report on the effectiveness of land degradation policies and programs (Australia 1989) summarized the assistance given to soil conservation over the years, including land use policy, financially supported policies and programs and tax concessions, concluding that soil conservation policy has failed, as evidenced by the existing high level of land degradation. Despite the knowledge about soil degradation problems, some government agencies still see their primary role as promoting economic development with little regard to environmental costs (Australia 1996; Chisholm and Dumsday 1987).

The Australian Standing Committee on Soil Conservation had an influential role with the states and territories on soil conservation policy, but was generally restrained by constitutional inadequacies (Boer and Hannam 1992). As a result, there remains an inconsistency between the states and territories in the adoption of criteria for, and implementation of soil conservation policy. This body was, until 1985, administratively linked to the Australian Agricultural Council, a link which forged an agricultural rather than an ecological view of soil conservation. The committee's soil conservation policy, involving the study of community benefits of, and finance for soil conservation (Dept. of Primary Industry 1971), examined the activities by which soil conservation organizations contributed toward conservation and improvement of soil resources since the 1940s and presented a plan for soil conservation achievements to the year 2000. A further study examined the "benefits of soil conservation" by using land productivity criteria and economic aspects of land use.

Collaborative study—a chance for new direction

A comprehensive Commonwealth-States' Collaborative Study of 1978 (DEHCD 1978) provided a good basis for the different levels of government in Australia to develop consistent soil conservation and sustainable land and water management policies and programs. The main limitation of this study was its restriction to broad policy issues, rather than outlining new policy direction in conjunction with specific policies for state soil conservation agencies to adopt. There is no clear evidence to indicate that the states took the opportunity to turn the findings of this study into effective policy programs, nor implement them within the guidelines suggested by global policies which appeared around that time (FAO 1982; Olembo 1983; UNEP 1982). It is argued that the Commonwealth could have used the collaborative study to initiate a comprehensive philosophical, ideological, and strategic change in soil conservation policy and legislation in Australia.

National soil conservation strategy

A national soil conservation strategy was prepared by the Australian Soil Conservation Council as a statement of the way in which land degradation would be tackled in Australia (ASCC 1989), and in response to the soil conservation "elements" of the National Conservation Strategy for Australia (Australia 1983). Its purpose was to provide nationally agreed policies and priority actions for land degradation prevention and control and rehabilitation of affected areas and to sustain Australia's economic prosperity for the needs and aspirations of its population. Its principles generally showed some concern for ecological aspects of soil, including the use of land within its capability, proper evaluation and planning of land, and land management practices which maintain or improve soil qualities. The strategy gave recognition to the intergenerational responsibility to prevent and mitigate against land degradation, and to the integrated management of soil, water, flora and fauna. Although not a Commonwealth government strategy, it was a combined strategy based on the land management roles of the Commonwealth, state and territory governments. In keeping with other national soil conservation policies, it was left up to each party to determine the way in which the strategy would be implemented and consequently there is little evidence of this occurring.

Integrated catchment management policy

In the past decade a significant amount of policy material has appeared in Australia aimed at managing land, water and other biophysical resources under the concept of "integrated catchment management" (Blackmore et al. 1995). This policy group broadly considers stability and productivity of soils, sustainable native vegetation management, maintaining a satisfactory yield of high quality water, and minimizing adverse environmental effects due to development. Some of the policies have been very specific (SCS of New South Wales 1987c; 1991), and supported by a special piece of legislation, e.g., the New South Wales Catchment Management Act 1989. Ecologically, this has been a significant environmental law initiative for soil because it is a logical way to approach and solve land related environmental problems. The implementation of the "integrated resource" policy is actually quite difficult as it relies on the effective coordination of government departments, authorities, private companies and individuals who have the land management responsibilities within catchments (Burton 1986). However, despite wide community acceptance of this policy, progress in actual implementation has been hesitant and unsystematic. At its most sophisticated level, the Commonwealth and three States have devised an integrated catchment management approach to identify and solve natural resource degradation issues in the Murray-Darling Basin, which is the most important agricultural catchment in southeast Australia (Australia 1996b; Blackmore et al. 1995).

In most cases, coordination of the individual catchment-based policies is left to multidisciplinary committees (Booth and Teoh 1989). In rural areas the committees are predominantly represented by rural landowners who can veto any proposal for punitive action, and the catchment management legislation usually does not contain interventionist or regulatory powers. The principal source areas of the "integrated" policies still reside in the single-issue resource legislation (e.g., soil, water, forestry, and environmental planning and assessment acts, etc.). These acts, which are the responsibility of the authorities represented on the committees (Hannam and Watkins 1989), in effect, make up the catchment management process. Public agencies usually have fragmented and shared responsibilities either from one level of government to another or among agencies at the same level of government, which is a barrier to effective integrated decision-making.

State soils policy

In 1987 New South Wales released the first 'State Soils Policy' in Australia (SCS of New South Wales 1987a). It was released with the triad of new natural resources policy including catchment management policy and a vegetation management policy (SCS of New South Wales 1987b;c). It was also viewed as an integral component for the successful implementation of the new "integrated resource management policy" in southeast Australia. This soil policy is a major soil conservation policy initiative in Australia but is weakened by not being a direct outcome of the New South Wales Soil Conservation Act, 1938. Key aspects of the policy are neither linked to the objects, or any specific provisions of the act. The policy covers multiple land use objectives, development and conservation objectives, and admits past misuse of soil. It lacks a sustainable ethic (Hannam 1992) and proposes environmental protection and continuing economic growth as mutually compatible and not necessarily as conflicting objectives (Australia 1983). A further weakness of this policy is its failure to provide firm direction to change attitudes toward soil, or to nominate priority land degradation control actions that could be systematically treated over nominated time periods. However, it does recognize soil as finite and nonrenewable by its low rate of formation (Edwards 1991)—a key ecological standard which should be central to the implementation of the policy. Having recognized that the extent of soil erosion and land degradation affects this nonrenewable position, the policy optimistically argued that the social welfare and economic independence of the population in southeast Australia were greatly dependent on an ecologically sustainable soil resource.

National landcare policy

The nationally used term of landcare originated from rural community groups who were voluntarily undertaking land conservation activities (Australia 1995b). It is now Australia's most ambitious land conservation program, expanding into a set of ideas and values relating to sustainable land management and ecologically sustainable development (Alexander 1995; Campbell and Siepen 1994). A Decade of Landcare was inaugurated in 1990 by the Australian Soil Conservation Ministers following joint representation to the Commonwealth Government by the National Farmers Federation and Australian Conservation Foundation concerning the seriousness of the degradation of Australia's natural resources. Landcare has developed various meanings but the policy, being quite varied, encourages an integrated approach to solving land degradation by government and community groups working under individual state plans and a national program (Australia 1996b). Since it became recognized by governments there has been a trend for the states to amalgamate land, water and forest management agencies into land conservation departments—a move which conjunctively leads to the disappearance of some specialist land degradation management agencies and in some cases, specialist scientific expertise (Boer and Hannam 1992). There has also been a noticeable decline in the development of specific soil conservation policy. Most of the policy attention is now diverted to general land conservation policy, focusing on community group activities to deal with land degradation (Douglas et al. 1995; Roberts 1992). While landcare unquestionably raises public awareness for integrated land conservation management, it has done this with reluctance to use regulatory land use control for unsustainable land use practices. Governments are less inclined to use punitive measures and take the view that peer pressure within and between the community groups will alone resolve land degradation. A further criticism of landcare is the length to which the community can intervene in the operational responsibilities of specialist land conservation institutions. A dichotomous situation now exists whereby the urban community is demanding that governments increase legislative capabilities to deal with land degradation matters, while the rural community, the group widely represented as "landcare," advocates nonregulatory approaches.

The general decline of state funding to the agencies, and the diversion of departmental staff to service the 2,200 community-based groups is reducing the capability for agencies to undertake long-term research into the development of indicators and attributes of ecologically sustainable land management. (Standing Committee on Agriculture and Resource Management 1996). The amalgamation of numerous, formerly autonomous specialist organizations into a central integrated program is constructive in an institutional sense. It has substantially improved community awareness of specialist land conservation programs and capabilities, including the structure and allocation of funding. However, conversely, there has been less attention given by state governments and the Commonwealth government to the specific impacts of the departmental amalgamations on the effectiveness to deal with land degradation. In keeping with the position documented in the 1980s, the landcare program is producing a similar form of policy with an equally low interest on the role of, and reform of, soil conservation legislation (Bradsen 1994).

Soil conservation reform

Ecologically sustainable soil

There is now substantial argument for movement away from the legal and policy processes, which have led to unsustainable use of soil, to a new policy and legislative paradigm that is based on ecologically sustainable concepts and provisions. Under the new global environmental guidelines, individual nations are not only encouraged to make a serious commitment to the formulation of national sustainability policies and strategies, but to support these with appropriate environmental law (Boer 1995; Harding 1994; UN 1992a,b). Various attempts have been made to raise the consciousness about soil sustainability in Australia (Bradsen 1991; Roberts 1992), and there is now adequate strategic material and guidelines to support these aims (Australia 1992a,b). Reforming soil conservation law under the concept of land conservation would give special consideration to a sustainable land use objective (Bradsen 1988; Gardner 1994). Australian soil conservation law and policy would adopt the critical elements of globally established ecologically sustainable concepts (UN 1992a,b; UNEP 1992a) as well as regional and state ecological issues (Bradsen 1992; Gardner 1994). A fundamental change in attitude toward the soil is critical (Bradsen 1994; Hannam 1993). It should also embrace the concept of soil as part of our natural heritage, especially the links to our cultural heritage (Boer 1991; Boer and Hannam 1992). Emphasis would be placed on developing attributes to measure land and water quality to sustain agriculture and to introduce sustainable land management practices capable of being enforced and measured over the long term (Fowler 1993; Standing Committee on Agriculture and Resource Management 1996).

Policy and institutional aspects

The first level of action would involve a shift in attitude from the commodity-oriented view of the soil to a social ecological view (IUCN 1991). Legislation alone will not be sufficient to achieve this and it must be supported by the continuation and expansion of "landcare" type community education programs (Alexander 1995). Existing state soil policy is not adequate and it is critical to

504

redraft some of this material with a "duty of care" theme, including a firm commitment to the development and implementation of ecologically sustainable land use systems, priorities for biodiversity maintenance, education, and reappraisal of the application of some existing ecological concepts, such as "protected land" (Hannam 1992, 1995). Secondly, ethical rules must be established for soil ecological sustainability (Beatley 1991) as well as rules that recognize the need to research and establish physical and chemical limits for soil and land use enterprises (Department of Primary Industry and Energy and Australian Soil Conservation Council, 1988; Olembo 1983), including guidelines to establish and assess the value of ecologically sustainable standards/indicators and a review of the ability of soil conservation technology to reach and maintain ecologically sustainable standards, and indicators to measure the progress of ecologically sustainable land use and sustainable land management (Hamblin 1991; Hannam 1992). This involves substantial commitment to research and investigation (Edwards 1991; Standing Committee on Agricultural and Resource Management 1996).

The evaluation of ecological soil standards for sustainable land use against existing institutional and legal standards will indicate whether philosophical and research objectives are achievable. It will also depend on reassessment of government and community attitudes to soil conservation, and must address the ethical and practical aspects of property rights and responsibility (Gardner 1994). Factors requiring attention include Commonwealth and state responsibilities for soil conservation/soil environmental protection, and commitment to implement enforcement powers to ensure that ecological standards of soil sustainability are met (Bradsen 1994). Major economic and sectoral agencies of government must be made directly responsible and fully accountable for ensuring that policies, programs, and budgets support land use and development that is ecologically sustainable (Australia 1992a; Carew-Reid et al. 1994).

Soil legislation

There is an urgent need to reform Australian soil conservation legislation. The basic philosophy should focus on prevention of ecological deterioration of soil, rehabilitating degraded land and ecosystems, and establishing decision-making systems that safeguard against unsustainable and degrading land uses. Some of the basic mechanisms to achieve soil sustainability include concepts that prescribe behavior and establish a duty-of-care ethic, create the right to take jurisdiction over soil, establish rules and criteria for ecological soil standards, establish mechanisms to create cooperation in soil management (environmental assessment and planning), decide ultimate or primary respon-

505

sibility for various groups, and determine the circumstances for intervention (Gardner 1994).

Definitions

If the current 'agriculturally-based' definition and perception of soil conservation is maintained, an ecological objective for soil is not possible. Therefore, in a reform process, the move from a soil conservation act to an act 'for the soil' means that the implementation of soil conservation would move from a technocentric role to an ecocentric role, i.e., recognizing a moral duty of humans to protect the ecological integrity of soil. Provisions dealing with practical soil conservation methods would then become just one of many aspects of legislation. A definition of this type could accommodate both the ecological and economic objectives for viable agriculture.

By adopting an ecologically sustainable land use objective, the definition of land degradation, i.e., "decline in quality of natural resources, commonly caused through improper use of the land by humans" (Houghton and Charman 1986), would be modified to place an obligation on humans to rectify existing degradation of land, vegetation, and water, to prevent degradation occurring and take soil conservation action to restore an ecosystem "as close as possible to its predisturbed state" (recognizing that many of Australia's natural ecosystems have suffered irreversible changes). It is also suggested that the term soil degradation be developed to give attention to the specific ecological needs of the soil.

Public participation in policy and legislation

Sustainable soil legislation would include provisions to invite public opinion or action on any area of soil, by constituting a community-based soil advisory council; requiring the soil conservation authority to prepare state soil strategies: publicly exhibiting all soil management plans with a legal obligation to take public responses into account; providing for soil conservation agreements that contain ecologically sustainable techniques; including a range of enforcement powers to control degrading land use activities; and including a standing provision so an action may be brought against an individual or corporation if an activity threatens or potentially threatens the ecological integrity of soil.

Rights and responsibilities

All nations should recognize their responsibility to an adequate environment for present and future generations (Australia 1992a). This is an important step toward the broader societal goal of ecological sustainability identified in

506

the global sustainable development literature on the rights and responsibilities of individuals and states (Chisholm and Dumsday 1987; Carew-Reid 1994; The Commission on Global Governance 1995; UN 1992a,b; Vigod 1989). An Australian law of this type would not only be a bold step forward for the environment in general, but it would help raise the ecological consciousness of soil. However, it is recognized that various constitutional difficulties for soil would probably still remain (Crawford 1991) until the constitution could be amended to include an environmental power. Major rights for individuals relevant to the soil are considered to be effective access to judicial and administrative proceedings, including redress and remedy in exercising their rights and obligations to soil legislation; the right of any person to take legal action against another person for causing soil degradation, the right for any member of the public to participate in the planning process for soil, including the making of regulations; the right of access to soil environmental information; the right to a healthy environment; and all persons have the duty to protect and conserve the environment for the benefit of present and future generations. A general obligation would give individuals the right to participate in decision-making activities where developments are likely to have a significant effect on the ecological integrity of soil, and the right to legal remedies and redress for those whose soil environment has been or may be seriously affected.

Conclusion

There is sufficient grounds to advocate that Australia should seek an alternative paradigm for soil—aimed at conserving its ecological integrity (Australia 1996b; Boer 1995; Harding 1994). It is difficult to see that it can be reached without significant changes within the Australian social and cultural fabric, particularly the current attitudes towards development and the preservation of "private property rights" for agricultural land. A willingness to adopt changes may lead to a change in the condition of the soil as well as other changes in the biophysical system. This would be a measure of the effectiveness of ecologically sustainable techniques. The final standard will be the reduction in the types and levels of soil degradation, and long-term viability of agriculture (Australia 1996b; Standing Committee on Agriculture and Resource Management 1996).

The European occupation of Australia has destroyed much of the Aboriginal culture and in doing so, failed to recognize the benefits of their human-nature bond to the ecological sustainability of the continent (McTanish and Boughton 1993). Environmental degradation has become an entrenched feature of European land use in Australia, particularly soil degradation. The implementation of soil conservation cannot be left as a choice alone by the individual. The results of past institutions allowing this to happen, together with its promotion of soil conservation as a necessity for "production" have not been successful (Boer and Hannam 1992). The landcare and catchment management programs, while advocating solutions to land degradation, in their current form offer little more than what the past policy has offered. Building these concepts into broad-based ecologically sustainable policy and legislation with the specific provisions for their implementation and for ongoing support (including incentives), would be a substantially better approach.

The most effective way that a change in thinking and ultimately the method of soil decision-making will come about is by a general change in attitude toward the soil in the community. A change of attitude must be based on ecological respect for soil and it should work toward the establishment of a code of ecological ethics for soil. Soil must be seen as an integral element of the major global, national, and state biodiversity campaigns, particularly its important role in building a sustainable society as advocated in the *Global Biodiversity Strategy* (UNEP 1992b) and *Agenda 21* (UN 1992a). At the Commonwealth level, the challenge involves a greater commitment to the basic obligations of national environmental strategies, in particular the *Intergovernmental Agreement on the Environment* (Australia 1992b; Gardner 1994), to the soil. This action will make clear the responsibilities of the Commonwealth, the state and local governments to soil conservation policy and legislation, and it will ultimately provide a better basis for cooperative conservation programs.

References

Alexander, H. 1995. A Framework for Change, the State of the Community Landcare Movement in Australia, A National Landcare Facilitator Project. Annual Report. National Landcare Program.

Australia. 1983. National Conservation Strategy for Australia. Living Resource Conservation for Sustainable Development. Australian Government Publishing Service. Canberra.

Australia. 1984. Land Use Policy in Australia, Senate Standing Committee on Science, Technology and the Environment. Australian Government Publishing Service. Canberra.

Australia. 1986a. Economic and Rural Policy. A Government Policy Statement. Australian Government Publishing Service. Canberra.

Australia. 1986b. The Constitution as Altered to 31 October, 1986. Australian Government Publishing Service. Canberra.

Australia. 1987. Fiscal Measures and the Achievement of Environmental Objectives. Report of the House of Representatives Standing Committee on Environment and Conservation. Australian Government Publishing Service. Canberra.

Australia. 1989. The Effectiveness of Land Degradation Policies and Programs. Report of the House of Representatives Standing Committee on Environment, Recreation and the Arts. Australian Government Publishing Service. Canberra.

Australia. 1992a. National Strategy for Ecologically Sustainable Development. Australian Government Publishing Service. Canberra.

Australia. 1992b. Intergovernmental Agreement on the Environment.

Australia. 1995a. Native Vegetation Clearance, Habitat Loss and Biodiversity Decline. An overview of recent native vegetation clearance in Australia and its implications, Biodiversity Series. Paper No. 6. Biodiversity Unit. Department of the Environment, Sport and Territories.

Australia. 1995b. Landcare Information, Land, Water, and Vegetation Programs. 1995-1996. Department of the Environment, Sport and Territories, Department of Primary Industries and Energy. Australian Nature Conservation Agency.

Australia. 1996a. Outlook 96: Proceedings of the National Agricultural and Resources Outlook Conference, Canberra. Vol. 2. Agriculture. ABARE.

Australia. 1996b. State of the Environment. Australia. An Independent Report Presented to the Commonwealth Minister for the Environment by the State of the Environment Advisory Council. Department of the Environment, Sport and Territories.

Australian Soil Conservation Council. 1989. National Soil Conservation Strategy. Department of Primary Industries and Energy. Canberra.

Aveyard, J.M., and P.E.V. Charman. 1991. Soil Degradation and Productivity - A Concluding Perspective in P.E.V. Charman, and B.W. Murphy (eds.). Soils, Their Properties and Management. A Soil Conservation Handbook for New South Wales. Sydney University Press.

Balderstone, J.S. 1982. Agricultural Policy Issues and Options for the 1980s. Working Group Report to the Minister for Primary Industry. September 1982. Australian Government Publishing Service. Canberra.

Bates, G. M. 1992. Environmental Law in Australia. Butterworths. Sydney.

Beatley, T. 1991. A Set of Ethical Principles to Guide Land Use Policy. Land Use Policy 8:3.

Blackmore D.J., S.W. Keyworth, F.L. Lynn, and J.R. Powell. 1995. The Murray-Darling Basin Initiative: A case study in integrated catchment management. in Sustaining the Agricultural Resource Base. Office of the Chief Scientist. Department of the Prime Minister and Cabinet 61-67.

Blyth, M.J., M.G. Kirby. 1985. The Impact of Government Policy on Land Degradation in the Rural Sector in A.J. Jakeman, D.G. Day, and A.K. Dragun (eds). Policies for Environmental Quality Control. CRES. Australian National University. Canberra.

Boer, B.W. 1991. Sustaining the Heritage, in Our Common Future. J.M., Behrens, and B.M. Tsamenyi (eds). Environmental Law and Policy Workshop. Papers and Proceedings: Faculty of Law, University of Hobart, Australia. 88-102.

Boer, B.W. 1995. Institutionalising Ecologically Sustainable Development: The Roles of National, State, and Local Governments in Translating Grand Strategy into Action. Willamette Law Review 31:2:307-358.

Boer, B.W., and I.D. Hannam. 1992. Agrarian Land Law in Australia in W. Brussaard and M. Grossman. (eds). Agrarian Land Law in the Western World. C.A.B. International. Wallingford, Oxon. 212-233.

Booth, C.A., and C.H. Teoh. 1989. Total Catchment Management in NSW as a Community Involvement Process in Proceedings of Total Catchment Management Workshop. University of Wollongong.

Bradsen, J.R. 1987. Land Degradation: Current and Proposed Legal Controls. EPLJ 4:113.

Bradsen, J.R. 1988. Soil Conservation Legislation in Australia. Report for the National Soil Conservation Program. University of Adelaide.

Bradsen, J.R. 1991. Perspectives on Land Conservation. EPLJ 8:16.

Bradsen, J.R. 1992. Biodiversity Legislation: Species, Vegetation, Habitat. EPLJ 9:175.

Bradsen, J.R. 1994. Natural Resource Conservation in Australia: Some Fundamental Issues, in T.L. Napier, S.M. Camboni, and S.A. El-Swaify (eds). Adopting Conservation on the Farm, An International Perspective on the Socioeconomics of Soil and Water Conservation, Soil and Water Conservation Society. Ankeny, Iowa. 435-460.

Burton, J. 1986. The Total Catchment Concept and its Application in New South Wales. Hydrology and Water Resources Symposium. Brisbane.

Campbell, A., and G. Siepen. 1994. Landcare: Communities Shaping the Land and the Future. Allen and Unwin.

Carew-Reid, J., R. Prescott-Allen, S. Bass, and B. Dalal-Clayton. 1994. Strategies for National Sustainable Development. A Handbook for their Planning and Implementation. IUCN: The World Conservation Union. Gland. Switzerland.

Charman, P.E.V., and B.W. Murphy. (eds). 1991. Soils, Their Properties and Management. A Soil Conservation Handbook for New South Wales. Sydney University Press.

Chisholm, A., and R. Dumsday (eds). 1987. Land Degradation, Policies and Programs. Cambridge University Press. Sydney.

Commonwealth Scientific, Industrial and Research Organisation. 1996. Division of Soils. Figures supplied.

Constitutional Commission. 1987. Report of the Advisory Committee to the Constitutional Commission on Trade and Economic Management.

Crawford, J. 1991. The Constitution and the Environment. Sydney Law Review 13 (1):11.

Dendy, T., and J. Murray (eds). 1996. From Conflict to Conservation. Native Vegetation Management in Australia. South Australian Department of Environment and Natural Resources.

Department of Environment, Housing and Community Development. 1978. A Basis for Soil Conservation Policy in Australia, Commonwealth and State Government Collaborative Soil Conservation Study, 1975-1977. Australian Government Publishing Service. Canberra.

Department of Primary Industry. 1971. Study of Community Benefits of and Finance for Soil Conservation. Standing Committee on Soil Conservation. Canberra.

Department of Primary Industry and Energy and Australian Soil Conservation Council. 1988. Report of the Working Party on the Effects of Drought Assistance Measures and Policies on Land Degradation. Australian Government Publishing Service. Canberra.

Douglas, J., H. Alexander, and B. Roberts. 1995. Sustaining the Agricultural Resource Base: Community Landcare Perspective in Sustaining the Agricultural Resource Base, Office of the Chief Scientist. Department of the Prime Minister and Cabinet. 68-76.

Edwards, K. 1991. Soil Formation and Erosion Rates in P.E.V. Charman and B.W. Murphy (eds). Soils Their Properties and Management, A Soil Conservation Handbook for New South Wales. Sydney University Press, 36-47.

Fisher, M.P. 1986. Western Land Administration in New South Wales and Possible Guidelines for Australia in The Mulga Lands. North Quay Qld. Royal Society of Queensland, 155-159.

Food and Agriculture Organisation of the United Nations. 1982. World Soil Charter.

Fowler, R.J. 1984. Environmental Law and its Administration in Australia. EPLJ 1:10.

Fowler, R.J. 1993. A Brief Review of Federal Legislative Powers with Respect to Environment Protection. Australian Environmental Law News 1:51-64.

Gardner, A. 1994. Developing Norms of Land Management in Australia. The Australian Journal of Natural Resources Law and Policy 1:127-165.

Graham, O.P. 1992. Survey of Land Degradation in New South Wales, Australia. Environmental Management 2:205.

Hamblin, A. 1991. Sustainability: Physical and Biological Considerations for Australian Environments. Working paper No. WP 19/89 (revised edition). Bureau of Rural Resources. Department of Primary Industry and Energy. Canberra.

Hannam, I.D. 1992. The Concept of Sustainable Land Management and Soil Conservation Law and Policy in Australia in P. Henriques, (ed). Proceedings of International Conference on Sustainable Land Management. International Pacific College, Palmerston North. New Zealand 153-168.

Hannam, I.D. 1993. The Policy and Law of Soil Conservation. Unpublished Doctor of Philosophy. Macquarie University. Sydney.

Hannam, I.D. 1995. Environmental Law and Private Property Forest Management in New South Wales. Defending the Environment. Second Public Interest Environmental Law Conference. Australian Centre for Environmental Law. Adelaide.

Hannam, I.D., and W. Watkins. 1989. Concepts of Total Catchment Management and Future Directions, The Interdepartmental Committee and the Statutory Framework for Total Catchment Management in New South Wales. Proceedings of Total Catchment Management Workshop. University of Wollongong.

Harding, R. 1994. Interpretation of the Principles for the Fenner Conference on the Environment. Sustainability-Principles to Practice. Unisearch. University of New South Wales.

Houghton, P.D., and P.E.V. Charman. 1986. Glossary of Terms Used in Soil Conservation. Soil Conservation Service of New South Wales and the Standing Committee on Soil Conservation.

International Union for the Conservation of Nature and Natural Resources, United Nations Environment Program and World Wildlife Fund. 1991. Caring For The Earth. A Strategy for Sustainable Living. Gland. Switzerland,

McDonald, G.T., and T.J. Hundloe, 1993. Policies for Sustainable Future in G.H. McTainsh, and W.C. Boughton (eds). Land Degradation Processes in Australia. Longman. Cheshire. Melbourne 347-383.

McTainsh, G.H., and W.C. Boughton (eds). 1993. Land Degradation Processes in Australia. Longman. Cheshire. Melbourne.

Morgan, R.P.C. 1986. Soil Erosion and Conservation. Longman Scientific and Technical. Hong Kong.

Musgrave, W., and R.A. Pearse. 1984. Soil Management Policy in Australia: Institutions, Criteria and Socioeconomic Research. In E.T. Craswell, J.V.

Remenyi, and L.G. Nallana (eds). Soil Erosion Management. Proceedings of a Workshop held at PCARRD. Los Banos. Philippines. Australian Centre for International Agricultural Research Bureau of Soils.

New South Wales Department of Land and Water Conservation. 1996. Native Vegetation Protection and Management in NSW. Discussion of Options and Reform.

Olembo, R.J. (ed). 1983. Environmental Guidelines for the Formulation of National Soil Policies. United Nations Environment Program. Nairobi.

Parliament of New South Wales. 1984. Reports of the Joint Select Committee of the Legislative Council and Legislative Assembly to Inquire into the Western Division of New South Wales. Government Printer. New South Wales.

Prately, J.E., and D.L. Rowell. 1987. From the First Fleet - Evolution of Australian Farming Systems in P.S. Cornish, and J.E. Prately (eds). Tillage, New Directions in Agriculture. Inkata Press. Sydney.

Roberts, B. 1992. Landcare Manual. New South Wales University Press. Kensington. NSW. Australia.

Shelly, E. 1990. Degradation of Marginal Cropping Lands. Soil Conservation Service of New South Wales. Technical Report No 24.

Soil Conservation Service of New South Wales. 1987a. State Soils Policy.

Soil Conservation Service of New South Wales. 1987b. State Tree Policy.

Soil Conservation Service of New South Wales. 1987c. Total Catchment Management Policy.

Soil Conservation Service of New South Wales. 1989. Land Degradation Survey New South Wales 1987-1988.

Soil Conservation Service of New South Wales. 1991. Discussion Paper. Total Catchment Management.

Standing Committee on Agriculture and Resource Management of the Agricultural and Resource Management Council of Australia and New Zealand. 1996. Indicators for sustainable agriculture: Evaluation of pilot testing. A report prepared for the Sustainable Land and Water Resources Management Committee.

The Commission on Global Governance. 1995. Our Global Neighbourhood. The Report of the Commission on Global Governance. Oxford University Press, Inc. New York.

United Nations. 1992a. Agenda 21.

United Nations. 1992b. The Rio Declaration.

United Nations Environment Program. 1982. World Soils Policy. Nairobi.

United Nations Environment Program. 1992a. Convention on Biological Diversity.

United Nations Environment Program. 1992b. Global Biodiversity Strategy. Guidelines for Action to Save, Study, and Use Earth's Biotic Wealth Sustainably and Equitably.

Vigod, T. 1989. Legal Tools for Implementing Sustainable Development. Address to the University of Western Ontario. Faculty of Law Interdisciplinary Conference on Planet out of Balance: Is Sustainable Development the Solution, Canadian Environmental Law Association. Ontario.

Wadham, S.M., and G.L. Wood. 1939. Land Utilisation in Australia. Melbourne University Press.

Woods, L.E. 1983. Land Degradation in Australia. Department of Home Affairs and Environment. Australian Government Publishing Service. Canberra.

33

Landcare Policies and Programs in Australia

John R. Bradsen

Landcare has its origin in rural community groups. These groups began to emerge in several rural communities in the mid-1980s in response to concern about both land degradation[1] and rural community decline,[2] and encouraged by financial assistance. The term Landcare came to describe these groups and the land caring activities they undertook.[3]

The name Landcare was first formally recognized when it was registered in Victoria in 1985 as the name of a state government program to assist voluntary community groups. The word Landcare has been commonly used since the introduction of Landcare as a federal program in 1989, to refer to the groups, their activities and the entire Landcare program. This emerging nomenclature is a useful commentary on the evolutionary, rather chameleon-like character of the Landcare movement.

The emergence and development of community Landcare groups is a remarkable phenomenon which would seem to be explained largely by the three factors mentioned above. First, these groups revealed a landholder concern about degradation and demonstrated a grass roots capacity to act (Alexander 1995), which sets an empowering, nonthreatening example for other rural communities to follow.

Secondly, given difficult rural conditions and rural communities in decline, there was an urge to recreate a sense of community, and Landcare groups and the Landcare movement have continued to meet this need. The expression of this urge in the form of Landcare groups may well reflect the Australian ethos of mateship in hard times. Thirdly, the availability of financial assistance has been significant. Groups received federal funding in 1985 under the National Soil Conservation Program (NSCP) (Bradsen 1988), and from 1992 under the National Landcare Program (NLP).

Perhaps the most significant feature of the Landcare movement is the group concept which has changed the whole conservation dynamic in both its physical and social dimensions. Landholders often saw degradation next door but not their own. Many struggled alone with degradation which extended beyond property boundaries. The group structure has ensured a more open, objective acknowledgment of the problems, a more open discussion about the issues, a wider breadth of vision, and has helped bolster the activity of flagging individuals. In short, the group focus has resulted in greater awareness, action, commitment, and support.[4]

In looking at Landcare, group numbers have been something of a touchstone. When Landcare became an official national program—in 1989 there were some 200 rural groups in existence (LAL 1990)—it was assumed that this number would stabilize at about 1,000-1,200. In fact, group numbers have continued to climb and had risen to some 2,728 as at late 1995 (LAL 1995). Only about 30 percent of landholders, however, are members of groups.

This explanation for this remarkable and continued growth in group numbers is that Landcare, fostered by a major awareness campaign and an increasingly environmentally conscious population, has expanded well beyond its original rural focus. Rural groups still lie at the heart of Landcare, and it is at them that core programs like the Property Management Planning process applies. Groups are being formed, however, to grapple with a wide range of urban, coastal, and other environmental issues in the fields of land and water management and nature conservation. The breadth of issues involved is illustrated by the wide range of their broadly self-explanatory names such as "Bushcare," "Coastcare," "Dunecare," "Friends of Grasslands," and "Park Care."

Other groups with a monitoring focus have names such as "Streamwatch," "Saltwatch," "Rivers of Blue," and "Frogwatch." Urban groups, including school-based groups, are proliferating. A good example of the latter is the Kingscliff High Enviro Club which describes itself as a "foundation" member of Landcare.[7] And there is now even "Officecare" which encourages city people to join Landcare by changing their work habits.

These groups are linked together in that they appear to see themselves as part of the Landcare movement, though many of them appear to understand this in broad environmental terms. They do not fall within what may be termed "institutional Landcare" and may be hard-pressed to say exactly what they mean by Landcare.

There is enormous value in the group concept and value in it being sufficiently widely defined to encourage general community participation (Alexander 1995). However the diffusion of Landcare is now so considerable that it may come to hinder its capacity to provide the structure and the rigorous consideration of sustainability that is necessary to ensure that the use of natural resources is truly sustainable.

Landcare program

The Landcare proposal was put to the federal government jointly by environmental antagonists, the Australian Conservation Foundation and the National Farmers Federation (NFF and ACF 1989). It is hardly surprising that the proposal, consisting of 13 elements, all of equal emphasis,[8] was found to be politically irresistible. This origin and the continued cooperation of these two groups has been a key factor in Landcare's development.[9]

Landcare came into existence as a national program in July 1989 when the prime minister issued a statement on the environment entitled *Our Country Our Future*. The Landcare Program was one of three elements in a package of measures to attack the problem of land degradation. The three were: first, the Year (1990) and the Decade of Landcare (to the year 2000); second, a review of policies; and third, an expansion of the National Soil Conservation Program (NSCP). Landcare was said to be the federal government's contribution to the National Soil Conservation Strategy (NSCS) introduced in 1987 (LAL 1990).

To implement these measures it was proposed that each state and territory should prepare a Decade of Landcare Plan, that the federal government should do likewise, and that these component plans should together form the National Decade of Landcare Plan.[10] These plans were progressively completed by August 1992.

Landcare's institutional evolution, clearly influenced by its evolutionary development at a grass roots level, has also been remarkable. When the National Landcare Program (NLP) was first launched in 1989, it was a program within the National Soil Conservation Strategy (NSCS) and introduced to operate alongside the National Soil Conservation Program (NSCP) and the Federal Water Resources Assistance Program (FWRAP).

Other separate programs also introduced in 1989 included One Billion Trees (OBT) and Save The Bush (STB).[11] These two programs were based on the recognition that much natural resource degradation began with clearance of native vegetation. Hence they sought respectively to plant one billion trees in the Decade of Landcare and encourage the retention of remaining native vegetation.

Another related program is the Murray Darling Basin Natural Resources Management Strategy (NRMS). It was introduced to deal on a total catchment management basis with the enormous degradation problems in the basin which is Australia's agricultural heartland.[12]

National Landcare Program

By 1992, the need for greater integration of natural resource management had become apparent. Landcare had been such a success that it was given the preeminent policy position with the creation of the National Landcare Program (NLP) as the umbrella program within which others were located.[13] The National Soil Conservation Program (NSCP) and the Federal Water Resources Assistance Program (FWRAP) disappeared by name to be amalgamated as the core land and water elements of the NLP, while other programs, the OBT, STB, and NRMS, became distinct components of the National Landcare Program.

Additional components of the NLP now include the Waterwatch Program (WP), also created in 1992 to provide a national focus for community groups to monitor water quality on a coordinated basis, and the Drought Landcare Program (DLP). This component is a special short-term initiative introduced in 1995 to stimulate protective Landcare activity when natural resources are vulnerable in time of drought.

Two other associated national programs are Landcare and Environment Action Program (LEAP) and Regional Environmental Employment Program (REEP). They seek to marry Landcare with employment and training to provide work for around 15,000 and 10,000 unemployed people respectively, particularly on Landcare/local community projects.

There are other programs which should be mentioned. The River Murray Corridor of Green Program (COGP) is a 3-year program to develop vegetation corridors within the 100 kilometer wide, 2500 kilometer long river corridor. The Rural Adjustment Scheme (RAS) assists landholders to move away from inappropriate land use. Many agricultural enterprises in Australia are of dubious viability and readjustment should play a more significant role, but the scheme is small and largely confined to cases of manifest nonviability.

Other programs could be mentioned. Landcare information lists some 60 organizations or programs which have objectives or functions related to Landcare, some of which may come to be part of Landcare as it continues to evolve as has occurred with community groups.

Several other important institutional aspects of Landcare which are provided for in the National Decade of Landcare Plan must be mentioned. The Commonwealth Plan in particular, which establishes the overall strategic framework for Landcare, provides for three categories of coordinator and two categories of facilitator. Coordinators play a key role with individual Landcare groups, while facilitators operate at the regional and national level.

Group coordinators are employed full or part time by individual groups to organize and run group meetings and activities. Their role is invaluable for the effective functioning of groups.[14] State Landcare coordinators are employed by state and territory governments to develop policies for Landcare groups and act as a focal point for group activities within state agencies. Decade of Landcare coordinators are similarly employed to promote awareness in coordination with Landcare Australia Limited.

Regional facilitators are employed by state and territory governments to stimulate group development within regions and to promote their coordination on a regional basis. The National Facilitator is engaged by the federal government with terms of reference which include evaluating, monitoring, and providing advice on Landcare on a national basis.

These institutions, from local group coordinators to the national facilitator, are crucial features of Landcare without which it could not function as it does. Improvements could be made, however, including better training and accreditation especially at the regional level where Landcare is weakest.[15]

The NLP Regional Initiatives Program seeks to promote the development of board scale regional structures and other programs such as COG and NRMS support projects at the subregional scale (Alexander 1995). Moreover the Regional and State Assessment Panel process lends the structure a regional thrust. But a clear, long-term strategic framework for identifying regions and for the regional implementation of Landcare is lacking at a national level.[16]

Varying regional structures are established under state legislation. Examples are Land Conservation Districts in Western Australia.[17] District Soil Boards in South Australian,[18] Total Catchment Management Committees in New South Wales,[19] and Catchment and Land Protection Boards in Victoria.[20] Each seeks to integrate natural resource management on a regional basis most commonly the catchment. However, the extent to which these operate effectively with Landcare is variable.

It should be said that although Landcare now involves an elaborate institutional framework, it is still not implausibly spoken of as a "movement." Here the distinction again needs to be referred to between the program/policy aspects of "Landcare," that is the official policy of component programs which go under the heading National Landcare Program, and the activities aspects of Landcare, that is the wide range of "landcare" activities which may not have a formal role within the program but which are nonetheless seen by the community as part of Landcare. The fact that they are so seen is no doubt in considerable degree testament to the extraordinary success of the Landcare awareness campaign.

Landcare and awareness

Generating awareness was seen from the outset as critical to the success of the Landcare program. This obviously applied to landholders, but because significant public funding was required and wider community support sought, awareness needed to cover the urban population. To this end, in October 1989 the federal government created a nonprofit corporation limited by guarantee, called Landcare Australia Limited. It was given two broad functions, which it has summarized as creating awareness of Landcare, including a Landcare ethic, and raising nongovernment funds to support Landcare.[21]

Landcare Australia Limited became fully functional in March 1990 and focused its early efforts on promoting awareness. This included establishing extensive contact with the media where it now has well over 1,000 outlets. It also obtained sponsorship in the form of media space or time (LAL 1995).

Central to the awareness campaign was the development of Landcare awards. Eight categories of awards were introduced in 1991 in each state and territory[22] (now increased to ten),[23] with the winners being eligible for the national awards. These have attracted considerable prestige and publicity with awards being made by the prime minister in the Great Hall of Parliament before some 500 distinguished guests (LAL 1994). They have also received considerable publicity with the 1994 awards resulting in the publication of 146 articles.

Landcare Australia has also sought awareness through promoting March as the National Landcare Month with themes such as "catchments" and "Landcare is good business."[24] The highlight of the month is the national awards. A National Landcare Day in September[25] has also received considerable publicity.

Awareness has also been sought through sponsorship and promotion of particular events. Perhaps the best known example of the former is the Uncle Toby's Company "Let's Landcare Australia" campaign run yearly since 1992, while among the latter the "Angry Anderson Challenge" stands out. This media personality obtained major TV coverage for a project aiming to plant 1 million trees on 14 sites in the River Murray area (LAL 1995).

Landcare Australia has employed a leading research organization[26] to monitor public awareness on a regular basis. The results show that the national awareness of Landcare has risen from 22 percent in 1991 to 66 percent in 1994, 10 percent to 57 percent in the capital cities.[27]

Several comments may be made about this campaign. It has clearly raised awareness and no doubt stimulated considerable activity in support, especially in the cities. But whether it demonstrates the "fostering of a Landcare ethic" as claimed (LAL 1993) is less clear. More particularly, being aware of Landcare and its logo does not indicate any real grasp of the meaning and implications of sustainable resource use. Moreover, there has been very considerable awareness in the past which has come to remarkably little, a matter discussed below.

There is no doubt that a high level of awareness has been achieved. But it should not be seen as achieving, or even as a way of achieving, sustainable resource use. It should be seen as presenting an opportunity, perhaps a "one off opportunity," for real change. If it is squandered it may well set back the move towards sustainability for a considerable time to come.

Landcare and funding

Landcare groups were first funded in 1985 under the NSCP[28] and financial assistance has been one of the mainsprings of Landcare groups, "a tremendous catalyst" (Campbell 1992). Nevertheless they were initially formed and functioned in good measure without external funding. Indeed much Landcare activity is still landholder and community funded. One condition of NLP funding is that recipients contribute at a rate of at least $1 for each $2 of assistance. It has been said that the figure is really about $4 for $149. Other estimates suggest that the figure is much higher.[29] This may be consistent with an NLP goal to have groups, once established, bear more of their Landcare costs. Be this as it may the Landcare movement could not have developed as it has without public funding.

When Landcare became official policy in 1989, the federal government indicated that it would provide the total sum of $320 million to be made available over the period of the Decade of Landcare which was widely applauded as a long-term commitment. This funding was divided among the various programs existing at that time.

From 1992, when the NSCP and FWRAP became the core land and water elements of the NLP, the funds that were available under those programs were transferred accordingly. These funds are divided into these three components: first, a community Landcare component, which deals with the community group aspects of the various activities and programs; secondly, a Commonwealth/state/territory component, which is concerned with the functions undertaken jointly by the Commonwealth and states or territories and which are implemented through Partnership Agreements; and thirdly, a national component, which focuses on activities that have national significance.

Other NLP programs such as OBT, STB, and NRMS, which also make funds available to the community, continued to be funded separately though under the general umbrella of the National Landcare Program. Funds also became available under other programs such as COG and DLP.

The way in which Landcare funds are available to the community is somewhat controversial. The original Landcare proposal was that funds were to be directed at the underlying causes of degradation with the broader picture and longer time frame in mind, not at the on-ground symptoms. Funds were to be "catalytic" with the general purpose of promoting future sustainable land use[30] rather than for dealing with the symptoms of past unsustainable use. Individual landholders were therefore expected to be responsible for on-ground works with such costs to be incorporated into their economic strategies.

Such funding remains available for groups but not for individuals, and for activities like natural resource mapping and planning, for demonstrations and trials and for the employment of a coordinator, but not for on-ground works. The focus is community and change. Hence it is said that there is a "trend within the NLP to support projects that satisfy the criteria: 'It is community driven, it has a strategy, it is about change'" (Alexander 1995).

Landholders have expressed some frustration with the lack of funds for on-ground work as many have the natural inclination to deal with the more obvious and immediate on-ground problems. This concern is obviated to some degree, however, in that funds are available to individuals for on-ground works under other components of the NLP such as the OBT, STB, NRMS, DLP, and COG programs. Moreover, although community funding is not lavish with most NLP grants being under $5,000, the ceiling of $20,000 per grant has been lifted (Alexander 1995).

An issue of concern is the availability of funding at the regional level. Funds are available under the State/Commonwealth component of the NLP for regional initiatives and under NRMS, COOP, and DLP in the context of a regional plan (Alexander 1995). The confusion which tends to prevail where funding is concerned is high at the regional level—a situation not helped by the range of government departments involved.

The processing of applications has been improved in that applications are assessed by Regional and State Assessment Panels which provide joint community/government forums for assessing priorities and projects.[31] This increases regional coherence though the system is under strain from the large number of small and short-term applications. Funds are now available under the *One-Stop-Shop* process (Alexander 1995), whereby, from the landholder perspective, all applications for funding have a common entry and exit point.

The emphasis on federal funding should not obscure the fact that Landcare is also supported by considerable state government funds. However, some funds such as those previously provided for extension services, for example, have been rerouted to become Landcare funding. Detailed information cannot be given here, however, as there is no comprehensive register of state Landcare funding.

It should also be mentioned that public support for individuals is also available through tax deductibility, at least for those landholders fortunate enough to be liable for tax (Alexander 1995).

The Landcare Foundation is also a significant source of funding. The second function of Landcare Australia Limited was to raise funds from the private sector and it created the foundation early in 1994 as its fund-raising arm. In addition to general donations, it created three categories of giver: gold, silver, and bronze, exceeding $200,000, $100,000, and $50,000, respectively. Fuji Xerox with $800,000 launched Officecare. The Foundation is nearly halfway to its target of $10 million after 3 years (LAL 1996). The first major Foundation Project is the 1 million tree Angry Anderson Challenge project for which it is providing $850,000.

Landcare and history

Landcare lacks a sense of history. Indeed it tends to deny history with the view that the past is the past and that we simply go forward from here.[32] Unfortunately, we cannot. It does seem that without an awareness of history we repeat mistakes. This does not just mean repeating a particular act, but includes, for example, being led astray by following similar mindsets.

Historically, in Australia, attitudes toward land use have been dominated by myths. A classic example is the myth that land degradation in Australia was the result of adopting British methods of agriculture with which Australian soils could not cope. In fact, until relatively recently, British agriculture, both in practice and in its values, placed considerable emphasis on conservation, adopting methods of land use appropriate to each area. Its "method" of agriculture was to conserve.

Australian agriculture, in contrast, quickly developed an aggressive land use ethos and practices to match what was most often inappropriately adapted to particular areas. Its "method" of land use was to exploit. It is not suggested that this is commonly the case today. The point is that the myth was crucial in precluding Australian land users and others from seeing what they were doing.[33] The critical lesson from history for Landcare is not so much that we may still have that particular myth, but rather to ask what are the present myths and how might they be leading us astray?

There is another question posed by history. Landcare emphasizes the importance of awareness. It is not suggested that this is inappropriate but it needs to be remembered that over 50 years ago there was such widespread awareness of damage to land that a member in parliament could say that, "During the last few years the question of soil conservation has been on the lips of practically everyone. So much publicity has been given to it that the public has become erosion conscious" (Bradsen 1988). The lesson from history is to ask why so much awareness was, in the long run, apparently of such limited effect, and whether in our own way and time Landcare might be repeating the pattern.

Similar comments can be made about the use of knowledge. It was well known in the 1940s for example, that salinity was caused by the over-clearance of native vegetation (Bradson 1988). Yet massive, foolish clearance was allowed, indeed encouraged, including in western Australia, where in Parliament salinity was explained and where we now have the greatest salinity problem. Yet, at the time, people still managed to feel justified in what they were doing. The question which history poses here is whether there are aspects of Landcare where we fail to apply what we know, or we are blind, foolish, or lacking in courage.

The history of land degradation, including the archeology of land degradation, shows how seriously we must take nonrenewable resource conservation issues. It may be that in various respects we are repeating mistakes and not taking the issue seriously enough. One way in which to explore this issue is to ask what view history will be likely to take of the Decade of Landcare. The historic pattern does seem to be one of crisis followed by action followed by crisis and so on. Will history look back and see that Landcare fits this pattern? Will it be seen as really effective or as only effective enough to delay and thus magnify the next crisis?

Landcare and philosophy

The effectiveness of a program such as Landcare turns on philosophical as well as practical considerations. The group ethos which lies at the heart of Landcare is an excellent case in point and, although the groups are practical, the group ethos is Landcare's great philosophical strength.

The case for a landcare ethic has been well argued for some time (Roberts 1984), and its importance for Landcare is highlighted by the first goal of Landcare Australia Limited, which was to develop awareness of Landcare and of a Landcare ethic. There appears to be no doubt that this approach has engendered awareness and a broad community support, but it is suggested that too much is expected of it and that the debate needs to be more specific and focused.

Establishing a Landcare ethic requires more than a general awareness and involves some real understanding of the issues. An awareness of Landcare, which is essentially what surveys test, says little about an underlying appreciation of sustainable resource use or the implications of its adoption. Indeed such a campaign may create an idealized or superficial view of sustainability.

This general awareness approach may be contrasted with the approach to resource use which sustained indigenous peoples. They did not rely on a broad ethic. Their ethic may have provided a philosophical framework, but at the core of their sustainability system were quite specific, hard-edged obligations, typically part of the kinship system. The ethic had no efficacy without the obligations. We must ask whether the same might be the case for us.

There are also several specific respects in which philosophical debate about land care needs to be more focused. One issue in need of further exploration is the relationship between the profitable and the sustainable. Landcare is permeated with the notion that land care must be both profitable and sustainable.

It is true that private landholders/resource users cannot continue to operate while making a loss. But they cannot make a profit in the long run if the production system is not sustained. Even more fundamentally, the community cannot survive in the long run if its essential resource base is not sustained. Thus one cannot boldly insist on both profitability and sustainability either for the individual or for the community by pushing the resource base harder.

It is not the resource, such as the land, but the enterprise which needs to be profitable. To this end enterprises must be restructured, or alternative activities or value adding adopted. In the case of Australian landholders, for example, this may be practically difficult given the average age of nearly 60 and given a widespread lack of education, skill, and confidence.

These issues give rise to philosophical problems including questions about how the community views its most fundamental resources as against the value it places on individuals in time of change. But these problems are no reason for not looking things in the eye. Both profitability and sustainability cannot be insisted upon if this allows the resource base to be pushed too hard.

A similar question arises concerning the relationship between the private and the public. The notion broadly underlying Landcare is that individual landholders should bear environmental costs where the benefits are private, or go to the landholder, whereas the community should bear the cost where the benefits are public. Present funding arrangements very broadly seek to reflect this distinction.

To a degree this may be a question of fact. But in good measure it is also a philosophical question which includes asking about the extent to which present landholders should bear the costs of repairing resources or restructuring enterprises where that need results from exploitation which was set in train at an earlier time, often at the behest of governments, and which underlies the present wealth of city populations.

These two relationship issues tend to merge when one adds the question of time. One could argue that sustainability is inherently a long-term, and hence, public issue, while profitability is inherently a short-term private issue. Thus expenditure going to the former is in the public interest. In general, there is confusion about these issues in that resource users are asked to play both a stewardship and a commercial role at the same time with poor guidance about the drawing of boundaries.

There is a one other issue requiring consideration which has a philosophical aspect. It concerns the lack of an overall, strategic framework or "shared vision" (Alexander 1995) for Landcare and sustainable development, especially in rural Australia. Landcare is largely an amalgam of local responses to local problems and local community decline. If it has a unifying philosophy it would seem to be the group ethos. But there also needs to be a national long-term philosophy or shared vision. And this must in good measure turn on the question of sustainability.

Landcare and sustainability

The fundamental goal of Landcare, when first introduced in 1989, was that all land would be used within its sustainable capacity by the end of the Decade of Landcare. Such a clear-cut goal may have been politically useful, but it is so unattainable that it suggests a failure to understand the nature and extent of the problem of sustainable resource use and of the difficulties in framing effective solutions.

We are now halfway through the decade yet "it is not possible to say that Landcare group activity is responsible for the greater adoption of sustainable land use practices. Moreover, compared with other developed nations, Australia has only rudimentary information on the productive capacity of its land resources, and the potential hazards associated with their use" (Australia report 1996). This is surely remarkable not only after 5 years of Landcare but also after all the effort that went into the development of the National Strategy for Ecologically Sustainable Development.[34]

In physical terms the problems are considerable and sustainability questions remain unanswered. Nevertheless, there is the knowledge and the pressing need to develop sustainability indicators. Where there is uncertainty we must apply the precautionary principle, now formally adopted as policy in Australia. Moreover, applying concepts such as degradation or conservation to single resources like soil or water is too narrow a way to think about sustainability. The holistic consideration of all resources in their context and on a regional basis, such as a catchment, is necessary.

Sustainability also has a conceptual aspect which means that it can never have a fixed or final meaning. Such a view is like that of a student who believes the professor has the "right answers" to fundamental questions hidden in a drawer. Sustainability is a way of thinking about the world as well as about physical reality. It will change as circumstances and our understandings change. For instance the adoption of the precautionary principle must shift our way of thinking about sustainability. Hence, our search for physical or on-ground solutions must be matched by a constant examination of our understandings.

So far as Landcare is concerned, the approach to sustainability is naturally influenced by Landcare groups and tends to reflect the level of awareness and concern of local communities. The result is that the more glaring problems and those having the most immediate economic consequences are attended to, rather than the underlying sustainability issues. Thus, groups focus on things like gully erosion, conspicuous weeds, contour banks, salinity, waterlogging, and especially tree planting.

Revegetation is important. Much is being achieved, and one may get the impression that Australia is being revegetated. Yet, despite some laws prohibiting clearance, there is still far more vegetation being cleared than planted. Indeed, the program to plant 1 billion trees in a decade pales against the billions of native plants being cleared every year. The revegetation efforts are somewhat reminiscent of the Civilian Conservation Corps in the United States (Mealing 1995), but one would hope that Landcare's tree planting will leave a longer legacy.

Meanwhile more insidious and fundamental issues affecting "underlying production systems" (Alexander 1995), such as soil structure decline, acidification, declining organic carbon, poor vegetation management, and loss of biodiversity, receive limited attention. Therefore it is not surprising that with half the Decade of Landcare gone we "have (still) only taken the first tentative steps toward sustainability." Nor is it surprising that "in many areas the economic and environmental viability of current practice is on a downward spiral" (Alexander 1995).

It is time to take a much harder look at sustainability and its implementation. It is a complex, evolving concept having physical, conceptual, and social/organizational aspects, and there is much to learn. But there is also much that we know. It may be that some present uncertainty reflects our wish to press our resources closer to their limits. As long ago as the 1930s it was said that the effective implementation of existing knowledge would be a major step in dealing with our problems. This is still the case, and it is time to take a much harder look at sustainability and its effective implementation.

Landcare and legislation

The prime minister's statement launching Landcare said that "Effective legislation is the foundation of a comprehensive soil conservation program" (LAL 1990). This has not occurred. Indeed, sensible debate has not taken place, as it is quickly derailed by the emotive use of words like "regulation" and by assumptions that the law should only be used to penalize serious offenders.[35] The latter goes without saying; hence, if a landholder threatened a legitimate Landcare

527

inspector, a penalty would be available. But the law should not be thought of as if its only purpose is to lay down detailed rules governing the use of resources like land which are then enforced against individuals.

Before outlining the use of law it is appropriate to comment on the justification for its enactment since this should be the subject of community discussion. Its essential basis is that the community has a strong and legitimate interest in the sustainable use of its natural resources which greatly exceeds the interest of any particular resource user. There are two forces which dominate recource use and management. The first is economics and the second is law. Other factors like attitudes and policy play a role, but they are subservient to these forces. Market forces will not ensure sustainable use.[36] This leaves the law.

The primary law should not set out detailed prescriptive rules. It should provide for a comprehensive framework for Landcare and a comprehensive program for sustainable land use.

The organizational framework requires an effective national focus. A Landcare Commission has been recommended to oversee implementation of the next phase of Landcare. This is consistent with a truly long-term commitment to Landcare and with the development of a shared vision and coherent program. Where a body has a holistic oversight over a range of resources previously, the preserve of a range of ministers and departments, questions of ministerial responsibility and departmental jurisdiction will arise. But the present central structure is complex and confusing.

There also needs to be a stronger regional planning structure within which local group activity can be effectively coordinated. Again the present picture is complex and confusing. The need is for a more coherent structure which leaves room for continued evolution.

The community group process should remain flexible and able to meet the needs of local communities without more formal structure than is necessary to provide for accountability. Where local planning is involved, it needs to be consistent with regional arrangements. Resource users cannot, of course, be required to join groups. Hence, if Landcare is to be made comprehensive, this needs to be ensured at the individual property level.[37] Thus, it should become the norm for all resource users to have and follow a property plan. Methods of accreditation and the greater use of computer technology could be used, for example, to obviate the need for extensive administration.

At the substantive level, primary law should ensure the development of sustainability indicators or best bet practices. This should occur at both a regional and local level and involve local people and communities. But it needs to be based on rigorous sustainability criteria. There should also be maximum reliance on local responsibility to operate monitoring mechanisms.

Landcare conclusion

In assessing Landcare one would have thought that there would be an agreed basis on which to do so. But this is not the case. Several approaches have been suggested, including the number and health of Landcare groups and "works on the ground." It is suggested that such specific tests will invariably prove to be inadequate, while the ultimate goal of actually managing resources sustainably can only be applied retrospectively, if at all. Thus, as things stand, the best assessment of Landcare is the extent to which it amounts to a comprehensive, rigorous system which will provide the greatest likelihood of ensuring the sustainable management of Australia's nonrenewable natural resources within reasonable time. Despite its remarkable achievements, as yet Landcare falls short.

Notes

1. Only one comprehensive study has been undertaken into the state of Australia's land. Commonwealth and State Governments (1978), A Basis for Soil Conservation Policy in Australia Report No. 1 AGPS Canberra. The most recent study is Australia State of The Environment (1996), an independent report prepared by expert reference groups and presented to the Commonwealth Minister for the Environment. 1996 CSIRO Adelaide.
2. Farm numbers in Australia have dropped dramatically from 189,400 in 1970 to 124,975 in 1990 and may drop to 75,000 by 2000. Hassall & Assocs P/L 1994 Economic Study for the "Review of the Landcare Group" AUC 494 cited in Alexander fn 4 p. 3. Along with other factors there has been a serious decline of some rural communities.
3. For an outline of the origin of Landcare groups see S. Ewing. It's in Your Hands: An assessment of the Australian Landcare Movement, Ph.D. Thesis UOM 1995 p. 85.
4. Their rapid growth reflected a need as well as a concern. Sustaining the Agricultural Resource Base. Independent paper prepared for the prime minister's Science and Engineering Council. AGPS 1995.
5. Landcare has helped create a culture in which it is acceptable, even expected to consider sustainability issues. fn 4 p. 14.
6. It has tended to place heavy burdens, however, on more committed members and "There is concern about burnout in key group members..." fn 4 p. 31.

7. Detailed information is available on the Internet.
8. These were 1) Landcare groups; 2) property plans; 3) technical support and advice; 4) taxation rebates; 5) incentives for conservation farming; 6) direct government funding; 7) state support; 8) national assessment; 9) land capability; 10) economic data; 11) legislation; 12) conservation incentives; 13) education.
9. It is also significant that the committees responsible for the preparation and review of Landcare plans included wider community representation.
10. There is a separate document entitled Decade of Landcare Plan which is a very slim document.This is in contrast to the components, see, e.g., the 55-page Commonwealth Decade of Landcare Plan.
11. These programs were a response to the aggressive overclearance of native vegetation which was well recognized as the precursor to much of Australia's natural recourse degradation.
12. The basin produces over 40 percent of Australia's agricultural output, including half the sheep flock, half the cropland, and three quarters of the irrigated land. fn 9 p. 61.
13. Each of the programs closely associated with the National Landcare Program are briefly described in fn 4.
14. There are now over 150 group coordinators nationwide. fn 5 p. 8.
15. fn 5 There a number of points in this document that refer to the theme of regional inadequacy.
16. All levels of government and the community are looking to the catchment and regional level for the implementation phase of Landcare fn 5 p. 65.
17. Soil and Land Conservation Act 1945 (WA).
18. Soil Conservation and Land Care Act 1989 (SA).
19. Catchment Management Act 1989 (NSW).
20. Catchment and Land Protection Act 1994 (Vic).
21. The company's seven objects as prescribed in its Memorandum of Association are set out in Landcare Australia Limited Second Annual Report 1991 p. 1.
22. The categories are: Landcare primary producer; Landcare individual (non primary producer); community Landcare group; media; research (or innovative technology); local government; education; business. falO p. 7.
23. With the addition of the BP Landcare Catchment Award and the Landcare Nature Conservation Award sponsored by the Australian Nature Conservation Agency. fn 12 p. 5.
24. For the fourth and fifth months, respectively. fn 39.
25. fn 39 p. l2 for an example of activities undertaken.
26. Roy Morgan Research Centre

27. The LAL Annual Reports also show that the national recognition of the Landcare Logo reached 62 percent in 1994. fn 12 p. 5.
28. Soil Conservation (Financial Assistance) Act, 1985 C/W
29. Alexander puts the figure at from $5 to $12 for $1. fn 5 p. 68.
30. Commonwealth Decade of Landcare Plan component p. 27, 30.
31. fn 5 p. 71. This process has been praised for its role in integrating community and departments and for effectively locating local priorities within a national framework. It is under strain, however, because of the need to process many small applications and because of short time frames.
32. "Landcare has succeeded because it is forward looking." Fischer T MP Deputy PM fn 39.
33. This is not unlike the U.S. myth about rates of soil formation which comforted those who applied the USLE.
34. The core objectives of the strategy are 1) to enhance individual and community well-being by following a path of economic development that safeguards the welfare of future generations; 2) to provide for equity within and between generations; and 3) to protect biological diversity and maintain essential ecological processes and life support systems. Seven guiding principles follow. Council of Australian Governments 1992 AGPS Canberra. The strategy searches for a balance between sustaining resources and sustaining the economy. The short term is hard to resist.
35. Such discussions in Australia are also bedevilled by argument about states' rights. It is time to focus the discussion on the resources and on the measures best able to ensure their sustainable use.
36. Indeed they may be bound to do otherwise. Where economic forces are shaped by legislation through targeted subsidies or cross compliance, then in truth this is not the market at work but a legislative program. Australia cannot afford costly schemes of this sort. 78 fn 6 p. 75.
37. Ewing says that "Landcare has succeeded, probably beyond even the most optimistic of expectations, in mobilizing community groups." p. 253. Nevertheless, she also points out that "Landcare is not the all-embracing program that the rhetoric would have us believe..." p. 249.

References

4 Landcare Information Land, Water and Vegetation Programs 1995-1996 3rd Ed p. 2 DPIE Canberra.

Alexander, H. 1995 A Framework for Change The State of the Community Landcare Movement in Australia. The National Landcare Facilitator Annual Report. NLFP, Canberra.

Bradsen J.R. 1988. Soil Conservation Legislation in Australia Report for the National Soil Conservation Program p. 148, UOA Adel.

Landcare Australia Limited. 1990. First Annual Report LAL Syd. p. 21

fn 12 p. 7.

Landcare Australia Limited. 1995. Sixth Annual Report pS LAL Syd.

fn 4 p. 2.

fn 12 p. 5. 16fn 5 p. vii.

NFF and ACF A National Land Management Program A joint submission by the ACF and NFF to the Federal Government of Australia. Feb 1989.

The Hon RJL Hawke MP 20 July 1989 21 fn 10 p. 20.

fn 5 p 67.

fn 12 p 15.

Landcare Australia Limited. 1994 5th Ann. Rep. p. 9.

Landcare Feedback No 13 May 1996 Landcare Australia limited.

fn 12 p. 15.

Landcare Australia Limited Fourth Annual Report 1993 p. 7.

fn 5 p. 79.

Campbell, A. 1992. Landcare in Australia — Taking the Long View in Tough Times National Landcare Facilitator Third Annual Report p. vii NSCP Canberra

fn 3 p. 182.

fn 5 p. 72.

fn 5 p. 35.

fn 5 p. 35.

fn 5 p. 72.

fn 5 p. ix Sec 75D Taxation Act C/W.

fn 39

fn 3 p. 76.

Melrose, A.J. SA Parliament. 1945. Cited in J.R. Bradsen. In 7 pl 1.

Wise, F. MP PM WA Parliament. 1945. cited in Bradsen. In 7

Roberts, B. See for example Land Ethics — A Necessary Addition to Australian Values ANU Canberra 1984

fn 5 p. 51.

fn 3 p. 215.

Australia State of The Environment Report1996 p. 32-3.

Intergovernmental Agreement on the Environment. 1992. Commonwealth/ state/ local government agreement. 70 fn 3 p. 215.

Mealing, R.D. The Civilian Conservation Corps Unpublished Paper 1995. 72fn 5 p. 24.

fn 5 p. 76.

fn 5 p. 24.

fn 10 p. 21. The focus on soil is too narrow but this does not vitiate the statement.

34

Successful Soil and Landscape Conservation in Australia

John Cary

Judgments about successful soil and landscape conservation often ignore the bigger pictures of environmental change and changes in aspects of human valuation of the environment. In the 200 years since European settlement, Australia's land has changed—it has lost some properties and gained others. When humans dwell on land it will always be changed. Some land has become seriously degraded; other land is in a more productive state than in times past.

At the time of European settlement, early white explorers and settlers commonly described much of the land along Australia's eastern and southwest margins as "park like." As was the case in the U.S., much of this was due to the use of vegetation-burning regimes by indigenous inhabitants. They created pastoral grasslands and landscapes which Europeans likened to noblemen's parks. The land management practices of the indigenous inhabitants provided the landscape that was so attractive to waves of new settlers intent on pastoral and farming activity. In Australia, the relatively sparse rainfall on much of the land, subsequently converted to pastoral and farming land, also contributed to the park-like savanna landscapes that characterized about 30 percent of Australia's land area.

The commander of the Beagle during its voyage in 1837–1843, reflected a common view of the potential for the dominance of Australia's land (by Europeans) when he expressed "the feeling of pride engendered by the thought that we are in any way instrumental to the extension of man's influence over the world which has been given him to subdue" (Stokes 1846). This instrumental view overlay an aesthetic response to the attractiveness of much of Australia's landscape. Such a view was more commonly expressed by the more affluent and land-owning classes amongst the earlier settlers. The less fortunate and the "battlers" did not gain access to the attractive land and were often overwhelmed by the challenges of settling less hospitable landscapes, which often turned out to be unsustainable bases for farming activity.

In the early period of white settlement of Australia, the European impact on the Aboriginal inhabitants was dramatic and deadly, but the relative impact on the landscape was of relatively smaller significance. Compared to the small European settlement, the Australian land was vast, overwhelming, and intimidating. No single view prevailed among early European inhabitants. While some were effusive about the utilitarian beauty and pastoral potential of the landscape, others were daunted. The complexity of Australia's ecosystems was generally unrecognized by most early settlers who were dominated by their need for survival in an often threatening and inhospitable landscape.

Early conservationists in the U.S., such as Theodore Roosevelt, were concerned with the prudent use of the earth and its resources as a means of economic growth and progress rather than preservation (Fox 1981). In the 1880s similar sentiments held sway in Australia. In Victoria and in other states, there were campaigns for more conservative forest management and against the reckless practices of saw millers, pioneer farmers, and mining companies. Such campaigns principally reflected utilitarian views, but also aesthetic and ecological values.

The more recent perceptions of the ecological destruction of white settlement also have tended to be focused on destruction of forests and forest woodlands; however, the most devastated ecosystems in Australia have been native grasslands and grassy woodlands. The grassland habitat was the home of many of Australia's extinct or endangered small mammals. Very little of that habitat remains, because the Australian reserve and park systems have been based upon forests.

There are many areas in Australia where the removal of trees has been an important cause of land degradation; however, this is not always the case. Some of Australia's worst land degradation is caused by irrigated salinity in parts of the Murray Darling Basin, where the removal of trees from irrigated land has been a minor destabilizing factor in a complex hydrogeological system destabilized by adding too much water to an inappropriate land system. There is a widespread public perception that the establishment of trees will ameliorate land degradation. Such a perception is independent of other motives for planting trees such as for aesthetic enhancement. However, there may be good reason to focus on the aesthetic. And, indeed, the aesthetic may be a biologically determined preference determined by natural selection determined by evolution.

In this chapter, broad elements that underlie conservation behavior will be considered within the Australian context. The focus is the place of economic, psychological, and social factors which influence specific conservation activity. Particular attention will be given to the place of trees and apparent innate

536

human preference for trees in landscapes. The place of the Landcare phenomenon in Australia and the linkages with tree planting and other farm conservation behavior will also be considered.

Conservation on Australian farms

There are differences among the Australian, European, and U.S. contexts for soil and landscape conservation. In the European context, conservation commonly is concerned with the prevention of nonpoint source pollution or the prevention of further production intensification. Conservation behavior in such cases is generally motivated and compensated (for profit foregone) by government or nongovernment agencies representing external beneficiaries. Farm conservation programs in the U.S. often have similar characteristics with cross compliance programs and programs such as land "set asides," compensating landholders for profits foregone. In the Australian context, compensation to individual landholders for engaging in safe environmental behavior is relatively uncommon. Land degradation such as deterioration of soil structure or soil salinization commonly impinges directly on the profit of an individual landholder on whose property the degradation exists. In Australia, with the exception of salinity associated with large-scale irrigation schemes, efforts to ameliorate land degradation have tended to focus on encouraging landholders to respond to the on-farm production and profit possibilities which might be secured by conservation behavior. Soil and landscape conservation programs in Australia rely predominantly on voluntary compliance or the informed self-interest of land users. This occurs against a background of extensive government research programs, a declining public extension service, and extensive efforts to influence public and private norms towards more desirable practices of land use.

Market failures and public goods

Most forms of land degradation lead to external costs over time and these may be large or small. For example, the decline of soil structure or the decline of soil fertility, which are characterized by on-site costs, often give rise to other forms of soil erosion that subsequently may have off-site costs. Other forms of land degradation, such as dryland soil salinity, may have both on-site and off-site effects that will vary for different properties.

The economic costs to a landholder of at least some conservation practices may exceed the on-farm benefits on a short-term and possibly long-term basis. The lack of immediate financial incentive in a dynamic farm economy may result in many landholders not adopting these practices. For forms of land degradation which are entirely internal to the boundaries of individual farm properties, profits attributable to conservation practices can potentially be captured by individual property owners.

In situations involving externalities, noninstrumental motives (such as stewardship or an environmental orientation) could be expected to become more important in influencing the use of conservation practices, when no regulation is applied or no external subsidization is available. In such cases, without appropriate policy instruments, a self-interested perception of profitability will not be sufficient to produce an optimal level of adoption of such technologies.

Stewardship

Colman (1995), in discussing agricultural stewardship as one of many forms of ethical behavior in relation to economic transactions, suggested it should be encouraged by policy. There is an obvious need to understand the relative importance of stewardship amongst the other factors which may influence individual adoption of conservation practices. The notion of stewardship, however, is a problematic factor in determining conservation behavior.

Rural landholders in Australia have both followed and led changing community values concerning land management for conservation (Cary and Barr 1994). However, the beliefs of farmers concerning stewardship and the management of land resources are generally multifaceted and complex, and often in conflict with each other. Dynamic perceptual mapping to monitor changes in landholders' beliefs has established farmers' beliefs about conservation-orientated management systems which are often in conflict with, and outweighed by, beliefs about the need for management systems to be profitable, simple to use, and assured in their outcome (Wilkinson and Cary 1992 and 1993).

Tree planting has been an important focus of recent land conservation activity in Australia. In many areas of Australia, landholders see no good financial reason to plant large numbers of trees on their farms, yet many of them plant relatively small numbers for aesthetic reasons or for shade and shelter. The nature of the environmental choices of such landholders is not uncharacteristic of the environmental choices facing the wider population. Appropriate environmental behavior often has no immediate financial gain and may often involve immediate financial or other disadvantage. The perceived benefits from individual action accrue in the future and are likely to be shared with others.

For most people there is a discrepancy between beliefs about the environment and behavior related to the environment. A solution to this apparent psychological inconsistency is for individuals to publicly espouse the relevant social norms but engage only in token conservation behavior, sufficient to provide apparent cognitive consistency for the individual (Cary 1993). Instrumental beliefs about the environment, related to self-interest, thus are likely to be more powerful than such publicly espoused symbolic beliefs in influencing environmental behavior.

The tenuous link between conservation values, sometimes expressed as stewardship, and conservation behavior has been observed in a number of studies in Australia, as well as in other countries. Attempts to establish linkages between measures linked to stewardship and conservation behavior related to crop farming practices in Australian research studies have generally been unsuccessful. Harvey and Hurley (1990) in a study of crop farmers in Victoria found no statistical relationship between perception of erosion, or concern about erosion, and use of the conservation tillage techniques. A study of wheat producers in New South Wales found that adherence to the conservation ethic did not significantly differ between the adopters and nonadopters of conservation cropping (Sinden and King 1988). Vanclay (1988) found farmers adhering to a conservation ethic were less likely to adopt conservation farming. These findings may reflect the measurement difficulties associated with operationalizing "stewardship" in survey interviews; however, it is more likely for cropping farmers to go with cropping management decisions that are essentially determined on the instrumental grounds of convenience and cost.

Predisposing factors for conservation behavior

Economic

While most environmental practices can be expected to enhance productivity and economic profitability in the longterm, many environmental practices (for example, conservation cropping) have productivity benefits within a shorter time span. The most successful soil and land conservation practices in Australia have been practices that confer large, and privately captured, economic benefits within a short time (Barr and Cary 1992). Given the observed importance of profitability in decisions about using conservation practices, the use of environmental technologies which do not produce immediate financial gain or which produce benefits of an ambiguous nature has to be facilitated by factors other than those that are primarily economic.

Psychological

The delineation between many psychological and social factors influencing behavior towards the environment is somewhat arbitrary given that attitudes to the environment and values associated with the environment and conservation tended to be socially learned. The small or insignificant influence of stewardship variables or personal environmental orientations in predicting environmental behavior for many rural conservation practices has already been discussed.

Enduring landscape preferences. There is another group of psychological influences on environmental behavior, related to preference for landscapes, and particularly the place of trees. Elements of these preferences will be socially determined, as well as influenced by familiarity. However, there also appears to be an underlying element of preference which is more innate and determined by natural selection associated with the evolution of the human brain.

Human brains evolved to process landscape information over a long time. At some time, up to 5 million years ago, the ancestors of the human species came down from the trees and were living as terrestrial primates in mainly open grassland. The species, *Homo erectus,* existed in African savanna between one million to 300,000 years ago. Through most of this time man and his species ancestors have been a hunters of large game in savanna environments, consisting of grass and isolated trees or groups of trees, and the most significant evolutionary adaptation was a large increase in brain size. Also, there has consistently been a need to remain fairly close to bodies of water, because humans need a constant supply of fresh water. These facts seem to provide a plausible explanation for modern man's apparent desire to create park-like landscapes and to live in view of water when given the opportunity (Bourassa 1991).

The evolutionary basis of such preference is linked to brain adaptation for survival in a savanna landscape, particularly to "read" such landscapes and speedily process information related to them. The interpretive process of natural landscapes relies heavily on the past—upon information gathered at previous times. In the evolutionary context such information allowed quick response (for hunting or flight) without the need to contemporaneously process all the information present.[1]

The apparent consistency across cultures of human preference for "savanna type" configurations of grassland and trees probably reflects the importance in human evolution of effective cognitive functioning within such categories of landscape (Kaplan and Kaplan 1978; Kaplan 1987; Tuan 1974). Landscape preference assessments appear to be influenced by the potential for human functioning in a setting. Research on the psychology of perception of landscape has established that while preference is subjective, the "subjectivity is often shared to a remarkable degree" (Kaplan 1975, 123). Wide-open, undifferentiated vistas and dense, impenetrable forests both fail to provide appropriate information about one's whereabouts, and both frequently appear as distinct perceptual categories (Kaplan and Kaplan 1989). Similar findings have been confirmed in Australia and New Zealand for rural pastoral landscapes (Cary 1995).

Social

While utilitarian orientations, where an individual's interests are maximized, tend to dominate behavior toward the environment, such orientations are modified by socially transmitted norms, whereby an expectation of behavior is transmitted by members of a group that its members ought to behave in a certain way with respect to the environment. Potentially, where strong pro-environmental norms are present, the social desirability of pro-environmental behavior can temper utilitarian predominance.

The magnitude of individual pro-environmental behavior, particularly for behaviors where utilitarian attributes are less strongly present, will be greater where intimate or immediate groups provide social support and reinforcement for such behaviors. As social norms are often transmitted and reinforced via groups we will consider two successful Australian conservation programs, Landcare and Soilcare, which employ a group approach.

Landcare membership and conservation behavior

When the Australian Government declared the Decade of Landcare in 1990 it was expected there would be 1,200 Landcare groups by the year 2000. The target was exceeded: the number of groups increased from about 100 in 1985 to 900 groups in 1991, and 2,700 groups in 1995. In 1995, 37 percent of Australia's farmers were members of Landcare groups. The enthusiastic embrace of landholders for the landcare movement is a phenomenon that has no apparent equivalent in other western countries. Such a social movement, the development of which reflects a series of special and fortuitous circumstances, is unlikely to be easily transferable to other countries. Indeed, Landcare is not a universal extension mechanism because not all landholders are members and not all Landcare members are active members. As well, not all farm enterprise types are equally represented; more pastoral farmers are Landcare members than are cropping farmers or dairy farmers (Cary and Wilkinson 1992). The Landcare movement does provide some lessons of the relative importance of social factors in influencing desired environmental behavior.

Landcare groups tended to be started by the most environmentally oriented farmers and such groups reinforced and legitimized these environmental values. The environmental outcomes for such early group membership can be inferred from an examination of conservation behaviors, related to tree planting and the planting of deep-rooted pasture, undertaken by Landcare members in Central Victoria (Table 1).[2] Initially, such groups did not appeal to production-oriented landholders. As more farmers joined

Landcare groups the scope of the groups widened, and they eventually become more attractive to production-oriented farmers.

Characteristics associated with Landcare success

A longitudinal study of Landcare membership in central Victoria presents some insights that reflect on the influence of economic, psychological, and social factors on successfully implementing soil and landscape conservation in Australia. The discussion will focus on two aspects of conservation decisions about two conservation practices—the decision to use a practice and the extent of use of a practice. The two practices were tree planting and the planting of deep-rooted pasture. In the area for which these results are reported poor land management practices, particularly the removal of deep-rooted vegetation, have caused water-tables to rise, contributing to increased local and regional land salinization of the slopes and plains. Revegetation of district farms using deep-rooted species such as trees or improved pastures can reduce both on-site as well as off-site soil salinity.[3]

Perception of long-term profit was an important predictor of the decision to use both practices, but was more important for the decision to plant trees. Trees were planted irrespective of whether salinity was perceived by the landholder as an environmental problem. In the case of phalaris pasture, recognition of salinity as an environmental problem increased the likelihood of planting. Membership of a Landcare group was a significant, but not substantive, predictor of the decision to plant deep-rooted pasture and was not significant for predicting the decision to plant trees.

When we consider the more significant decision—the extent of planting of trees and of deep-rooted pasture—recognition of salinity as an environmental

Table 1. Length of membership in a Landcare group and extent of planting and pasture planting between 1988 and 1991

Earliness of joining a Landcare group	Extent of conservation behavior	
	Mean number trees planted	Mean area of deep-rooted pasture sown (ha)
Not a member	191	9.1
Joined between 1988–1991	465	21.8
Joined before 1988	1537	35.2
F ratio (farm size)	0.78	18.11***
F ratio (earliness of joining)	16.73***	4.00*

* $p<0.05$; **$p<0.01$; ***$p<0.001$

problem did not affect the extent of tree planting or of deep-rooted pasture planting. Membership of a Landcare group was a very significant influence in explaining the extent of tree planting. The extent of tree planting, relatively, was much more dependent on Landcare membership than on perceived profitability. The extent of planting deep-rooted pasture was dependent only on perceived profitability (which was a more significant factor than it was for tree planting), not on Landcare membership.

One explanation of these findings is that the presence of some trees (i.e., a landholder engaging in some planting), in contrast to extensive planting or dense plantations of trees, meets aesthetic or innate psychological preferences discussed above. As low-density tree plantings are generally desirable, they attract a capital premium (extra profitability) for the land on which they are planted. In contrast, denser plantings within the study area had an opportunity cost of income lost from other farming activity. Thus, membership of a Landcare group became the influential factor for increased tree planting.

The transmission of pro-environmental norms via membership of a conservation group provided a means of encouraging proenvironmental behavior, such as extensive tree planting, which was perceived as less profitable. Where a practice was generally perceived as profitable, as in the case of planting deep-rooted pasture, there appeared to be little need for other means to encourage the appropriate behavior. However technical impediments, such as the lack of appropriate cultivating equipment for planting deep-rooted pasture on smaller sized properties, suggests the need for strategies such as group ownership of relevant equipment.

"Soilcare" and conservation behavior

Learning from the past

Soilcare, like Landcare, was first established in Victoria. An earlier attempt at a large scale soil conservation program provided an important lesson which has led to a more effective *modus operandi* for such conservation programs involving landholders. In the 1960s, the Eppalock soil conservation project involved joint programs between the former Soil Conservation Authority and farmers in a program of pasture improvement and on-farm engineering works to protect the eroding catchment of a water reservoir. While the "partnership" involved joint investment activity in pasture improvement and some consultation with landholders, the on-farm engineering works were determined independently of the respective landholders. After completion of the project, landholders felt they had little or no responsibility for the maintenance of the erosion control structures established by Soil Conservation Authority; these structures subsequently deteriorated, reducing the longer-term effectiveness of the project (Barr and Cary 1992). In subsequent projects attempts have been made to involve landholders more comprehensively. This lesson was applied in the Soilcare program.

Soilcare

Soilcare is a group extension activity which has successfully implemented the use of conservation cropping practices amongst farmers. Soilcare encourages farmers to engage in minimum tillage cropping or, preferably, direct drill cropping, where all tillage is eliminated. The most complex form of conservation farming is crop stubble retention, where the burning of crop stubble is eliminated and crops are drilled directly into the soil while the stubble is still standing. Trash farming aims to gain all the benefits of direct drilling as well as the added benefit of increasing soil organic matter by retaining the wheat stubble to break down in the soil and to reduce erosion risk.

Scientists promoting the uptake of conservation cropping in Victoria initially relied upon journal articles and field days to promote and extend the message of reduced costs, timely preparation, and improved soil structure associated with conservation cropping. A revised strategy involved farmers in evaluation and adaptation of the conservation technology in their own districts. The project was called "Soilcare." Demonstrations on paddocks owned by local farmers became a focus for groups of local farmers. The group members discussed their concerns about conservation cropping and, with advisory and research officers, decided on the experiments to be run on the demonstration blocks with the aim of solving cropping problems specified by the group. A small experimental plot, in the corner of the demonstration block, helped farmers and advisers understand how much each individual practice was contributing to the success or failure of the package being demonstrated on the whole paddock. This created local ownership of the demonstration and the applicable findings and reinforced belief changes about conservation cropping (Wilkinson and Cary 1993).

Soilcare differs from Landcare in some important respects. With the possible exception of soil erosion, the various soil degradations being faced by Soilcare do not have the level of externalities exhibited by the land degradation being tackled by Landcare groups. The soil problems tend to be confined to individual farms. This essential difference allowed Soilcare groups to be established in a different way from Landcare groups. Whereas Landcare group boundaries are based on geographic areas, Soilcare group boundaries were chosen on the basis of social, soil type, and geographic similarities. For each Soilcare group a few "catalytic" individuals were approached, who then suggested names of other farmers who might be interested. In all cases Soilcare group members selected themselves. Soilcare group boundaries thus tended to be social boundaries, rather than the geographic boundaries.

Concluding thoughts

Today there are a number of environmental problems associated with Australian land for which, currently, there are no adequate solutions. Sometimes there may be solutions available but they do not possess the characteristics for spontaneous and widespread on-farm use. Currently, practical, low-cost, and profitable solutions that can be easily incorporated into farming systems do not exist for problems such as irrigation-induced rising groundwater, soil acidification, regionally induced dryland salinity, woody weeds in the rangelands, vertebrate pests in the rangelands, regional dieback of native remnant vegetation, and nutrient contamination of watercourses and wetlands. In cases where there are technically feasible solutions, and the problem is considered to be a serious problem in the public domain, incentives or regulation will be required to bring about changed land management.

Establishment of deep-rooted species and reestablishment of native vegetation on rural land has come to be seen as increasingly important in Australia. Revegetation programs frequently have not been well thought through in terms of appropriateness or in terms of wider and longer-term landscape planning. Tree planting as an environment-enhancing measure appears to interact with innate landscape preference for open or park-like woodland, rather than denser plantings. Such innate preference is a natural facilitator of low density, but not higher density, tree planting. Similarly, denser understoreys often associated with natural remnant vegetation may not satisfy inherent human landscape preference.

The more general lessons from conservation programs which have been successful are that, ideally, the practice is profitable within a short time frame. Otherwise there need to be strong social pressures present or exerted through conservation group membership. While many landholders espouse stewardship values only a small group of landholders is sufficiently concerned about land degradation to make significant changes to management programs in situations where there is cost without significant economic gain. Conservation groups, in an instrumental sense, need to provide tangible benefits for membership and then be able to exert social influence to encourage behavior where the benefits tend to be longer term rather than short term. Not all landholders will be influenced by such group activity. In the case of Landcare, membership is often seen as socially desirable within a local community, and it may also allow access to

545

relatively small government subsidies provided to some groups, but not all members are committed to implementing more significant environmental activity. In the case of Soilcare the benefit of membership was often to jointly learn with other group members ways of overcoming some of the technical difficulties associated with implementing conservation cropping.

In the most recent history of soil and landscape conservation in Australia, rural action to ameliorate land degradation has not been held back by urban reticence to commit resources to such programs. The Australian government has committed $1 billion to projects related to sustainable resource management in rural lands. The commitment includes an environmental trust fund, a national land and water audit, and additional funding for the Landcare program.[4] This chapter has broadly established the relative influence of economic, psychological, and social factors in determining conservation activity. The wise expenditure of large government outlays on privately owned lands will take account of the relative influence of these factors, particularly the need for economic advantage to be present. When subsidy payments are used in situations where there is no short-term economic benefit associated with conservation behavior it will be unwise to depend on the stewardship ethic as a concomitant for widespread implementation of conservation behavior. For more universal uptake of noneconomic conservation behavior, regulation of such behavior may be necessary.

Notes

1. We experience the environment not as a series of snapshots of what is going on immediately in front of us, but rather as a construction. This construction is made up of a good deal of prior knowledge and only a sampling of current information. Despite the incompleteness of the information, this construction tends to feel both clear and definite (Kaplan and Kaplan 1978).
2. The details of this study are described in Cary and Wilkinson (1997).
3. A more detailed analysis of results for decision to use and the extent of use of both of these practices is presented, respectively, in Cary and Wilkinson (1997) and Cary (1994).
4. This commitment was contingent upon funds being made available from the privatization of the national telecommunications provider.

References

Appleton, J. 1975. The Experience of Landscape. London: Wiley.

Barr, N.F., and J.W. Cary. 1992. Greening a Brown Land: The Australian Search for Sustainable Land Use. Melbourne: Macmillan.

Bourassa, S.C. 1991. The Aesthetics of Landscape. London: Belhaven Press.

Cary, J.W. 1993. The nature of symbolic beliefs and environmental behavior in a rural setting, Environment and Behavior 25: 555-576.

Cary, J.W. 1994. The adoption of conservation practices in Australia: An exploration of commercial and environmental orientations. In Adopting Conservation on the Farm: An International Perspective on the Socioeconomics of Soil and Water Conservation. T.L. Napier, S.M. Camboni, and S.A. El-Swaify (eds.). Ankeny, IA: Soil and Water Conservation Society Press.

Cary, J.W. 1995. An Analysis of Perceptions of High Country Landscapes. Lincoln, N.Z.: Landcare Research New Zealand Ltd.

Cary, J. W., and R.L. Wilkinson. 1992. The Provision of Government Extension Services to the Victorian Farming Community. Parkville, Vic: School of Agriculture and Forestry, University of Melbourne.

Cary, J. W., and R.L. Wilkinson. 1997. Perceived profitability and farmers' conservation behavior. Journal of Agricultural Economics 48: 13-21.

Cary, J.W., and N.F. Barr. 1994. The browning of landcare. The Independent Monthly 5 (7): 46-47.

Colman, D.R. 1995. Ethics and externalities: Agricultural stewardship and other behavior. Journal of Agricultural Economics 45: 299-311.

Fox, S. 1981. John Muir and His Legacy: The American Conservation Movement. Boston: Little, Brown & Co.

Harvey, J.T., and F.T. Hurley. 1990. Cropping and Conservation. Ballarat, Vic.: Regional Studies Unit, Ballarat University College.

Kaplan, R. 1975. Some methods and strategies in the prediction of preference. In Landscape Assessment: Values, Perceptions and Resources. E.H. Zube, R.O. Brush and J.G. Fabos (eds.). Stroudsburg, PA: Dowden, Hutchinson & Ross Inc.

Kaplan, R., and S. Kaplan. 1989. The Experience of Nature: A Psychological Perspective. Cambridge: Cambridge University Press.

Kaplan, S. 1987. Aesthetics, effect, and cognition: environmental preference from an evolutionary perspective. Environment and Behavior 19: 3-32.

Kaplan, S., and R. Kaplan. 1978. Humanscape: Environments for People. Ann Arbor, MI: Ulrich's Books.

Sinden, J.A., and D.A. King. 1988. Who adopts conservation practices. Aust. JSWC 1 (1): 32-6.

Stokes, J.L. 1846. Discoveries in Australia: With an Account of the Coasts and Rivers Explored and Surveyed During the Voyage of H.M.S. Beagle in the Years 1837-43. London: T&W Boone.

Tuan, Y. 1974. Topophilia: A Study of Environmental Perception. Englewood Cliffs, NJ: Prentice-Hall.

Vanclay, F. 1988. Socio-Economic Characteristics of Adoption of Soil Conservation. MEnvS Thesis. Queensland: Griffith University.

Wilkinson, R.L., and J.W. Cary. 1992. Monitoring Landcare in Central Victoria. Parkville, Vic: School of Agriculture and Forestry, University of Melbourne.

35

Down the Track From Rio:
Lessons from the Australian Landcare Program

Sarah Ewing

Agenda 21, as adopted by the UNCED Summit meeting in June 1992, represents the current international consensus on actions necessary to move towards
the goal of "sustainable development" (Kelly 1992). One of several key themes
running through the Agenda 21 document, is that of community involvement.
For policymakers, this demands not only a renewed emphasis on local policy
and action but also recognition of a need for greater "democratization" and
involvement in policymaking. The new environmental agenda views an expression of greater democratization to be integral to the achievement of an environmentally sustainable future, encompassing ideas of subsidiarity, decentralization, empowerment, and participation (Agyeman and Evans 1994). Using the example of the Australian Landcare program, this paper considers the
implications of this apparently universal embrace of "community" and the "local," for the way in which we think about, and assess, soil and water conservation policy.

The paper is divided into two sections. It opens with a brief historical overview of the policy setting for Landcare. This is followed by a review of the idea
of "community" as it is employed in Landcare and the implications this has for
the way in which the success of the program is measured.

The nature of the problem

The agricultural production processes that have both created Australia's
wealth and provided Australians with jobs, have cost the country dearly. It is
estimated that most areas of cropland and improved pasture in Australia are
affected by land degradation of one form or another (State of the Environment
Advisory Council 1996); soil acidity, for example, is affecting about 26m hectares (64m acres) or 35 percent of the nation's prime cropping country, and

Australia's croplands are commonly losing between 100 and 200 tonnes of topsoil per hectare each year (40-80 tons per acre). The total cost of land and water degradation to the Australian community is now estimated to be in the order of A$1410m per annum, a cost compounded by profound economic and social problems as unreliable rainfall and declining terms of trade accelerate rural adjustment in Australia (Alexander 1995; Vanclay and Lawrence 1995).

Serious national concern for land degradation was first aroused in the 1930s, largely as a result of the widespread publicity surrounding the U.S. "Dust Bowl." The leading role taken by the U.S. government in developing a model of federal responsibility and intervention was adopted in Australia soon after. After the crises of the 1930s and a postwar flurry of activity, concern for the degradation problem diminished and it was not until 1985 that a national approach was realized.

Like its U.S. counterpart, the Australian government has been reluctant to invoke legislation in response to land degradation, preferring instead to place the emphasis on voluntary compliance and attitudinal change. The rhetoric of Landcare, the most recent government policy initiative in this area, relies heavily upon the ethic of stewardship and upon voluntary cooperation between landholders. It brings together the private property interests of participating farmers, the stewardship assumptions of group conservation, and the notion of the public or national good.

"A revolution in the bush"

Landcare has been described variously as "a revolution in the bush," "an inspired piece of public policy," and "undoubtedly the most exciting and significant development in land conservation in Australia." It is certainly true that, *prima facie*, the Landcare program has succeeded beyond even the most optimistic of expectations. From humble beginnings in Victoria in 1986, Landcare has been embraced by governments throughout Australia, at both State[1] and federal levels, as a model for community-led action towards more sustainable use of the nation's land and water resources. Such is its strength, Landcare has now gained a reputation beyond Australia. Specifically, the Landcare concept has been incorporated into the United Nations Convention to Combat Desertification and the Program of Action for the Sustainable Development of Small Island Developing States, adopted in 1994 (Williams et al. 1995; Wensley 1994).

Landcare first came to national prominence in 1989, when the National Farmers' Federation and the Australian Conservation Foundation put forward a joint proposal to the Federal Government, urging that a concerted national effort be made to tackle land degradation (Toyne and Farley 1989). Their proposal, or at least part of it, won government endorsement and was launched publicly in the Prime Minister's *Statement of the Environment.* On the banks of the Murray River in the Murray-Darling Basin, the site of some of the most devastating land degradation in the country, the Prime Minister declared that the 1990s be known as the "Decade of Landcare" (Hawke 1989). In 1990, Commonwealth, State and Territory Ministers responsible for soil conservation agreed to develop a national "Decade of Landcare Plan" (DOLP), integrating action by governments, individuals and the community to address land degradation in Australia. Plans for each State and Territory and the Commonwealth have since been developed and now combine to form the National Decade of Landcare Plan (Commonwealth of Australia 1991).

Briefly, the Commonwealth component of the DOLP outlines five goals, as follows:

- to have the *whole community* aware of the problem of land degradation and the benefits of sustainable land use;
- to ensure continuing development and implementation of sustainable land use principles and practices;
- to have all public and private land users and land managers understanding the principles of sustainable land use and applying them in their use and management decisions;
- to have *all* Australians *working together in partnership* for sustainable land use; and
- to have effective and appropriate economic, legislative, and policy mechanisms in place to facilitate the achievement of sustainable land use.[2]

In Decade of Landcare documents, frequent use is made of these ideas of the community, working together in partnership with the state—images now used to market Landcare to the world.

Community Landcare

The community Landcare movement forms the cornerstone of the Landcare program and is made up of voluntary, community-based groups. These groups are comprised mainly of rural landholders, who come together to tackle local land management issues. There are currently in the order of 2,200 Landcare groups across the country, involving 30 percent of the nation's farmers. Landcare groups vary in size from small groups of neighbors with a local problem to larger groups with wider, more complex concerns.

Landcare groups also vary widely in the diversity of issues they tackle, the diversity of environments in which they operate and in their diversity of approaches to a problem. Initially, groups tend to focus on a local problem related to land degradation, such as salinity, the control of pest plants and animals or a decline in remnant vegetation or water quality. Over time, as groups' experiences and confidence expands, they tend to embrace a broader suite of concerns, for example, property and catchment planning and, in cooperation with other groups, issues such as regional development.

Measuring success

The success of Landcare rests squarely on optional action by these voluntary groups, based upon the assumption that if landholders have an awareness of, and concern for land degradation, they will, in turn, change their land management practices. This is one reason why the success of Landcare is often equated with the number of groups. For policymakers, such a measure is seductive and is often used in political discourse as "proof" that things are happening in Landcare "on the ground" and, presumably, changing the face of the nation. National surveys, however, have yet to establish any causality between participation in Landcare and the adoption of appropriate land management practices (Mues et al. 1994). As Bradsen (1994) puts it, we have shown an "unfortunate tendency" to get "too excited" by the number of groups and by specific success stories.

The first tentative step towards a less crude assessment of community Landcare was in the establishment of the National Landcare Facilitator Project to provide an independent national overview of Landcare activities (Campbell 1992; Alexander 1995). As a starting point for the Project, Campbell (1992) assumed that the primary role for Landcare groups, and thus the expectation against which their effectiveness should be assessed, is that of involving people more actively in land conservation in order to generate a greater commitment towards land conservation at the individual and community level. According to Campbell, it is, therefore "unfair to judge the effectiveness of Landcare groups against the criterion of sustainable land use, as too many other players outside the influence of Landcare groups are involved" (Campbell 1992: iii).

Campbell suggests that an alternative way in which the potential and effectiveness of Landcare group activity may be assessed, is through what he terms indicators of "group health" and "group impact," such as leadership, credibility, and involvement. In Victoria, Curtis and his colleagues have elaborated this idea in an "index of group effectiveness," used to help identify characteristics of more (or less) effective groups (for example, Curtis and DeLacy 1995).

Generalizations like these have often been uncritically accepted in the Landcare literature. This is not to deny their contribution to an overall evaluative framework for Landcare. But, such broad-scale frameworks tend to disconnect Landcare, and Landcare groups, from the local context in which state policy and practice is played out. We need now, more than ever, to understand the inner workings of these programs at the *local* level, since their effectiveness depends so much on the communities themselves.

Several evaluation studies have shown that an understanding of the *local* demands different indicators and different methodologies. Hinchcliffe et al. (1995), for example, in measuring the effectiveness of participatory watershed programs, concluded that:

> Understanding the economic, social and environmental impacts of watershed development means expanding the indicators beyond "kilometres of terraces built" or "numbers of farmers trained." It means understanding the benefits and costs as perceived by local men and women (Hinchcliffe et al 1995: 6).

Similarly the experience of Reij (1994) in Mali who, given a mandate to measure the success of certain technical interventions, found that initiatives that met the criteria of being sustainable and replicable resulted from the combination of social, financial, and technical factors:

> We therefore placed increased emphasis on the *observational* content of the site visit reports. That is, we did not want to see only what we were looking for, but rather to understand the underlying factors at play as fully as possible (Reij 1994).

It is against this theoretical background that I chose recently to adopt a case study approach to an assessment of Landcare (Ewing 1995). The study drew upon the experience of Landcare by farmers of the Dundas Tablelands in the Glenelg River catchment of Victoria's Western District (Figure 1). Through in-depth interviewing and participant observation across seven Landcare groups, I sought to understand how Landcare works "on the ground." In the remainder of this chapter, I will draw upon examples from the Tablelands, to illustrate the way in which the farming community engages with the Landcare program, particularly the way in which the notion of community functions in a local setting. For example, to what extent is Landcare a community-driven rather than state-driven program? Does it work in, and for the community? Or is the idea of "community" just part of state rhetoric and legitimation?

Briefly, the Glenelg catchment drains an area of 12,000km² (4,600 square miles) and has one of the longest records of accelerated soil erosion since European settlement in Victoria. Despite concerted effort, by both landholders and government over a period of 30 years, significant land degradation problems persist including dryland salinity, stream and soil erosion, declining water quality, reduced biodiversity, and the invasion of pest plants and animals. Landcare formally arrived in the area in 1986, with the establishment of one of the first Landcare groups in Victoria. There are now in the order of 16 groups on the Tablelands, with membership varying from 8 to 120 members. The Tablelands also embrace one of Victoria's prime wool growing districts, a legacy of its early settlement by squatters[3] and wealthy pastoralists. The region's dependence on wool production for such a large proportion of its income has made it particularly prone to the effects of the depressed Australian wool market in recent years.

The bureaucratization of community

Structures for participation

One way in which the government has sought to provide for community in Landcare, is through participatory planning. The development of Victoria's

Figure 1. Location map of the Glenelg River catchment southwest of Victoria

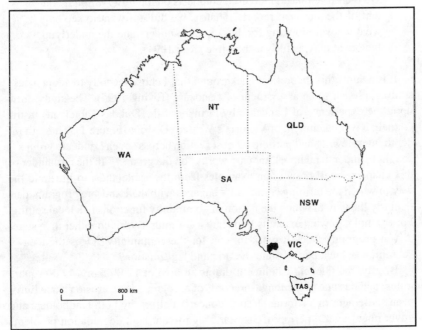

Decade of Landcare Plan (DOLP) began formally in 1990 and, after a period of public consultation, was launched in 1992. Consistent with the Commonwealth component of the Decade Plan, it supported an emphasis on a coordinated and integrated approach to land management, with a community/government partnership at its core. For planning purposes, the plan identified nine Landcare regions across the State, each of which required a *Regional* Landcare Plan to be developed as a collaborative exercise between participating government agencies and non-government regional representatives. A hierarchy of Landcare planning was thus established, from the Commonwealth DOLP, to the Victorian DOLP and finally, at the local level, to the Regional Landcare Plans. Representatives of the nine regional communities presented their "final" plans to government in June 1993, at a formal reception at Victoria's parliament house in Melbourne. The occasion was significant for the symbolism of the community meeting the state, albeit in a place where the politicians were likely to feel more at home than the visiting "community."

While the Regional Landcare Plan process had been heralded in the Victorian DOLP, it was perceived by regional communities as being imposed upon them. In trying to allow for the development of comprehensive plans and community participation, the process quickly became "bureaucratized" and a tension emerged between the original idea and its enactment. In the Glenelg region, it prompted a wearied response to yet *more* planning, as typified by this farmer's comments:

> I'm a little bit concerned with some of these issues that we can have too many surveys and plans ... Every time something comes up, "oh we must do a survey on that" and another so many thousands of dollars goes into that survey and it's almost a copycat of the previous survey that's been done but it's just a little bit wider or a little bit something else. And then something else comes up and then "oh we'll take a survey of that." So they do another survey and you never heard the end result ... money is just being spent on administration and these wonderful plans. But does it get back to the people that are going to put them into practice?

Respondents were quick to make the distinction between "them" and "us," between government experts and the lay community. And, as the following extract illustrates, it is a distinction that implies not only different types of knowledge, but also that each group is differently empowered in the planning process:

> We saw that with [the government planner] when he brought out the [draft Glenelg Regional Landcare] Plan ... and I was horrified about that, and you probably heard all about that. I wrote letters and all sorts of things. And he wanted to come up and talk to me, and I wrote back and said "I don't want you to talk to me. I want you to listen to me." And that's what worries me. If you get the wrong person in there who wants to talk to you, they're the ones with the pens.

A little more than half of the respondents knew of the Glenelg Regional Landcare Plan despite them all being Landcare group members and despite the local community having been invited to participate in its development over the previous eight months. Of those who had seen the Plan, only a handful had submitted a written response to the draft version. This is not surprising, since the document is uninviting in its complexity, verbosity, and liberal use of technical jargon. Several respondents felt disillusioned after earlier experience of a community planning exercise for salinity:

> And then you've got the Salinity Strategy which was really presented as a *fait accompli* in a beautiful glossy document. And people were asked for comment, after a launch of the strategy ... most people didn't even realise they were meant to comment on it. There's still people ... there's still individuals that don't know that any plan ever came out of it ... because that's an important document that should have been out in paperback for six months before the community.

The issues raised here about the dominance of "expert" discourse, of consultation upon plans which appear as if already in operation and of participation being equated with consultation are, of course, well-rehearsed issues in planning literature. Community ideas, if they are heard at all, appear to get waylaid into long-term strategic planning activities which then get delayed, get "operationalised" elsewhere and simply do not seem to appear on the ground.

There is not only the question of *if* "the community" is heard, but also by whom the community's views are represented. In Victoria, the overwhelming majority of community appointees to regional Landcare planning groups were known to have previous experience and expertise in natural resource planning (Curtis et al. 1993). The process, therefore, favored already advantaged groups in the community, those with resources and political aptitude, and did little to foster widespread community involvement in the political process. Landcare practice is, then, vulnerable to charges of "elitism"; those with political experience, who know how to work with bureaucracy or who can leave the farm work to someone else, are the ones most likely to be involved in Landcare planning. Policymakers clearly need to acknowledge the many social and situational differences within the farming "community." As argued by Iris Marion Young (1990), the perception of anything like the common good can only be the outcome of public interaction that expresses rather than suppresses these particularities.

Community grants

Another way in which the structures of Landcare allow for community involvement in decision-making, is in the allocation of community grants, the main source of federal financial support to Landcare groups. Under the National Landcare Program (NLP), applications for community grants are called for once a year. Any community group involved in natural resources management can apply. The allocation of NLP grants is devolved to community-based Regional Assessment Panels (RAPs). RAPs assess, rank, and make recommendations upon both community and agency grants in their region. The devolution of decision-making in this way is consistent with the rhetoric of the state working in partnership with community. In practice, however, it has become clear that community members generally have a more limited understanding of the complexities of the NLP's administration than do agency representatives, with the result that they are quite differently empowered in the task. Thus, accounts emerge such as this one, from a RAP representative in Queensland:

> "The State blokes came along to the assessment meeting to push their own submissions—all the big guns—and I can tell you, we were cowed. We agonized for an hour over a $6,000 grant for one group and passed a State submission for $70,000 in a blink. It's this sort of subtle politics that we community groups can't handle" (*Landline,* September 18, 1994, ABC Television).

Groups resent the difficulty they have in obtaining funds for worthwhile and small scale community projects, when agencies appear to win support with relative ease. This sense of the community process being constrained by the interests of government is a common concern. One Tablelands farmer sums it up well, in this recollection of his remarks to a State Landcare conference and the effect it had on the audience:

> "What was worrying me about Landcare, was the fact that the first couple of years we were in it, it was very much a local departmen-

Figure 2. Letter to the editor, Sydney Morning Herald, April 4, 1994

Catch 22 of gully management

As a new Landcare member, I am only just beginning to understand how to apply for Government funding. Correct me if I am wrong, but it seems to go something along these lines.

Year 1: Application for funding of $500 to aid in the purchase of fencing materials and trees to control gully erosion.

Result: Application rejected — however, please find enclosed a cheque for $100 to assist in Incorporation and Establishment Fees of a new Landcare Group.

Year 2: We have just established the Washaway Gully Landcare Group. Can we now get $500 to aid in gully erosion control?

Result: Application rejected — however, find enclosed a cheque for $1,000/member for a Whole Farm Plan

and $800 for a Total Catchment Management (TCM) Policy. This will enable your group to define the boundaries and assist in detecting other potential degradation problems which may impinge on the future sustainability of your catchment area.

Year 3: We have just completed our Whole Farm Plan and TCM policy. Is it now possible to get $500 for gully control?

Result: Application rejected — however, we have agreed to appoint a Landcare co-ordinator to help you and six other groups fill in your application forms correctly [approx cost $40,000 per annum].

Year 4: We thank you for the Landcare co-ordinator and you will note that all our forms are now correctly filled in. Can we please have $500 for the gully?

Result: Application rejected — but we enclose $1,000 for a Gully Control Expert to assess your gully and to recommend a series of strategic control measures to be implemented.

Year 5: We have just had our gully checked by J. Blow (Diploma of Gullies). He says we have a C3 gully with a 4 score erosion component and he suggests we fence it off and plant trees. Can we now *please* have $500 for the gully?

Result: Application rejected — however, we have agreed to give you $5,000 to make it a Demonstration Gully Control Site so other Landcare co-ordinators, soil conservationists, agronomists, CALM experts, etc can see how to control a gully.

John Hindmarsh,
Nangus.

April 4

tal Landcare group operation. And I said, 'Mr. Minister, it's worrying me that you are just creating an employment bureau for so many people, well for your bureaucrats.' And I said that in our little group, we only had nine people in it, we had a wool bale full of correspondence in the last twelve months. I tipped it out in our lounge room and I said 'that is ridiculous.' I said 'we want trees and pastures and Landcare work done. We don't want jobs and we don't want paperwork.' And with that, the four hundred people all stood up and clapped me like mad. I felt ten foot tall."

Groups are likewise frustrated at the bureaucracy associated with grants, most of which are for less than A$5,000. This is exemplified in a recent letter to the editor of the *Sydney Morning Herald* (Figure 2). The correspondent succeeds in sketching an unflattering, albeit unlikely, scenario in which a Landcare group receives government funding to do anything other than the project for which the original application was made. One farmer, when I asked her to explain to me her understanding of Landcare, said, "I think about the paddock—that's where Landcare is. Out on the farm. But it's not that. It's getting off the rails."

She went on to articulate a widely held view that, increasingly, there are forces moving Landcare away from "the local." It is certainly true that community Landcare is being nested amongst an increasingly complex legislative and operative framework. In its success, Landcare has found many friends in the bureaucratic structures of government and many now wish to claim an association with it. The bureaucratic Landcare "family" grows ever larger, embracing the interests not only of primary industries and nature conservation, but also of the long-term unemployed, water supply, regional development, and large corporations. It seems that the original concept of Landcare, as one built upon coordinated, local community action, is being increasingly put aside, in a rush for government and sponsor support funds and the identity associated with a successful program (Poussard 1994; Vanclay 1994).

A dose of economic rationalism?

The espoused rationale for a group-based approach to land degradation relates to the ethical arguments for community participation. But, there is also a pragmatic concern for public extension agencies unable to meet a demand for their services at a time of belt-tightening by government. According to Cary (1993), the declining importance of agriculture in Australia has prompted a reappraisal of the nature and extent of government-funded extension. Government involvement in agricultural extension is a responsibility of the States, and

Cary (1993) notes that at least four State governments have recently undertaken reviews of their extension services. With a trend towards smaller government, there has been a closer questioning of the justification for channelling public funds into services, the benefits of which might be seen to accrue largely to farmers.

In Victoria, for example, since coming to power in 1992, the State Government has sought to rationalize its services, the effects of which have been particularly hard-felt in rural communities. Schools, hospitals, railway stations, and government offices have been closed. The government's rationale has been that the user-pays principle should be applied to the provision of government services. Of course, the user-pays approach has never been popular with farmers who argue that, as beneficiaries, consumers have for too long been subsidised by the farming community. So, when the Department of Natural Resources and Environment announced plans to contract for soil conservation, revegetation, salinity control, and vermin and noxious weed control services, in line with the State's privatization program, the reaction was predictable. The editorial in Victoria's major rural newspaper urged thus:

> Farmers must make it clear they will not tolerate any weakening of the Landcare movement by the Victorian government's moves to privatise its vital catchment and land management unit ... The government has a big job to convince farmers this privatization is about delivering a better, cheaper service and not an excuse to fob-off the entire cost of Landcare to farmers. Because in the end if Landcare fails, the whole community pays *(Weekly Times,* 8 March 1995: 2).

Whatever the arguments for or against privatization of extension services, it remains that it is perceived by the rural community as another burden upon an already disadvantaged group. Historically, political, economic, and statutory power has tended to be vested in State Government in State urban capitals. This combined with a rural decline and a still growing proportion of rural people in poverty means that the tradition of self-help and self-suffciency in rural communities is being sorely stretched (Lawrence and Killion 1994). One Tablelands farmer described to me the demands now being made on rural communities, in this way:

> "I know that eventually they want us to become independent and self-sufficient, but I think that they ... I don't know, but I think there's too much strain on country communities as it is. *If it's not*

Landcare, it's the schools. If it's not the schools, it's health. If it's not health, it's the local church and ... if it's not that, it's the fire brigade. And three to four nights a week you'd often be out to meetings trying to organise and get things running."

At a broader level, it might be, as suggested by Martin et al. (1992), that in encouraging the development of Landcare, the state is not so much adopting an ethical stance in favor of community empowerment, but rather is seeking to devolve responsibility for land degradation management to the community. Effectively, government services can be rationalised under the guise of involving community in Landcare. In the discourse of economic rationalists, community becomes "a 'thing' to be modernized, professionalized and commodified" (Kenny 1994:9).

Local ideas of community

What is this community we hear of night and day?
It must be something special, or its name would fade away.
Am I part of this phenomenon, does it need the likes of me?
Or will I stand there on the sidelines and view it critically?

Perhaps I fit in somewhere, as a farmer, yes I should ...

Michael Ryan, Farmer
Dookie, Victoria

Community is one of those words which seems to cause little difficulty in everyday speech but which, when imported into the literature of planners and social scientists, becomes slippery and difficult to define. Rather than offer a structural definition of "community," as was once the case, many social scientists now set out to deal with community as it is symbolically constructed—that is, with the emphasis on the *meaning* of community (Cohen 1985). This more open approach to defining community has succeeded in demonstrating that "structures do not, in themselves, create meaning for people" (Cohen 1985: 9). Iris Marion Young suggests that, for most people, insofar as they consider themselves members of communities at all, "a community is a group that shares a specific heritage, a common self-identification, a common culture, and a set of norms" (Young 1990: 234). Indeed, it is often this idea of the homogenous farming community, with longstanding ties to the land and a locality, which forms the unstated community model of Landcare rhetoric.

The history of land conservation in Australia, and particularly in Victoria, has argued a case for two broad approaches to land management. The first is based upon the physical logic of the catchment and the landscape and the second upon the cultural logic of the idea of community. This inherited understanding of the physical and social organization of locally-based land conservation activities suggests that the "ideal" Landcare group would be based on a community spanning the area of a catchment. But the task of meeting the expectations of a diverse community even within a catchment creates difficulties of its own. For example, one of the Landcare groups in my study, the membership of which is defined by a subcatchment boundary, embraces three distinct land systems; the type and extent of land degradation varies between each system, as does the land use. This group, therefore, encompasses farmers of diverse interests and competing concerns who are physically and socially distant from others in the group. The implications of this are described by one member in this way:

> "You know a classic was that tree pruning day where they all knew where a group member's block was, but I didn't have a clue and there was no one there to tell me and no signs. I suppose you've got the blokes down the bottom end of the group at the river ... You know, you tend to have your 'squattocracy' for want of a better term at that end of the group higher up the catchment and the smaller farmers down the other end of the group. Within the group, you're sort of small—probably socially and economically and in every other way, we're miles apart."

Clearly, this farmer's group does not conform to the idea of a community-based upon mutual recognition and identification. If it did, "signs" would not be needed and a distinction would not be drawn so readily, between the "squattocracy" and "the blokes down the bottom end."

Another example of internal division which persists, in spite of the supposedly unifying logic of the physical landscape, is afforded by the experience of an umbrella group formed to represent the interests of Landcare groups on the Dundas Tablelands. The formation of the combined group gives its membership a strategic advantage in the competition for community grants. Through strength of numbers and the suggestion of united community support, the group has been able to argue a strong case for funding assistance from government. However, early on, some groups and landholders did not identify with this idea of a homogenous Tablelands community, the limits of which were prescribed by land management units as defined many years earlier, by a government soil

562

scientist, on a map. This "new" community, formed *because* of Landcare and in order to attract government funds, was faced with internal conflict and was very nearly disbanded. Rather than the group's energy being directed towards on-ground Landcare works, much of it has been taken up with its attempt to align with the government's ideal of a what a Landcare community should look like.

For many farmers the ideas of community and stewardship that are central to Landcare are at odds with a rural culture which, for so long, has placed a high value on the right of individual landowners to determine land use as he or she thinks fit. Many respondents indicated that prior to the advent of Landcare, with its farm walks and field days, the only part of their neighbor's farms that they knew, was the track from the road to the shearing shed. Through Landcare, they now not only know their neighbors but also their neighbors' farms and their neighbors' problems. So, whilst Young's (1990) ideal of a "self-identifying" community in Landcare does not necessarily hold, there is evidence that Landcare may *generate* a sense of community, in a form or place where it did not exist before.

Summary

In the opening to this paper, I referred to Agenda 21 and to a renewed emphasis on "the local" and the role of the community in environmental management and decision-making. This chapter has examined Landcare, considered, locally at least, to be one of the most successful community-based programs that governments in Australia have ever implemented. There is little doubt that Landcare has been a remarkable achievement. Now stretching across 2,200 local communities nationwide, community Landcare groups have mobilized community action to an unprecedented degree and are having a fundamental impact upon community attitudes towards land management.

In this chapter, I have considered just one aspect of the program, its ideological and practical dependence on the notion of "community." I have argued that, to the extent that this idea of community underpins Landcare, it is important that it functions as the rhetoric suggests, and that the structures of Landcare bureaucracy work to the advantage of local Landcare practice. The only way in which we can properly gain an insight into this important interface between the state and the community, is to look backstage, behind that which is commonly looked for in Landcare evaluation. In arguing a case for the inclusion of the *local* in Landcare evaluation, I am not wanting to suggest that the evaluative frameworks already established for Landcare should be abandoned. In assessing the program's outcomes, we must clearly attend to the question of whether, after all the attention to community, there is indeed measurable change in the

563

physical extent of the problems we first set out to tackle.

Faced with the euphoria induced by community Landcare, it easy to get distracted from the processes that may yet cause the program to "come off the rails," such as its reliance on voluntarism, the absence of other actions relating to public policy issues (such as private property rights and taxation), and the fact that there are still 70 percent of the nation's farmers, including farming women, not yet involved. If we accept that an emphasis on "the local" is an important maxim which will ultimately secure more sustainable use of our land and water resources, we must ensure that the national and regional frameworks put in place invest it with substance.

Notes

1. The lowercase "state" is used as a generic term to refer to the institutions and governance of advanced capitalist societies. The upper case "State" is used in specific reference to Australian States, e.g., Victoria.
2. Emphasis added.
3. The term "squatter" was first used as a derogatory term for those, often ex-convicts, who occupied land without authority. Eventually the term was applied to all large pastoralists in Australia. They became a very powerful group socially, economically, and politically.

For comments on earlier versions of this chapter, grateful thanks are due to Dr. Brian Finlayson, Dr. Lesley Hodgson, and Mr. Brian Williams at the University of Melbourne. The research on which this chapter is based was funded by the Rural Industries Research and Development Corporation.

References

Agyeman, J. and B. Evans (eds.) 1994. Local Environmental Policies and Strategies, Longman, Harlow.

Alexander, H. 1995. A Framework for Change: The State of the Community Landcare Movement in Australia, National Landcare Facilitator Annual Report, National Landcare Program, Canberra.

Bradsen, J. 1994. Natural resource conservation in Australia. In: T.L. Napier, S.M. Camboni, and S.A. El-Swaify (eds.), Adopting Conservation on the Farm: An International Perspective on the Socioeconomics of Soil and Water Conservation, Soil and Water Conservation Society, Ankeny, Iowa, pp. 435-460.

Campbell, A. 1992. Landcare in Australia: Taking the Long View in Tough Times, National Landcare Facilitator Third Annual Report, National Landcare Program, Canberra.

Cary, J. 1993. Changing foundations for government support of agricultural extension in economically developed countries. Sociologica Ruralis 33 (3/4):336 -347.

Cohen, A. 1985. The Symbolic Construction of Community, Routledge, London.

Commonwealth of Australia. 1991. Decade of Landcare Plan: Commonwealth Component, Australian Government Publishing Service, Canberra.

Curtis, A., J. Brickhead, and T. DeLacy. 1993. Community Participation: The Regional Landcare Action Plan Process Victoria, 1992/93, The Johnstone Centre, Charles Sturt University, Albury.

Curtis, A. and T. DeLacy. 1995. Evaluating landcare groups in Australia: how they facilitate partnership between agencies, community groups, and researchers. Journal of Soil and Water Conservation 50(1): 15-20.

Ewing, S. 1995. It's in Your Hands: An Assessment of the Australian Landcare Movement, unpublished Ph.D. Thesis, University of Melbourne.

Hawke, R.J.L. 1989. Our Country, Our Future: Statement on the Environment, Australian Government Publishing Service, Canberra.

Hinchcliffe, F., I. Guijt, J.N. Pretty, and P. Shah. 1995. New Horizons: The Economic, Social and Environmental Impact of Participatory Watershed Development, International Institute for Environment and Development, London.

Kelly, R. 1992. Report on the Earth Summit. The UN Conference on Environment and Development (UNCED), Australian Government Publishing Service, Canberra.

Kenny, S. 1994, Unsettling community. Australian Sociological Association Conference, Deakin University, Geelong, 7 - 10 December.

Landline (video recording) 18 September 1994, Australian Broadcasting Corporation, Sydney.

Lawrence, G. and F. Killion. 1994. Beyond the eastern seaboard: what future for rural and remote Australia? Australian Council of Social Services National Congress, Brisbane, 26 - 28 October.

Martin, P., S. Tarr, and S. Lockie. 1992. Participatory environmental management in New South Wales: policy and practice. In: Lawrence, G., F. Vanclay, and B. Furze (eds.), Agriculture, Environment and Society: Contemporary Issues for Australia, Macmillan Melbourne, pp. 184-207.

Mues, C., H. Roper, and J. Ockerby. 1994. Survey of Landcare and Land Management Practices, 1992-93, Australian Bureau of Agricultural and Resource Economics, Canberra.

Poussard, H. 1994. Letter to the editor. Australian Journal of Soil and Water Conservation 7(3):2.

Reij, C. 1994. Defining success. In: Hurni, H., H. Liniger, T. Wachs, and K. Harweg (eds.), Workshop Proceedings: 2nd International WOCAT Workshop, Group for Development and Environment, Institute of Geography, University of Berne, p. 32.

State of the Environment Advisory Council. 1996. Australia: State of the Environment 1996, Commonwealth Scientific and Industrial Research Organisation (CSIRO) Publishing, Melbourne.

Toyne, P. and R. Farley. 1989. A national land management program. Australian Journal of Soil and Water Conservation 2(2):6-9.

United Nations Conference on Environment and Development, 1992. Agenda 21: Action Plan for the Next Century, UNCED.

Vanclay, F., 1994. Hegemonic Landcare: further reflections from the National Landcare Conference. Rural Society 4(3/4):45-47.

Vanclay, F. and G. Lawrence. 1995. The Environmental Imperative: Eco-Social Concerns for Australian Agriculture, Central Queensland University Press, Rockhampton.

Wensley, P. 1994. Creating the future. In: Defenderfer, D. (ed.), Proceedings of the 1994 Australian Landcare Conference v. 2, Department of Primary Industries and Fisheries, Hobart, pp. 87-99.

Williams, M., M. McCarthy, and G. Pickup. 1995. Desertification, drought and Landcare: Australia's role in an international convention to combat desertification. Australian Geographer 26 (1):23-32.

Young, I.M. 1990. Justice and the Politics of Difference, Princeton University Press, Princeton.

36

Water Allocation Policy in Australia

David Farrier

Low or temporally variable rainfall in many areas combined with high rates of evaporation, have led to a significant focus within Australia on measures to harvest available water supplies. Large dams built in all states with the exception of South Australia are the primary source of water for domestic purposes and irrigation (Industry Commission 1992: 22; Ecologically Sustainable Development Working Group Chairs 1992: 111). Irrigated agriculture (broadacre crops such as cotton, sugarcane, oilseeds, and rice; vineyards and orchards; pasture; and horticulture) account for about 75 percent of water use in Australia. In 1988–1989, half of this comprised irrigated pasture, with most of this in New South Wales (NSW) and Victoria. But in terms of the value of irrigated production (4.6 billion in 1988–1989), irrigated pasture accounted for only 20 percent. Surface irrigation (e.g., flood and furrow) is the usual method of application, but there is growing use of pressurized systems (e.g., spray, micro-spray, and drip), primarily for horticultural production (Industry Commission 1992: 193-194).

The Murray-Darling Basin covers most of inland southeastern Australia, and includes most of the country's best farmland. The value of agricultural production in the Basin is in excess of $10 billion, and irrigation accounts for $3 billion of this (Murray-Darling Basin Ministerial Council 1995: 2). Around 75 percent of total surface water use in Australia occurs in the Basin, and it is estimated that 85 percent of the available surface water supply is used (Industry Commission 1992: 22). Of water extracted between 1988/89 and 1992/93, over 95 percent was for irrigation. An audit of water use in the Basin showed that between 1988 and 1994, water diversions grew by about 8 percent, and that this was continuing at a rate of 1 percent per year. The greatest increases have been in northern NSW and Queensland because of the high returns available from irrigated cotton. The increase in Queensland has been 89.3 percent, but from a small base. In NSW, these increases have not been fueled by the grant of new water allocations for consumptive uses (compare Queensland where some new licences have been granted) but by operating within the flexibilities of the existing allocations system, discussed below. Moreover, there was still scope for even further diversions under the existing management system (Murray-Darling Basin Ministerial Council 1995: 6, 9-10).

The environmental problems

A wide range of environmental problems have resulted from poor land and water management in the Basin (Murray-Darling Basin Ministerial Council 1990 and 1995; Knights et al. 1995):

- dryland salinity stemming from rising water tables caused by land clearing for dryland agricultural development;
- water quality problems caused by diffuse runoff (fertilizers and pesticides) from agricultural operations, and saline water discharges from irrigation areas, as well as the impact of urban communities;
- degradation of instream values, in particular damage to ecosystems, resulting from river regulation and water extraction for consumptive uses reducing the frequency of high flows, shifting flows from spring to summer and autumn, when irrigation water is needed, and changing flow variability;
- changes in river flow characteristics resulting, on the one hand, in permanent inundation of some floodplain wetlands and, on the other, reductions in the flooding of others in spring, impacting on fish and bird breeding (NSW EPA 1995: 43-44);
- loss of up to 50 percent of the freshwater/inland wetlands since European settlement (NSW EPA 1995: 43), with consequent loss of ecosystem services, such as nutrient cycling, flood mitigation, water filtration, and sediment trapping;
- reductions in river flow, contributing to the conditions for the growth of blue-green algal blooms, with 115 occurrences in NSW in the drought year of 1993-1994, 54 of which were serious (NSW EPA 1995: 49);
- irrigation salinity, caused by rising water tables and waterlogging of typically saline sediments.

While dryland salinity and diffuse runoff have only a tenuous connection with extraction of water from watercourses, other problems are connected. Some of them, such as blue-green algae and salinity, pose immediate and transparent threats to human self-interest in terms of health impacts and loss of productivity, while in the case of others—loss of ecosystem services and reductions in biodiversity—the threats posed are long-term and cumulative, and, consequently, less real to those whose behavior contributes to their existence.

The legal framework for the extraction of water for consumptive uses

Although the formal position under the Australian Constitution remains unclear (Fisher 1995: 31-32), in practice, the regulation and management of water has been treated as a matter for the States, with the Commonwealth Government's role substantially confined to one of moral and financial persuasion. In 1992, the primacy of the States in natural resource management was confirmed by the *Intergovernmental Agreement on the Environment.* This was a formal agreement between State, Territory, and Commonwealth governments. The agreement established jurisdictional boundaries in the environmental context. This, however, recognizes the Commonwealth's interest in meeting its obligations under international conventions. These conventions include the Ramsar Convention which focuses on wetlands and the United Nations Convention on Biological Diversity.

There are two modifications to this general position: the Murray-Darling Basin Agreement and the Dumaresq-Barwon Border Rivers Agreement. These set up consensual mechanisms involving the Commonwealth and the relevant State governments relating to inter-state rivers.

Although the Murray-Darling Basin Agreement had its origins in the River Murray Waters Agreement of 1915 (Clark 1983), providing for the regulation of the river and the sharing of water between the States concerned, it has over the years taken on a broader focus. The current functions of the Murray-Darling Basin Ministerial Council, set up in 1985, now include the determination of major policy issues of common interest to the parties "concerning effective planning and management for the equitable efficient and sustainable use of the water, land and other environmental resources" of the Basin as a whole (Agreement cl 9(a)). Resolutions of the Council must, however, be unanimous (Agreement cl 12(3)), and implementation is left to individual States. For example, a resolution by the Ministerial Council to cap diversions from rivers in the Basin at 1994 levels, discussed below, is creating some controversy because of the varying approaches taken by the States to implementation.

Following the substantial rejection at the end of the nineteenth century of the doctrine of riparianism, exported from England but found to be too inhibiting to development in a country where water was scarce, each of the Australian States adopted an administrative system for allocating water. This was based on assumption of a right of primary access by the state, with water allocated for consumptive uses through a system of licences and other authorizations granted for specified periods (Fisher 1995; Bartlett 1996). In theory, at least, this system allowed for considerable flexibility (Clark and Renard 1972: 15). As developed in New South Wales, the main features were (Farrier 1993) as follows:

- Assumption by the state of the exclusive right to use and control water.
- The annexation of water allocations for consumptive uses to particular areas of land, such that transfer of the land involved transfer of the water allocation, and the water allocation could not be transferred independently. In combination with low water charges (see below), this has led to water allocations being capitalized into land values even where they have not been used. In addition, the annexation of water allocations to land has led to significant technical legal problems in relation to attempts to amend the legislation to allow separate transfer of water allocations (Bond and Farrier 1996).
- Water charges traditionally only covering part of the cost of delivering water from dams to the point of extraction ("running the rivers"), with irrigators making no contribution to capital, maintenance, and refurbishment costs of water storages, or to the costs of managing the resource (including planning, resource evaluation, and granting permits). In September 1995 there was an interim "resource management charge" of $1.35 per megalitre was introduced in NSW.
- Initially, indirect control of water use through restrictions on the area of land which could be irrigated rather than on the amount of water used. This approach has now been replaced on rivers regulated by head storages by volumetric water allocations schemes. These allow the irrigation of any area within assigned volumetric limits.
- No automatic loss of rights for failure to use allocations. Although there is the theoretical possibility that they might be terminated or not renewed for lack of beneficial use, this has not occurred in practice. The result is that there are a considerable number of so-called "sleeper" licences, held by nonirrigator farmers or graziers, and "dozer" licences, held by small scale or retired irrigators who do not have the resources to fully utilize their allocation.
- Allocation of water to irrigators in a context where they were guaranteed very limited security of supply.

The prior appropriation doctrine which governs water allocation for consumptive uses in some U.S. states guarantees priority of access to water based on order of historical usage, provided that beneficial use of the quantity claimed has not been abandoned. If there is not enough water to go around, junior appropriators must give way to senior appropriators (Sax and Abrams 1986). By contrast, those with water allocations under Australian administrative systems generally share the pain. Aside from the priority given to restricted riparian rights for domestic supply/stock watering, town water supplies and "high security" supplies (originally designed for permanent plantings, such as orchards and vineyards, but more recently made more widely available in return for certain trade-offs), irrigators in particular catchments are equally vulnerable to water shortages in any particular year. This may see irrigators in some valleys receiving only an across-the-board percentage of their notional allocation. This is principally the case with certain valleys in the north of the Basin where security of supply can be as low as 35 percent (on average, 100 percent of the notional allocation can be expected in only 35 percent of the years). While from one perspective this lack of security of supply is an inevitable feature of natural systems, it stems fundamentally from human optimism in notionally allocating a resource which is not guaranteed to be available. Apart from this seasonal insecurity, the security of those with existing allocations is inevitably devalued by the grant of further allocations, for urban water supply purposes, for example. The uncertainty generated by these factors has led cotton irrigators in some valleys, encouraged in the past by the NSW water agency, to invest in large off-river water storages, that they have filled during declared "off-allocation" periods—flood events or natural runoff not intercepted by the dam—when the flow in the river is judged to exceed immediate requirements and irrigators are allowed to divert water above and beyond their allocation under licence. This has further interfered with instream flows.

More recently, threats to security of supply have emerged in the form of arguments claiming that some proportion of existing water allocations and water traditionally taken during "off allocation" periods may have to be reclaimed from holders. The reason is so that more water can be left in watercourses to protect in-stream values, such as the protection of biodiversity and the maintenance of ecosystems.

Ecologically sustainable development

Natural resources policy in Australia is now driven by a concern, not simply to prevent and remediate land/water degradation for productive purposes, but to achieve *ecologically sustainable development* (ESD). ESD represents Australia's response to calls for sustainable development embodied in the 1987 Report of the World Commission on Environment and Development, *Our Common Future* (Brundtland). In 1990 the Commonwealth government published a discussion paper (Commonwealth of Australia 1990), and subsequently set up nine working groups, to discuss implementation of ecologically sustainable development within particular sectors (agriculture, energy production, energy use, fisheries, forests, manufacturing, mining, tourism, and transport), as well as cross-sectorally, with representatives from government, industry, environment and consumer groups, trade unions, and science. These groups reported in November 1991, and, building on these, in December 1992 the Commonwealth government released a *National Strategy for Ecologically Sustainable Development,* developed by an intergovernmental committee, and endorsed by the Council of Australian Governments (Commonwealth of Australia 1992).

The core objectives of the strategy are as follows (Commonwealth of Australia 1992: 8):

- to enhance individual and community well-being and welfare by following a path of economic development that safeguards the welfare of future generations;
- to provide for equity within and between generations;
- to protect biological diversity and maintain essential ecological processes and life support systems.

The guiding principles include the integration of economic, environmental, social, and equity considerations in decision-making processes, the precautionary principle, the need to develop a strong, growing, and diversified economy, the use of cost effective and flexible policy instruments, such as improved valuation and pricing and incentive mechanisms, and community involvement. The strategy emphasizes the need for a balanced approach, with no objective or principle predominating over the others (Commonwealth of Australia 1992: 9).

The concept of ESD has now begun to appear in environmental and resource management legislation enacted by the Australian States. In NSW, the Protection of the Environment Administration Act 1991, provides that the Environment Protection Authority (EPA) should have regard to the "need to maintain ecologically sustainable development," which requires "the effective integration of economic and environmental considerations in decision-making processes" (s 6). The principles which the legislation indicates can be used to achieve ESD are set out in the following terms (s 6):

(a) The precautionary principle, namely, that if there are threats of serious or irreversible environmental damage, lack of full scientific certainty should not be used as a reason for postponing measures to prevent environmental degradation.

(b) Intergenerational equity, namely, that the present generation should ensure that the health, diversity and productivity of the environment is maintained or enhanced for the benefit of future generations.

(c) Conservation of biological diversity and ecological integrity.

(d) Improved valuation and pricing of environmental resources.

Currently, the Department of Land and Water Conservation is the water resource manager in NSW. It operates under older legislation which makes no reference to ESD (Water Act 1912; Water Administration Act 1986). The EPA, however, has a broad charter to protect, restore, and enhance the quality of the environment in NSW, and a direct interest in instream flows because of the interrelationship between water quantity and water quality. Under its legislation, the EPA has the power to direct the Department of Land and Water Conservation to do anything within its powers which will contribute to environment protection (s 12). This would include the adoption of principles of ecologically sustainable development which it is clearly within the Department's powers to apply. In practice, however, government policy has led to the development of a cooperative relationship between the two agencies to identify both water quality and quantity objective for NSW watercourses.

A number of points relevant to the subsequent discussion can be made briefly here. In the first place, ESD contains a commitment to protect biodiversity and to maintain ecosystems. Secondly, the precautionary principle outlaws arguments that we should delay taking action until there is overwhelming scientific evidence of adverse impact on instream biodiversity and ecosystems, provided that we have enough evidence to indicate that there are "threats of serious or irreversible damage." Thirdly, although this version of the precautionary principle has nothing to say about what weight should be given in the overall decision making calculus to nonmarket environmental values, as against traditional economic values, there is a strong emphasis in ESD on ensuring that natural resources are appropriately valued, rather than simply being costless externalities in production processes. This has significant implications for arguments examined below that the most efficient way of redistributing what are now scarce water resources is to create so-called "property rights" in allocations for ex-stream uses and allow them to be traded in the marketplace. The market has a miserable record when it comes to factoring public costs stemming from environmental degradation into private decision-making processes.

Finally, there is a strong emphasis in the *National Strategy for Ecologically Sustainable Development* not only on intergenerational equity, but intragenerational equity. Surprisingly, this is missing from the definition which appears in the NSW legislation discussed above. When it comes to water resource management in the Murray-Darling Basin, significant equity questions are raised by arguments examined below that, in order to provide for ecologically sustainable instream flows, water will have to be "clawed back" from extractors. Historical water allocations have been factored into land values, and some irrigators have invested heavily in infrastructure (e.g., large off-river water storage) in the expectation that there would be no radical changes to government policy. Indeed, until recent times, government policy has been to actively encourage such investment.

In the end, ESD principles require decision-makers to "integrate" environmental factors which the precautionary principle places firmly on the agenda, with socioeconomic factors. The aim is to secure a "balance," in which neither prevails over the other. Whether this can be achieved in practice within a well-established socioeconomic tradition and culture which has been historically driven by developmental imperatives, is quite another question.

From water allocation to resource management

As it became clear that water resources were fully committed and, arguably overcommitted, Australian water resources policy has shifted from a development phase to one emphasizing the management of existing supplies (modifying demands and increasing efficiency) (Pigram and Hooper 1994: 5). The introduction of administrative, and later statutory, embargoes on the issue of new allocations on streams within the NSW section of the Basin, commenced in the 1970s, has gradually been extended so that they are now virtually comprehensive. The focus is no longer on facilitating extraction of water from the stream, but upon making better use of existing supplies and improving the quality of supplies.

A key tool in the quest for economic efficiency which has emerged is the water transfer. Permanent transfer was allowed in South Australia as early as 1982, in Queensland from 1989, and in Victoria from 1991 (Murray-Darling Basin Ministerial Council 1995: 37). In NSW, temporary transfers were permitted in the 1983-1984 season, and legislation allowing permanent transfers commenced in 1989 (Bond and Farrier 1996). Even though the system of administrative allocation was theoretically flexible enough to allow government to reclaim water from existing holders of allocations and reassign it to those putting it to higher value uses, the argument was that the market would do this more efficiently. "Grandfathering" allocations to existing holders did not attract any of the political pain or social disruption which would have been associated with modifications to allocations made through administrative intervention.

In terms of both permanent and temporary transfers, by far the greatest amount of activity has occurred in New South Wales, in part because water is particularly scarce as a result of the combined effect of climatic factors and the fact that irrigation development pressures, particularly involving cotton, have been greatest here. In 1993-1994, 29,145 megalitres were transferred permanently, and in 1994-1995, a further 31,600. But activity on the temporary market has been much greater, albeit varying significantly from year to year, with as much as 896,300 megalitres transferred in 1994-1995 (Hill).

The expectation was that the operation of the free market would ensure that water would be transferred to higher value uses, for example, away from irrigated pasture to cotton production. The problem is that when it comes to the determination of what are higher value uses, there is no mechanism in the market-place for factoring in the differential public environmental costs associated with various land and water uses. A water allocation transferred upstream can mean a reduced flow in the intermediate section of the river with subsequent impact on the environment. Temporary transfers are likely to be heaviest at particular times of the year, such as at the end of growing seasons, putting extra pressure on instream flows. Water transferred to irrigate a different crop may result in extractions at different times of the year, for example, irrigation of pasture is relatively even throughout the year, whereas cotton irrigation is concentrated in the summer growing season. Cotton production results in other costs to the community which are not factored into the market price. It relies on heavy applications of fertilizers and pesticides, creating problems of diffuse pollution. Moreover, the high value of the product has made it economical to construct large off-river water storages to harvest off-allocation the floods and freshes vital to a healthy river system.

In spite of the commitment, as one of the principles of ecologically sustainable development, to improve the valuation and pricing of environmental resources, there has been no attempt to factor the costs to the community of environmental externalities into the price charged by the NSW government for water, although in NSW this is now being considered by the Independent Pricing and Regulatory Tribunal as part of its review of rural water pricing. So far as ensuring that externalities associated with transfer are accounted for, the NSW legislation does provide that government approval for transfer is required. It only refers specifically to the social and economic effects of transfer as potentially relevant considerations in the decision as to whether to grant approval, but the agency can take into account environmental matters to the extent that "it thinks fit" (Water Act 1912 s 20AI(6)). In practice, it appears that this regulatory system does not present a significant barrier to transfers, although the tribunal which reviews agency decisions has recently taken a more aggressive stance by refusing approval to a transfer 150 kilometres upstream, on the grounds of both environmental impact and impact on the ability of intermediate users to take water (Bond and Farrier 1996).

The transfer system, then, must cater for the fact that extraction of water from the stream imposes environmental costs on society through loss of instream values. This has become crucial with dawning recognition that protection of instream values is not simply an obstacle to consumptive use, but a precondition. This message was first driven home to the wider community in the early 1990s with the realization that the quantity of water in the stream was a factor in deterioration of water quality through increasing occurrence of blue-green algal outbreaks (Verhoeven 1993).

These failings of the transfer system have been exacerbated by the fact that there has been a disparity between water notionally allocated by the water resource manager and water actually extracted by consumptive users. The allocation system has been administered with the expectation that a proportion of the water allocated in any one year will not be taken up (Murray-Darling Basin Ministerial Council 1995: 6-8, 35; Bain et al. 1996). Data shows that in the Murray-Darling Basin between 1988/89 and 1992/93 only 63 percent of the water made available annually under the allocations system was taken up by irrigators. Some part of this deficit is attributable to "sleeper" and "dozer" licences. Although there are provisions in the legislation which would allow these licences to be cancelled, without compensation, for lack of beneficial use (Water Act 1912 ss 17A(l)(e), 20H(l)(e)), this has not occurred in practice. Other reasons for the gap between notional allocations and actual use, relevant during wetter periods, are

- "off-allocation" policy which allows irrigators to get access to water during periods of high flow without drawing on their allocation under licence;
- agencies allocating more water than is actually available in the knowledge that not all entitlement holders will use their full allocation (the aim being to encourage those who do have the capacity to use water to use it to its full).

The result is that the administrative allocation system only operates to restrict extractions in drought periods (Bain et al. 1996). During wetter years it actually encourages diversions. It is factors other than the licensing system which have restricted diversions in practice, including low levels of on-farm storage in the south, river channel capacity, the undeveloped market for water entitlements, and low returns from irrigation (Murray-Darling Basin Ministerial Council 1995: 35).

All this could change dramatically as the transfer system takes hold. For the allocated but unused water is available for sale on the short-term transfer market. The likely result is that the water market will be flooded, driving down prices and making it unlikely that water will be retired from low value uses, such as irrigated pasture, for transfer to higher value uses. Increased sales will simply result in increased water use:

"The introduction of water trading, now well underway, and other changes such as technical development of new crops and marketing opportunities, are generating considerable pressures for the slack in the allocation system to be taken up." (Murray-Darling Basin Ministerial Council 1995: 35)

This has made inevitable changes to announced allocations and off-allocation policy with a view to bridging the gap between notional allocations and historical use. In mid-1995, the Murray-Darling Basin Ministerial Council announced a decision to "cap" diversions between July 1995 and June 1997 by holding them at the 1993-1994 level of development and, thereafter, following consultation with the community, to impose a final cap which will balance diversions and the protection of instream values. Closely linked to this, in September 1995, the NSW government announced a water reform package which included the development of river flow objectives to provide benchmarks to guide catchment planning. A moratorium on the permanent transfer of unused portions of allocations was announced. Temporary transfer remains available, but at present only for up to 3 years. Percentage allocations announced annually as available within catchments are being deliberately reduced to bring them closer to expected use, and changes to off-allocation policy are certain. As a result, heavier users are becoming increasingly dependent on the short-term transfer market, and it seems likely that the 3-year restriction on temporary transfers will have to be removed. Measures are also being put in place in Victoria to stop increases in diversions by preventing unused allocations being taken up, but Queensland has reserved the right to allow this, while undertaking not to issue any new allocations (Bain et al. 1996).

The traditional approach in land use planning law has been to deny compensation where new regulatory impositions have defeated mere expectations. Only existing uses of land, not unrealized aspirations, have been protected against the introduction of new regulatory requirements with which they conflict (Farrier 1991). This is reflected in the water context in the doctrine of prior appropriation's insistence on continuing beneficial use if water rights are to be preserved. By analogy and a fortiori in the context of an administrative system of water allocation, the modification of water allocations not fully required to support existing development would have been quite legitimate prior to the introduction of a transfer system. Close analysis of the legislation indicates that there is no requirement for payment of compensation for modification of permits, as distinct from outright revocation, and certainly not for government failure to declare water as being available "off-allocation." Nor is there any provision requiring compensation to be paid for takings of private property rights in the NSW Constitution, or the constitutions of any of the other States, although there is a provision in the Australian Constitution requiring the "acquisition of property on just terms" (s 51 (xxxi)) which might apply in the unlikely event that the Commonwealth took the initiative.

At the same time, it is crucial to distinguish between legal theory and practical realities. Landholders clearly regard themselves as having de facto property rights, particularly in a context where cheap irrigation water has been capitalized into land values (Industry Commission 1992: 201) and investment decisions have been made on the basis of assumptions of water availability encouraged by governments. Introduction of transferability—a key indicator of private property—has confirmed these expectations, and further marginalized those provisions in the legislation that provides administrative flexibility in the allocation of water. The NSW government which, in introducing the transfer system, grandfathered existing water allocations free of charges is now likely to face demands for compensation where water is reclaimed from permit holders. The suggestion that "a full-scale review of entitlements undertaken within an open and consultative planning process" (ARMCANZ 1995: 9) will obviate the need for compensation, may well prove over optimistic. However, the absence of any legal requirement for compensation in situations falling short of licence revocation gives government the flexibility to look to the future by offering a financial package designed to facilitate restructuring, rather than one which looks backwards to compensation.

Protecting the instream environment

Having put the cart before the horse, by facilitating transfers before identifying the prerequisites for maintaining a healthy river system and assessing the cumulative effect that transfers might have, government is now trying to bolt the stable door. While remaining committed to enhancing the transfer system as a management tool, one of the fundamental thrusts of contemporary policy is that recognition of property rights in water as a vital component of facilitating transfer should only take place against a backdrop of catchment planning in which we first assess and specify environmental needs (ARMCANZ 1995). The first principle of a recent policy position paper on the establishment of a national framework for implementation of property rights in water, developed in the context of a commitment by the Council of Australian Governments to implement water reform (COAG 1994), states "That all consumptive and nonconsumptive water entitlements be allocated and managed in accordance with comprehensive planning systems and based on full basin-wide hydrologic assessment of the resource." (ARMCANZ 1995: 5)

There is, however, a strong suspicion that the property rights horse may have already bolted. Developmental imperatives appear once again to have run ahead of careful environmental planning.

Our understanding of what is required to ensure that the development of water resources is ecologically sustainable has become considerably more sophisticated in recent times. The instream environment is no longer viewed simply as another competitor for water allocations, alongside consumptive out-of-stream uses, as suggested by the earlier discourse of "water allocations for the environment." There is growing recognition that sustainability of the water resource through "protection of instream values" is a core requirement: "what is necessary to sustain the environment is also necessary for the long-term sustainability of human uses" (NSW EPA 1995: 62). Moreover, the focus is no longer solely on the provision of *quantities* of water to satisfy instream needs but on approximating natural *variations* in seasonal flow.

The impracticability of returning to a "natural" flow regime is acknowledged. The aim is to "mimic" natural flows. Specific environmental contingency allocations from stored water are only part of a package of responses. The package must also incorporate passing flows through storages to reduce their impact, environmentally sympathetic water supply delivery rules and weir management to reintroduce flow variability, and substantial adjustments to current off-allocation management to modify the interception of flood flows (Knights et al.1995). During past drought periods water users have accepted the need for substantial restrictions on use, but they will have much more difficulty in accepting that restrictions should apply during periods of high flow.

Knights et al (1995: 253) have suggested a risk management approach to the question of environmental flows "which recognizes that there is no scientifically correct solution, but that some of the damage caused by river regulation and water abstraction can be minimized or redressed through a process of trial and adjustment." Decisions are not simply scientific ones, but contain a substantial value component. Consistent with the demand for intragenerational equity as one of the principles of ecologically sustainable development, social and economic considerations arising from disturbance of expectations will inevitably play a significant part in the initial readjustment of allocations to protect instream values.

To acknowledge the role that social and economic factors will play in reaching a final resolution is not to diminish the substantial role that science must play in setting the parameters. It is clear, however, that currently available scientific information is quite inadequate to allow us to make decisions guaranteeing a high degree of resource security. Application of the precautionary principle clearly indicates that lack of full scientific certainty in the face of "threats" of serious environmental harm must not be used as a reason for delaying remedial responses. This does not, however, require once-and-for-all responses that are incapable of adjustment as new scientific information becomes available, whether this indicates that the initial response has gone too far, or not far enough. The flexibility to make adjustments in the future through a process of adaptive management based on ongoing monitoring and adjustment, is a crucial component of any package (Knights et al.1995). The fundamental question is how this acknowledged need for flexibility can be aligned with commitments to establish property rights in water.

579

Framing "property" rights

In the current context of a fully committed water resource, a perception that water is not being used efficiently, and an understandable lack of political will to make the necessary adjustments through regulation, the task ahead is seen to be one of framing allocations for productive uses in such a way as to facilitate their voluntary transfer. This, and not the creation of property rights as an end in itself, is the objective. The notion of a "property right" is already intimately associated by the community with rights of private ownership in relation to land. Its use in the context of water law reform is unfortunate because ownership of land is of unlimited duration, and private property in rural contexts is popularly associated with claims to unrestricted use.

To facilitate transfer of water allocations, it has been argued that rights must be *explicit, exclusive, enforceable,* and *transferable* (Hill et al. 1993). So far as the first of these requirements is concerned, the argument is that people will not trade if they do not have an acceptable degree of certainty about what they are getting. This is intimately related to questions of security (Millington 1991). Rights must be *"well specified in the long-term sense* —the market can interpret and depend on what the rights really mean" (ARMCANZ 1995: 4). The first trick which must be performed, then, is to provide this certainty and security in a context where it is acknowledged that instream values must be protected and where we may need to adjust the balance between consumptive and nonconsumptive uses as more scientific information becomes available:

In specifying ownership of entitlements, the individual right holder's security of tenure should generally be maximised subject to the degree to which resource availability and environmental provisions within a catchment can be defined. Ownership tenure should be perpetual, but with conditions of access associated with entitlements that are subject to reviewability within an open planning system as discussed in Principle 1. However, to ensure an acceptable level of security over entitlements for water users, such reviews should be specified as occurring after periods of several years, with sufficient notice being given of impending reviews (ARMCANZ 1995: 8).

Whether these limited guarantees will be sufficient to bring about the efficient operation of the transfer market remains to be seen. Extractors already cope with insecurity resulting from climatic variability, and one argument is that as long as they are given clear advance warning through this additional element of insecurity being built into the definition of their rights, they will not be discouraged from operating in the transfer market (Millington 1991).

The ARMCANZ approach does not go as far as meeting landholder claims for property rights in terms of "resource security," with government having to purchase any additional water required to protect instream values in the transfer market or having to provide compensation for compulsory claw backs of water if scientific research in the future shows this to be necessary. The concern must be that this conception of property rights will overwhelm the more fluid, and, from the extractor's perspective, less secure notion envisaged by ARMCANZ, and that a government which grandfathered water allocations free of charge will ultimately have to buy them back. Apart from the administrative reallocation envisaged by ARMCANZ and the alternative of purchasing water in the open market, another technique for reclaiming allocated water involves "taxing" or discounting volumes of water on transfer and reallocating the amount reclaimed for instream purposes (Collins and Scoccimarro 1995; Doolan and Fitzpatrick 1995).

Questions also arise in relation to the matter of *enforceability* of water rights. Enforceability is intimately related to *exclusivity*. The objective is to ensure that only the holder of the property right benefits from the resource, the corollary being that others are excluded. This requires regulation, traditionally provided in common law jurisdictions through the criminal law and the law of torts. What many economists have failed to recognize in drawing a crude distinction between command and control regulation and fiscal instruments, such as tradeable water rights, is that transferable rights are ultimately dependent on this command and control regulation.

In the past there was no criminal offence of trespass to land, in spite of optimistic threats from landholders on signposts that "trespassers will be prosecuted." There was no need to have outside agencies involved, because landholders themselves could police their boundaries. Protecting mooted property rights in water by preventing illegal interception by intermediate landholders is much more difficult. There have been significant problems with unauthorized diversions in the past, although modern metering technology and the sheer capacity of some of the pumps involved makes detection increasingly easier. The question which arises is whether property right holders themselves should be responsible for protecting their rights through selfhelp and civil proceedings, or whether proposed water rights should be guaranteed by the community under the criminal law, backed-up by a specialist enforcement agency, as is the case at present. If the latter, then the transaction costs of enforcing exclusivity through provision of a specialist enforcement agency must be factored into any assessment of the efficiency of reallocation through the market mechanism. The situation is complicated by the fact that current initiatives designed to protect instream values will result in more water being left in the stream, providing significant temptation to irrigators in a context where neighbours have no direct self-interest in taking action. In these circumstances, the involvement of an outside enforcement agency appears inevitable.

Finally, there is the question of *transferability*. The need for some level of regulation of transfers to address the question of the public costs associated with transfer, is acknowledged in general terms in the literature (ARMCANZ 1995: 8-9; Industry Commission 1992: 143-144), but the potential ramifications do not seem to have been appreciated. If the environmental costs associated with upstream and interbasin transfers, and transfers that will involve different patterns of water use, are to be taken much more seriously than they have been under the existing transfer system, this could involve significant constraints on the operation of the water market.

I would like to thank Don Geering, Manager, Natural Resources and Environmental Policy Branch, NSW Department of Urban Affairs and Planning, and Hugh Milner, Senior Hydrologist, NSW Department of Land and Water Conservation for their assistance in locating relevant material, and for comments made on an earlier draft.

References

ARMCANZ (Agriculture and Resource Management Council of Australia and New Zealand). 1995. Water Allocations and Entitlements: A National Framework for the Implementation of Property Rights in Water. Task Force on COAG Water Reform Occasional Paper No 1.

Bain, D., A. Close, H. Milner, and A. Hee. 1996. Managing the Water Resources of the Murray-Darling Basin Beyond the Year 2000. Paper presented at the Twenty-Third Hydrology and Water Resources Conference (Institute of Engineers, Australia), Hobart 21-24 May.

Bartlet, R.H. 1995. The Development of Water Law in Western Australia. In R.H. Bartlet, A. Gardner, and B. Humphries, Water Resources Law and Management in Western Australia: 43-116.

Bond, M. and D. Farrier. 1996. Transferable Water Allocations - Property Rights or Shimmering Mirage. Environmental and Planning Law Journal 13(3): 213-224.

Clark, S.D. 1983. Intergovernmental Quangos: The River Murray Commission. Australian Journal of Public Administration XLII(1): 154-172.

Clark, S.D. and I.A. Renard. 1972. The Law of Allocation of Water for Private Use (Australian Water Resources Council Research Project 69/16).

COAG (Council of Australian Governments). February 1994. Communique.

Collins, D. and M. Scoccimarro. 1995. Economic Issues in the Creation of Environmental Water Allocations. In Proceedings of the National Agricultural and Resources Outlook Conference, Australian Bureau of Agricultural and Resource Economics, Canberra: 241-251.

Commonwealth of Australia. 1990. Ecologically Sustainable Development: A Commonwealth Discussion Paper.

Doolan, J.M. and C.R. Fitzpatrick. 1995. Protecting Our Rivers Through Property Rights and Water Trading. In Proceedings of the National Agricultural and Resources Outlook Conference, Australian Bureau of Agricultural and Resource Economics, Canberra: 262-269.

Commonwealth of Australia. December 1992. National Strategy for Ecologically Sustainable Development.

Ecologically Sustainable Development Working Group Chairs. 1992. Intersectoral Issues Report.

Farrier, D. 1991. Vegetation Conservation: The Planning System as a Vehicle for the Regulation of Broadacre Agricultural Land Clearing. Melbourne University Law Review 18: 26-59.

Farrier, D. 1993. Environmental Law Handbook: Planning and Land Use in New South Wales. 2nd edition: 326-343.

Fisher D. 1995. Water. In Laws of Australia 14.9: 1-210.

Hill, C., B. Fitzgerald, and D. Cleary. 1993. Property Rights in New South Wales Water Management.

Hill, C., Senior Economist, Department of Land and Water Conservation. 1996. Personal communication.

Knights, P., B. Fitzgerald, and R. Denham. 1995. Environmental Flow Policy Development in NSW. In Proceedings of the National Agricultural and Resources Outlook Conference, Australian Bureau of Agricultural and Resource Economics, Canberra: 252-261.

Industry Commission. 1992. Water Resources and Waste Water Disposal. Report No. 26.

Millington, P. 1991. The Water Market — A NSW Case Study. In Proceedings of the New Environmentalism Conference, Sydney.

Murray-Darling Basin Ministerial Council. 1990. Natural Resources Management Strategy.

Murray-Darling Basin Ministerial Council. 1995. An Audit of Water Use in the Murray-Darling Basin.

NSW Environment Protection Authority. 1995. State of the Environment 1995.

Pigram, J. and B.P. Hooper. 1990. Transferability of Water Entitlements.

Pigram, J. and B.P. Hooper. 1994 Water Resource Management in an Environment of Change. Australian Journal of Soil and Water Conservation 7(1): 4-8.

Pigram, J. and B.P. Hooper. 1992. Water Allocations for the Environment. Seminar Proceedings, Centre for Water Policy Research, UNE, Armidale.

Sax, J. and R.H. Abrams. 1986. Legal Control of Water Resources: Cases and Materials: 278-285.

Verhoeven, J. 1993. Implementing the New South Wales Algal Management Strategy. Australian Journal of Soil and Water Conservation 6(3): 30-34.

37

Policies to Promote Sustainable Land Management in New Zealand

Wayne Bettjeman

New Zealand's economy is largely an agricultural one. More than two-thirds of its land area is used for agricultural purposes. It is therefore inevitable that agriculture has significant impacts on the environment. These impacts are not new but are the culmination of land management practices over the last 150 years.

Land management in New Zealand today is influenced by the country's geographical setting, its patterns of settlement, land use and development, and by recent wide-ranging economic and environmental reforms. This chapter examines in turn each of these influences and their effects on sustainable land management in New Zealand. It then sets out the key points of the government's recent strategy for sustainable land management.

Geographic and historic influences on land management

New Zealand lies in the southwest Pacific Ocean some 1,900 kilometers southeast of Australia. It consists of two main islands (the imaginatively named North Island and South Island), and a number of smaller islands, whose combined area of 270,500 square kilometers is similar to the size of Japan or the British Isles. It is a small country inhabited by 3.6 million people.

New Zealand straddles two tectonic plates which are moving against each other. The resulting earth movements have produced hilly and mountainous terrain over two-thirds of the land area, with frequent earthquakes in most parts of the country and a zone of volcanic and geothermal activity in the central North Island. Much of the land is still undergoing significant uplift. Land slopes are close to the maximum. Three-quarters of New Zealand's landmass consists of geologically young sedimentary rocks. All this makes parts of the country particularly prone to erosion.

The removal of native bush and destruction of other ecosystems by Maori and European settlers has significantly increased the erosion rate. The indigenous Maori migrated to New Zealand from the Pacific over a period between 650 and 900 years ago. During their occupation, the natural environment was modified significantly with large tracts of forests in the dry eastern region of the South Island destroyed by fire. Less than 200 years ago, European colonists arrived and settled in New Zealand. Large numbers arrived after the signing of a treaty with Maori in 1840.

In less than 150 years, European settlers radically changed the landscape and its fragile ecosystems. The native temperate forests, which once covered about 85 percent of the land area have been reduced to 23 percent of the land and confined largely to economically unproductive mountain areas. Settlers cut or burned much of the remaining forest and replaced it with European introduced pasture grasses for the production of sheep and cattle.

Applying European farming techniques to a wet, steep, and deforested landscape has led to soil erosion, siltation of rivers, and flooding. These problems, which resulted largely from forest clearances first became apparent in the 1920s. Other difficulties encountered by settler farmers included inadequate trace minerals in some soils, a problem later corrected by applying trace elements with fertilizer.

Land management issues in New Zealand today

New Zealand's history of land management has resulted in soil erosion, silted-up streams, rivers and estuaries, and waterways polluted by agricultural runoff. Among the more serious issues of concern are:

- Accelerated erosion and nutrient decline as a result of deforestation. About 10 percent of New Zealand is classed as severely eroded and the loss of soil through erosion and transport by rivers to the sea is estimated to be 400 million tons per year. Voluntary and government-subsidized soil conservation programs over the last few decades have achieved only a localized reduction in soil erosion on hill country farmlands. The loss of soil-stabilizing vegetation has resulted in soil degradation on North Island hill farms and South Island tussock country, as well as in upper catchment indigenous forests, along streams and river banks and in coastal dune lands.

- The impacts of pests (particularly possums) and weeds on agricultural systems and natural ecosystems. Some of the plants and animals European settlers brought with them have quickly developed into pests which have proven difficult to control. Pests and weeds, together with the pressures of livestock and burning off, are a major cause of degradation of soil and tussock cover in the South Island high country.

- The degradation of highly productive, but relatively scare "elite" soils, through urban encroachment and through intensive cropping and grazing which can cause compaction, loss of soil structure, and fertility decline.
- The costs of natural disasters and adverse climatic events. Between 1986 and 1990, natural disasters and adverse climatic events cost the government more than $175 million in direct assistance to primary producers. The costs to individual businesses and to communities were considerably greater. Protected areas, especially forests and wetlands, play an important role in limiting vulnerability to flooding, erosion, and other damage.
- The contamination of water from agricultural activities. Though waterways in sparsely developed areas of New Zealand are in good condition, lowland river reaches in agriculturally developed catchments are in surprisingly poor condition. Some small creeks and streams in dairying areas are in very poor condition, as a result of diffuse and point-source agricultural wastes. It has also been found that lakes fed by catchments in which more than 50 percent of the land is used for agriculture show symptoms of eutrophication. Eroded sediment from deforested land also contributes to water quality problems. Finally nitrate contamination of shallow groundwater is an important issue in some intensively used catchments in New Zealand.

A less tangible, but no less important issue is the effects of land management practices on the ecological cultural, historical, and traditional values associated with the land. As a natural ecosystem, the New Zealand landscape is home to a unique array of birds, plants, and other biota. It is this landscape that local recreationalists and tourists come to visit and enjoy.

Maori have a strong affinity with land and water. The health or *mauri* of a tribal river or lake is a cornerstone to the *mana* (status or prestige) of an *iwi* (tribe). Rivers and lakes which are valued highly by Maori are often located in agricultural catchments which are affected by point source discharges and diffuse sources of agricultural runoff. The desecration of *waahi tapu* (sacred sites), *urupa* (burial grounds) and other sites of significance through land development or land degradation is also of primary concern to Maori.

A further issue is New Zealand's international reputation. New Zealand makes strong statements internationally about its environmental status. The country's reputation for "quality products from a quality environment" depends on environmentally sustainable land use practices. Sound land management is increasingly recognized as offering a competitive advantage, particularly in the global market.

Many issues listed above have been apparent in different parts of the country for some time. But they have not always been recognized as matters of concern. It could be argued that these are the result of the pursuit of economic and social goals without due regard to environmental needs.

Land management in New Zealand has also, until relatively recently, been characterized by a lack of knowledge of the processes at work behind land degradation problems, and lack of coordinated response to the effects of the problems. It is clear now that unless these issues are addressed in a more systematic way, underpinned by effective science, a large part of New Zealand's export income, will be put at risk, and the well-being of future generations of New Zealanders jeopardized.

However, the picture is not all bad. Considerable progress has been made towards achieving sustainable land management, although some problem areas, such as the North Island hill country and South Island high country, remain. Many of New Zealand's land management issues have been resolved by changing technology and management practices, and by the institutional reforms of the last decade.

Institutional reform in New Zealand

In comparison to most other developed countries, a particular characteristic of economic, agricultural, and environmental reform in New Zealand over the past decade has been the rapid pace of change. The speed of change has been facilitated by New Zealand's current system of government, which comprises a single legislative chamber based on the "first past the post" electoral system. New Zealand's form of government is similar to that of the United Kingdom except that there is no upper house, no "senate." There is no written constitution and no federal system of states or provinces. The executive can therefore make major changes to laws relatively quickly should it decide to do so. Recently, the country decided to implement a new system of proportional representation which changed the way the executive can exercise power.

Agricultural and economic reforms

Prior to New Zealand's economic reforms in the mid-1980s, agricultural policy focused on development and largely ignored the need for consistency between economic and environmental policy objectives. Policies on economic development, resource use, and disaster relief emphasized short-term economic considerations. These policies encouraged farmers to take greater risks with climatic factors by adopting more intensive and less diversified management systems. The environmental consequences and costs of agricultural production were ignored. While soil erosion in particular was seen as a problem, few saw the connection with agricultural support policies. Environmental policies were subordinated to development objectives, and were largely oriented to cleaning up the most obvious adverse effects of agricultural development.

In early 1985, the government abolished financial assistance for agricultural products such as meat and wool. Subsequently, fertilizer and other input subsidies were abolished as were investment and land development concessions. In addition, tax concessions for farmers were withdrawn. Free government services for farmers were eliminated. Producer boards had their access to concessionary Reserve Bank funding withdrawal; they now have no access to taxpayer funds. Starting in 1987, central government subsidies for soil conservation, flood control, and drainage schemes were substantially eliminated, although some local funding continued.

Changes in the general economy also affected agriculture. Farmers expected the floating of the New Zealand dollar in 1985 would lead to devaluation given New Zealand's high inflation. But with a tight monetary policy in place to control inflation, high interest rates attracted money into the country. The exchange rate rose between 1985 and 1988, lowering returns to farmers in New Zealand dollars and making it more difficult for the agricultural sector to adjust to the changing economy.

Assistance to New Zealand pastoral agriculture over the past 15 years, in terms of Producer Subsidy Equivalents (PSEs), declined from an average PSE of 24 percent in 1979-1986 to 4 percent in 1995. The result is that New Zealand farmers are now fully exposed to world market forces. This is in marked contrast to most other developed countries.

The elimination of agricultural subsidies caused a number of changes with environmental implications. The removal of subsidies occurred at the same time as high interest rates and an overvalued exchange rate, which were associated with macroeconomic reforms. These factors contributed to financial stress for many farmers, which in some cases led to short-term exploitation of the resource base. On the positive side, livestock numbers declined, forestry plantings continued to increase, the development of marginal lands virtually ceased, and the use of fertilizers and other agricultural chemicals decreased. These changes lessen the likelihood of off-site contamination and of severe degradation of marginal lands.

Despite these changes, environmental challenges remain. The New Zealand experience has confirmed that, while the removal of agricultural subsidies is beneficial, specific environmental policies are necessary to address the environmental effects of agriculture.

The development of these policies is well underway. The New Zealand government no longer provides significant disaster relief payments to farmers, but instead requires farmers to manage their land in consideration of these risks.

User-pays and polluter-pays principles are well established in New Zealand. Agricultural inspection and extension services formerly funded by government are now on a user-pays basis. Pest management programs are funded largely by levies or special property taxes on landholders who benefit from them.

Furthermore, farmers must obtain environmental permits for a range of activities and often pay the administration and monitoring costs associated with the permit. Local authorities are encouraging landowners to take collective responsibility for devising solutions to problems such as nonpoint source pollution. Although there is still some way to go, New Zealand is moving towards internalization of environmental costs in order to encourage the efficient and sustainable use of natural resources. Reform of environmental laws has been a key part of that process.

Environmental reforms

The Resource Management Act, which came into force on October 1, 1991, is the corner stone of the reform of New Zealand's environmental laws. It is a major and fundamental, if not unique, regulatory reform. The Resource Management Act is based on the single and over-arching purpose of promoting sustainable management. Central to sustainable management is the idea that a sustainable economy, especially a resource based one, is reliant on a sustainable environment. Sustainable management is now the sole objective for a law which replaces 59 resource and planning statutes, including major legislation dealing with town and country planning and water and soil management.

Other new environment laws which complement the Resource Management Act are the Biosecurity Act 1993, which provides for pest control operations, and the Hazardous Substances and New Organisms Act 1996, which addresses various environmental risks not fully addressed under the Resource Management Act. New legislation is also being developed to cover the registration and use of agricultural chemicals.

The key features of the Resource Management Act are as follows:
- It sets *environmental limits* by specifying tough duties and responsibilities that are built on in the management system of policies, plans and resource consents (permits). For example, no one may discharge a pollutant into water except in ways that are provided for through the law. Everyone has a duty to avoid, remedy or mitigate any adverse effect on the environment. The setting of environmental limits clarifies the extent to which trade-offs may be made between environmental quality and development;

- It requires an *ecosystem approach* in which decision makers consider all impacts of a proposal for the use or development of resources. Pollutants are managed in a cross media way and impacts on land, air and water are considered together. The aim is to minimize adverse effects on the environment;
- It does not presume *regulation and related rule making is the only or necessarily the best way* to manage the environment. It requires decision-makers to be clear about environmental outcomes and use the best means to achieve them; and
- It requires a focus on the *effects* activities have on the environment. This ensures that decision-makers concentrate on environmental outcomes and that resource users have more freedom in how they comply with environmental rules.

In New Zealand, major natural resources users generally support the approach and requirements of the new law. They consider that environmental quality is an important part of product quality. They believe that sustainable management is an important component of being able to use resources. And they see a positive and compatible connection between environmental and economic sustainability. For these reasons, many resource users are developing their own environmentally rigorous codes of practice which will comply with and help further develop the requirements of the new law.

So what effect is the Resource Management Act having on sustainable land management in New Zealand? It is too early to judge its success or otherwise, at least in terms of measuring its impact on the sustainability of resources. But initial indications suggest that it is having an impact. There is a considerable and widespread awareness, especially by business and industry and by many in the community, of its purpose and its liability provisions. The latter, of course, encourage compliance with the relatively tough requirements of the act compared to the previous law.

The act is already stopping or reducing environmental damage and exposing the environmental effects of development proposals to public scrutiny. There will be no more substandard landfills or new discharges of raw or crudely treated sewage or leachate into water and coastal systems. Dairy shed effluent direct discharges to water have now virtually ceased—a direct result of rigorous monitoring and inspection of discharge permits by regional councils.

Integrated management still has a long way to go but it is beginning to work. There will be no more adverse side effects on media such as water or land caused by trying to manage effects on another medium, such as air. Monitoring environmental performance is increasing and self-monitoring is developing. And nonregulatory techniques such as information and participatory management through landcare programs are developing.

A start is being made on defining environmental quality baselines and setting out ways of improving the state of the environment. Once more policies and plans are prepared and implemented under the act, they will help to achieve the wider objectives of the Resource Management Act. The result should be a progressive improvement in environmental quality.

Local government reform

New Zealand has a system of local government that is largely independent of the central executive government. Local authorities have their own source of income independent of central government based on local taxes on property. The functions and powers of local authorities are conferred only by parliament, and are prescribed by local government legislation and other acts, such as the Resource Management Act. Local government in New Zealand is in no way comparable with state governments in countries such as Australia or the U.S. in terms of scope or powers. However, local authorities do have significant and vital roles in a number of areas, including the management of natural resources.

A fundamental reform of local government in New Zealand was carried out in 1989. One of the most obvious results of the reform has been the reduction of local authorities from over 800 (including the large number of special purpose, ad hoc authorities) to around 90. This has increased efficiency in local government and reduced overlapping jurisdiction.

The reform of local government was closely coordinated with the environmental reform. Decisions taken in the local government reform process have therefore contributed to better environmental management. For example, regional councils now have the principal responsibility for water, soil, pollution, and coastal management. Regions are based on catchment boundaries, thereby ensuring effective catchment-based water and related pollution management.

Recent government environmental policy initiatives

Following the 1992 Earth Summit in Rio de Janiero, the New Zealand Government implemented two environmental policy initiatives.

The first of these is the *Environment 2010 Strategy* released in September 1995. This strategy identifies the key issues and sets out the broad goals for protecting the environment. It identifies 11 priority areas for action: land resources; water resources; air quality; indigenous habitats and biological diversity; pests, weeds, and diseases; fisheries; energy; transport; waste (including contaminated sites and hazardous substances); climate change; and ozone layer protection.

Environment 2010 stated that the government will give priority to developing and implementing a *Sustainable Land Management Strategy* for New Zealand.

The Sustainable Land Management Strategy is the second major environmental policy initiative, and was released in June 1996. The purpose of the strategy is to enable those contributing to or affected by sustainable land management to work together more effectively.

The strategy defines land management problems, sets out the priorities for action, and states what the government aims to achieve. The strategy complements the initiatives which regional councils have already undertaken on sustainable land management as they exercise their responsibilities under the Resource Management Act.

The focal point for the strategy is the land user. Land users are encouraged to continually improve the way that they operate and in doing so take into account the effects their businesses have on the environment and vice versa.

Two principal means by which the strategy promotes sustainable land management are through the following:
- direct help to land users to achieve desired changes, and
- indirect assistance by improving research, coordination and other support from government and industry.

The principal activities that the government will undertake in each of these areas are presented in the strategy. In summary, these actions are:
- financial support to the *New Zealand Landcare Trust* to train a national network of landcare and community group facilitators and to support the establishment of landcare groups;
- the preparation, in conjunction with industry sectors, of *best management practices* for land users;
- the establishment of a *National Science Strategy Committee* to provide advice to government and where possible to coordinate publicly and privately funded sustainable land management research at a regional and national level;
- the development of *standards and/or guidelines* for water quality and soil characteristics;
- the promotion of *voluntary and market incentives*; and
- the development of on-site indicators and a *set of national environmental indicators*.

The strategy identifies three priority areas for action—high country degradation, agricultural impacts on aquatic ecosystems, and hill country erosion. The priority areas will be addressed through Action Plans developed by the Ministry for the Environment in cooperation with other departments and in agreement with major stakeholders.

The strategy provides a national framework and statement of what central government intends to do to encourage environmental improvements on commercially used land. But this is only part of the picture. Initiatives by industry sectors, local government and land users themselves will be crucial to achieving the long term and permanent changes, effective environmental management, and sustainable land use require.

Conclusions

Past and current land management practices in New Zealand have had considerable impacts on native ecosystems, soils, and waterways. These impacts which result from past and current land management practices have proven difficult to address over the years because of pervasive institutional problems that have resisted change. A lot has been done by successive governments over the last ten years to address these institutional problems. Legislative reform, such as the Resource Management Act, local government reform and economic reforms have all contributed to improved environmental management. These reforms are relatively new and their full impact and effect cannot yet be measured in quantitative terms.

However, institutional reform alone will not achieve sustainable land management. Access to information is also important. Many land users do not recognize the problems they face and have inadequate tools or information to make changes. Furthermore, the research, advisory, and regulatory systems of central and local government are not adequately coordinated or targeted. The Sustainable Land Management Strategy recognizes and addresses these problems.

Already some encouraging progress towards sustainable management is being made. Some land users have made significant advances in the way that they operate their businesses to reduce effects on the surrounding environment. Recent initiatives by forestry and agricultural organizations also suggest that attitudes are changing. But, even so, short-term economic motives, rather than longer term environmental concerns, continue to drive many land management decisions.

References

Biosecurity Act 1993. Published under the Authority of the New Zealand Government -1993.

Castles, F.G., R. Gerritsen, and J. Vowles. (eds.) 1996. *The Great Experiment: Labour Parties and public policy transformation in Australia and New Zealand.* Auckland University Press.

Gow, L.J.A . 1995. *Implementing Sustainability: New Zealand's Experience with its Resource Management Act.* Address to the World Resources Institute (WRI)/NZ Embassy Seminar. Washington, D.C. June.

Hazardous Substances and New Organisms Act 1996. Published under the Authority of the New Zealand Government -1996.

Macsarenhas, R.C. 1996. *Government and the Economy in Australia and New Zealand: The Politics of Economic Policy Making.* Austin and Winfield. Maryland.

Ministry of Agriculture. 1996. *Environmental Effects of Removing Agricultural Subsidies: The New Zealand Experience.* New Zealand paper for the OECD Seminar on Environmental Benefits of Sustainable Agriculture: Issues and Policies. Helsinki. September.

Ministry for the Environment. 1995. *Environment 2010 Strategy. A Statement of the Government's Strategy on the Environment.* Wellington. September.

Ministry for the Environment. 1996. *Sustainable Land Management. A Strategy for New Zealand.* Wellington. June.

Ministry of Research, Science and Technology 1995. *Science for Sustainable Land Management: Towards a New Agenda and Partnership.* Wellington. November.

Resource Management Act 1991. Published under the Authority of the New Zealand Government -1994 (Reprint).

Statistics New Zealand. 1996. *New Zealand Official Yearbook.* 99th Edition. Wellington

Taranaki Regional Council 1995. *Approaches to Managing Discharges to Land and Water in Taranaki. A Discussion Document.* Stratford. November.

38

Soil and Water Conservation Policies and Programs: Successes and Failures: A Synthesis

Ted L. Napier, Silvana M. Napier, and Jiri Tvrdon

Several conclusions were derived from discussions presented in previous chapters. Conclusions are presented followed by a brief discussion of the implications for soil and water conservation policies and programs.

Degradation of soil and water resources associated with agricultural production is a universal problem on this planet. All geographical regions of this planet report environmental degradation from soil erosion associated with the production of agricultural products. However, environmental degradation is most problematic in economically disadvantaged countries that have a large number of subsistence farmers who are attempting to survive on limited arable farmland. Use of highly erodible land for crop production has significantly increased soil erosion in many countries of the world. Often marginal land is brought into agricultural production to feed increasing human populations and the result is often environmental disaster in terms of displacement of soil resources.

Widespread adoption of technology-intensive agricultural production systems has resulted in degradation of soil and water resources throughout the world via soil displacement, compacting of soil, and elimination of long fallow periods. Extensive use of agricultural chemicals to increase and/or maintain high levels of production has contributed to degradation of water resources. Adoption of a limited number of food and fiber crops combined with specialization in large animal production has also contributed to environmental degradation by agricultural production sectors in all societies.

The major implication of these conclusions is that societies must implement soil and water conservation policies and programs in the near future to prevent further degradation of vital natural resources required to produce food and fiber to feed and clothe domestic populations. Societies must develop soil and water conservation policies and programs that are relevant to the socioeconomic and political situations in their respective countries, because the evidence provided in previous chapters demonstrates that some policies and programs have been successful in certain situations but not in others. While conservation policies and programs implemented in other societies may be useful as guides for the development of conservation policies and programs in societies wishing to formulate such initiatives, it must be recognized that conservation efforts that have been very successful in some societies may fail miserably in another sociocultural context.

Another implication of these conclusions is that societies should examine methods embraced to achieve national agricultural production goals. Overemphasis on technology-intensive and chemical-intensive production systems will have severe environmental consequences in the form of soil erosion and reduced water quality.

Voluntary conservation programs that emphasize the provision of information and education combined with partial economic subsidies were shown to be successful during the Dust Bowl era for reducing the incidence of soil erosion in North America; however, such approaches have been shown to be less effective during the past 20 years.

This conclusion suggests that traditional approaches used to develop and to implement soil and water conservation policies in the U.S. and Canada need to be reassessed in the context of behavioral outcomes. If contemporary conservation policies and programs are not effective in motivating landowner/operators to adopt and use relevant soil and water conservation production systems, then new policy approaches should be explored. Evidence presented in the previous chapters strongly suggests that future soil and water conservation policies and programs in the U.S. and Canada will probably employ more command and control approaches such as those used in Europe. It is highly likely that future conservation policies and programs in the U.S. and Canada will continue to emphasize voluntary programs where possible. It is generally agreed that contemporary soil and water conservation policies and programs will not be adequate to address future environmental problems.

The major implication of this conclusion is that policy-makers in North America must have the courage to develop and implement soil and water conservation policies and programs that will not be politically popular. It also means that landowner/operators must be willing to include the preferences of nonfarm populations in on-farm production decisions. Both of these outcomes will be difficult to achieve.

Soil and water conservation policies and programs in Europe have employed command and control approaches. Evidence suggests that more emphasis will be placed on the provision of economic incentives by local and national governments to encourage voluntary adoption of conservation production systems in the future.

Unlike conservation policies implemented in North America, European approaches to soil and water conservation have emphasized command and control. While these approaches have been partially successful, consideration is now being given to the use of economic inducements to motivate landowner/operators to adopt and use soil and water protection practices. Future policies and programs in Europe will probably combine command and control with economic incentive systems. It is highly likely that preference will continue to be given to command and control over voluntary economic incentive conservation approaches, due to the economic costs of the latter approach. This is especially true in Eastern and Central European countries where public economic resources for investment in soil and water conservation programs are very limited.

The implications of this observation are that countries that have traditionally relied on command and control approaches to achieve environmental objectives will be required to appropriate limited public resources to finance future conservation programs. Such programs have been shown to be extremely expensive in societies such as the U.S. which has traditionally employed economic incentives to motivate landowner/operators to adopt and use soil and water conservation production systems. It is highly likely that governments electing to reallocate public financial resources to soil and water conservation programs will be severely criticized by sectors that experience reduced funding.

Point pollution remains a very serious environmental problem in Eastern and Central Europe. Given that public economic resources are very limited in many Eastern and Central European countries, governments within these geographic regions will be required to make very difficult decisions relative to priorities given to specific environmental problems. It must be recognized that governments in Eastern and Central Europe countries can address only a very limited number of environmental problems with the economic resources at their command. It is highly likely that governments will choose to allocate limited economic resources to the most severe environmental problems. If that is the rationale used to make funding allocation decisions, it is highly likely that agricultural pollution will be ranked lower in priority than other environmental problems. The highest priority for action will probably be given to point source pollution.

Point pollution from cities and industries demands immediate attention due to the threat to public health and the potential long-term negative consequences to the physical environment. Soil erosion and degradation of water resources from agricultural sources cannot be ignored, but action may have to be postponed until point pollution is at least partially controlled. It is highly likely that greater environmental benefits will be achieved per economic unit input by investing in control of point pollution.

The provision of conservation information and education will constitute important elements of soil and water conservation policies and programs in Central and Eastern Europe in the near term, because many land operators are not aware of environmental issues and techniques to prevent environmental degradation. While public economic resources are being initially focused on control of point pollution, public investments in information and education designed to foster adoption of low-cost and easily implemented soil and water conservation practices at the farm level appear to be justified.

Soil and water conservation policies and programs require institutional structures to be effectively implemented. Evidence presented in this volume clearly demonstrates that institutional structures are essential for effective implementation of soil and water conservation policies and programs. Societies that have authorized and funded institutional structures to implement conservation policies have been much more successful addressing soil and water conservation issues than societies that have not authorized and funded such institutions. Societies that anticipate formulating comprehensive soil and water conservation policies without establishing institutional structures to implement such initiatives will be terribly disappointed when the policies fail to achieve expected objectives.

Given the technical skills required to implement effective soil and water conservation programs at the farm level, institutional organizations commissioned to implement conservation policy must be staffed with very capable professionals. This means that societies that aspire to address soil and water conservation problems using policy instruments will be required to allocate extensive financial resources to pay for professional salaries and to finance training programs for landowner/operators. The creation and maintenance of soil and water conservation agencies is extremely expensive; however, they are requisites for effective implementation of conservation policy.

The implication of this conclusion is that investment of public resources in the formation and maintenance of conservation agencies is essential for the protection of soil and water resources. Societies that fail to invest in conservation agencies will continue to suffer significant degradation of soil and water resources. Formation of public conservation agencies in economically poor countries will require committed political leaders who will be pressured to allocate limited economic resources to pressing social needs.

Adoption of soil and water protection practices at the farm level is strongly influenced by the profitability of the practices being considered. While environmental ethics has a role to play in the decision-making process about adoption of soil and water conservation production practices at the farm level, the major determining factor of whether or not farmers will adopt or reject a specific farming practice is profitability. If a conservation production practice is not profitable, landowner/operators will not adopt it without the use of some type of incentive. Unfortunately, there is general agreement among contributors to this volume that most soil and water conservation practices at the farm level are not profitable in the short term and often not in the long term.

This means that conservation agency personnel are now required to convince landowner/operators to modify existing production systems without guarantees of increased economic benefits associated with adoption. Conservationists are forced to ask farmers to introduce higher levels of risk into their farm businesses without being able to state without qualification that landowner/operators adopting conservation production systems will be adequately compensated by greater farm level income. If conservationists are totally honest with farmers, they would have to inform farmers that adoption of conservation production systems will probably not increase farm income. Conservationists can legitimately state that adoption of conservation production systems will produce many benefits for society. Such arguments will probably not convince many farmers to change existing production systems particularly in economically poor countries.

The implications of these observations is that governments must provide incentives to farmers to encourage them to adopt and to continue use of soil and water conservation production systems. The policy options to create incentives to adopt range from economic subsidies to regulations. Subsidies compensate land operators for any economic losses incurred when conservation practices are employed. Regulations impose costs in the form of penalties. Penalties associated with the use of degrading production systems decrease the profitability of using such practices and make the adoption of conservation production systems more attractive.

Agricultural production policies and land use policies can be effectively used to reduce or prevent degradation of soil and water resources. Agricultural production policies that encourage landowner/operators to shift production to less erosive crops contribute maintaining environmental quality. Agricultural production policies that encourage shifts to animal production rather than row crops on cultivated farmland often have substantial environmental benefits. Conversion of cultivated cropland to pasture significantly reduces soil loss even on highly erodible land. Thus, environmental quality is often improved when landowners shift production from row crops to animal production. This assumes, of course, that waste control programs are implemented at the farm level to ensure that animal wastes are used in an environmentally benign manner. It also assumes that the number of animal units per unit of land does not exceed the "carrying capacity" of the land resource. Too many animal units per unit of land will result in removal of ground cover and increase rates of soil displacement.

Agricultural production policies that encourage shifts in the types of crops produced on cultivated land can also result in improved environmental quality.

Shifts to wheat production rather than row crops, such as corn or soybeans, can reduce soil loss. Shifts of production to crops that require lesser amounts of chemical fertilizers can improve water quality because fewer nutrients are available for transport to waterways.

Another policy approach that can reduce degradation of soil and water resources is imposition of stringent land-use controls designed to prevent conversion of forests, prairies, and wetlands to alternative uses. Several societies are presently experimenting with land-use policies that prohibit shifts of forest and agricultural land to nonagricultural uses. Such policy approaches could prevent degradation of soil and water resources. Other natural resources such as forests and wildlife could benefit from use of land-use controls.

The primary implication of these observations is that societies have many policy options that can be effectively employed to address soil and water pollution problems. Several of the policy options are not directly related to the improvement of environmental quality. While land-use controls can be effectively used to protect natural resources from degradation, it must be recognized that use of these policy instruments demands considerable foresight on the part of policy-makers. Use of land-use policies to control soil and water pollution necessitates proactive action by national, regional, and local governments. Land-use controls must be applied prior to the emergence of environmental problems for them to be effective in maintaining or improving environmental quality. It is often difficult to motivate populations to act on potential soil and water resource problems until the problems are perceived as being problematic. Often by the time environmental problems are recognizable, the opportunity to employ land-use control as a means of preventing degradation of soil and water resources is past.

Conclusion

The contributors to this volume have discussed some of the soil and water conservation experiences from their respective countries in North America, Europe, and Australia. Some of the policies and programs were successful while others were not. The authors have identified some of the factors that have contributed to the success or failure of the conservation efforts. While there are similarities among successful conservation initiatives, there are also many differences.

Successful conservation policies and programs tend to be most successful when they are relevant to the local situation, are integrated and consistent with national programs of socioeconomic development, have continuity over time, are adequately funded, and have the institutional support to be effectively implemented at the farm level. Conservation initiatives that have not been successful tend to deviate from the characteristics of successful policies and programs.

One of the implications of this conclusion is that societies wishing to increase the probability of successful conservation policies and programs must develop conservation initiatives that address relevant environmental problems that may be unique to specific geographic regions. Over-reliance on conservation policies and programs shown to be effective in other countries and/or geographic regions should be avoided. While experiences in other societies are useful to provide a basis for development of conservation policies and programs, it must be recognized that societies should develop conservation initiatives that are relevant to their situation.

A second implication is that conservation of soil and water resources is economically very costly. It is unrealistic to expect conservation policies and programs to be successful unless they are adequately funded. A significant portion of the cost will be associated with development and maintenance of institutional structures to implement conservation policies. Societies must also recognize that commitment of economic resources to conservation must be long term, or investments made in conservation will be lost due to landowners reverting to use of environmentally degrading production systems. Economic costs of soil and water conservation must be assumed by domestic governments. Reliance on external sources of conservation funding is unsound, because those resources can be terminated whenever the funding source elects to withdraw economic resources. This discussion strongly suggests that conservation must not become a residual claimant on societal resources.

The final implication of this conclusion is that soil and water conservation must become closely integrated with national development goals. This demands that political decision-makers and members of society must place considerable value on the protection of natural resources and be willing to accept the constraints placed on socioeconomic development by the collective decision to protect soil and water resources for future generations. This is probably the most significant barrier to the formation and implementation of soil and water conservation programs. Most people on this planet believe that technology-intensive and energy-intensive industrial and agricultural production systems are the best means of achieving life styles that are perceived to be highly desirable. Such growth models for societal development are certain to result in the abuse of natural resources. One of the best means of lessening the adverse environmental impacts of socioeconomic development is to incorporate conservation policies into a national development program. Protection of natural resources should be emphasized even if the rate of socioeconomic growth is reduced. Rapid increases in socioeconomic growth at the cost of the environment will probably prove to be very costly in the long term.

Contributing Authors

Chapter 1
Ted L. Napier, Silvana M. Napier, and Jiri Tvrdon
Ohio Agricultural Research and
  Development Center of
  The Ohio State University
Columbus, Ohio

Chapter 2
Hans Hurni
Bern, Switzerland

Chapter 3
Maurizio G. Paoletti
University of Padova
Padova, Italy

Chapter 4
Paul W. Johnson
Former Chief
USDA-NRCS
Washington, D.C.

Chapter 5
**Thomas A. Weber
and Gary A. Margheim**
USDA-NRCS
Washington, D.C.

Chapter 6
Daniel Conrad
USDA-NRCS
Washington, D.C.

Chapter 7
Max Schnepf
USDA-NRCS
Ankeny, Iowa

Chapters 8 and 9
Ted L. Napier and Silvana M. Napier
Ohio Agricultural Research and
  Development Center of
  The Ohio State University
Columbus, Ohio

Chapter 10
J. Dixon Esseks, Steven E. Kraft, and Douglas M. Ihrke
Northern Illinois University
DeKalb, Illinois

Chapter 11
**Dana L. Hoag,
Jennie S. Hughes-Popp,
and Paul C. Huszar**
Colorado State University
Fort Collins, Colorado

Chapter 12
**Jennie S. Hughes-Popp,
Paul C. Huszar, and Dana L. Hoag**
Winnipeg, Manitoba, Canada

Chapter 13
Andrew P. Manale
U.S. Environmental Protection Agency
Washington, D.C.

Chapter 14
**David R. Cressman, Scott N. Duff,
Paul H. Brubacher, and Jim Arnold**
Ecologistics International LTD
Waterloo, Ontario, Canada

Chapter 15
D. Peter Stonehouse
University of Guelph
Guelph, Ontario, Canada

Chapter 16
Jill S. Vaisey, Ted W. Weins, and Robert J. Wettlaufer
Prairie Farm Rehabilitation
Agriculture and Agri-Food Canada
Regina, Saskatchewan, Canada

Chapter 17
Michael A. Fullen
School of Applied Sciences
The University of Wolverhampton
Wolverhampton, United Kingdom

Chapter 18
Alex Dubgaard
Royal Veterinary and
   Agricultural University
Copenhagen, Denmark

Chapter 19
Jesper S. Schou
Copenhagen, Denmark

Chapter 20
Paul Vedeld and Erling Krogh
Agricultural University of Norway
Norway

Chapter 21
**Peter Weingarten
and Klaus Frohberg**
Institute of Agricultural Development
   in Central and Eastern Europe
Halle an der Saale, Germany

Chapter 22
**Monika Frielinghaus
and Hans-Rudolf Bork**
Centre for Agricultural Landscape and
   Land Use Research
Muncheberg, Germany

Chapter 23
Rita Kindler
Berlin, Germany

Chapter 24
A. Klik and O.W. Baumer
University of Agricultural Sciences
Vienna, Austria

Chapter 25
Andrzej Sapek and Barbara Sapek
Institute for Land Reclamation and
   Grassland Farming
Raszyn, Poland

Chapter 26
**Walter E. Foster
and Vilija Budvytiené**
U.S. Environmental Protection Agency
Kansas City, Kansas

Chapter 27
**Stanimir C. Kostadinov,
Miodrag D. Zlatic,
and Nenad S. Rankovic**
Belgrade University
Belgrade, Yugoslavia

Chapter 28
Jon Hron
President
Czech University of Agriculture
Prague, Czech Republic

Chapter 29
**F. Dolezal, M. Janecek, J. Lhotsky,
and J. Nemecek**
Research Institute for Soil and Water
   Conservation
Praha, Zbraslav, Czech Republic

Chapter 30
Jiri Tvrdon
Czech University of Agriculture
Prague, Czech Republic

Chapter 31
**Miloslav Lapka, J. Sanford Rikoon,
and Eva Cudlinova**
Institute of Landscape Ecology
  Academy of Sciences of the
  Czech Republic
Ceske Budejovice, Czech Republic

Chapter 32
Ian Hannam
Department of Land and
  Water Conservation
Parramatta, NSW, Australia

Chapter 33
John R. Bradsen
Law School
The University of Adelaide
South Australia

Chapter 34
John Cary
University of Melbourne
Parkville, Victoria, Australia

Chapter 35
Sarah Ewing
University of Melbourne
Parkville, Victoria, Australia

Chapter 36
David Farrier
University of Wollongong
NSW, Australia

Chapter 37
Wayne Bettjeman
Wellington, New Zealand

Chapter 38
**Ted L. Napier, Silvana M. Napier,
and Jiri Tvrdon**
Ohio Agricultural Research and
  Development Center of
  The Ohio State University
Columbus, Ohio

Index

V